Gmelin Handbuch der Anorganischen Chemie

Achte Auflage und Ergänzungswerk

Main Series, 8th Edition, and New Supplement Series

The following Index volumes for individual System Numbers of Gmelin Handbook are published up today:

System Number 3 „Sauerstoff"	Formula Index and Alphabetical Index (comprised in Section 8)
System Number 13 „Boron Compounds"	Formula Index
System Number 16 „Phosphor"	Formula Index and Alphabetical Index
System Number 21 „Natrium"	Formula Index and Alphabetical Index
System Number 34 „Quecksilber"	Formula Index and Alphabetical Index (comprised in Section B 4)
System Number 48 „Vanadium" System Number 49 „Niob" System Number 50 „Tantal"	Formula Index and Alphabetical Index (commonly for System Numbers 48, 49, 50, as well as for New Supplement Series volume 2)
System Number 71 „Transuranium Elements"	Alphabetical Index of Subjects and Substances

Other Indexes are comprised in various volumes of Main Series and New Supplement Series

Zu einzelnen System-Nummern des Gmelin Handbuchs sind bisher folgende Register erschienen:

System-Nr. 3 „Sauerstoff"	Formel- und Schlagwortregister (in Lieferung 8 enthalten)
System-Nr. 13 „Borverbindungen"	Formula Index
System-Nr. 16 „Phosphor"	Formel- und Schlagwortregister
System-Nr. 21 „Natrium"	Formel- und Schlagwortregister
System-Nr. 34 „Quecksilber"	Formel- und Schlagwortregister (in Lieferung B 4 enthalten)
System-Nr. 48 „Vanadium" System-Nr. 49 „Niob" System-Nr. 50 „Tantal"	Formel- und Schlagwortregister, gemeinsam für die System-Nr. 48, 49, 50 sowie für Band 2 des Ergänzungswerks zur 8. Auflage
System-Nr. 71 „Transurane"	Alphabetical Index of Subjects and Substances

Weitere Register sind in verschiedenen Bänden des Haupt- und Ergänzungswerks enthalten.

Gmelin Handbuch der Anorganischen Chemie

Achte völlig neu bearbeitete Auflage

BEGRÜNDET VON Leopold Gmelin

ACHTE AUFLAGE BEGONNEN im Auftrag der Deutschen Chemischen Gesellschaft von R. J. Meyer

FORTGEFÜHRT VON E. H. E. Pietsch und A. Kotowski
Margot Becke-Goehring

HERAUSGEGEBEN VOM Gmelin-Institut für Anorganische Chemie der Max-Planck-Gesellschaft zur Förderung der Wissenschaften
Direktor: Ekkehard Fluck

Springer-Verlag
Berlin · Heidelberg · New York 1980

Gmelin Handbuch der Anorganischen Chemie

Achte Auflage und Ergänzungswerk

Main Series, 8th Edition, and New Supplement Series

Index

Formula Index

Volume 12

O–Zr

Elements 104 to 132

AUTHORS (BEARBEITER) | Ramona Becker, Helga Hartwig, Herbert Köppe, Hans Vanecek, Paul Velić, Rudolf Warncke, Anna Zelle

EDITOR (REDAKTEUR) | Rudolf Warncke

Springer-Verlag
Berlin · Heidelberg · New York 1980

The volumes of the Main Series are evaluated up to the end of 1974, the volumes of the New Supplement Series up to the end of 1973.

Es sind die Bände des Hauptwerkes bis zum Erscheinungstermin Ende 1974, die des Ergänzungswerkes bis Ende 1973 berücksichtigt.

Die vierte bis siebente Auflage dieses Werkes erschien im Verlag von Carl Winter's Universitätsbuchhandlung in Heidelberg

Library of Congress Catalog Card Number: Agr 25-1383

ISBN 3-540-93413-8 Springer-Verlag, Berlin · Heidelberg · New York
ISBN 0-387-93413-8 Springer-Verlag, New York · Heidelberg · Berlin

Wiesbadener Graphische Betriebe GmbH, Wiesbaden

Introduction

Volume 12 completes the alphabetical Formula Index. The index covers all volumes of the Eighth Edition through 1974 for the Main Series and through 1973 for the New Supplement Series. In all, there are more than 142,000 entries.

The index includes all definite elements, compounds, ions, and systems discussed in the volumes covered. The substances are arranged by their empirical formulas. In the second column the usual formula is given. The elements are entered by their symbols. Transuranium elements that do not yet have an official IUPAC name are listed separately at the end of this volume.

Supplements will keep this index up-to-date. The first supplement, which will cover the years 1974 through 1979, is already in preparation.

Frankfurt/Main, January 1980

Rudolf Warncke

Vorwort

Mit dem vorliegenden Band 12 ist das alphabetische Formelregister zum Gmelin Handbuch abgeschlossen, das mit nahezu 142000 Einträgen alle Bände der 8. Auflage des Handbuchs umfaßt, soweit sie bis Ende 1974 (Hauptwerk) bzw. Ende 1973 (Ergänzungswerk) erschienen sind.

Das Formelregister enthält alle in den vorgenannten Bänden besprochenen definierten Elemente, Verbindungen, Ionen und Systeme. Als Haupt-Ordnungsbegriff dient die Summenformel; neben jeder Summenformel ist die entsprechende Formel in üblicher Schreibweise aufgeführt. Die Elemente selbst sind unter ihrem Symbol eingeordnet; diejenigen Transurane, die noch keinen von der IUPAC anerkannten Namen besitzen, sind im Anhang des vorliegenden Bandes zusammengestellt.

In einer Ergänzung zum Formelregister werden die in den letzten Jahren erschienenen Handbuch-Bände, die im vorliegenden Register noch nicht berücksichtigt werden konnten, erfaßt. Die Ergänzung für die Jahre 1974 bis 1979 ist in Vorbereitung.

Frankfurt/Main, Januar 1980

Rudolf Warncke

Introduction to Volume 1

The extraordinary quantity of material included in the Gmelin Handbook of Inorganic Chemistry can make it extremely difficult to find a specific compound using only a general understanding of the basic rules for the arrangement of the subject matter (the so-called "System of the Last Position"). Consequently, subject indexes have been provided for the past few years to individual Handbook volumes, covering only a single section, or in some cases, a complete system number. During this period a comprehensive subject index–the Gmelin Index–was being prepared, whose first volume is now available.

The broad distribution and extensive utilization enjoyed by the Gmelin Handbook in English-speaking countries has led to a decision to publish the Gmelin Index in the English language. This decision was partly based on the fact that since the English language today occupies a preeminent position in the field of chemistry, a German-speaking user could utilize an English text much more readily than would be the case for an English-speaking user and a German text.

The core of the Gmelin Index is formed by a Formula Index which includes all of the definite elements, compounds, ions, and systems which are discussed or mentioned anywhere in the entire Eight Edition of the Gmelin Handbook. All volumes of the Main Series which were published before the end of 1974 are included, as well as volumes 1 through 12 of the New Supplement Series. The Formula Index will consist of about 12 volumes which will appear at intervals of four to six months.

Subsequent index volumes will contain accessible entries for items, which could not be considered in the Formula Index, for example, minerals, alloys, trivial names, and names of classes of compounds.

The Gmelin Index will be printed–for the first time in the history of the Gmelin Handbook–not by using lead type but rather by optical printing. It has thus become possible to store the Index text on magnetic tape, thereby having it available for future expansion when newly-published Handbook volumes can be readily included. It also has become possible to collect subject areas from various specific points of view.

It is not possible to include in a single published index *all* of the material which is scattered throughout the Handbook text. Such an undertaking would vastly exceed the capacity of a printed index, and could only be done utilizing the capabilities of an electronic computer system. At the present time, the Gmelin Institute is endeavoring to work out a project of this type.

The Gmelin Index should give an user better access to the information contained in the Gmelin Handbook, and should put him in position better to utilize the advantages of the Handbook than heretofore. These advantages, for example, include a review of the entire literature of chemistry going back to the very origins of this field and a collection not merely of literature citations but also of the data themselves–critically evaluated, sifted, and arranged in exhaustive monographs.

We wish to thank the Ministry for Research and Technology for their financial support during development and preparation of the Gmelin Index.

Frankfurt/Main, August 1975

Rudolf Warncke

Vorwort zu Band 1

Die überaus große Fülle des im Gmelin Handbuch dargebotenen Materials macht es selbst bei Kenntnis des zugrundeliegenden Ordnungsprinzips (»System der letzten Stelle«) mitunter recht schwierig, eine bestimmte Verbindung aufzufinden. Es wurden daher schon seit mehreren Jahren einigen Handbuch-Bänden Register beigegeben, die jeweils eine Lieferung oder auch einzelne System-Nummern umfaßten, und gleichzeitig wurde ein Gesamtregister, das Gmelin Register, in Angriff genommen, dessen erste Lieferung heute vorliegt.

Die große Verbreitung, die das Gmelin Handbuch im angelsächsischen Sprachraum gefunden hat, führte zu dem Entschluß, das Gmelin Register in englischer Sprache herauszugeben. Damit wird zugleich der Tatsache Rechnung getragen, daß im Bereich der Chemie heute die englische Sprache an erster Stelle steht und daß infolgedessen ein deutschsprachiger Benutzer leichter einen englischen Text zu lesen vermag als dies umgekehrt der Fall ist.

Das Kernstück des Gmelin Registers bildet ein Formelregister, das alle in der 8. Auflage des Gmelin Handbuchs besprochenen bzw. erwähnten definierten Elemente, Verbindungen, Ionen und Systeme umfaßt. Berücksichtigt sind dabei alle Bände des Hauptwerks, soweit sie bis zum Ende des Jahres 1974 erschienen sind, sowie die Bände 1 bis 12 des Ergänzungswerks. Das Formelregister wird etwa 12 Lieferungen umfassen, die in Abständen von 4 bis 6 Monaten erscheinen sollen.

In zusätzlichen Register-Bänden sollen Angaben zugänglich gemacht werden, die im Formelregister nicht berücksichtigt werden können, z.B. Mineralien, Legierungen, Trivialnamen und Verbindungsgruppen-Namen.

Die Herstellung des Gmelin Registers erfolgt – erstmalig für das Gmelin Handbuch – nicht im Bleisatz, sondern im Lichtsatz. Damit ist es möglich geworden, den Register-Text auf Magnetband zu speichern und für eine spätere Erweiterung unter Einbeziehung der in der Zwischenzeit neu erschienenen Handbuch-Bände zur Verfügung zu halten. Zugleich wird damit die Möglichkeit geschaffen, Stoffzusammenstellungen nach bestimmten Gesichtspunkten vorzunehmen.

Es ist nicht möglich, alle im Handbuch-Text verstreuten Angaben durch ein Register zu erfassen. Ein derartiges Vorhaben würde den Rahmen eines gedruckten Registers bei weitem überschreiten und wäre nur in der Form eines elektronischen Informationsspeichers durchführbar. Das Gmelin-Institut ist zur Zeit bemüht, die Voraussetzungen für ein derartiges Projekt herauszuarbeiten.

Das Gmelin Register soll dem Benutzer einen besseren Zugang zu den im Gmelin Handbuch enthaltenen Informationen geben und ihn in die Lage versetzen, die Vorteile des Handbuchs – zum Beispiel Berücksichtigung der gesamten chemischen Literatur, Zurückgehen bis in die Anfänge der Chemie, Zusammenstellung nicht nur von Literaturstellen, sondern von Daten, die kritisch gesichtet und in größere Zusammenhänge eingefügt werden – besser als bisher zu nutzen.

Dem Bundesministerium für Forschung und Technologie danken wir für die finanzielle Unterstützung bei der Entwicklung und Vorbereitung des Gmelin Registers.

Frankfurt am Main, August 1975

Rudolf Warncke

Instructions for the user

The first column of the Formula Index contains the empirical formulas which provide the principal basis for the sequence in which the entries are arranged. The formulas are written in their more normal form in the second column, together with extensions and subdivisions. The citations themselves are given in the third column.

In the *empirical formula* listing (first column), the element symbols for each compound are arranged in alphabetical order; C and H are not treated separately here.

H_2O is included in these empirical formulas only when it is bound in the compound as a complex, as shown in the formulations in the second column. In all other cases, H_2O is not considered.

Isotopes are not shown in the empirical formulas, but are replaced by the normal symbol for the element. However, ionic charges are included in the empirical formulas.

For the elements themselves, the symbol of the element is used as the empirical formula–specifically, in the monatomic form (e.g., Br); entries which definitely refer to polyatomic molecules can be found under the corresponding empirical formula (e.g., Br_2, Br_3).

In the empirical formulas the letters x, y, and z are used for indefinite subscripts. High polymers of the type $(AB)_n$ are listed under the empirical formula of the monomeric compound.

Multicomponent systems (solid solutions, melts, etc.) can be found under the empirical formula of their inorganic components, while solutions are only listed under the formula of the dissolved substance.

The arrangement of the empirical formulas follows in alphabetic sequence and in terms of increasing subscripts; indefinite subscripts follow at the end of each subscript sequence (e.g., $Al_2Br_{x+6}Cu_x$ comes after $Al_2Br_{31}Sb_5$). Ions follow the corresponding neutral compound, and are arranged sequentially in terms of increasing positive, and then negative, charges. The transuranium elements for which no element symbol has been assigned can be found under their atomic number at the end of the Formula Index.

The formulas used in the *second column* of the Index are generally those which are employed in the body of the Handbook text; if more than one formulation is given for a single compound, these are all shown in the Index. In some cases–especially for isomeric compounds–due to limited space formulas are used which deviate from the Handbook text style so as to better illustrate the structure of the compound.

Compounds with the same empirical formula are arranged in the second column in the following sequence: the compound–isotopic compounds–polymeric compounds–hydrates. Isomeric compounds, if any, then follow in the same sequence, and, finally, multicomponent systems corresponding to this empirical formula. For example, the following arrangement can be found in the second column under the empirical formula ClNa:

NaCl
$Na^{37}Cl$
$(NaCl)_2$
$NaCl \cdot 2\,H_2O$
...
NaCl glasses:
  $NaCl–B_2O_3$
...

NaCl solid solutions:
Na(Br,Cl)
...
NaCl solutions:
$NaCl–H_2O$
...
NaCl systems:
$NaCl–As_2O_3–H_2O$
...

If it is not possible to distinguish the isomers of a given composition by different formulas, they are labelled by successive numbers.

The term "system" is used in a very narrow sense in the Index; it includes those equilibria which are characterized by phase diagrams. Ionic systems are treated as systems of the corresponding compounds. Data on solutions are given under the formula of the compound itself for those compounds which have only rather short text passages.

For the elements, the name of the element is utilized in the second column (e.g., "Silver"). The previously given sequence also applies here, but it has been modified so that the ions (in general, i.e., "Silver ions") and the isotopes (i.e., "Silver isotopes") are inserted before the multicomponent systems.

The hydrates are sequenced in terms of increasing water content.

In the case of multicomponent systems, the components are listed in the sequence: inorganic components–organic components–H_2O; the entries then follow in alphabetic order, but the organic components are considered in order of increasing C and H content as well. For elements appearing as components in multicomponent systems, the monatomic element symbol is always used. Isotopic compounds as components are sequenced directly behind the corresponding usual compounds.

Elements and compounds for which the associated Handbook text is voluminous, are subdivided in the Index by subject headings; the degree of subdivision depends on the amount of text material involved.

The concepts "solubility", "solutions", and "systems" partly overlap; in this case, all three positions must be searched. The same is true for the concepts "diffusion" and "systems" as well as for "sorption" and "systems".

With respect to the *citations* (third column of the Index), the following abbreviations are utilized:

Hb	=	Main Series Volume
Eb	=	Supplement Volume to Main Series
Ab	=	Appended Volume to Main Series
Erg. W.	=	New Supplement Series Volume.

For the first three of the above cases, the citation includes the System Number, the system element, the nature of the volume (Main Series Volume, Supplement Volume, or Appended Volume to Main Series), the Part or Section (if any), and then, finally, following a hyphen, the page or pages involved. All volumes of the Main Series which are not Supplement Volumes or Appended Volumes to Main Series, are designated as Main Series volumes in the citation. The citation "61 (Ag): Hb/A3–70/106" has the following meaning: pages 70 to 106 in Section 3 of Part A of System Number 61, Silver. The System Number 1, Rare Gases, and the System Number 39, Rare Earth Elements, are designated by "1 (EG)" and "39 (SE)" respectively.

Volumes of the New Supplement Series are cited only using the volume number, for example "Erg. W. 12–53" means: page 53 in volume 12 of the New Supplement Series.

The Supplement Volumes to Sections 3, 4, and 5 of Part A for System Number 59 "Iron" (Gmelin-Durrer) are designated as "59 (Fe): Eb/GD1a", "59 (Fe): Eb/GD1b", etc.; the volumes on magnetic materials for this same System Number are designated as "59 (Fe): D/1. Eb" or "59 (Fe): D/2. Eb".

In the case of *"see..." or "see also..." referrals,* reference is made to entries in both the first and second columns of the Index; for example, *"see* $Al_2Na_2O_4$ ... $Na_2O \cdot Al_2O_3$". In the case of such referrals within the subjects headings, only a specific subject heading is given, for example, *"see Deposits"*.

In this index the volume numbers where an entry may be found are based on the traditional system numbers for the Main Series and the volume numbers for the New Supplement Series. Since January 1978, all handbook volumes, both the Main Series and the New Supplement Series, have been arranged by the symbols of the elements. However, this change should cause no difficulty in the use of the index since the symbol is given in parenthesis immediately after the system number. For example, the volumes 61 (Ag) are sought under Ag. There are only four exceptions:

1 (EG) corresponds to He
8 (J) corresponds to I
39 (SE) corresponds to Sc
69 (Ma) corresponds to Tc

The volumes of the New Supplement Series, denoted by "Erg. W.", are found as follows:

Erg. W. 1	He
Erg. W. 2	V
Erg. W. 3	Cr
Erg. W. 4, 7a, 7b, 8	Np
Erg. W. 5, 6	Co
Erg. W. 9, 12	F
Erg. W. 10	Zr
Erg. W. 11	Hf

Volumes 2 and 3 are bound together, as are Volumes 10 and 11.

Hinweise für den Benutzer

Das Formelregister enthält in seiner ersten Spalte die als Haupt-Ordnungsbegriff dienenden Summenformeln. In der zweiten Spalte sind die Formeln in der üblichen Schreibweise nebst Ergänzungen und Unterteilungen aufgeführt, und in der dritten Spalte finden sich die Band- und Seitenangaben.

In der *Summenformel* (erste Spalte) sind die Element-Symbole der Verbindung in alphabetischer Reihenfolge angeordnet, eine Voranstellung von C und H erfolgt nicht.

H_2O wird in die Summenformel nur einbezogen, wenn es entsprechend der Formulierung in der zweiten Spalte komplex gebunden ist. In allen anderen Fällen wird H_2O nicht berücksichtigt.

Isotope werden in der Summenformel nicht angeführt, sondern durch das normale Element-Symbol ersetzt. Die Ladung von Ionen wird in der Summenformel angegeben.

Bei den Elementen selbst wird als Summenformel das Element-Symbol verwendet, und zwar in einatomiger Schreibweise (z. B. Br); Angaben, die speziell das mehratomige Molekül betreffen, sind zusätzlich bei der entsprechenden Summenformel (z. B. Br_2, Br_3) zu finden.

Für unbestimmte Indices werden in der Summenformel die Buchstaben x, y und z verwendet. Hochpolymere vom Typ $(AB)_n$ sind unter der Summenformel der monomeren Verbindung eingereiht.

Mehrstoffsysteme (Mischkristalle, Schmelzen usw.) sind bei den Summenformeln ihrer anorganischen Komponenten zu finden, Lösungen jedoch nur bei derjenigen des gelösten Stoffes.

Die Reihung der Summenformeln erfolgt in alphabetischer Folge und nach aufsteigenden Indexzahlen; unbestimmte Indexzahlen stehen dabei jeweils am Ende einer Indexreihe (z. B. $Al_2Br_{x+6}Cu_x$ hinter $Al_2Br_{31}Sb_5$). Ionen folgen hinter der entsprechenden neutralen Verbindung und sind in sich nach steigenden positiven und danach negativen Ladungen geordnet. Transurane, denen noch kein Element-Symbol zugeordnet ist, finden sich unter ihrer Ordnungszahl am Schluß des Formelregisters.

Für die in der *zweiten Spalte* des Registers aufgeführten Formeln wird meist die Schreibweise verwendet, die im Handbuch-Text zu finden ist; sind dort für eine Verbindung mehrere Formulierungen aufgeführt, so werden diese übernommen. In vielen Fällen – besonders beim Vorliegen von isomeren Verbindungen – wird jedoch im Rahmen des verfügbaren Platzes eine vom Handbuch-Text abweichende Schreibweise gewählt, die die Struktur der Verbindung besser zum Ausdruck bringt.

Haben Verbindungen die gleiche Summenformel, so werden sie in der zweiten Spalte in folgender Reihenfolge aufgeführt: Verbindung – isotope Verbindungen – polymere Verbindungen – Hydrate. Danach folgen ggf. isomere Verbindungen mit der gleichen Folge und anschließend die zu der Summenformel gehörenden Mehrstoffsysteme. Beispielsweise ist unter der Summenformel ClNa in der zweiten Spalte folgende Anordnung zu finden:

NaCl
$Na^{37}Cl$
$(NaCl)_2$
$NaCl \cdot 2\, H_2O$
...

NaCl glasses:
$NaCl–B_2O_3$
...

NaCl solid solutions:
  Na(Br,Cl)
  ...
NaCl solutions:
  $NaCl-H_2O$
  ...
NaCl systems:
  $NaCl-As_2O_3-H_2O$
  ...

Können zu einer Summenformel gehörende Isomere nicht durch ihre Schreibweise unterschieden werden, so werden sie fortlaufend numeriert.

Der Begriff »Systeme« wird im Register im engen Sinne verwendet; er umfaßt Gleichgewichte, wie sie durch das Zustandsdiagramm charakterisiert sind. Ionensysteme sind als Systeme der entsprechenden Verbindungen erfaßt. Angaben über Lösungen sind bei Verbindungen mit geringem Textumfang unter der Formel der Verbindung selbst zu finden.

Bei Elementen wird inder zweiten Spalte der Name des Elements eingesetzt (z. B. »Silber«). Auch hier gilt die oben angeführte Reihenfolge, jedoch mit der Änderung, daß vor der Gruppe der Mehrstoffsysteme die Ionen (allgemein, z. B. »Silber-Ionen«) und die Isotope (z. B. »Silber-Isotope«) eingereiht werden.

Hydrate werden in sich nach steigendem Wassergehalt gereiht.

Bei Mehrstoffsystemen werden die Komponenten in der Reihenfolge »Anorganische Komponenten – Organische Komponenten – H_2O« aufgeführt; die Reihung erfolgt dann in alphabetischer Folge, wobei jedoch bei organischen Komponenten der C-Gehalt und danach der H-Gehalt maßgeblich ist. Für Elemente als Komponenten in Mehrstoffsystemen wird stets das einatomige Element-Symbol verwendet. Isotope Verbindungen als Komponenten sind unmittelbar hinter den entsprechenden normalen Verbindungen eingeordnet.

Elemente und Verbindungen, bei denen der zugehörige Handbuch-Text umfangreich ist, werden im Register durch Sachverhaltsangaben unterteilt; der Grad der Unterteilung richtet sich dabei nach dem Textumfang.

Die Begriffe »Löslichkeit«, »Lösungen« und »Systeme« überlappen sich teilweise; hier muß ggf. bei allen drei Stellen nachgelesen werden. Dasselbe gilt für die Begriffe »Diffusion« und »Systeme« sowie für »Sorption« und »Systeme«.

Bei den *Band- und Seitenangaben* (dritte Spalte des Registers) werden folgende Abkürzungen verwendet:

Hb	Hauptband des Hauptwerks der 8. Auflage
Eb	Ergänzungsband des Hauptwerks der 8. Auflage
Ab	Anhangband des Hauptwerks der 8. Auflage
Erg. W.	Ergänzungswerk zur 8. Auflage.

Für die Bände des Hauptwerks wird zunächst die System-Nummer und das System-Element angegeben, danach folgt die Angabe der Art des Bandes (Hauptband, Ergänzungsband oder Anhangband) und gegebenenfalls des Teils und der Lieferung und schließlich hinter einem Bindestrich die Seitenzahl bzw. Seitenzahlen. Alle Bände des Hauptwerks, die nicht Ergänzungsband oder Anhangband sind, werden dabei als Hauptband bezeichnet. Die Angabe »61 (Ag): Hb/A3–70/106« bedeutet beispielsweise: Seiten 70 bis 106 in der Lieferung 3 des Teils A der System-Nummer 61 »Silber«. Die System-Nummer 1 »Edelgase« wird mit »1 (EG)«, die System-Nummer 39 »Seltenerdelemente« mit »39 (SE)« bezeichnet.

Hinweise für den Benutzer

Die Bände des Ergänzungswerks werden nur mit der Bandnummer zitiert, es bedeutet beispielsweise »Erg. W. 12–53«: Seite 53 im Band 12 des Ergänzungswerks.

Die Ergänzungsbände zu den Lieferungen 3 bis 5 des Teils A der System-Nummer 59»Eisen« (Gmelin-Durrer) werden mit »59 (Fe): Eb/GD1a«, »59 (Fe): Eb/GD1b« usw. zitiert, die Bände »Magnetische Werkstoffe« der gleichen System-Nummer mit »59 (Fe): D/1. Eb« bzw. »59 (Fe): D/2. Eb«.

Bei *Verweisungen* wird auf die Angaben in der ersten und der zweiten Spalte des Registers verwiesen, also z. B. »*see* $Al_2Na_2O_4$ ... $Na_2O \cdot Al_2O_3$«. Bei Verweisungen innerhalb von Sachverhaltsangaben wird lediglich der Sachverhalt angeführt, z. B. »*see Deposits*«.

Im Formelregister beziehen sich die Angaben der Handbuch-Textstellen auf die alte Anordnung der Handbuch-Bände nach System-Nummern (Hauptwerk) bzw. laufenden Nummern (Ergänzungswerk). Nach der im Januar 1978 eingeführten neuen Anordnung werden die Handbuch-Bände, Hauptwerk und Ergänzungswerk gemeinsam, alphabetisch nach den Element-Symbolen abgestellt. Für die im Formelregister angeführten Zitate bedeutet dies:

1. Hauptwerk:
   Jeder Band steht bei dem Element-Symbol, das im Zitat hinter der System-Nummer angeführt ist, also z.B. der Band 61 (Ag) beim Symbol Ag.
   Ausnahmen:
   Es gehören die Bände mit dem Zitat
   1 (EG) zum Symbol He
   8 (J) zum Symbol I
   39 (SE) zum Symbol Sc
   69 (Ma) zum Symbol Tc
2. Ergänzungswerk:
   Die mit „Erg. W." zitierten Bände gehören zu folgenden Element-Symbolen:

Erg. W. 1:	He
Erg. W. 2:	V
Erg. W. 3:	Cr
Erg. W. 4, 7a, 7b, 8:	Np
Erg. W. 5, 6:	Co
Erg. W. 9, 12:	F
Erg. W. 10:	Zr
Erg. W. 11:	Hf

Die Bände Erg. W. 2 und 3 sowie 10 und 11 sind zusammengebunden.

Formula Index

O–Zr

Elements 104–132

Formula Index O–Zr

$O_{0.03}Ta$ . . . . $TaO_{0.03}$ . . . . 50 (Ta): Hb/B1–30
$O_{0.4}Sm$ . . . . $SmO_{0.4}$ . . . . 39 (SE): Hb/C1–314
$O_{0.53}V$ . . . . $VO_{0.53}$ . . . . 48 (V): Hb/B1–20
$O_{0.6}Sm$ . . . . $SmO_{0.6}$ . . . . 39 (SE): Hb/C1–314

O . . . . Oxygen
see subdivisions:
O ... Air
O ... Oxygen
O ... Ozone

–. . . . . Air:
Acoustic properties . . . . 3 (O): Hb/4–949/56
Boiling point . . . . 3 (O): Hb/4–861/3
Chemical reactions . . . . 3 (O): Hb/4–819
Composition . . . . 3 (O): Hb/4–819/20
Compressibility . . . . 3 (O): Hb/4–827/41
Critical constants . . . . 3 (O): Hb/4–858/9
Density . . . . 3 (O): Hb/4–820/7
Detection and determination. . . . . . 3 (O): Hb/4–819
Diffusion. . . . . 3 (O): Hb/4–912/21
Electrical properties . . . . 3 (O): Hb/4–967/76
Entropy . . . . 3 (O): Hb/4–891/6
Equation of state . . . . 3 (O): Hb/4–837/41
Expansion coefficient. . . . . 3 (O): Hb/4–853/6
Expansion effects . . . . 3 (O): Hb/4–841/53
Heat content. . . . . 3 (O): Hb/4–887/91
Heat of vaporization . . . . 3 (O): Hb/4–863/5
Internal energy . . . . 3 (O): Hb/4–876, 884/7
Isotherms . . . . 3 (O): Hb/4–827/37
Joule-Thomson effect . . . . 3 (O): Hb/4–842/52
Liquefaction . . . . 3 (O): Hb/2–182/98
Magnetic properties . . . . 3 (O): Hb/4–966
Mechanical properties . . . . 3 (O): Hb/4–820/948
Melting point . . . . 3 (O): Hb/4–863
Optical properties. . . . . 3 (O): Hb/4–956/66
Pressure coefficient. . . . . 3 (O): Hb/4–856/8
Rectification . . . . 3 (O): Hb/2–191/3
Solubility . . . . 3 (O): Hb/4–921/35
Sorption . . . . 3 (O): Hb/4–935/48

See the "Instructions for the user"
at the beginning of the volume

O . . . . Air:

Specific cohesion . . . . 3 (O): Hb/4–903/4
Specific heat . . . . 3 (O): Hb/4–865/84
Spectra . . . . 3 (O): Hb/4–963/6
States of aggregation. . . . . 3 (O): Hb/4–819/20
Surface tension . . . . 3 (O): Hb/4–903/4
Thermal conductivity . . . . 3 (O): Hb/4–896/903
Thermal properties . . . . 3 (O): Hb/4–820/903
Vapor pressure . . . . 3 (O): Hb/4–859/61
Viscosity. . . . . 3 (O): Hb/4–904/12

–. . . . . Oxygen:

Absorption of light . . . . 3 (O): Hb/3–527/30
– of X- and gamma rays . . . . 3 (O): Hb/3–559/61
Acoustic properties . . . . 3 (O): Hb/3–522/5
Active oxygen. . . . . 3 (O): Hb/4–976/84
Association . . . . 3 (O): Hb/3–362
Atom. . . . . 3 (O): Hb/3–319/47
Atomic nucleus . . . . 3 (O): Hb/3–319
Atomic radius . . . . 3 (O): Hb/3–324/33
Atomic scattering factors. . . . . . . 3 (O): Hb/3–374/80
Atomic volume . . . . 3 (O): Hb/3–333/42
Atomic weight . . . . 3 (O): Hb/3–342/6
Behavior at the anode . . . . 3 (O): Hb/3–614/21
– at the cathode. . . . . 3 (O): Hb/3–621/3
– at the dropping mercury electrode . . . . 3 (O): Hb/3–621/3
Birefringence, electrical . . . . 3 (O): Hb/3–531/2
–, magnetic . . . . 3 (O): Hb/3–530/1
Boiling point . . . . 3 (O): Hb/3–409/11
Boiling point elevation . . . . 3 (O): Hb/3–411
Bond energy . . . . 3 (O): Hb/3–364/5
Breakdown voltage . . . . 3 (O): Hb/3–586
Cells . . . . 3 (O): Hb/3–604/14
Chemical constant . . . . 3 (O): Hb/3–430/1
Chemical reactions . . . . 3 (O): Hb/4–819
– with hydrogen
see O . . . Oxygen mixtures: O–H

Color. . . . . 3 (O): Hb/3–525
Combustion process, history . . . . . 3 (O): Hb/1–1/33
Compressibility . . . . 3 (O): Hb/3–387/98
Compton scattering. . . . . 3 (O): Hb/3–560/1
Cosmochemistry . . . . 3 (O): Hb/2–91/6, 129
Critical constants . . . . 3 (O): Hb/3–401/3
Crystal form . . . . 3 (O): Hb/3–372
Crystallographic properties . . . . . . 3 (O): Hb/3–371/80
Debye temperature . . . . 3 (O): Hb/3–430
Density . . . . 3 (O): Hb/3–381/7
Depolarization of light . . . . 3 (O): Hb/3–533

Hb = Main Series Volume; GD = Gmelin-Durrer; Ab = Appended Volume to Main Series;
Eb = Supplement Volume to Main Series; Erg. W. = New Supplement Series Volume

O	Oxygen:		
	Detection	3 (O):	Hb/4–819
	– in iron	59 (Fe):	Hb/FI–42, 57/8
	Determination	3 (O):	Hb/4–819
	– in iron	59 (Fe):	Hb/FI–41/103
			Hb/FII2–344
	Dielectric constant	3 (O):	Hb/3–584/5
	Diffusion	3 (O):	Hb/3–446/55
	Discovery	3 (O):	Hb/1–49/77
	Dissociation	3 (O):	Hb/3–365/71
	Ebullioscopic constant	3 (O):	Hb/3–411/2
	Effusion	3 (O):	Hb/3–446
	Electrical conductivity	3 (O):	Hb/3–586/7
	Electrical moment	3 (O):	Hb/3–363/4
	Electrical properties	3 (O):	Hb/3–584/93
	Electrochemical behavior	3 (O):	Hb/3–593/623
	Electron affinity	3 (O):	Hb/3–588/90
	Electron configuration	3 (O):	Hb/3–319/20
	Electronegativity	3 (O):	Hb/3–347
	Electrostriction	3 (O):	Hb/3–585/6
	Entropy	3 (O):	Hb/3–427/30
	Equation of state	3 (O):	Hb/3–391/5
	Expansion coefficient	3 (O):	Hb/3–399/400
	Expansion effects	3 (O):	Hb/3–398/9
	Faraday effect	3 (O):	Hb/3–530/1
	Free energy	3 (O):	Hb/3–426/7
	Geochemistry	3 (O):	Hb/2–96/101, 111/81
	Handling	3 (O):	Hb/2–225
	Heat content	3 (O):	Hb/3–426/7
	Heat of fusion	3 (O):	Hb/3–415/6
	– of sublimation	3 (O):	Hb/3–415
	– of transformation	3 (O):	Hb/3–416
	– of vaporization	3 (O):	Hb/3–413/5
	History	3 (O):	Hb/1–1/82
			Hb/2–183/4
			Hb/3–343/5
	Industrial hazards	3 (O):	Hb/2–226
	Internal energy	3 (O):	Hb/3–426
	Internal pressure	3 (O):	Hb/3–439
	Ionization potentials	3 (O):	Hb/3–590/2
	Joshi effect	3 (O):	Hb/3–587/8
	Kerr effect	3 (O):	Hb/3–531/2
	Lattice structure	3 (O):	Hb/3–372/4
	Liquefaction	3 (O):	Hb/3–314
	Magnetic moment	3 (O):	Hb/3–324, 363
	Magnetic properties	3 (O):	Hb/3–578/84
	Magnetostriction	3 (O):	Hb/3–583/4
	Mass absorption coefficient	3 (O):	Hb/3–559

See the "Instructions for the user"
at the beginning of the volume

O . . . . Oxygen:

Mass fluorescence coefficient	3 (O):	Hb/3–559/60
Mass scattering coefficient.	3 (O):	Hb/3–560
Mechanical properties	3 (O):	Hb/3–381/525
Melting point	3 (O):	Hb/3–412/3
Molar volume	3 (O):	Hb/3–360/1
Molecular radius	3 (O):	Hb/3–354/9
Molecular structure.	3 (O):	Hb/3–348/51
Molecular weight	3 (O):	Hb/3–361/2
Molecule.	3 (O):	Hb/3–348/71
Moment of inertia.	3 (O):	Hb/3–362/3
Occurrence, extraterrestrial	3 (O):	Hb/2–91/6, 129
–, terrestrial	3 (O):	Hb/2–96/181
– in the atmosphere	3 (O):	Hb/2–96/101
– in the biosphere.	3 (O):	Hb/2–179/81
– in the hydrosphere	3 (O):	Hb/2–111/9, 161/79
– in the lithosphere	3 (O):	Hb/2–119/28, 129/38, 145/61
Optical properties.	3 (O):	Hb/3–525/78
Oxygen electrode	3 (O):	Hb/3–595/604
Oxyhydrogen gas electrode	3 (O):	Hb/3–604
Parachor.	3 (O):	Hb/3–439/41
Phlogiston theory.	3 (O):	Hb/1–33/49
Photoelectron emission	3 (O):	Hb/3–593
Polarizablitity	3 (O):	Hb/3–320/3, 364
Polymorphism	3 (O):	Hb/3–371/2
Position in the electrochemical series	3 (O):	Hb/3–593/4
– in the periodic system	3 (O):	Hb/3–346
Potentials	3 (O):	Hb/3–594/5
Preparation	3 (O):	Hb/3–301/14
–, industrial	3 (O):	Hb/2–182/226
–, –, from air.	3 (O):	Hb/2–182/202
–, –, from substances rich in oxygen .	3 (O):	Hb/2–221/4
–, –, from water	3 (O):	Hb/2–202/21
Pressure coefficient.	3 (O):	Hb/3–400/1
Purification	3 (O):	Hb/2–224 Hb/3–311/3
Purity testing	3 (O):	Hb/2–224/5 Hb/3–313
Refraction	3 (O):	Hb/3–525/7
Rheochor	3 (O):	Hb/3–446
Scattering of light.	3 (O):	Hb/3–533
– of X- and gamma rays	3 (O):	Hb/3–559/61
Solubility	3 (O):	Hb/3–455/94
– in gels and hydrosols.	3 (O):	Hb/3–481
– in ice	3 (O):	Hb/3–473
– in inorganic solvents	3 (O):	Hb/3–475/82

O	Oxygen:		
	Solubility in iron	59 (Fe):	Hb/A5–856 Eb/GD1a–57/8 Eb/GD1b–11
	– in natural waters	3 (O):	Hb/3–473/5
	– in organic solvents	3 (O):	Hb/3–481/94
	– in steam	3 (O):	Hb/3–473
	– in tin	46 (Sn):	Hb/B–351
	– in water	3 (O):	Hb/3–456/75
	Sorption on carbon	3 (O):	Hb/3–495/508
	– on metal oxides	3 (O):	Hb/3–511/6
	– on organic substances	3 (O):	Hb/3–521
	– on palladium	65 (Pd):	Hb/2–257/9
	– on platinum	68 (Pt):	Hb/C1–26/41
	– on silica gel	3 (O):	Hb/3–508/10
	– on other inorganic substances	3 (O):	Hb/3–516/21
	Sound absorption	3 (O):	Hb/3–524/5
	Sound velocity	3 (O):	Hb/3–522/4
	Specific heat	3 (O):	Hb/3–416/24
	Spectra:		
	Arc spectrum	3 (O):	Hb/3–534/46
	Band spectrum	3 (O):	Hb/3–563/78
	Continuous spectrum	3 (O):	Hb/3–578
	Microwave spectrum	3 (O):	Hb/3–562/3
	Raman spectrum	3 (O):	Hb/3–533/4
	Spark spectra	3 (O):	Hb/3–546/57
	X-ray spectrum	3 (O):	Hb/3–557/9
	Standard potential	3 (O):	Hb/3–593
	Structure	3 (O):	Hb/3–372/4
	Surface tension	3 (O):	Hb/3–436/9
	Susceptibility	3 (O):	Hb/3–578/83
	Thermal conductivity	3 (O):	Hb/3–431/6
	Thermal properties	3 (O):	Hb/3–387/436
	Thermochor	3 (O):	Hb/3–441
	Toxicity	3 (O):	Hb/2–298
	Transformation points	3 (O):	Hb/3–413
	Triple point	3 (O):	Hb/3–408/9, 412/3
	Valence	3 (O):	Hb/3–346/7
	Vapor pressure	3 (O):	Hb/3–403/9
	Vibration frequencies	3 (O):	Hb/3–354, 430
	Viscosity	3 (O):	Hb/3–441/6
	Working dangers	3 (O):	Hb/2–226
–	$O_2 \cdot xH_2O$	3 (O):	Hb/5–1191
–	Ozone:		
	Chemical reactions *see also Decomposition*	3 (O):	Hb/4–1147/66
	– with air	3 (O):	Hb/4–1147

See the "Instructions for the user"
at the beginning of the volume

O . . . . Ozone:

	Chemical reactions with elements . . .	3 (O):	Hb/4–1148/57
	– with inorganic compounds	3 (O):	Hb/4–1147/8, 1157/66
	– with organic compounds	3 (O):	Hb/4–1166
	– with water	3 (O):	Hb/4–1147/8
	Decomposition	3 (O):	Hb/4–988/1082
	– in corpuscular radiation	3 (O):	Hb/4–1077/82
	– in electrical discharges	3 (O):	Hb/4–1038/77
	– in photochemical processes	3 (O):	Hb/4–1026/38
	– in thermal processes	3 (O):	Hb/4–999/1019
	Decomposition equilibria	3 (O):	Hb/4–988/94
	Detection	3 (O):	Hb/4–819
	Determination	3 (O):	Hb/4–819
	Electrical properties	3 (O):	Hb/4–1139
	Electrochemical behavior	3 (O):	Hb/4–1139/47
	Formation	3 (O):	Hb/4–988/1100
	–, formation equilibria	3 (O):	Hb/4–988/94
	–, thermodynamic data	3 (O):	Hb/4–1109/10
	– in chemical reactions	3 (O):	Hb/3–640/1, 778, 782 Hb/4–1092/9
	– in corpuscular radiation	3 (O):	Hb/4–1077/82
	– in electrical discharges	3 (O):	Hb/4–1038/77
	– in electrolytic processes	3 (O):	Hb/4–1082/91
	– in photochemical processes	3 (O):	Hb/4–1020/5
	– in thermal processes	3 (O):	Hb/4–995/9
	– in other processes	3 (O):	Hb/4–1099/100
	History	3 (O):	Hb/4–984/7
	Magnetic properties	3 (O):	Hb/4–1139
	Mechanical properties	3 (O):	Hb/4–1117/24
	Molecule	3 (O):	Hb/4–1111/7
	Occurrence	3 (O):	Hb/2–101/11
	Optical properties	3 (O):	Hb/4–1124/39
	Preparation	3 (O):	Hb/4–1100/9
	–, industrial	3 (O):	Hb/2–226/32
	Solubility	3 (O):	Hb/4–1121/3
	Sorption	3 (O):	Hb/4–1123/4
	Spectra	3 (O):	Hb/4–1127/39
	Thermal properties	3 (O):	Hb/4–1117/24
	Toxicity	3 (O):	Hb/2–298/300
	Uses	3 (O):	Hb/2–232
–.	$^{16}O^{16}O^{18}O$	3 (O):	Hb/4–1111
–.	$^{16}O^{18}O^{16}O$	3 (O):	Hb/4–1111
–.	$^{16}O^{18}O^{18}O$	3 (O):	Hb/4–1111
–.	$^{18}O_3$	3 (O):	Hb/4–1111
–.	$O_3 \cdot H_2O = H_2O_4$	3 (O):	Hb/8–2626/7

–. . . . . Oxygen ions
see O^{n+} ... O^{n+}, etc.

Hb = Main Series Volume; GD = Gmelin-Durrer; Ab = Appended Volume to Main Series;
Eb = Supplement Volume to Main Series; Erg. W. = New Supplement Series Volume

O	Oxygen isotopes:		
	Atomic nucleus	3 (O):	Hb/3–319
	Determination in water	3 (O):	Hb/6–2019/23
	Enrichment and separation	3 (O):	Hb/3–314/8
	Occurrence	3 (O):	Hb/2–83/91
–	Oxygen mixtures:		
	Air–Ar–CO	14 (C):	Hb/C2–105
	Air–CH_4	3 (O):	Hb/4–902/3
	Air–CH_4–CO	14 (C):	Hb/C2–99, 106
	Air–CH_4–CO–H	14 (C):	Hb/C2–99, 107
	Air–CO	3 (O):	Hb/4–902/3
		14 (C):	Hb/C1–276
			Hb/C2–94/7, 103/4
	Air–CO–CO_2	14 (C):	Hb/C2–105
	Air–CO–C_2H_2	14 (C):	Hb/C2–106
	Air–CO–C_2H_4	14 (C):	Hb/C2–99, 106
	Air–CO–C_2H_4–H	14 (C):	Hb/C2–107
	Air–CO–C_2H_6	14 (C):	Hb/C2–99, 106
	Air–CO–C_3H_8	14 (C):	Hb/C2–99
	Air–CO–H	14 (C):	Hb/C2–98, 106
	Air–CO–He	14 (C):	Hb/C2–105
	Air–CO–N	14 (C):	Hb/C2–105
	Air–CO_2	3 (O):	Hb/4–886/7, 890/1
		14 (C):	Hb/C1–528/31
	Air–C_2H_2	3 (O):	Hb/4–902/3
	Air–C_2H_4	3 (O):	Hb/4–912
	Air–C_2H_5OH	3 (O):	Hb/4–896
	Air–CH_3COCH_3	3 (O):	Hb/4–825, 884, 903
	Air–$(C_2H_5)_2O$	3 (O):	Hb/4–825, 884
	Air–C_6H_6	3 (O):	Hb/4–825
	Air–C_8H_{18}	3 (O):	Hb/4–896
	Air–C_8H_{18}–H_2O	3 (O):	Hb/4–896
	Air–H	3 (O):	Hb/4–825, 884/7, 890/1, 903, 912
	Air–H_2O	3 (O):	Hb/4–886/7, 890/1, 896, 902/3, 912
	Air–He	3 (O):	Hb/4–845
	Air–NH_3	3 (O):	Hb/4–902/3
	O–Ar–CO	14 (C):	Hb/C2–109
	O–Ar–N	4 (N):	Hb/1–151
	O–CO	3 (O):	Hb/3–510, 512, 515
		14 (C):	Hb/C1–268/9
			Hb/C2–75/111
	O–CO–CO_2	14 (C):	Hb/C2–95/6, 108/9
	O–CO–C_2H_4	14 (C):	Hb/C1–284
	O–CO–Cl	14 (C):	Hb/C2–109
	O–CO–H	14 (C):	Hb/C2–97/8
	O–CO–H_2O	14 (C):	Hb/C2–98/9

See the "Instructions for the user"
at the beginning of the volume

O . . . . Oxygen mixtures:

O–CO–He	14 (C):	Hb/C2–109
O–CO–N	4 (N):	Hb/1–143
	14 (C):	Hb/C2–95/6, 108
O–CO–N–H_2O	14 (C):	Hb/C2–99
O–CO_2	3 (O):	Hb/3–510, 515
	14 (C):	Hb/C1–520/1
O–CO_2–H–N	14 (C):	Hb/C1–528
O–CO_2–N	14 (C):	Hb/C1–527/8
		Hb/C3–79
O–C_2H_4	3 (O):	Hb/3–395, 446
O–H:		
Compression coefficient	3 (O):	Hb/5–1190
Heat conductivity	3 (O):	Hb/3–432, 436
		Hb/5–1190
Hydrogen flames	3 (O):	Hb/3–768/803
–, flame temperature	3 (O):	Hb/3–783/9
–, flame velocity and flame propagation	3 (O):	Hb/3–781/803
–, general properties	3 (O):	Hb/3–768/70
–, spectrum and chemism	3 (O):	Hb/3–770/82
Ignition limits	3 (O):	Hb/3–804/16
Ignition temperature	3 (O):	Hb/3–816/8
Oxyhydrogen gas coulometer	3 (O):	Hb/3–620/1
Oxyhydrogen gas electrode	3 (O):	Hb/3–604
Photochemical reaction	3 (O):	Hb/3–639/51
Reaction in presence of electrical discharges	3 (O):	Hb/3–633/9
– – of foreign O or H atoms	3 (O):	Hb/3–625/30
– – of ionizing radiation	3 (O):	Hb/3–630/3
Sound velocity	3 (O):	Hb/3–523
Surface reaction	3 (O):	Hb/3–651/92
– on metals	3 (O):	Hb/3–655/85
– on oxides	3 (O):	Hb/3–685/91
– on other materials	3 (O):	Hb/3–691/2
Thermal reaction	3 (O):	Hb/3–692/768
–, detonation limits	3 (O):	Hb/3–695/737
–, detonation peninsula	3 (O):	Hb/3–702/6
–, reaction mechanism	3 (O):	Hb/3–737/59
–, sensitizers and inhibitors	3 (O):	Hb/3–759/68
–, steady reaction	3 (O):	Hb/3–714/31
Viscosity	3 (O):	Hb/3–445/6
O–H–N	4 (N):	Hb/1–142
O–H_2O	3 (O):	Hb/3–523
		Hb/6–1750
O–He	3 (O):	Hb/3–523
		Hb/5–1185
O–N	3 (O):	Hb/3–510, 512
	4 (N):	Hb/1–142, 148, 168

Hb = Main Series Volume; GD = Gmelin-Durrer; Ab = Appended Volume to Main Series; Eb = Supplement Volume to Main Series; Erg. W. = New Supplement Series Volume

O	Oxygen mixtures:		
	$O-NH_3$	4 (N):	Hb/3-639/84
–.	Oxygen solutions:		
	Ozone-CO_2	3 (O):	Hb/4-1167
	Ozone-Cl	3 (O):	Hb/4-1167
	Ozone-H_2O	3 (O):	Hb/4-1167/83
	Ozone-inorganic solvents	3 (O):	Hb/4-1183/4
	Ozone-N	3 (O):	Hb/4-1167
	Ozone-NH_3	3 (O):	Hb/4-1167
	Ozone-O	3 (O):	Hb/4-1166/7
–.	Oxygen systems:		
	O-Ag	61 (Ag):	Hb/B1-33/53
	O-Ag-Ag_2O-B_2O_3	61 (Ag):	Hb/B3-249
	O-Ag-Cr_2O_3	61 (Ag):	Hb/B4-288
	O-Ag-Cu	61 (Ag):	Hb/B4-438/9
	O-Ag-Mn	61 (Ag):	Hb/B4-359
	O-Ag-S.	9 (S):	Hb/B1-185
		61 (Ag):	Hb/B3-61/2
	O-Al-Cu	60 (Cu):	Hb/B3-1169/70
	O-Al-Ta	50 (Ta):	Hb/B2-197
	O-Am	Erg. W. 4-32/4	
	O-Am-Cm	Erg. W. 4-79	
	O-Ar.	3 (O):	Hb/5-1185/8
	O-Ar-CO-N	14 (C):	Hb/C1-276
	O-Ba-Cu	60 (Cu):	Hb/B3-1139
	O-Ba-Ni	57 (Ni):	Hb/B3-1125/6
	O-Ba-S.	9 (S):	Hb/B1-183
	O-BaO-BaO_2.	30 (Ba):	Hb-93/4
	O-Be-Ta	50 (Ta):	Hb/B2-172
	O-Bi.	19 (Bi):	Eb-464
	O-Bi-Cu-Fe	60 (Cu):	Hb/B3-1315
	O-Bi-Nb	49 (Nb):	Hb/B1-422/3
	O-Bi-Pb	47 (Pb):	Hb/C3-966
	O-Bi-S	9 (S):	Hb/B1-182
		19 (Bi):	Eb-745
	O-Bi-Ta	50 (Ta):	Hb/B1-275
	O-Bi-Te	19 (Bi):	Eb-817
	O-C-Fe	59 (Fe):	Hb/A5-861/74
			Hb/A8-1703/16
			Hb/B1-21/6
			Eb/GD1a-58/64
			Eb/GD1b-11/2
	O-C-Mo	53 (Mo):	Hb-196
	O-C-N-Nb.	49 (Nb):	Hb/B1-387
	O-C-N-Ta	50 (Ta):	Hb/B1-250
	O-C-Nb	49 (Nb):	Hb/B1-381/2
	O-C-Pu	Erg. W. 4-224/5	
	O-C-Ta	50 (Ta):	Hb/B1-248

O . . . . Oxygen systems:

O–C–Ti	41 (Ti):	Hb–368
O–C–V	48 (V):	Hb/B1–335/6
O–C–Zn	32 (Zn):	Eb–977
O–CO_2	14 (C):	Hb/C1–516/20
O–Ca–Ni	57 (Ni):	Hb/B3–1118
O–Ca–S	9 (S):	Hb/B1–182/3
	28 (Ca):	Hb/B3–657/60
O–CaS	28 (Ca):	Hb/B3–658
O–Cd	33 (Cd):	Eb–418
O–Ce–N	39 (SE):	Hb/C2–179/80
O–Cl–S–Ta	50 (Ta):	Hb/B1–178
O–Cm	Erg. W. 4–35	
O–Co	58 (Co):	Eb/A–475
O–Co–CoO	58 (Co):	Eb/A–475/6
O–Co–Co_3O_4	58 (Co):	Eb/A–476
O–Co–Cu	60 (Cu):	Hb/B3–1263
O–Co–H	58 (Co):	Hb/A2–224
O–Co–S	9 (S):	Hb/B1–183
	58 (Co):	Eb/A–626
O–Co–V	58 (Co):	Eb/A–873
O–Co–W	58 (Co):	Eb/A–877
O–CoO–Co_3O_4	58 (Co):	Eb/A–476
O–Cr	52 (Cr):	Hb/B–7/9
O–Cr–Cu	60 (Cu):	Hb/B3–1215/7
O–Cu	60 (Cu):	Hb/B1–23/34
O–Cu–Fe	60 (Cu):	Hb/B3–1270
O–Cu–Fe–Mn	60 (Cu):	Hb/B3–1330
O–Cu–Fe–S	60 (Cu):	Hb/B3–1295
O–Cu–Fe_2O_3	60 (Cu):	Hb/B3–1270
O–Cu–Mn	60 (Cu):	Hb/B3–1242/3
O–Cu–Mn–S	60 (Cu):	Hb/B3–1246
O–Cu–Mo	60 (Cu):	Hb/B3–1230
O–Cu–Ni	60 (Cu):	Hb/B3–1250/2
O–Cu–P	60 (Cu):	Hb/B2–918
O–Cu–Pb	60 (Cu):	Hb/B3–1197/8
O–Cu–S	9 (S):	Hb/B1–184/5
O–Cu–U	60 (Cu):	Hb/B3–1240
O–Cu–W	60 (Cu):	Hb/B3–1235
O–Cu–Zn	60 (Cu):	Hb/B3–1145/6
O–Er	39 (SE):	Hb/C1–414/5
O–Eu	39 (SE):	Hb/C1–323/5
O–Eu–N	39 (SE):	Hb/C2–180
O–Fe	59 (Fe):	Hb/A5–856/60
		Hb/A8–1692/8
		Hb/B1–12/5
		Eb/GD1a–53/7
		Eb/GD1b–9/11

Hb = Main Series Volume; GD = Gmelin-Durrer; Ab = Appended Volume to Main Series;
Eb = Supplement Volume to Main Series; Erg. W. = New Supplement Series Volume

O . . . . Oxygen systems:

O–Fe–H	59 (Fe):	Hb/A8–1700/2
		Hb/B1–17/21
		Eb/GD1a–58/64
		Eb/GD1b–11/2
O–Fe–Mn	59 (Fe):	Hb/A5–881/923
O–Fe–Ni	59 (Fe):	Hb/A5–942/3
O–Fe–Ni–Si	59 (Fe):	Hb/A5–943
O–Fe–P	59 (Fe):	Hb/A5–923/35
O–Fe–S	9 (S):	Hb/B1–183/4
O–Fe–Si	59 (Fe):	Hb/A5–876/81
O–Ga–Si–V	48 (V):	Hb/B2–592
O–Ga–V.	48 (V):	Hb/B2–591/2
O–Gd–N	39 (SE):	Hb/C2–181
O–H	3 (O):	Hb/5–1190/1
O–H–Nb–H_2O	49 (Nb):	Hb/B1–33, 52
O–H_2O.	3 (O):	Hb/6–1750
O–He	3 (O):	Hb/5–1185
O–Hf–Nb	49 (Nb):	Hb/B3–218
O–Hf–Ni	57 (Ni):	Hb/B1–118
O–K–Nb	49 (Nb):	Hb/B2–1/2
O–Kr.	3 (O):	Hb/5–1188/9
O–La–N.	39 (SE):	Hb/C2–179
O–Li–Ta.	50 (Ta):	Hb/B2–1, 126
O–Mg–Ta.	50 (Ta):	Hb/B2–172
O–Mn	56 (Mn):	Hb/C1–7/14
O–Mn–H_2O	56 (Mn):	Hb/C1–367/8, 414
O–Mn–Si	59 (Fe):	Hb/A5–941/2
O–Mo–Ni.	57 (Ni):	Hb/B3–1217
O–N–Nb	49 (Nb):	Hb/B1–127
O–N–Pm	39 (SE):	Hb/C2–180
O–N–Sm	39 (SE):	Hb/C2–180
O–Na–Ta	50 (Ta):	Hb/B2–141
O–Nb	49 (Nb):	Hb/B1–14/23
O–Nb–Ta	50 (Ta):	Hb/B2–245/6
O–Nb–Ti	49 (Nb):	Hb/B3–178/9
O–Nb–Zn	49 (Nb):	Hb/B3–94
O–Nb–Zr	49 (Nb):	Hb/B2–128/9, 171/2
		Hb/B3–209
O–Nd	39 (SE):	Hb/C1–296/7
O–Ne	3 (O):	Hb/5–1185
O–Ni.	57 (Ni):	Hb/B2–377/8
O–Ni–S	9 (S):	Hb/B1–183
	57 (Ni):	Hb/B2–676/7
O–Ni–Ti.	57 (Ni):	Hb/B1–97
O–Ni–V	57 (Ni):	Hb/B1–159
		Hb/B3–1198
O–Ni–W	57 (Ni):	Hb/B3–1225

O . . . . Oxygen systems:
O–Ni–Zr . . . . 57 (Ni): Hb/B1–114
O–Np–Sc . . . . Erg. W. 4–45/6
O–ozone . . . . 3 (O): Hb/4–1166/7
O–Pb . . . . 47 (Pb): Hb/C1–51/2
O–Pb–S . . . . 9 (S): Hb/B1–183
47 (Pb): Hb/C2–533/4
O–Pb–Sb . . . . 47 (Pb): Hb/C3–926/9
O–Pb–Ta . . . . 50 (Ta): Hb/B2–232
O–Pb–Te . . . . 47 (Pb): Hb/C2–678/9
O–Pd–S . . . . 9 (S): Hb/B1–185
O–PrO_x . . . . 39 (SE): Hb/C1–275/8
O–Pt . . . . 68 (Pt): Hb/C1–26/41
O–Pu . . . . Erg. W. 4–10/2
O–Pu–H_2O . . . . Erg. W. 4–10/2
O–Pu–U . . . . Erg. W. 4–59/60
O–Ru . . . . 63 (Ru): Eb–232/3
O–Ru–H_2O . . . . 63 (Ru): Eb–232/3
O–S–Sb . . . . 9 (S): Hb/B1–182
O–S–Sr . . . . 9 (S): Hb/B1–183
O–S–Ti . . . . 41 (Ti): Hb–343
O–Sc . . . . 39 (SE): Hb/C1–180
O–Si–Ta . . . . 50 (Ta): Hb/B1–263
O–Sm . . . . 39 (SE): Hb/C1–313/4
O–Sn . . . . 46 (Sn): Hb/C1–24/5
O–Ta . . . . 50 (Ta): Hb/B1–17/23
O–Ta–Th . . . . 50 (Ta): Hb/B2–229
O–Ta–Ti . . . . 50 (Ta): Hb/B2–221
O–Ta–Ti–Zr . . . . 50 (Ta): Hb/B2–229
O–Ta–V . . . . 50 (Ta): Hb/B2–244
O–Ta–Y . . . . 50 (Ta): Hb/B2–205
O–Ta–Zr . . . . 50 (Ta): Hb/B2–228
O–TbO_x . . . . 39 (SE): Hb/C1–391/4
O–Ti . . . . 41 (Ti): Hb–211/5
O–Ti–V . . . . 48 (V): Hb/B2–658
O–U . . . . 55 (U): Hb–82
O–V . . . . 48 (V): Hb/B1–11/2, 18/9
O–Xe . . . . 3 (O): Hb/5–1190
O–Y . . . . 39 (SE): Hb/C1–188/9
O–Yb . . . . 39 (SE): Hb/C1–426
O–Zn . . . . 32 (Zn): Eb–809
O–Zr . . . . 42 (Zr): Hb–208/12

Ozone–KOH . . . . 22 (K): Hb/1–245/6
Ozone–O . . . . 3 (O): Hb/4–1166/7

O^{n+} . . . . O^{n+} (n = 1 to 7):
Electron configuration . . . . 3 (O): Hb/3–320
Formation . . . . 3 (O): Hb/3–370/1

Hb = Main Series Volume; GD = Gmelin-Durrer; Ab = Appended Volume to Main Series;
Eb = Supplement Volume to Main Series; Erg. W. = New Supplement Series Volume

O^{n+} . . . . O^{n+} (n = 1 to 7):
Ionic radius . . . . 3 (O): Hb/3–326/7
Ionic volume . . . . 3 (O): Hb/3–335
Ionization potentials . . . . 3 (O): Hb/3–590/2
Magnetic moment . . . . 3 (O): Hb/3–324
Spectra . . . . 3 (O): Hb/3–546/57

O^- . . . . O^- . . . . 3 (O): Hb/3–320, 324, 335, 557

O^{2-} . . . . O^{2-} . . . . 3 (O): Hb/3–320, 321/3, 324/33, 335/42

OOs . . . . OsO . . . . 66 (Os): Hb–25
–. . . . . $OsO \cdot H_2O$ . . . . 66 (Os): Hb–25
OP. . . . . PO . . . . 16 (P): Hb/C–64/6
–. . . . . $(PO)_x$ . . . . 16 (P): Hb/C–67
OP^+ . . . . $[PO]^+$ . . . . 16 (P): Hb/C–66
OP_2 . . . . P_2O . . . . 16 (P): Hb/C–59
OP_2Zn. . . . . ZnP_2O . . . . 32 (Zn): Hb–276
OP_3 . . . . P_3O . . . . 16 (P): Hb/C–59
OP_4 . . . . P_4O . . . . 16 (P): Hb/C–58/64
–. . . . . $P_4O \cdot 2\,H_2O$ . . . . 16 (P): Hb/C–62
OPb . . . . PbO:
Chemical reactions . . . . 47 (Pb): Hb/C1–105/19
Colloidal forms . . . . 47 (Pb): Hb/C1–62/3
Crystallographic properties . . . . . . 47 (Pb): Hb/C1–68/74
Electrical properties . . . . 47 (Pb): Hb/C1–83/95
Electrochemical behavior . . . . 47 (Pb): Hb/B2–414
Hb/C1–98/105
Formation . . . . 47 (Pb): Hb/C1–57/61, 65/7
History. . . . . 47 (Pb): Hb/A1–100/11
Magnetic properties . . . . 47 (Pb): Hb/C1–84/5
Mechanical properties . . . . 47 (Pb): Hb/C1–74/6
Molecule. . . . . 47 (Pb): Hb/C1–67/8
Optical properties. . . . . 47 (Pb): Hb/C1–95/8
Preparation . . . . 47 (Pb): Hb/C1–57/65
–, industrial . . . . 47 (Pb): Hb/C1–63/5
– of special forms. . . . . 47 (Pb): Hb/C1–61/3
Purity testing . . . . 47 (Pb): Hb/C1–64/5
Spectra . . . . 47 (Pb): Hb/C1–68
Thermal properties . . . . 47 (Pb): Hb/C1–76/83

–. . . . . $PbO \cdot 0.33\,H_2O = 3\,PbO \cdot H_2O = Pb_3O_2(OH)_2$ . . . . 47 (Pb): Hb/C1–121/2
–. . . . . $8\,PbO \cdot 3\,H_2O$ . . . . 47 (Pb): Hb/C1–119, 121
–. . . . . $PbO \cdot 0.4\,H_2O = 5\,PbO \cdot 2\,H_2O$ . . . . . . 47 (Pb): Hb/C1–121
–. . . . . $PbO \cdot 0.5\,H_2O = 2\,PbO \cdot H_2O = Pb_2O(OH)_2$ . . . . 47 (Pb): Hb/C1–120
–. . . . . $PbO \cdot 0.6\,H_2O = 5\,PbO \cdot 3\,H_2O$ . . . . . . 47 (Pb): Hb/C1–120
–. . . . . $PbO \cdot nH_2O$ . . . . 47 (Pb): Hb/C1–119/22

OPb . . . . PbO glasses:

$PbO-AlF_3-B_2O_3$	47 (Pb):	Hb/C4–1295/6
$PbO-Al_2O_3-B_2O_3$	47 (Pb):	Hb/C4–1295
$PbO-Al_2O_3-B_2O_3-CeO_2$	47 (Pb):	Hb/C4–1375
$PbO-Al_2O_3-B_2O_3-Na_2O$	47 (Pb):	Hb/C4–1302
$PbO-Al_2O_3-SiO_2$	47 (Pb):	Hb/C4–1297/300
$PbO-B_2O_3-Bi_2O_3$	47 (Pb):	Hb/C3–978
$PbO-B_2O_3-Cs_2O$	47 (Pb):	Hb/C3–1156
$PbO-B_2O_3-K_2O-SiO_2$	47 (Pb):	Hb/C3–1109/10
$PbO-B_2O_3-La_2O_3$	47 (Pb):	Hb/C4–1375
$PbO-B_2O_3-Li_2O$	47 (Pb):	Hb/C3–992/3
$PbO-B_2O_3-Na_2O-SiO_2$	47 (Pb):	Hb/C3–1047
$PbO-B_2O_3-SiO_2$	47 (Pb):	Hb/C3–858/60
$PbO-BaO-SiO_2$	47 (Pb):	Hb/C3–1208/10
$PbO-Bi_2O_3$	47 (Pb):	Hb/C3–978
$PbO-Bi_2O_3-P_2O_5$	47 (Pb):	Hb/C3–978
$PbO-Bi_2O_3-SiO_2$	47 (Pb):	Hb/C3–978
$PbO-CaO-SiO_2$	47 (Pb):	Hb/C3–1191/2
$PbO-CdO-SiO_2$	47 (Pb):	Hb/C4–1257
$PbO-GeO_2$	45 (Ge):	Eb–501
	47 (Pb):	Hb/C4–1475/6
$PbO-K_2O-Li_2O-SiO_2$	47 (Pb):	Hb/C3–1116
$PbO-K_2O-Na_2O-SiO_2$	47 (Pb):	Hb/C3–1120/2
$PbO-K_2O-SiO_2$	47 (Pb):	Hb/C3–1102/6
$PbO-La_2O_3-SiO_2$	47 (Pb):	Hb/C4–1376
$PbO-Li_2O-Na_2O-SiO_2$	47 (Pb):	Hb/C3–1052
$PbO-Li_2O-SiO_2$	47 (Pb):	Hb/C3–993
$PbO-Li_2O-TeO_2$	47 (Pb):	Hb/C3–992/3
$PbO-Na_2O-SiO_2$	47 (Pb):	Hb/C3–1039/46
$PbO-Na_2O-TeO_2$	47 (Pb):	Hb/C3–1034
$PbO-P_2O_5$	47 (Pb):	Hb/C3–864/5
$PbO-P_2O_5-ThO_2$	47 (Pb):	Hb/C4–1471
$PbO-Pr_2O_3-SiO_2$	47 (Pb):	Hb/C4–1376
$PbO-Rb_2O-SiO_2$	47 (Pb):	Hb/C3–1144/5
$PbO-Sb_2O_3-SiO_2$	47 (Pb):	Hb/C3–941
$PbO-SiO_2$:		
Chemical reactions	47 (Pb):	Hb/C3–835/6
Enamel	47 (Pb):	Hb/C3–836
Formation and preparation	47 (Pb):	Hb/C3–824/5
Physical properties	47 (Pb):	Hb/C3–827/34
Solder glass	47 (Pb):	Hb/C3–836
Structure	47 (Pb):	Hb/C3–825/6
$PbO-SiO_2-SrO$	47 (Pb):	Hb/C3–1199
$PbO-SiO_2-ThO_2$	47 (Pb):	Hb/C4–1471
$PbO-SiO_2-TiO_2$	47 (Pb):	Hb/C4–1392/4
$PbO-SiO_2-ZnO$	47 (Pb):	Hb/C4–1231
$PbO-TeO_2$	47 (Pb):	Hb/C2–678/9

Hb = Main Series Volume; GD = Gmelin-Durrer; Ab = Appended Volume to Main Series;
Eb = Supplement Volume to Main Series; Erg. W. = New Supplement Series Volume

OPb . . . . PbO melts:
$PbO-B_2O_3-Na_2O-SiO_2$ . . . . 47 (Pb): Hb/C3-1047
$PbO-B_2O_3-Tl_2O$ . . . . 47 (Pb): Hb/C4-1354
$PbO-MgO$ . . . . 47 (Pb): Hb/C3-1175
$PbO-Na_2B_4O_7$ . . . . 47 (Pb): Hb/C3-1034
$PbO-Na_2O$ . . . . 47 (Pb): Hb/C3-1004/5
$PbO-SiO_2$ . . . . 47 (Pb): Hb/C3-810/23
$PbO-SiO_2-Tl_2O$ . . . . 47 (Pb): Hb/C4-1354

–. . . . . PbO phosphors:
$PbO-SiO_2-ZnO$ . . . . 47 (Pb): Hb/C4-1231

–. . . . . PbO solid solutions:
$PbO-Al_2O_3-Ga_2O_3$ . . . . 47 (Pb): Hb/C4-1306
$PbO-BaO-HfO_2-Nb_2O_5$ . . . . 49 (Nb): Hb/B3-319
$PbO-Fe_2O_3$ . . . . 59 (Fe): D/2. Eb-441
$PbO-K_2O-Nb_2O_5-Ta_2O_5$ . . . . 50 (Ta): Hb/B2-265
$PbO-Na_2O-Nb_2O_5-Ta_2O_5$ . . . . 50 (Ta): Hb/B2-265
$PbO-Sb_2O_5-SnO_2$ . . . . 47 (Pb): Hb/C4-1595

–. . . . . PbO solutions:
$PbO-NaOH-ZnO-H_2O$ . . . . 47 (Pb): Hb/C4-1230

–. . . . . PbO systems:
$PbO-Ag_2O-CuO$ . . . . 61 (Ag): Hb/B4-441
$PbO-Ag_2O-Cu_2O$ . . . . 61 (Ag): Hb/B4-441
$PbO-Al_2O_3$ . . . . 47 (Pb): Hb/C4-1291
$PbO-Al_2O_3-B_2O_3-Y_2O_3$ . . . . 47 (Pb): Hb/C4-1375/6
$PbO-Al_2O_3-SiO_2$ . . . . 47 (Pb): Hb/C4-1296/7
$PbO-As_2O_3$ . . . . 47 (Pb): Hb/C3-889
$PbO-As_2O_3-H_2O$ . . . . 47 (Pb): Hb/C3-889
$PbO-As_2O_5$ . . . . 47 (Pb): Hb/C3-890/1
$PbO-As_2O_5-H_2O$ . . . . 47 (Pb): Hb/C3-890/1
$PbO-B_2O_3$ . . . . 47 (Pb): Hb/C2-685/91
$PbO-B_2O_3-BaO-TiO_2$ . . . . 47 (Pb): Hb/C4-1412/3
$PbO-B_2O_3-Bi_2O_3-CdO$ . . . . 47 (Pb): Hb/C4-1258
$PbO-B_2O_3-CdO$ . . . . 47 (Pb): Hb/C4-1256/7
$PbO-B_2O_3-Na_2O$ . . . . 47 (Pb): Hb/C3-1034
$PbO-B_2O_3-Nb_2O_5$ . . . . 49 (Nb): Hb/B4-313/4
$PbO-B_2O_3-PbTiO_3-PbZrO_3$ . . . . . . 47 (Pb): Hb/C4-1454
$PbO-B_2O_3-SO_3$ . . . . 47 (Pb): Hb/C2-697
$PbO-B_2O_3-SiO_2$ . . . . 47 (Pb): Hb/C3-857/60
$PbO-B_2O_3-TeO_2$ . . . . 47 (Pb): Hb/C2-697
$PbO-B_2O_3-TiO_2$ . . . . 47 (Pb): Hb/C4-1391
$PbO-B_2O_3-ZnO$ . . . . 47 (Pb): Hb/C4-1227/8
$PbO-BaO-MgO-SiO_2$ . . . . 47 (Pb): Hb/C3-1211
$PbO-BaO-P_2O_5-TiO_2$ . . . . 47 (Pb): Hb/C4-1413
$PbO-BaO-SiO_2$ . . . . 47 (Pb): Hb/C3-1207/8
$PbO-Bi_2O_3$ . . . . 47 (Pb): Hb/C3-966/7
$PbO-Bi_2O_3-TiO_2$ . . . . 47 (Pb): Hb/C4-1394

OPb . . . . PbO systems:
$PbO-Bi_2O_3-V_2O_5$ . . . . 48 (V): Hb/B2-573
$PbO-CO_2$ . . . . 47 (Pb): Hb/C2-697/702
$PbO-CaO-SiO_2$. . . . . 47 (Pb): Hb/C3-1191/2
$PbO-CaO-SiO_2-ZnO$ . . . . 47 (Pb): Hb/C4-1231
$PbO-CrO_3-H_2O$ . . . . 52 (Cr): Hb/B-931
$PbO-Cr_2O_3$ . . . . 52 (Cr): Hb/B-910
$PbO-CuO$ . . . . 60 (Cu): Hb/B3-1198
$PbO-Cu_2O$ . . . . 60 (Cu): Hb/B3-1198
$PbO-Fe_2O_3$ . . . . 59 (Fe): Hb/B5-1130/1
$PbO-GeO_2$ . . . . 47 (Pb): Hb/C4-1474/5
$PbO-GeO_2-Ta_2O_5$ . . . . 50 (Ta): Hb/B2-243
$PbO-KBr$ . . . . 47 (Pb): Hb/C3-1086
$PbO-KCl$ . . . . 47 (Pb): Hb/C3-1081
$PbO-KF$ . . . . 47 (Pb): Hb/C3-1068
$PbO-KI$ . . . . 47 (Pb): Hb/C3-1092
$PbO-KNaC_4H_4O_6-H_2O$ . . . . 47 (Pb): Hb/C3-1120
$PbO-KNaC_4H_4O_6-NaOH-H_2O$ . . . . 47 (Pb): Hb/C3-1120
$PbO-K_2CO_3$ . . . . 47 (Pb): Hb/C3-1097
$PbO-K_2C_4H_4O_6-H_2O$ . . . . 47 (Pb): Hb/C3-1098
$PbO-K_2O-Nb_2O_5-H_2O$ . . . . 49 (Nb): Hb/B3-248/9
$PbO-K_2O-SiO_2$. . . . . 47 (Pb): Hb/C3-1100/1
$PbO-K_2SO_4$ . . . . 47 (Pb): Hb/C3-1096
$PbO-La_2O_3$ . . . . 47 (Pb): Hb/C4-1373
$PbO-La_2O_3-Nb_2O_5-TiO_2$ . . . . 49 (Nb): Hb/B3-302
$PbO-Li_2CO_3$. . . . . 47 (Pb): Hb/C3-993
$PbO-Li_2SO_4$. . . . . 47 (Pb): Hb/C3-992
$PbO-MgO-SiO_2$ . . . . 47 (Pb): Hb/C3-1177
$PbO-MoO_3$ . . . . 53 (Mo): Hb-307
$PbO-N_2O_5-H_2O$ . . . . 47 (Pb): Hb/C1-243
$PbO-NaBr$. . . . . 47 (Pb): Hb/C3-1026
$PbO-NaCl$. . . . . 47 (Pb): Hb/C3-1025
$PbO-NaCl-NaOH$ . . . . 47 (Pb): Hb/C3-1025
$PbO-NaCl-PbCl_2-H_2O$ . . . . 47 (Pb): Hb/C3-1025
$PbO-NaF$ . . . . 47 (Pb): Hb/C3-1017
$PbO-NaOH$ . . . . 47 (Pb): Hb/C3-1005/6
$PbO-NaOH-Nb_2O_5-H_2O$ . . . . 49 (Nb): Hb/B3-248
$PbO-NaPO_3$. . . . . 47 (Pb): Hb/C3-1047/9
$PbO-Na_2CO_3$ . . . . 47 (Pb): Hb/C3-1034
$PbO-Na_2C_4H_4O_6-H_2O$ . . . . 47 (Pb): Hb/C3-1037
$PbO-Na_2O-H_2O$ . . . . 47 (Pb): Hb/C3-1006/7
$PbO-Na_2O-SiO_2$ . . . . 47 (Pb): Hb/C3-1038/46
$PbO-Na_2O-SiO_2-TiO_2$ . . . . 47 (Pb): Hb/C4-1392/3
$PbO-Na_2O-ZnO-H_2O$ . . . . 47 (Pb): Hb/C4-1229/30
$PbO-Na_2SO_4$ . . . . 47 (Pb): Hb/C3-1030
$PbO-Nb_2O_5$ . . . . 49 (Nb): Hb/B3-223/5
$PbO-Nb_2O_5-P_2O_5$ . . . . 49 (Nb): Hb/B4-361
$PbO-Nb_2O_5-TeO_2$ . . . . 49 (Nb): Hb/B4-307

Hb = Main Series Volume; GD = Gmelin-Durrer; Ab = Appended Volume to Main Series; Eb = Supplement Volume to Main Series; Erg. W. = New Supplement Series Volume

OPb	PbO systems:		
	$PbO-Nb_2O_5-TiO_2$	49 (Nb):	Hb/B3-284
	$PbO-P_2O_5$	47 (Pb):	Hb/C3-863/4
	$PbO-P_2O_5-H_2O$	47 (Pb):	Hb/C3-870
	$PbO-P_2O_5-SiO_2$	47 (Pb):	Hb/C3-885/6
	$PbO-P_2O_5-V_2O_5$	48 (V):	Hb/B2-572
	$PbO-P_2O_5-ZrO_2$	47 (Pb):	Hb/C4-1431
	PbO-Pb	47 (Pb):	Hb/C1-52/4
	$PbO-PbBr_2$	47 (Pb):	Hb/C2-369/70
	$PbO-PbBr_2-H_2O$	47 (Pb):	Hb/C2-371
	$PbO-Pb(CH_3COO)_2-H_2O$	47 (Pb):	Hb/C2-771/2
	$PbO-PbCl_2$	47 (Pb):	Hb/C1-333
	$PbO-PbCl_2-H_2O$	47 (Pb):	Hb/C1-336
	$PbO-PbCrO_4$	52 (Cr):	Hb/B-928
	$PbO-PbF_2$	47 (Pb):	Hb/C1-280
	$PbO-PbI_2$	47 (Pb):	Hb/C2-404
	PbO-PbS	47 (Pb):	Hb/C2-534
	$PbO-PbSO_4$	47 (Pb):	Hb/C2-582/3
	$PbO-PbSO_4-SiO_2$	47 (Pb):	Hb/C3-856
	PbO-PbSe	47 (Pb):	Hb/C2-632
	$PbO-PbSeO_3$	47 (Pb):	Hb/C2-634/5
	$PbO-Pb_3(AsO_4)_2$	47 (Pb):	Hb/C3-898/9
	$PbO-Pb_3O_4$	47 (Pb):	Hb/C1-54
	PbO-RbCl	47 (Pb):	Hb/C3-1142
	$PbO-SO_2-H_2O$	47 (Pb):	Hb/C2-535
	$PbO-SO_3-TeO_2$	47 (Pb):	Hb/C2-681
	$PbO-Sb_2O_3$	47 (Pb):	Hb/C3-927/8
	$PbO-Sb_2O_3-Sb_2S_3$	47 (Pb):	Hb/C3-939
	$PbO-Sb_2O_4$	47 (Pb):	Hb/C3-928/9
	$PbO-Sb_2O_5$	47 (Pb):	Hb/C3-929
	$PbO-SeO_2-TeO_2$	47 (Pb):	Hb/C2-684
	$PbO-SiO_2$	47 (Pb):	Hb/C3-808/10
	$PbO-SiO_2-ZrO_2$	47 (Pb):	Hb/C4-1431
	$PbO-SnO_2$	47 (Pb):	Hb/C4-1581
	$PbO-Ta_2O_5$	50 (Ta):	Hb/B2-232/3
	$PbO-TiO_2$	47 (Pb):	Hb/C4-1378/9
	$PbO-TiO_2-V_2O_5$	48 (V):	Hb/B2-574
	$PbO-TiO_2-ZrO_2$	47 (Pb):	Hb/C4-1439/40
	$PbO-V_2O_5$	48 (V):	Hb/B2-567/8
	$PbO-WO_3$	54 (W):	Hb-317
	$PbO-ZrO_2$	47 (Pb):	Hb/C4-1423
OPb_2	Pb_2O	47 (Pb):	Hb/C1-106
OPb_2^{2+}	$[Pb_4O_2]^{4+}$	47 (Pb):	Hb/C1-199
OPd	PdO	65 (Pd):	Hb/2-260/3
–	$PdO \cdot H_2O$	65 (Pd):	Hb/2-263/4
–	$PdO \cdot xH_2O$	65 (Pd):	Hb/2-263/5
OPd_2	Pd_2O	65 (Pd):	Hb/2-259
OPm	PmO	39 (SE):	Hb/C1-312

See the "Instructions for the user" at the beginning of the volume

Formula	Compound	Volume	Reference
OPo	PoO	12 (Po):	Hb–133
OPo^{2+}	$[PoO]^{2+}$	12 (Po):	Hb–102
OPr	PrO	39 (SE):	Hb/C1–87/92, 96, 98/102
OPr^{+}	PrO^{+}	39 (SE):	Hb/C1–102
OPt	PtO	68 (Pt):	Hb/C1–26/8, 42
–	$PtO \cdot H_2O$	68 (Pt):	Hb/C1–26, 42/5
–	$PtO \cdot 2\,H_2O$	68 (Pt):	Hb/C1–26, 42/5
OPtS	PtOS	68 (Pt):	Hb/C1–109
–	$PtOS \cdot 0.5\,H_2O$	68 (Pt):	Hb/C1–109
–	$PtOS \cdot H_2O$	68 (Pt):	Hb/C1–109
$OPu_{0.5}SU_{0.5}$	$U_{0.5}Pu_{0.5}OS$	Erg. W. 4–173	
OPu	PuO	Erg. W. 4–12, 19	
–	PuO solid solutions:		
	Pu(C,O)	Erg. W. 4–224/5	
	PuO–PuC–PuN	Erg. W. 4–226/8	
	PuO–(U,Pu)C–(U,Pu)N	Erg. W. 4–226/8	
OPuS	PuOS	Erg. W. 4–172/3	
OPuSe	PuOSe	Erg. W. 4–189, 191	
ORb_2	Rb_2O	24 (Rb):	Hb–105/6
–	Rb_2O glasses:		
	$Rb_2O–PbO–SiO_2$	47 (Pb):	Hb/C3–1144/5
–	Rb_2O melts:		
	$Rb_2O–V_2O_5$	48 (V):	Hb/B2–486
–	Rb_2O systems:		
	$Rb_2O–CO_2–CaO$	28 (Ca):	Hb/B3–1337
	$Rb_2O–Nb_2O_5$	49 (Nb):	Hb/B4–187/8
ORe	$ReO \cdot H_2O$	70 (Re):	Hb–96
ORe_2	$Re_2O \cdot 2\,H_2O$	70 (Re):	Hb–96
ORh	RhO	64 (Rh):	Hb–46
ORh_2	Rh_2O	64 (Rh):	Hb–45/6
ORu	RuO	63 (Ru):	Hb–25, Eb–234
ORu^{+}	$[RuO]^{+}$	63 (Ru):	Eb–233
ORu^{2+}	$[RuO]^{2+}$	63 (Ru):	Eb–248/9
ORu^{-}	$[RuO]^{-}$	63 (Ru):	Eb–233
ORu_2^{6+}	$[RuORu]^{6+}$	63 (Ru):	Eb–248/9
OS	SO	9 (S):	Hb/B1–58/9, 167/78
–	$(SO)_x$	9 (S):	Hb/B1–167
OS^{+}	SO^{+}	9 (S):	Hb/B1–262
OS^{2+}	SO^{2+}	9 (S):	Hb/B1–310
OSSi	SiOS	15 (Si):	Hb/B–752
OSTh	ThOS	44 (Th):	Hb–282/3
OSZn	ZnSO	32 (Zn):	Hb–219
$OSZn_2$	$ZnS \cdot ZnO$	32 (Zn):	Hb–219
OSZr	ZrOS	42 (Zr):	Hb–332/3
OS_2	S_2O	9 (S):	Hb/B1–167

Hb = Main Series Volume; GD = Gmelin-Durrer; Ab = Appended Volume to Main Series;
Eb = Supplement Volume to Main Series; Erg. W. = New Supplement Series Volume

Formula	Compound	Volume	Reference
OS_2Sb_2	$Sb_2OS_2 = 2\ Sb_2S_3 \cdot Sb_2O_3$	18 (Sb):	Hb/A3–305 Hb/B3–540
OS_2V^-	$[VOS_2]^-$	48 (V):	Hb/B1–281
OS_3	S_3O	9 (S):	Hb/B1–167
OS_3V^{3-}	$[VOS_3]^{3-}$	48 (V):	Hb/B1–281
OS_4Sr_2	$SrO \cdot SrS_4 \cdot 12\ H_2O$	29 (Sr):	Hb–171
OS_4Zn_5	$4\ ZnS \cdot ZnO$	32 (Zn):	Hb–219
OS_5Sb_4	$Sb_4OS_5 = 5\ Sb_2S_3 \cdot Sb_2O_3$	18 (Sb):	Hb/B3–540
$OS_6V_2^{4-}$	$[V_2OS_6]^{4-}$	48 (V):	Hb/B1–281
OSb	SbO	18 (Sb):	Hb/B2–348/9
OSb_2	Sb_2O	18 (Sb):	Hb/B2–348
OSc	ScO	39 (SE):	Hb/C1–85/92, 98/102, 180/1
OSc^+	ScO^+	39 (SE):	Hb/C1–102
OSc_2	Sc_2O	39 (SE):	Hb/C1–102/3
OSe	SeO	10 (Se):	Hb/B–23/5
OSi	SiO	15 (Si):	Hb/B–258/69
OSi^+	SiO^+	15 (Si):	Hb/B–267
OSm	SmO	39 (SE):	Hb/C1–90/2, 98/102, 106, 314/5
–	SmO solid solutions:		
	SmO_xN_{1-x}	39 (SE):	Hb/C2–180
OSn	SnO:		
	Chemical reactions	46 (Sn):	Hb/C1–42/8
	Electrochemical behavior	46 (Sn):	Hb/C1–41/2
	Formation	46 (Sn):	Hb/C1–25/8
	History	46 (Sn):	Hb/A–164/5
	Molecule	46 (Sn):	Hb/C1–28/35
	Physical properties	46 (Sn):	Hb/C1–31/41
	Preparation	46 (Sn):	Hb/C1–25/7
–	Sn_2O_2	46 (Sn):	Hb/C1–34/5, 104
–	Sn_3O_3	46 (Sn):	Hb/C1–34/5, 104
–	Sn_4O_4	46 (Sn):	Hb/C1–34/5, 104
–	$5\ SnO \cdot 2\ H_2O$	46 (Sn):	Hb/C1–49/54
–	$12\ SnO \cdot 5\ H_2O$	46 (Sn):	Hb/C1–49
–	$SnO \cdot 0.5\ H_2O$	46 (Sn):	Hb/C1–49
–	$3\ SnO \cdot 2\ H_2O$	46 (Sn):	Hb/C1–49
–	SnO solutions:		
	$SnO{-}NaOH{-}H_2O$	46 (Sn):	Hb/C1–55/7
–	SnO systems:		
	SnO–CoO	58 (Co):	Eb/A–871
	$SnO{-}H_2O$	46 (Sn):	Hb/C1–48
	$SnO{-}NbO_2$	49 (Nb):	Hb/B3–221
	$SnO{-}Nb_2O_5$	49 (Nb):	Hb/B3–221/2
	$SnO{-}Ta_2O_5$	50 (Ta):	Hb/B2–231
OSn_2^{2+}	$[Sn_2O]^{2+}$	46 (Sn):	Hb/B–384

See the "Instructions for the user"
at the beginning of the volume

OSr	SrO:		
	Chemical reactions	29 (Sr):	Hb–74/6
			Eb–159/61
	Electrochemical behavior	29 (Sr):	Eb–159
	Formation	29 (Sr):	Hb–71/2
			Eb–152/3
	Physical properties	29 (Sr):	Hb–72/4
			Eb–153/9
	Preparation	29 (Sr):	Hb–71/2
			Eb–152
	–, industrial	29 (Sr):	Hb–17/33
			Eb–84
	Purification	29 (Sr):	Hb–33
	Solubility	29 (Sr):	Hb–76/8
	Toxicity	29 (Sr):	Eb–146
	Uses	29 (Sr):	Eb–84
–	SrO glasses:		
	$SrO-PbO-SiO_2$	47 (Pb):	Hb/C3–1199
–	SrO solid solutions:		
	$SrO-BaO$	30 (Ba):	Eb–531/2
	$SrO-Fe_2O_3$	59 (Fe):	D/2. Eb–440/1
–	SrO systems:		
	$SrO-Al_2O_3-SiO_2$	35 (Al):	Hb/B2–603
	$SrO-As_2O_3-H_2O$	29 (Sr):	Eb–287
	$SrO-As_2O_5-H_2O$	29 (Sr):	Eb–289/90
	$SrO-B_2O_3$	29 (Sr):	Hb–188
			Eb–263
	$SrO-BaCl_2-BaO-SrCl_2$	30 (Ba):	Hb–388
	$SrO-BaO$	30 (Ba):	Eb–531/2
	$SrO-BaO-TiO_2$	41 (Ti):	Hb–457
	$SrO-Bi_2O_3$	29 (Sr):	Eb–295
	$SrO-C_{12}H_{22}O_{11}-H_2O$	29 (Sr):	Hb–208/9
	$SrO-CaCrO_4$	52 (Cr):	Hb/B–833
	$SrO-CaO$	29 (Sr):	Eb–303
	$SrO-CaO-SiO_2$	29 (Sr):	Eb–306
	$SrO-HCl-H_2O$	29 (Sr):	Hb–141
	$SrO-H_2O$	29 (Sr):	Hb–76/9
			Eb–161/2
	$SrO-MgO$	29 (Sr):	Eb–303
	$SrO-Nb_2O_5$	49 (Nb):	Hb/B3–26
	$SrO-P_2O_5-H_2O$	29 (Sr):	Hb–213
	$SrO-S-SO_2-SrS-SrSO_4$	29 (Sr):	Eb–257
	$SrO-SiO_2$	29 (Sr):	Hb–210
			Eb–280/1
	$SrO-SiO_2-H_2O$	29 (Sr):	Eb–282
	$SrO-SrBr_2-H_2O$	29 (Sr):	Hb–155
	$SrO-SrCl_2$	29 (Sr):	Hb–140
			Eb–217/8

Hb = Main Series Volume; GD = Gmelin-Durrer; Ab = Appended Volume to Main Series;
Eb = Supplement Volume to Main Series; Erg. W. = New Supplement Series Volume

OSr . . . . SrO systems:
$SrO-SrCl_2-H_2O$ . . . . 29 (Sr): Hb–140/1
$SrO-SrCl_2-SrS$ . . . . 29 (Sr): Eb–249
$SrO-SrCl_2-SrSe$ . . . . 29 (Sr): Eb–261
$SrO-SrI_2-H_2O$ . . . . 29 (Sr): Hb–164
$SrO-Ta_2O_5$ . . . . 50 (Ta): Hb/B2–181
$SrO-ZrO_2$ . . . . 42 (Zr): Hb–439

OSr_2 . . . . Sr_2O . . . . 29 (Sr): Hb–70/1
Eb–152

OTa . . . . TaO . . . . 50 (Ta): Hb/A1–8
Hb/A2–327, 498
Hb/B1–21, 26/8

–. . . . . TaO solid solutions:
TaO–TaN . . . . 50 (Ta): Hb/B1–81

OTa^+ . . . . $[TaO]^+$ . . . . 50 (Ta): Hb/A2–407
Hb/B1–59

OTa^{3+} . . . . $[TaO]^{3+}$ . . . . 50 (Ta): Hb/B1–59
OTa_2 . . . . Ta_2O . . . . 50 (Ta): Hb/B1–25/6, 30
–. . . . . Ta_2O solid solutions:
$Ta_2(O_xN_{1-x})$ . . . . 50 (Ta): Hb/B1–81

OTa_2^+ . . . . $[Ta_2O]^+$ . . . . 50 (Ta): Hb/A2–407
$OTa_{3.56}$ . . . . $Ta_{3.56}O = Ta_{32}O_9$ . . . . 50 (Ta): Hb/B1–24/5
OTa_4 . . . . Ta_4O . . . . 50 (Ta): Hb/B1–24, 51
OTb . . . . TbO . . . . 39 (SE): Hb/C1–87, 91/2, 97/102

OTe . . . . TeO . . . . 11 (Te): Hb–273/4
OTeTh . . . . ThOTe . . . . 44 (Th): Hb–295
OTh . . . . ThO . . . . 44 (Th): Hb–201/2
OTi . . . . TiO . . . . 41 (Ti): Hb–216/21
–. . . . . TiO solid solutions:
(Ti,V)O . . . . 48 (V): Hb/B2–558

–. . . . . TiO systems:
TiO–TiC . . . . 41 (Ti): Hb–368/9
TiO–VN . . . . 48 (V): Hb/B2–560

OTi^{2+} . . . . TiO^{2+} . . . . 41 (Ti): Hb–270, 471, 479
OTl . . . . TlO . . . . 38 (Tl): Hb/2–219
OTl_2 . . . . Tl_2O . . . . 38 (Tl): Hb/2–219/21
–. . . . . Tl_2O melts:
$Tl_2O-B_2O_3-PbO$ . . . . 47 (Pb): Hb/C4–1354
$Tl_2O-PbO-SiO_2$ . . . . 47 (Pb): Hb/C4–1354

–. . . . . Tl_2O systems:
$Tl_2O-B_2O_3$ . . . . 38 (Tl): Hb/3–377/8

OTm . . . . TmO . . . . 39 (SE): Hb/C1–87, 91/2, 97/100

OU^{2+} . . . . $[UO]^{2+}$ . . . . 55 (U): Hb–136

See the "Instructions for the user" at the beginning of the volume

OV	$VO = V_2O_2$:		
	Chemical reactions	48 (V):	Hb/B1–31
	Determination	48 (V):	Hb/A2–714
	Electrochemical behavior	48 (V):	Hb/A2–465/6, 498
	Formation and preparation	48 (V):	Hb/A2–20/3
	History	48 (V):	Hb/A1–7, 10
	Occurrence, extraterrestrial	48 (V):	Hb/A1–14
	Physical properties	48 (V):	Hb/A2–379/80, 391 Hb/B1–23/31
–	VO solid solutions:		
	(V,Ti)O	48 (V):	Hb/B2–558
–	VO systems:		
	VO–NbO	49 (Nb):	Hb/B3–321/2
	$VO–TiO_2–ZrO_2$	48 (V):	Hb/B2–564
	VO–V	48 (V):	Hb/B1–12/3
	VO–VN	48 (V):	Hb/B1–177
	$VO–V_2O_5$	48 (V):	Hb/B1–13
OV^+	VO^+:		
	Detection	48 (V):	Hb/A2–646
	Determination	48 (V):	Hb/A2–676/7
	Electrochemical behavior	48 (V):	Hb/A2–466/8, 486/7
	Transition reactions	48 (V):	Hb/A2–550/640
OV^{2+}	VO^{2+}:		
	Chemical reactions	48 (V):	Hb/A2–533/48 Hb/B1–161, 298
	Detection	48 (V):	Hb/A2–548, 646/7
	Determination	48 (V):	Hb/A2–676/7
	Electrochemical behavior	48 (V):	Hb/A2–465/9, 474/6, 498/505, 509/11
	Exchange reactions	48 (V):	Hb/A1–259 Hb/A2–549/50
	Properties of the ion	48 (V):	Hb/A2–340, 391 Hb/B1–70/2
	Sorption	48 (V):	Hb/A1–259 Hb/A2–549/50
	Transition reactions	48 (V):	Hb/A2–550/640
–	$(VO)_2^{4+}$	48 (V):	Hb/A2–601
–	$[VO(H_2O)]^{2+} = V(OH)_2^{2+}$	48 (V):	Hb/A2–534/5
–	$[VO(H_2O)_5]^{2+}$	48 (V):	Hb/A2–391, 534, 536 Hb/B1–71, 296/7
OV^{3+}	$[VO]^{3+}$	48 (V):	Hb/A2–511, 638 Hb/B1–157/60, 163, 236
OV_2	V_2O	48 (V):	Hb/B1–13
OV_2^{2+}	$[V_2O]^{2+}$	48 (V):	Hb/A2–475
OV_2^{3+}	$[V_2O]^{3+}$	48 (V):	Hb/A2–475
OV_2^{4+}	$[VOV]^{4+}$	48 (V):	Hb/A2–600
OV_3Zr_3	V_3Zr_3O	48 (V):	Hb/B2–673

Hb = Main Series Volume; GD = Gmelin-Durrer; Ab = Appended Volume to Main Series; Eb = Supplement Volume to Main Series; Erg. W. = New Supplement Series Volume

OV_4	V_4O	48 (V):	Hb/B1–21 Hb/B2–524
OW	WO	54 (W):	Hb–107, 111
OXe	XeO	Erg. W. 1–93	
OXe^+	XeO^+	Erg. W. 1–101	
OY	YO	39 (SE):	Hb/C1–85/93, 98/102, 188/9
–	Y_2O_2	39 (SE):	Hb/C1–91, 100
OY^+	YO^+	39 (SE):	Hb/C1–102
OY_2	Y_2O	39 (SE):	Hb/C1–102/3
OYb	YbO	39 (SE):	Hb/C1–91/2, 98/100, 426
–	YbO solid solutions:		
	YbO–EuO	39 (SE):	Hb/C1–430
OZn	ZnO:		
	Chemical reactions	32 (Zn):	Hb–131/6 Eb–809/17
	Commercial products	32 (Zn):	Eb–787/9
	Crystallographic properties	32 (Zn):	Hb–127 Eb–795/8
	Electrical properties	32 (Zn):	Hb–130 Eb–806/8
	Electrochemical behavior	32 (Zn):	Eb–808/9
	Formation	32 (Zn):	Hb–125, 127 Eb–772/3, 794/5
	Magnetic properties	32 (Zn):	Hb–130 Eb–805/6
	Mechanical properties	32 (Zn):	Hb–128 Eb–798/802
	Optical properties	32 (Zn):	Hb–128/30 Eb–804/5
	Pigments	32 (Zn):	Eb–788/9
	Preparation	32 (Zn):	Hb–125 Eb–773/6
	–, industrial	32 (Zn):	Hb–126 Eb–776/87
	Purification	32 (Zn):	Hb–92/3 Eb–789/91
	Purity testing	32 (Zn):	Eb–791/2
	Solubility	32 (Zn):	Hb–130/1 Eb–816/7
	Special forms	32 (Zn):	Eb–787/8, 792/4
	Thermal properties	32 (Zn):	Hb–128 Eb–802/3
–	$ZnO \cdot xH_2O$	32 (Zn):	Eb–833
–	ZnO glasses:		
	$ZnO-B_2O_3$	32 (Zn):	Eb–971/2
	$ZnO-Na_2O-SiO_2$	32 (Zn):	Eb–1008
	$ZnO-PbO-SiO_2$	47 (Pb):	Hb/C4–1231

See the "Instructions for the user"
at the beginning of the volume

OZn	ZnO melts:		
	$ZnO-B_2O_3-V_2O_5$	48 (V):	Hb/B2-536
–	ZnO phosphors:		
	$ZnO-PbO-SiO_2$	47 (Pb):	Hb/C4-1231
–	ZnO solid solutions:		
	$ZnO-CoO$	58 (Co):	Hb/A2-461 Eb/A-849
–	ZnO solutions:		
	$ZnO-NaOH-H_2O$	32 (Zn):	Eb-1004
	$ZnO-NaOH-PbO-H_2O$	47 (Pb):	Hb/C4-1230
–	ZnO systems:		
	$ZnO-Al_2O_3$	35 (Al):	Hb/B2-608
	$ZnO-Al_2O_3-CoO$	58 (Co):	Eb/A-866
	$ZnO-Al_2O_3-Na_2O-SiO_2$	35 (Al):	Hb/B2-611
	$ZnO-Al_2O_3-NiO$	57 (Ni):	Hb/B3-1162
	$ZnO-Al_2O_3-SiO_2$	35 (Al):	Hb/B2-611
	$ZnO-B_2O_3$	32 (Zn):	Eb-970/1
	$ZnO-B_2O_3-PbO$	47 (Pb):	Hb/C4-1227/8
	$ZnO-B_2O_3-SiO_2$	32 (Zn):	Eb-994
	$ZnO-CaO-PbO-SiO_2$	47 (Pb):	Hb/C4-1231
	$ZnO-CdO$	33 (Cd):	Eb-711
	$ZnO-CrO_3-H_2O$	52 (Cr):	Hb/B-856
	$ZnO-CuO-Fe_2O_3$	60 (Cu):	Hb/B3-1325
	$ZnO-GeO_2-Nb_2O_5$	49 (Nb):	Hb/B3-221
	$ZnO-H_2O$	32 (Zn):	Eb-818
	$ZnO-K_2O-SiO_2$	32 (Zn):	Eb-1015
	$ZnO-MgO$	32 (Zn):	Eb-1023
	$ZnO-NH_4Cl-H_2O$	32 (Zn):	Eb-874
	$ZnO-NH_4Cl-ZnCl_2-H_2O$	32 (Zn):	Eb-1017
	$ZnO-N_2O_5-H_2O$	32 (Zn):	Eb-843
	$ZnO-NaOH-H_2O$	32 (Zn):	Eb-1001/3
	$ZnO-Na_2O-H_2O$	32 (Zn):	Hb-294/5
	$ZnO-Na_2O-PbO-H_2O$	47 (Pb):	Hb/C4-1229/30
	$ZnO-Nb_2O_5$	49 (Nb):	Hb/B3-95
	$ZnO-Nb_2O_5-P_2O_5$	49 (Nb):	Hb/B4-361
	$ZnO-Nb_2O_5-SiO_2$	49 (Nb):	Hb/B4-359
	$ZnO-Nb_2O_5-V_2O_5$	49 (Nb):	Hb/B3-329
	$ZnO-NiO$	57 (Ni):	Hb/B3-1134/5
	$ZnO-NiO-TiO_2$	57 (Ni):	Hb/B3-1184/5
	$ZnO-P_2O_5-H_2O$	32 (Zn):	Hb-277
	$ZnO-P_2O_5-Ta_2O_5$	50 (Ta):	Hb/B2-349
	$ZnO-SiO_2$	32 (Zn):	Eb-989/90
	$ZnO-SiO_2-Ta_2O_5$	50 (Ta):	Hb/B2-349
	$ZnO-Ta_2O_5$	50 (Ta):	Hb/B2-193
	$ZnO-TiO_2$	41 (Ti):	Hb-460
	$ZnO-VO_2$	48 (V):	Hb/B2-533
	$ZnO-VO_2-V_2O_5$	48 (V):	Hb/B2-534

OZn	ZnO systems:		
	$ZnO-V_2O_5$	48 (V):	Hb/B2-534
	$ZnO-ZnCl_2-H_2O$	32 (Zn):	Eb-879
	$ZnO-ZnSO_4-H_2O$	32 (Zn):	Eb-963/4
	$ZnO-ZrO_2$	42 (Zr):	Hb-442
OZn_2	Zn_2O	32 (Zn):	Hb-124
OZr	ZrO	42 (Zr):	Hb-212/5
OZr^{2+}	ZrO^{2+}:		
	Chemical reactions	42 (Zr):	Hb-172/91
	– with elements	42 (Zr):	Hb-173
	– with inorganic compounds	42 (Zr):	Hb-173/80
	– with organic compounds	42 (Zr):	Hb-180/91
	Electrochemical behavior	42 (Zr):	Hb-157
$O_{1.27}V$	$VO_{1.27}$	48 (V):	Hb/B1-12/3, 31/2
$O_{1.41}Pb$	$PbO_{1.41} = Pb_{12}O_{17}$	47 (Pb):	Hb/C1-139/44
$O_{1.5}V$	$VO_{1.5}$		
	see O_3V_2 ... V_2O_3		
$O_{1.54}Pb$	$PbO_{1.54}$	47 (Pb):	Hb/C1-139/44
$O_{1.55}Pb$	$PbO_{1.55}$	47 (Pb):	Hb/C1-144/7
$O_{1.67}Pb$	$PbO_{1.67} = Pb_6O_{10}$	47 (Pb):	Hb/C1-144/7
$O_{1.67}V$	$VO_{1.67}$		
	see O_5V_3 ... V_3O_5		
$O_{1.71}Pr$	$PrO_{1.714} = Pr_7O_{12}$	39 (SE):	Hb/C1-284/5
$O_{1.71}Tb$	$TbO_{1.714} = Tb_7O_{12}$	39 (SE):	Hb/C1-387, 394/5
$O_{1.75}Tb$	$TbO_{1.75}$		
	see O_7Tb_4 ... Tb_4O_7		
$O_{1.75}V$	$VO_{1.75} = V_4O_7$	48 (V):	Hb/B1-13, 30, 50/1
$O_{1.76}Ta$	$TaO_{1.76}$	50 (Ta):	Hb/B1-30
$O_{1.78}Pr$	$PrO_{1.778} = Pr_9O_{16}$	39 (SE):	Hb/C1-285
$O_{1.8}Pr$	$PrO_{1.8} = Pr_{10}O_{18}$	39 (SE):	Hb/C1-285
$O_{1.8}V$	$VO_{1.80} = V_5O_9$	48 (V):	Hb/B1-13, 17, 50/1
$O_{1.82}Pr$	$PrO_{1.818} = Pr_{11}O_{20}$	39 (SE):	Hb/C1-285
$O_{1.82}Tb$	$TbO_{1.818} = Tb_{11}O_{20}$	39 (SE):	Hb/C1-387, 395/6
$O_{1.83}Pr$	$PrO_{1.833}$		
	see $O_{11}Pr_6$ ... Pr_6O_{11}		
$O_{1.83}Tb$	$TbO_{1.833} = Tb_6O_{11}$	39 (SE):	Hb/C1-387
$O_{1.83}V$	$VO_{1.833} = V_6O_{11}$	48 (V):	Hb/B1-13, 30, 50/1
$O_{1.86}V$	$VO_{1.86} = V_7O_{13}$	48 (V):	Hb/B1-13, 30, 50/1
$O_{1.87}V$	$VO_{1.87} = V_8O_{15}$	48 (V):	Hb/B1-13, 17, 50/1
O_2	O_2	3 (O):	Hb/3-348/51
	see also O ... Oxygen		
–	$O_2 \cdot xH_2O$	3 (O):	Hb/5-1191
O_2^+	O_2^+	3 (O):	Hb/3-351, 359/60, 367, 572/5, 638
O_2^{2+}	O_2^{2+}	3 (O):	Hb/3-592

See the "Instructions for the user" at the beginning of the volume

O_2^-	O_2^-	3 (O):	Hb/3–351/2, 359/61, 371, 590
			Hb/8–2626
		21 (Na):	Eb/6–338
O_2^{2-}	O_2^{2-}	3 (O):	Hb/3–360/1, 371, 590
O_2Os	OsO_2	66 (Os):	Hb–26/7
–	$OsO_2 \cdot H_2O$	66 (Os):	Hb–29
–	$OsO_2 \cdot 2\,H_2O$	66 (Os):	Hb–27/9
–	$OsO_2 \cdot 5\,H_2O$	66 (Os):	Hb–29
$O_2PS_2Tl_3$	$Tl_3PO_2S_2$	38 (Tl):	Hb/3–432
$O_2P_2Pb_3S_6$	$Pb_3(POS_3)_2$	47 (Pb):	Hb/C3–884
$O_2P_2S_3$	$P_2S_3O_2$	16 (P):	Hb/C–583
O_2Pb	PbO_2:		
	Chemical reactions	47 (Pb):	Hb/C1–189/95
	Crystallographic properties	47 (Pb):	Hb/C1–154/8
	Electrical properties	47 (Pb):	Hb/C1–160/2
	Electrochemical behavior	47 (Pb):	Hb/C1–163/89
	Estimation	47 (Pb):	Hb/C1–151
	Formation	47 (Pb):	Hb/C1–147/50
	History	47 (Pb):	Hb/A1–125/7
	Magnetic properties	47 (Pb):	Hb/C1–160
	Mechanical properties	47 (Pb):	Hb/C1–158/9
	Optical properties	47 (Pb):	Hb/C1–162/3
	Preparation	47 (Pb):	Hb/C1–147/54
	– of special forms	47 (Pb):	Hb/C1–151/4
	Purity testing	47 (Pb):	Hb/C1–151
	Thermal properties	47 (Pb):	Hb/C1–158/9
–	$PbO_2 \cdot xH_2O$	47 (Pb):	Hb/C1–195/6
–	PbO_2 systems:		
	$PbO_2–Cr_2O_3$	52 (Cr):	Hb/B–910
	$PbO_2–Pb_3O_4$	47 (Pb):	Hb/C1–55/6
	$PbO_2–Pb_{12}O_{19}$	47 (Pb):	Hb/C1–55
O_2Pb^{2-}	PbO_2^{2-}	47 (Pb):	Hb/B2–11
			Hb/C1–196/9
O_2PbSr	$SrPbO_2$	47 (Pb):	Hb/C3–1195
$O_2Pb_4^{4+}$	$[Pb_4O_2]^{4+}$	47 (Pb):	Hb/C1–199
O_2Pd	PdO_2	65 (Pd):	Hb/2–266
–	$PdO_2 \cdot 2\,H_2O$	65 (Pd):	Hb/2–266/7
–	$PdO_2 \cdot xH_2O$	65 (Pd):	Hb/2–266/8
O_2Po	PoO_2	12 (Po):	Hb–133
O_2Pr	PrO_2	39 (SE):	Hb/C1–103/4, 148, 290/4
–	$Pr_2O_4 \cdot nH_2O$	39 (SE):	Hb/C2–106/8, 112
–	$Pr_6O_{12} \cdot nH_2O$	39 (SE):	Hb/C2–106
–	PrO_2 solid solutions:		
	$PrO_2–CeO_2$	39 (SE):	Hb/C1–296
	$PrO_2–TbO_2$	39 (SE):	Hb/C1–404

O_2Pr	PrO_2 systems:		
	$PrO_2-Dy_2O_3$	39 (SE):	Hb/C1–295
	$PrO_2-Er_2O_3$	39 (SE):	Hb/C1–295
	$PrO_2-Gd_2O_3$	39 (SE):	Hb/C1–295
	$PrO_2-La_2O_3$	39 (SE):	Hb/C1–295
	$PrO_2-Nd_2O_3$	39 (SE):	Hb/C1–295
	$PrO_2-Sm_2O_3$	39 (SE):	Hb/C1–295
O_2Pr^+	PrO_2^+	39 (SE):	Hb/C1–104
O_2Pt	PtO_2	68 (Pt):	Hb/C1–26/9, 46
–	$PtO_2 \cdot H_2O$	68 (Pt):	Hb/C1–26, 46/7, 51/2
–	$PtO_2 \cdot 2\,H_2O = Pt(OH)_4$	68 (Pt):	Hb/C1–26, 46/7, 50/1
–	$PtO_2 \cdot 3\,H_2O$	68 (Pt):	Hb/C1–26, 46/7, 49/50
–	$PtO_2 \cdot 4\,H_2O$	68 (Pt):	Hb/C1–26, 46/9
O_2Pu	PuO_2:		
	Behavior as target	Erg. W. 8–287	
	Chemical reactions	Erg. W. 4–26/7	
	Compatibility with construction materials	Erg. W. 4–25/6	
	Criticality	Erg. W. 8–368/9, 372	
	Physical properties	Erg. W. 4–16/25	
		Erg. W. 8–298/300	
	Preparation	Erg. W. 4–14/5	
	Pressing and sintering behavior	Erg. W. 4–25	
	Sols	Erg. W. 8–244/5	
	Stoichiometry	Erg. W. 4–15	
–	$PuO_2 \cdot nH_2O$	Erg. W. 4–29	
–	PuO_2 fuels:		
	In-pile behavior	Erg. W. 8–259/77	
	Preparation	Erg. W. 8–242/8	
–	PuO_2 glasses:		
	$PuO_2-Na_2O-SiO_2$	Erg. W. 4–70	
–	PuO_2 solid solutions:		
	PuO_2-CeO_2	Erg. W. 4–57	
	PuO_2-CmO_2	Erg. W. 4–58	
	PuO_2-ThO_2	Erg. W. 4–58	
	$(Pu, U)O_2$	Erg. W. 4–60/70	
		Erg. W. 8–192	
	$PuO_2-UO_2-ZrO_2$	Erg. W. 4–71	
–	PuO_2 systems:		
	$PuO_2-Al_2O_3$	Erg. W. 4–56	
	PuO_2-BeO	Erg. W. 4–55	
	$PuO_2-EuO_{1.5}$	Erg. W. 4–57	
	$PuO_2-Fe_2O_3$	Erg. W. 4–72	
	$PuO_2-HoO_{1.5}$	Erg. W. 4–57	
	$PuO_2-LuO_{1.5}$	Erg. W. 4–57	
	PuO_2-MgO	Erg. W. 4–55/6	

See the "Instructions for the user"
at the beginning of the volume

O_2Pu	PuO_2 systems:		
	$PuO_2-PaO_{2.5}$	Erg. W. 4–58	
	$PuO_2-TmO_{1.5}$	Erg. W. 4–57	
	PuO_2-UO_2	Erg. W. 4–59	
	$PuO_2-U_3O_7$	Erg. W. 4–60	
	$PuO_2-YO_{1.5}$	Erg. W. 4–57	
	PuO_2-ZrO_2	Erg. W. 4–70/1	
O_2Pu^+	PuO_2^+	Erg. W. 8–69/70, 72	
O_2Pu^{2+}	PuO_2^{2+}	Erg. W. 4–99	
		Erg. W. 8–69/70, 74/5	
–	$[Pu_2O_4]^{4+}$	Erg. W. 4–28	
O_2Pu_2S	Pu_2O_2S	Erg. W. 4–172/3	
$O_2Pu_2S_3$	$Pu_2O_2S_3$	Erg. W. 4–172/3	
O_2Pu_2Se	Pu_2O_2Se	Erg. W. 4–189, 191	
O_2Pu_2Te	Pu_2O_2Te	Erg. W. 4–193	
O_2RbSc	$RbScO_2$	39 (SE):	Hb/C2–140/1
O_2RbTm	$RbTmO_2$	39 (SE):	Hb/C2–140
O_2RbY	$RbYO_2$	39 (SE):	Hb/C2–140/1
O_2RbYb	$RbYbO_2$	39 (SE):	Hb/C2–140
O_2Rb_2	Rb_2O_2	24 (Rb):	Hb–110
O_2Re	ReO_2	70 (Re):	Hb–89/90
–	$ReO_2 \cdot xH_2O$	70 (Re):	Hb–97
O_2Rh	RhO_2	64 (Rh):	Hb–48
–	$RhO_2 \cdot 2\,H_2O$	64 (Rh):	Hb–48/50
O_2Ru	RuO_2	63 (Ru):	Hb–25/7
			Eb–232/3, 234/8
–	$RuO_2 \cdot H_2O$	63 (Ru):	Eb–238/9
–	$RuO_2 \cdot nH_2O$	63 (Ru):	Eb–233, 238/9
O_2Ru^+	RuO_2^+	63 (Ru):	Eb–233
O_2Ru^-	RuO_2^-	63 (Ru):	Eb–233
O_2S	SO_2:		
	Acoustic properties	9 (S):	Hb/B1–247
	Behavior as solvent	9 (S):	Hb/B1–310/23
			Hb/B2–1122/30
	Chemical reactions	9 (S):	Hb/B1–262/96
	– in electrical discharges	9 (S):	Hb/B1–263
	– on heating	9 (S):	Hb/B1–262/3
	– with light	9 (S):	Hb/B1–263
	– with metal compounds	9 (S):	Hb/B1–288/94
	– with metals	9 (S):	Hb/B1–270/7
	– with nonmetal compounds	9 (S):	Hb/B1–277/88
	– with nonmetals	9 (S):	Hb/B1–263/70, 323/31, 336/7
	– with organic compounds	9 (S):	Hb/B1–294/6
	Crystallographic properties	9 (S):	Hb/B1–210/1
	Determination	9 (S):	Hb/A2–319, 423, 447
			Hb/B3–1145/9, 1222
	Electrical properties	9 (S):	Hb/B1–259/62

Formula	Entry	Volume	Reference
O_2S	SO_2:		
	Enrichment	9 (S):	Hb/A2–307/17 Hb/B3–1279/82
	Formation	9 (S):	Hb/B1–181/204
	History	9 (S):	Hb/A1–40/3
	Industrial hazards	9 (S):	Hb/A2–481/3
	Liquefaction	9 (S):	Hb/A2–318/9
	Magnetic properties	9 (S):	Hb/B1–259
	Mechanical properties	9 (S):	Hb/B1–211/6, 233/47
	Molecule	9 (S):	Hb/B1–204/10
	Optical properties	9 (S):	Hb/B1–247/59
	Oxidation *see also H_2O_3S ... H_2SO_3*	9 (S):	Hb/B1–323/31, 336/7
	–, chamber process	9 (S):	Hb/A2–426/34
	–, contact process	9 (S):	Hb/A2–324/51
	Preparation	9 (S):	Hb/B1–202/3
	–, industrial	9 (S):	Hb/A2–286/317
	–, –, concentration and purification	9 (S):	Hb/A2–307/18
	–, –, preparation from hydrogen sulfide	9 (S):	Hb/A2–288
	–, –, – from raw sulfur	9 (S):	Hb/A2–286/8
	–, –, – from sulfates	9 (S):	Hb/A2–300/7
	–, –, – from sulfidic ores	9 (S):	Hb/A2–288/300
	–, –, – from sulfites	9 (S):	Hb/A2–300
	–, –, – from waste gases	9 (S):	Hb/A2–307 Hb/B3–1279/82, 1418/9
	Purification	9 (S):	Hb/A2–317/8 Hb/B1–203
	Purity testing	9 (S):	Hb/A2–319
	Removal from waste gases	9 (S):	Hb/B3–1294, 1519
	Solubility in acid solutions	9 (S):	Hb/B3–1220/46
	– in organic materials	9 (S):	Hb/B1–298/310
	– in salt solutions	9 (S):	Hb/B3–1246/79
	– in technical absorbents	9 (S):	Hb/B3–1279/84
	– in water	9 (S):	Hb/B3–1131/219
	– in other inorganic substances	9 (S):	Hb/B1–296/8
	Sorption	9 (S):	Hb/B1–237/47 Hb/B3–1131/2, 1219, 1279/84
	Spectra	9 (S):	Hb/B1–251/9
	Storage	9 (S):	Hb/A2–319
	Thermal properties	9 (S):	Hb/B1–214/33
	Toxicity	9 (S):	Hb/A2–482, 506/9
	Uses	9 (S):	Hb/A2–319 Hb/B3–1294/5
–	$^{32}S^{16}O_2$	9 (S):	Hb/B1–206/7
–	$(SO_2)_2$	9 (S):	Hb/B1–209
–	$SO_2 \cdot H_2O$	9 (S):	Hb/B2–1121/2

See the "Instructions for the user" at the beginning of the volume

O_2S . . . . $SO_2 \cdot 6\,H_2O$:
- Composition . . . . 9 (S): Hb/B2–1094/9
- Formation and preparation . . . . . . 9 (S): Hb/B2–1099/108
- Mixed hydrates and hydrate solid solutions . . . . 9 (S): Hb/B2–1118/22
- Properties and behavior . . . . 9 (S): Hb/B2–1090/4, 1109/18

–. . . . . SO_2 mixtures:
- SO_2–CO_2 . . . . 14 (C): Hb/C1–546

–. . . . . SO_2 solid solutions:
- SO_2–Rn . . . . 9 (S): Hb/B1–210
- $SO_2 \cdot 6\,H_2O$–$Rn \cdot 6\,H_2O$. . . . . 9 (S): Hb/B2–1120/1

–. . . . . SO_2 solutions:
- SO_2–HCN . . . . 9 (S): Hb/B1–309
- SO_2–H_2O *see H_2O_3S ... H_2SO_3*
- SO_2–N. . . . . 9 (S): Hb/B1–296/7
- SO_2–O. . . . . 9 (S): Hb/B1–296
- SO_2–organic solvents . . . . 9 (S): Hb/B1–298/310
- SO_2–S. . . . . 9 (S): Hb/B1–297
- SO_2–SO_3 . . . . 9 (S): Hb/B1–366

–. . . . . SO_2 systems:
- SO_2–$Al_2(SO_4)_3$–H_2O . . . . 9 (S): Hb/B3–1282
- SO_2–BCl_3 . . . . 13 (B): Eb–200
- SO_2–BF_3. . . . . 9 (S): Hb/B1–298; 13 (B): Eb–186
- SO_2–BaO–BaS–$BaSO_4$–S. . . . . . . 30 (Ba): Eb–448
- SO_2–BaS–$BaSO_4$–S . . . . 30 (Ba): Eb–448
- SO_2–Br . . . . 9 (S): Hb/B1–297
- SO_2–CCl_4 . . . . 9 (S): Hb/B1–307
- SO_2–$CO(NH_2)_2$–H_2O . . . . 14 (C): Hb/D1–421
- SO_2–CH_3OH. . . . . 9 (S): Hb/B1–302
- SO_2–CO_2 . . . . 14 (C): Hb/C1–543/6
- SO_2–CH_3COOH–H_2O. . . . . 9 (S): Hb/B3–1245/6
- SO_2–$(CH_3)_2O$ . . . . 9 (S): Hb/B1–305
- SO_2–C_2H_5OH . . . . 9 (S): Hb/B1–302/3
- SO_2–CH_3COCH_3. . . . . 9 (S): Hb/B1–304
- SO_2–$(C_2H_5)_2O$ . . . . 9 (S): Hb/B1–305
- SO_2–$(CH_3)_4NI$–H_2O. . . . . 9 (S): Hb/B3–1251/3
- SO_2–$C_6H_4(OH)_2$–H_2O. . . . . 9 (S): Hb/B3–1251/4
- SO_2–$Ca(HSO_3)_2$–H_2O. . . . . 9 (S): Hb/B3–1262/79
- SO_2–CaO . . . . 28 (Ca): Hb/B3–657/8
- SO_2–CaO–H_2O . . . . 28 (Ca): Hb/B3–665/74
- SO_2–CaS . . . . 28 (Ca): Hb/B3–658
- SO_2–$CdBr_2$–H_2O . . . . 9 (S): Hb/B3–1248/50
- SO_2–$CdCl_2$–H_2O . . . . 9 (S): Hb/B3–1248/50
- SO_2–CdI_2–H_2O . . . . 9 (S): Hb/B3–1248/50
- SO_2–$CdSO_4$–H_2O. . . . . 9 (S): Hb/B3–1256/8

O_2S . . . . SO_2 systems:

SO_2-Cl	9 (S):	Hb/B1–297
$SO_2-Cl-SO_2Cl_2$	9 (S):	Hb/B3–1820
SO_2-GeCl_4	45 (Ge):	Eb–540
SO_2-HBr	9 (S):	Hb/B1–298
SO_2-HCl	9 (S):	Hb/B1–297/8
$SO_2-HSCN-H_2O$	9 (S):	Hb/B3–1245/6
SO_2-H_2O	9 (S):	Hb/B2–1046/88
		Hb/B3–1132/219
SO_2-H_2S	9 (S):	Hb/B1–298
$SO_2-H_2SO_4-H_2O$	9 (S):	Hb/B3–1220/45
$SO_2-HgBr_2-HgI_2$	34 (Hg):	Hb/B2–951
SO_2-I	9 (S):	Hb/B1–297
$SO_2-KBr-H_2O$	9 (S):	Hb/B3–1248/54
$SO_2-KCl-H_2O$	9 (S):	Hb/B3–1248/56
SO_2-KI	22 (K):	Hb/3–650/2
SO_2-KI-H_2O	9 (S):	Hb/B3–1248/54
$SO_2-KNO_3-H_2O$	9 (S):	Hb/B3–1256/8
SO_2-KSCN	22 (K):	Hb/4–912/3
$SO_2-KSCN-H_2O$	9 (S):	Hb/B3–1248/54
$SO_2-K_2SO_4-H_2O$	9 (S):	Hb/B3–1256/8
SO_2-LiI	20 (Li):	Hb–201
$SO_2-Mg(HSO_3)_2-H_2O$	9 (S):	Hb/B3–1262/4, 1273
$SO_2-MgSO_3-H_2O$	27 (Mg):	Hb/B2–209/10
SO_2-N	9 (S):	Hb/B1–297
SO_2-NH_3	9 (S):	Hb/B1–297
$SO_2-NH_3-H_2O$	23 (NH_4):	Hb/2–255/6
$SO_2-NH_4Br-H_2O$	9 (S):	Hb/B3–1248/50
$SO_2-NH_4Cl-H_2O$	9 (S):	Hb/B3–1248/50
$SO_2-NH_4HSO_3-H_2O$	9 (S):	Hb/B3–1273, 1276/9
SO_2-NH_4I	23 (NH_4):	Hb/1–231, 235
$SO_2-NH_4NO_3-H_2O$	9 (S):	Hb/B3–1256/8
SO_2-NH_4SCN	23 (NH_4):	Hb/2–387
$SO_2-NH_4SCN-H_2O$	9 (S):	Hb/B3–1248/50
	23 (NH_4):	Hb/2–387/8
$SO_2-(NH_4)_2SO_4-H_2O$	9 (S):	Hb/B3–1256/8
$SO_2-NaBr-H_2O$	9 (S):	Hb/B3–1248/50
$SO_2-NaCl-H_2O$	9 (S):	Hb/B3–1248/54
$SO_2-NaHSO_4-H_2O$	9 (S):	Hb/B3–1261, 1278/9
$SO_2-NaH_2PO_4-H_2O$	9 (S):	Hb/B3–1283/4
$SO_2-NaH_2PO_4-Na_2HPO_4-H_2O$	21 (Na):	Eb/5–2010
$SO_2-NaSCN-H_2O$	9 (S):	Hb/B3–1248/50
$SO_2-Na_2HPO_4-H_2O$	21 (Na):	Eb/5–2005
$SO_2-Na_2SO_4-H_2O$	9 (S):	Hb/B3–1256/61
$SO_2-PbO-H_2O$	47 (Pb):	Hb/C2–535
$SO_2-RbI-H_2O$	9 (S):	Hb/B3–1251/3
$SO_2-S-SrO-SrS-SrSO_4$	29 (Sr):	Eb–257
$SO_2-S-SrS-SrSO_4$	29 (Sr):	Eb–257
$SO_2-SO_2Cl_2$	9 (S):	Hb/B3–1820

See the "Instructions for the user"
at the beginning of the volume

O_2S . . . . SO_2 systems:
SO_2–$SiCl_4$ . . . . 15 (Si): Hb/B–692
SO_2–$SnBr_4$ . . . . 46 (Sn): Hb/C1–412
SO_2–$TiBr_4$ . . . . 41 (Ti): Hb–328
SO_2–$TiCl_4$ . . . . 41 (Ti): Hb–314/5
SO_2–$ZnSO_3$–H_2O . . . . 32 (Zn): Eb–935

O_2S^+ . . . . SO_2^+ . . . . 9 (S): Hb/B1–262
O_2SSb . . . . SbO_2S . . . . 18 (Sb): Hb/B3–541
O_2SSi . . . . SiO_2S . . . . 15 (Si): Hb/B–752
O_2SSr . . . . $SrSO_2$ . . . . 29 (Sr): Eb–249
O_2STl_2 . . . . Tl_2SO_2 . . . . 38 (Tl): Hb/3–346
O_2SU . . . . UO_2S . . . . 55 (U): Hb–143
O_2SW . . . . WO_2S . . . . 54 (W): Hb–186
O_2SZn . . . . $ZnSO_2$ . . . . 32 (Zn): Hb–219
O_2S_2 . . . . $S_2O_2 = (SO)_2$ . . . . 9 (S): Hb/B1–58/9, 167/78
$O_2S_2^{2-}$ . . . . $S_2O_2^{2-}$ . . . . 9 (S): Hb/B2–376
$O_2S_4U_3$ . . . . $UO_2 \cdot 2\ US_2$ . . . . 55 (U): Hb–143
$O_2S_5V_2^{4-}$ . . . . $[V_2O_2S_5]^{4-}$ . . . . 48 (V): Hb/B1–281
O_2S_x . . . . S_nO_2 (n = 8 to 16) . . . . 9 (S): Hb/B1–167
O_2Sb^- . . . . SbO_2^- . . . . 18 (Sb): Hb/B2–368/9
O_2Sb_3 . . . . Sb_3O_2 . . . . 18 (Sb): Hb/B2–348
O_2Se . . . . SeO_2:
Aerosols . . . . 10 (Se): Hb/B–27
Chemical reactions . . . . 10 (Se): Hb/B–34/40
Colloidal forms . . . . 10 (Se): Hb/B–27
Crystallographic properties . . . . . . 10 (Se): Hb/B–29/31
Electrical properties . . . . 10 (Se): Hb/B–34
Formation . . . . 10 (Se): Hb/B–25/6, 27
Mechanical properties . . . . 10 (Se): Hb/B–31
Molecule . . . . 10 (Se): Hb/B–28/9
Optical properties . . . . 10 (Se): Hb/B–31/3
Preparation . . . . 10 (Se): Hb/B–26/7
Spectra . . . . 10 (Se): Hb/B–28/9, 32/3
Thermal properties . . . . 10 (Se): Hb/B–31
Toxicity . . . . 10 (Se): Hb/A1–259

–. . . . . $SeO_2 \cdot H_2O = H_2SeO_3$ . . . . 10 (Se): Hb/A1–264/7, 273/81
Hb/B–44/51

–. . . . . SeO_2 solutions:
SeO_2–H_2F_2 . . . . 10 (Se): Hb/B–40
SeO_2–H_2O
see H_2O_3Se ... H_2SeO_3 solutions

SeO_2–H_2SO_4 . . . . 10 (Se): Hb/B–35, 40/1
SeO_2–H_2SeO_4–H_2O . . . . 10 (Se): Hb/B–41
SeO_2–organic solvents . . . . 10 (Se): Hb/B–41/3

–. . . . . SeO_2 systems:
SeO_2–$BaSeO_3$–H_2O . . . . 30 (Ba): Eb–454/5
SeO_2–HCl . . . . 10 (Se): Hb/B–145/6
SeO_2–HF . . . . 10 (Se): Hb/B–118

Hb = Main Series Volume; GD = Gmelin-Durrer; Ab = Appended Volume to Main Series;
Eb = Supplement Volume to Main Series; Erg. W. = New Supplement Series Volume

O_2Se . . . . SeO_2 systems:
SeO_2-H_2O . . . . 10 (Se): Hb/B–44/6
$SeO_2-Na_2O-H_2O$ . . . . 21 (Na): Eb/5–1906/7
$SeO_2-PbO-TeO_2$ . . . . 47 (Pb): Hb/C2–684

O_2SeU . . . . UO_2Se . . . . 55 (U): Hb–157

O_2Si . . . . SiO_2:
Absorption of light . . . . 15 (Si): Hb/B–349/51
– of X-rays . . . . 15 (Si): Hb/B–373
Adhesion . . . . 15 (Si): Hb/B–331
Behavior as electrode . . . . 15 (Si): Hb/B–403
Birefringence . . . . 15 (Si): Hb/B–361/8
Boiling point . . . . 15 (Si): Hb/B–339
Catalytic properties . . . . 15 (Si): Hb/B–577/82
Chemical reactions . . . . 15 (Si): Hb/B–539/82
– with acids . . . . 15 (Si): Hb/B–565/8
– with alkaline solutions . . . . 15 (Si): Hb/B–563/5
– with aqueous salt solutions . . . . . 15 (Si): Hb/B–571/6
– with metal compounds . . . . 15 (Si): Hb/B–549/63
– with metals . . . . 15 (Si): Hb/B–545/7
– with nonmetal compounds . . . . . 15 (Si): Hb/B–547/9
– with nonmetals . . . . 15 (Si): Hb/B–543/5
– with organic substances . . . . 15 (Si): Hb/B–576
– with water . . . . 15 (Si): Hb/B–569/71
Cleavage . . . . 15 (Si): Hb/B–310
Cohesion . . . . 15 (Si): Hb/B–331
Color . . . . 15 (Si): Hb/B–351/5
Compressibility . . . . 15 (Si): Hb/B–328/9
Crystallographic properties . . . . . . 15 (Si): Hb/B–305/20
Deformation . . . . 15 (Si): Hb/B–320
Density . . . . 15 (Si): Hb/B–321/3
Dielectric properties . . . . 15 (Si): Hb/B–376/80
Diffraction of X-rays . . . . 15 (Si): Hb/B–374/5
Dressing of sands and gravels . . . . . 15 (Si): Hb/B–271/7
Elasticity . . . . 15 (Si): Hb/B–323/9
Electrical conductivity . . . . 15 (Si): Hb/B–398/400
Electrical properties . . . . 15 (Si): Hb/B–376/400
Electrochemical behavior . . . . 15 (Si): Hb/B–400/8
Electron emission . . . . 15 (Si): Hb/B–400
Electrostriction . . . . 15 (Si): Hb/B–385
Enthalpy . . . . 15 (Si): Hb/B–341/5
Entropy . . . . 15 (Si): Hb/B–346
Faraday effect . . . . 15 (Si): Hb/B–371/2
Formation . . . . 15 (Si): Hb/B–269/70
– and preparation of cristobalite . . . . 15 (Si): Hb/B–283
– – of fused quartz . . . . 15 (Si): Hb/B–283/4
– – of quartz . . . . 15 (Si): Hb/B–277/80
– – of silica gel . . . . 15 (Si): Hb/B–458/99
– – of silicic acid sol . . . . 15 (Si): Hb/B 423/35

See the "Instructions for the user" at the beginning of the volume

O_2Si . . . . SiO_2:
Formation and preparation
of tridymite . . . . 15 (Si): Hb/B–280/3
– – of vitreous silica . . . . 15 (Si): Hb/B–284/5
– – of vitreous silica ware . . . . 15 (Si): Hb/B–286/8
– of special forms . . . . 15 (Si): Hb/B–270
Forms of state . . . . 15 (Si): Hb/B–298/306, 340/1
Glass electrode . . . . 15 (Si): Hb/B–864/920
Hardness . . . . 15 (Si): Hb/B–323
Heat capacity . . . . 15 (Si): Hb/B–341/5
Heat of fusion . . . . 15 (Si): Hb/B–340
– of transformation . . . . 15 (Si): Hb/B–340/1
Lattice energy . . . . 15 (Si): Hb/B–320
Lattice structure . . . . 15 (Si): Hb/B–310/20
Luminescence . . . . 15 (Si): Hb/B–372/3
Magnetic properties . . . . 15 (Si): Hb/B–376
Mechanical properties . . . . 15 (Si): Hb/B–321/34
Melting point . . . . 15 (Si): Hb/B–304/5
Modifications . . . . 15 (Si): Hb/B–298/306, 340/1
Molecule . . . . 15 (Si): Hb/B–307
Optical activity . . . . 15 (Si): Hb/B–369/71
Optical properties . . . . 15 (Si): Hb/B–348/76
Permeability . . . . 15 (Si): Hb/B–332/4
Physiological behavior . . . . 15 (Si): Hb/B–130/4
Piezoelectricity . . . . 15 (Si): Hb/B–380/97
Polymorphism . . . . 15 (Si): Hb/B–298/306, 340/1
Preparation
see Formation
Purification . . . . 15 (Si): Hb/B–276/7
Purity testing . . . . 15 (Si): Hb/B–277
Rayleigh scattering . . . . 15 (Si): Hb/B–360/1
Reflection of light . . . . 15 (Si): Hb/B–348/9
– of X-rays . . . . 15 (Si): Hb/B–373/4
Refraction . . . . 15 (Si): Hb/B–361/8, 375/6
Scattering of X-rays . . . . 15 (Si): Hb/B–375
Sound propagation . . . . 15 (Si): Hb/B–334, 395/6
Specific heat . . . . 15 (Si): Hb/B–341/5
Spectra . . . . 15 (Si): Hb/B–355/60
Strength . . . . 15 (Si): Hb/B–329/31
Surface tension . . . . 15 (Si): Hb/B–331/2
Susceptibility . . . . 15 (Si): Hb/B–376
Suspensions . . . . 15 (Si): Hb/B–288/98
Thermal conductivity . . . . 15 (Si): Hb/B–346/8
Thermal expansion . . . . 15 (Si): Hb/B–334/9
Thermal properties . . . . 15 (Si): Hb/B–334/48
Thermal radiation . . . . 15 (Si): Hb/B–372
Thermoelectric effects . . . . 15 (Si): Hb/B–400
Toxicity . . . . 15 (Si): Hb/B–130/4

Hb = Main Series Volume; GD = Gmelin-Durrer; Ab = Appended Volume to Main Series;
Eb = Supplement Volume to Main Series; Erg. W. = New Supplement Series Volume

O_2Si . . . . SiO_2:
Transformation points . . . . 15 (Si): Hb/B-300/3
Transmission, optical. . . . . 15 (Si): Hb/B-349/51
Ultrasound absorption . . . . 15 (Si): Hb/B-395/6
Ultrasound generation . . . . 15 (Si): Hb/B-397
Uses . . . . 15 (Si): Hb/B-396/7
Vapor pressure . . . . 15 (Si): Hb/B-339
Viscosity. . . . . 15 (Si): Hb/B-332
Wettability. . . . . 15 (Si): Hb/B-331/2

–. . . . . $SiO_2 \cdot 0.5\ H_2O = H_2Si_2O_5$ . . . . 15 (Si): Hb/B-412/3
–. . . . . $[2\ SiO_2 \cdot H_2O]_n$ . . . . 15 (Si): Hb/B-412/3
–. . . . . $SiO_2 \cdot H_2O = H_2SiO_3$. . . . . 15 (Si): Hb/B-411/2
–. . . . . $[SiO_2 \cdot H_2O]_6$ . . . . 15 (Si): Hb/B-411/2
–. . . . . $SiO_2 \cdot 1.5\ H_2O = H_6Si_2O_7$ . . . . 15 (Si): Hb/B-410/1
–. . . . . $SiO_2 \cdot 2\ H_2O = Si(OH)_4$ . . . . 15 (Si): Hb/B-409/10
–. . . . . $SiO_2 \cdot 2.5\ H_2O$ . . . . 15 (Si): Hb/B-409
–. . . . . $SiO_2 \cdot xH_2O$:
Aerogels. . . . . 15 (Si): Hb/B-496/8
Hydrosols:
Chemical reactions
see O_2Si ... SiO_2

Colloidchemical behavior . . . . . . 15 (Si): Hb/B-438/48
Constitution . . . . 15 (Si): Hb/B-435/8
Electrochemical behavior . . . . . . 15 (Si): Hb/B-453/6
Formation and preparation . . . . . 15 (Si): Hb/B-423/35
Physical properties. . . . . 15 (Si): Hb/B-448/53
Uses. . . . . 15 (Si): Hb/B-435

Organogels . . . . 15 (Si): Hb/B-498/9
Organosols . . . . 15 (Si): Hb/B-456/7
Silica gel:
Chemical reactions
see O_2Si ... SiO_2

Colloidchemical behavior . . . . . . 15 (Si): Hb/B-539/43
Constitution . . . . 15 (Si): Hb/B-414/7
Drying. . . . . 15 (Si): Hb/B-473/7, 486/7
Electrochemical behavior . . . . . . 15 (Si): Hb/B-537/9
Formation. . . . . 15 (Si): Hb/B-414, 458
Physical properties. . . . . 15 (Si): Hb/B-500/36
Preparation . . . . 15 (Si): Hb/B-458/77
–, industrial . . . . 15 (Si): Hb/B-477/89
– of special forms . . . . 15 (Si): Hb/B-489/96
Purification . . . . 15 (Si): Hb/B-473
Regeneration . . . . 15 (Si): Hb/B-499
Sorption . . . . 15 (Si): Hb/B-512/31

–. . . . . SiO_2 glasses:
$SiO_2-Ag_2O-B_2O_3$. . . . . 61 (Ag): Hb/B3-389
$SiO_2-Ag_2O-CaO-Na_2O$ . . . . 61 (Ag): Hb/B3-384/5

See the "Instructions for the user"
at the beginning of the volume

O_2Si . . . . SiO_2 glasses:
$SiO_2-Al_2O_3-CaO-Na_2O$ . . . . 35 (Al): Hb/B2-592/3
$SiO_2-Al_2O_3-PbO$ . . . . 47 (Pb): Hb/C4-1297/300
$SiO_2-B_2O_3$. . . . . 15 (Si): Hb/B-759/61
$SiO_2-B_2O_3-K_2O-PbO$ . . . . 47 (Pb): Hb/C3-1109/10
$SiO_2-B_2O_3-Li_2O$ . . . . 20 (Li): Eb-519
$SiO_2-B_2O_3-Li_2O-Na_2O$ . . . . 21 (Na): Eb/5-2242
$SiO_2-B_2O_3-Na_2O$. . . . . 21 (Na): Eb/5-1984/98
$SiO_2-B_2O_3-Na_2O-PbO$ . . . . 47 (Pb): Hb/C3-1047
$SiO_2-B_2O_3-PbO$ . . . . 47 (Pb): Hb/C3-858/60
$SiO_2-BaO-PbO$. . . . . 47 (Pb): Hb/C3-1208/10
$SiO_2-BeO-Na_2O$ . . . . 26 (Be): Hb-171
$SiO_2-Bi_2O_3$ . . . . 19 (Bi): Eb-844
$SiO_2-Bi_2O_3-PbO$ . . . . 47 (Pb): Hb/C3-978
$SiO_2-CaO-K_2O$. . . . . 28 (Ca): Hb/B3-1308/9
$SiO_2-CaO-K_2O-MgO-Na_2O$ . . . . . 28 (Ca): Hb/B3-1384
$SiO_2-CaO-K_2O-Na_2O$ . . . . 28 (Ca): Hb/B3-1318
$SiO_2-CaO-MgO$ . . . . 28 (Ca): Hb/B3-1370
$SiO_2-CaO-Na_2O$ . . . . 28 (Ca): Hb/B3-1275/8
$SiO_2-CaO-PbO$. . . . . 47 (Pb): Hb/C3-1191/2
$SiO_2-CdO-PbO$ . . . . 47 (Pb): Hb/C4-1257
$SiO_2-GeO_2-Na_2O$ . . . . 45 (Ge): Eb-501
$SiO_2-K_2O-Li_2O-PbO$ . . . . 47 (Pb): Hb/C3-1116
$SiO_2-K_2O-Na_2O-PbO$ . . . . 47 (Pb): Hb/C3-1120/2
SiO_2-K_2O-PbO. . . . . 47 (Pb): Hb/C3-1102/6
$SiO_2-La_2O_3-PbO$. . . . . 47 (Pb): Hb/C4-1376
SiO_2-Li_2O . . . . 20 (Li): Eb-518/9
$SiO_2-Li_2O-Na_2O$ . . . . 21 (Na): Hb-991
Eb/5-2240/2
$SiO_2-Li_2O-Na_2O-PbO$. . . . . 47 (Pb): Hb/C3-1052
SiO_2-Li_2O-PbO . . . . 47 (Pb): Hb/C3-993
SiO_2-NaBO_2 . . . . 21 (Na): Eb/5-1990, 1992
SiO_2-Na_2O:
Chemical reactions. . . . . 21 (Na): Eb/4-1505/9
Crystallographic properties . . . . . 21 (Na): Eb/4-1486/90
Devitrification . . . . 21 (Na): Eb/4-1484/5
Diffusion . . . . 21 (Na): Eb/4-1488
Electrical properties . . . . 21 (Na): Eb/4-1501/2
Electrochemical behavior . . . . . . 21 (Na): Eb/4-1503/5
Formation and preparation . . . . . 21 (Na): Eb/4-1482/4
Mechanical properties . . . . 21 (Na): Eb/4-1490/6
Optical properties . . . . 21 (Na): Eb/4-1498/501
Structure . . . . 21 (Na): Eb/4-1486/90
Thermal properties. . . . . 21 (Na): Eb/4-1496/9
$SiO_2-Na_2O-P_2O_5$. . . . . 21 (Na): Eb/5-2016
SiO_2-Na_2O-PbO . . . . 47 (Pb): Hb/C3-1039/46
$SiO_2-Na_2O-PuO_2$ . . . . Erg. W. 4-70
SiO_2-Na_2O-ZnO . . . . 32 (Zn): Eb-1008

Hb = Main Series Volume; GD = Gmelin-Durrer; Ab = Appended Volume to Main Series; Eb = Supplement Volume to Main Series; Erg. W. = New Supplement Series Volume

O_2Si . . . . SiO_2 glasses:
SiO_2-PbO:
Chemical reactions . . . . 47 (Pb): Hb/C3-835/6
Enamels . . . . 47 (Pb): Hb/C3-836
Formation and preparation . . . . . 47 (Pb): Hb/C3-824/5
Physical properties . . . . 47 (Pb): Hb/C3-827/34
Solder glasses . . . . 47 (Pb): Hb/C3-836
Structure . . . . 47 (Pb): Hb/C3-825/6
$SiO_2-PbO-Pr_2O_3$ . . . . 47 (Pb): Hb/C4-1376
$SiO_2-PbO-Rb_2O$ . . . . 47 (Pb): Hb/C3-1144/5
$SiO_2-PbO-Sb_2O_3$ . . . . 47 (Pb): Hb/C3-941
$SiO_2-PbO-SrO$ . . . . 47 (Pb): Hb/C3-1199
$SiO_2-PbO-ThO_2$ . . . . 47 (Pb): Hb/C4-1471
$SiO_2-PbO-TiO_2$ . . . . 47 (Pb): Hb/C4-1392/4
$SiO_2-PbO-ZnO$ . . . . 47 (Pb): Hb/C4-1231

– . . . . SiO_2 melts:
$SiO_2-B_2O_3-Na_2O-PbO$ . . . . 47 (Pb): Hb/C3-1047
$SiO_2-Li_2O-Na_2O$ . . . . 21 (Na): Eb/5-2240/2
SiO_2-Na_2O . . . . 21 (Na): Eb/4-1482/509
SiO_2-PbO:
Activity . . . . 47 (Pb): Hb/C3-812
Chemical reactions . . . . 47 (Pb): Hb/C3-823
Constitution . . . . 47 (Pb): Hb/C3-810/2
Electrochemical behavior . . . . . . 47 (Pb): Hb/C3-820/3
Physical properties . . . . 47 (Pb): Hb/C3-813/20
$SiO_2-PbO-Tl_2O$ . . . . 47 (Pb): Hb/C4-1354

– . . . . SiO_2 phosphors:
$SiO_2-PbO-ZnO$ . . . . 47 (Pb): Hb/C4-1231

– . . . . SiO_2 solid solutions:
$SiO_2-Ta_2O_5-ZrO_2$ . . . . 50 (Ta): Hb/B2-349

– . . . . SiO_2 solutions:
$SiO_2 \cdot xH_2O-H_2O$ . . . . 15 (Si): Hb/B-417/22

– . . . . SiO_2 suspensions . . . . 15 (Si): Hb/B-288/98
– . . . . SiO_2 systems:
SiO_2-Ag_2O . . . . 61 (Ag): Hb/B3-379
$SiO_2-Al_2O_3$ . . . . 35 (Al): Hb/B2-309/10
$SiO_2-Al_2O_3-BaO$ . . . . 35 (Al): Hb/B2-607
$SiO_2-Al_2O_3-BeO$ . . . . 35 (Al): Hb/B2-533
$SiO_2-Al_2O_3-CO_2-CaO-Na_2O$ . . . . . 35 (Al): Hb/B2-593
$SiO_2-Al_2O_3-CO_2-K_2O$ . . . . 35 (Al): Hb/B2-504
$SiO_2-Al_2O_3-CO_2-Na_2O$ . . . . 35 (Al): Hb/B2-426
$SiO_2-Al_2O_3-CaO$ . . . . 35 (Al): Hb/B2-576/81
$SiO_2-Al_2O_3-CaO-Fe_2O_3-MgO$ . . . . 35 (Al): Hb/B2-613
$SiO_2-Al_2O_3-CaO-Fe_2O_3$
$-MgO-Na_2O$ . . . . 35 (Al): Hb/B2-613
$SiO_2-Al_2O_3-CaO-K_2O$ . . . . 35 (Al): Hb/B2-594

See the "Instructions for the user" at the beginning of the volume

O_2Si . . . . SiO_2 systems:

System		Reference
$SiO_2-Al_2O_3-CaO-K_2O-Na_2O$	35 (Al):	Hb/B2–594/5
$SiO_2-Al_2O_3-CaO-MgO$	35 (Al):	Hb/B2–596/600
$SiO_2-Al_2O_3-CaO-MgO-Na_2O$	35 (Al):	Hb/B2–600
$SiO_2-Al_2O_3-CaO-Na_2O$	35 (Al):	Hb/B2–589/92
$SiO_2-Al_2O_3-Co_3O_4$	58 (Co):	Hb/A2–470
$SiO_2-Al_2O_3-Cr_2O_3-MgO$	52 (Cr):	Hb/B–897
$SiO_2-Al_2O_3-FeO-Fe_2O_3$	35 (Al):	Hb/B2–613
$SiO_2-Al_2O_3-K_2O$	35 (Al):	Hb/B2–489
$SiO_2-Al_2O_3-K_2O-H_2O$	35 (Al):	Hb/B2–503
$SiO_2-Al_2O_3-Li_2O$	35 (Al):	Hb/B2–356
$SiO_2-Al_2O_3-MgO$	35 (Al):	Hb/B2–540/1
$SiO_2-Al_2O_3-Na_2O$	35 (Al):	Hb/B2–388/90
$SiO_2-Al_2O_3-Na_2O-H_2O$	35 (Al):	Hb/B2–402
$SiO_2-Al_2O_3-Na_2O-ZnO$	35 (Al):	Hb/B2–611
$SiO_2-Al_2O_3-PbO$	47 (Pb):	Hb/C4–1296/7
$SiO_2-Al_2O_3-SrO$	35 (Al):	Hb/B2–603
$SiO_2-Al_2O_3-TiO_2$	41 (Ti):	Hb–464
$SiO_2-Al_2O_3-ZnO$	35 (Al):	Hb/B2–611
$SiO_2-Al_2O_3-ZrO_2$	42 (Zr):	Hb–444
$SiO_2-As_2O_5-NaF$	21 (Na):	Eb/5–2021
$SiO_2-As_2O_5-Na_2O$	21 (Na):	Eb/5–2021
SiO_2-BPO_4	16 (P):	Hb/C–642
$SiO_2-B_2O_3$	15 (Si):	Hb/B–759/61
$SiO_2-B_2O_3-BaO$	30 (Ba):	Eb–490
$SiO_2-B_2O_3-CaO$	13 (B):	Eb–135
	28 (Ca):	Hb/B3–1109/11
$SiO_2-B_2O_3-CaO-MgO-Na_2O$	28 (Ca):	Hb/B3–1382
$SiO_2-B_2O_3-CaO-Na_2O$	13 (B):	Eb–135
	28 (Ca):	Hb/B3–1282
$SiO_2-B_2O_3-K_2O$	22 (K):	Hb/5–979
$SiO_2-B_2O_3-Li_2O$	13 (B):	Eb–135
	20 (Li):	Eb–519
$SiO_2-B_2O_3-Na_2O$	13 (B):	Eb–135
	21 (Na):	Eb/5–1982/4
$SiO_2-B_2O_3-P_2O_5$	16 (P):	Hb/C–642
$SiO_2-B_2O_3-PbO$	47 (Pb):	Hb/C3–857/60
$SiO_2-B_2O_3-ZnO$	32 (Zn):	Eb–994
SiO_2-BaCO_3	30 (Ba):	Hb–331
SiO_2-BaO	30 (Ba):	Hb–331/2
		Eb–486
$SiO_2-BaO-CO_2-Na_2O$	30 (Ba):	Eb–517
$SiO_2-BaO-CaO-MgO-Na_2O$	30 (Ba):	Eb–531
$SiO_2-BaO-CaO-Na_2O$	30 (Ba):	Eb–531
$SiO_2-BaO-H_2O$	30 (Ba):	Eb–488
$SiO_2-BaO-K_2O$	30 (Ba):	Hb–376
		Eb–522/3
$SiO_2-BaO-K_2O-Na_2O$	30 (Ba):	Eb–523
$SiO_2-BaO-MgO$	30 (Ba):	Eb–527

Hb = Main Series Volume; GD = Gmelin-Durrer; Ab = Appended Volume to Main Series; Eb = Supplement Volume to Main Series; Erg. W. = New Supplement Series Volume

O_2Si . . . . SiO_2 systems:

$SiO_2-BaO-MgO-PbO$ . . . . 47 (Pb): Hb/C3-1211
$SiO_2-BaO-Na_2O$ . . . . 30 (Ba): Eb-515/7
$SiO_2-BaO-PbO$ . . . . 47 (Pb): Hb/C3-1207/8
$SiO_2-BeO-Na_2O$ . . . . 26 (Be): Hb-171
$SiO_2-Bi_2O_3$ . . . . 19 (Bi): Eb-844
SiO_2-C . . . . 15 (Si): Hb/B-857
SiO_2-CO_2-CaO . . . . 28 (Ca): Hb/B3-1112/4
$SiO_2-CO_2-CaO-K_2O$ . . . . 28 (Ca): Hb/B3-1312/3
$SiO_2-CO_2-CaO-MgO$ . . . . 28 (Ca): Hb/B3-1379
$SiO_2-CO_2-CaO-Na_2O$ . . . . 28 (Ca): Hb/B3-1282/5
$SiO_2-CO_2-K_2O$ . . . . 22 (K): Hb/5-979/80
$SiO_2-CO_2-Li_2CO_3-Li_2SiO_3$ . . . . . . 20 (Li): Eb-520
$SiO_2-CO_2-Li_2O$ . . . . 20 (Li): Hb-243
Eb-519/20
$SiO_2-CO_2-Na_2O$ . . . . 21 (Na): Hb-885/7
Eb/5-1982
$SiO_2-CaAl_2Si_2O_8$ . . . . 35 (Al): Hb/B2-579
$SiO_2-CaAl_2Si_2O_8-KAlSi_2O_6$ . . . . . . 35 (Al): Hb/B2-594
$SiO_2-CaO \cdot Al_2O_3 \cdot 2\ SiO_2$
$-K_2O \cdot Al_2O_3 \cdot 2\ SiO_2$
$-Na_2O \cdot Al_2O_3 \cdot 2\ SiO_2$ . . . . 35 (Al): Hb/B2-594
$SiO_2-CaO \cdot Al_2O_3 \cdot 2\ SiO_2-MgO$ . . . . 35 (Al): Hb/B2-596
$SiO_2-CaAl_2Si_2O_8-Mg_2SiO_4$ . . . . . . 35 (Al): Hb/B2-597/8
SiO_2-CaC_2 . . . . 28 (Ca): Hb/B3-1112
SiO_2-CaCl_2-CaO . . . . 28 (Ca): Hb/B3-1105/6
SiO_2-CaF_2 . . . . 28 (Ca): Hb/B3-1100
SiO_2-CaF_2-CaO . . . . 28 (Ca): Hb/B3-1100/2
SiO_2-CaF_2-KF . . . . 28 (Ca): Hb/B3-1310
SiO_2-CaF_2-MgO . . . . 28 (Ca): Hb/B3-1378/9
SiO_2-CaF_2-NaF . . . . 28 (Ca): Hb/B3-1281
$SiO_2-CaF_2-Na_2O$ . . . . 28 (Ca): Hb/B3-1281
$SiO_2-CaO \cdot MgO \cdot SiO_2$
$-2\ MgO \cdot SiO_2$ . . . . 28 (Ca): Hb/B3-1364
$SiO_2-CaO \cdot Nb_2O_5-CaO \cdot TiO_2$ . . . . . 49 (Nb): Hb/B4-360
SiO_2-CaO . . . . 28 (Ca): Hb/B3-1036/43
$SiO_2-CaO-CaS$ . . . . 28 (Ca): Hb/B3-1107
$SiO_2-CaO-CoO$ . . . . 58 (Co): Eb/A-844
$SiO_2-CaO-Cr_2O_3$ . . . . 52 (Cr): Hb/B-817/8
$SiO_2-CaO-FeO$ . . . . 59 (Fe): Hb/B5-1083
$SiO_2-CaO-Fe_2O_3$ . . . . 59 (Fe): Hb/B5-1086/7
$SiO_2-CaO-H_2O$ . . . . 28 (Ca): Hb/B3-1075/84
$SiO_2-CaO-K_2O$ . . . . 28 (Ca): Hb/B3-1307/9
$SiO_2-CaO-Li_2O$ . . . . 28 (Ca): Hb/B3-1247/8
$SiO_2-CaO-MgO$ . . . . 28 (Ca): Hb/B3-1362/70
$SiO_2-CaO-MgO-H_2O$ . . . . 28 (Ca): Hb/B3-1377
$SiO_2-CaO-MgO-Na_2O$ . . . . 28 (Ca): Hb/B3-1381/2
$SiO_2-CaO-MgO-NiO$ . . . . 57 (Ni): Hb/B3-1122
$SiO_2-CaO-NaCl-Na_2O$ . . . . 28 (Ca): Hb/B3-1281

O_2Si	SiO_2 systems:		
	$SiO_2-CaO-Na_2O$	28 (Ca):	Hb/B3-1272/8
	$SiO_2-CaO-Na_2O-H_2O$	28 (Ca):	Hb/B3-1280/1
	$SiO_2-CaO-Na_2O-TiO_2$	41 (Ti):	Hb-431
	$SiO_2-CaO-Nb_2O_5$	49 (Nb):	Hb/B4-357/8
	$SiO_2-CaO-Nb_2O_5-TiO_2$	49 (Nb):	Hb/B4-360
	$SiO_2-CaO-P_2O_5$	28 (Ca):	Hb/B1-204/6
			Hb/B3-1215/9
	$SiO_2-CaO-PbO$	47 (Pb):	Hb/C3-1191/2
	$SiO_2-CaO-PbO-ZnO$	47 (Pb):	Hb/C4-1231
	$SiO_2-CaO-SO_3$	28 (Ca):	Hb/B3-1107/8
	$SiO_2-CaO-SrO$	29 (Sr):	Eb-306
	$SiO_2-CaO-TiO_2$	41 (Ti):	Hb-430
	$SiO_2-CaTiO_3$	41 (Ti):	Hb-430
	$SiO_2-8\,CaO \cdot 7\,Nb_2O_5 \cdot 6\,TiO_2$	49 (Nb):	Hb/B4-360
	SiO_2-CoO	58 (Co):	Hb/A2-386/7
			Eb/A-735
	$SiO_2-CoO-MgO$	58 (Co):	Eb/A-840
	$SiO_2-Co_3O_4-Fe_2O_3$	59 (Fe):	Hb/B5-1166
	$SiO_2-Cr_2O_3$	52 (Cr):	Hb/B-435/6
	$SiO_2-Cr_2O_3-MgO$	52 (Cr):	Hb/B-791
	SiO_2-Cu_2O	60 (Cu):	Hb/B2-899
	$SiO_2-Fe-FeO-FeSi_2$	59 (Fe):	Hb/A8-1759/60
	$SiO_2-Fe-FeO-Ni-NiO$	59 (Fe):	Hb/B5-1160
	$SiO_2-Fe-FeO-Ni-NiO-S$	59 (Fe):	Hb/B5-1160
	SiO_2-FeO	59 (Fe):	Hb/A5-876/81
			Hb/B4-761/2
	$SiO_2-FeO-Fe_2O_3$	59 (Fe):	Hb/B4-762
	$SiO_2-FeO-MgO$	59 (Fe):	Hb/B5-1066
	$SiO_2-FeO-MnO$	56 (Mn):	Hb/C1-68
		59 (Fe):	Hb/B5-1153
	$SiO_2-Fe_2O_3-Na_2O$	59 (Fe):	Hb/B5-908/9
	$SiO_2-KAlSi_3O_8-NaAlSi_3O_8$	35 (Al):	Hb/B2-506
	$SiO_2-K_2O-Na_2O$	22 (K):	Hb/6-1222
	$SiO_2-K_2O-Nb_2O_5$	49 (Nb):	Hb/B4-356/7
	SiO_2-K_2O-PbO	47 (Pb):	Hb/C3-1100/1
	$SiO_2-K_2O-Ta_2O_5$	50 (Ta):	Hb/B2-348
	SiO_2-K_2O-ZnO	32 (Zn):	Eb-1015
	$SiO_2-K_2SiO_3$	22 (K):	Hb/5-973/4
	$SiO_2-K_2SiO_3-H_2O$	22 (K):	Hb/5-974/5
	$SiO_2-K_2SiO_3-Na_2SiO_3$	22 (K):	Hb/6-1222/4
	SiO_2-Li_2O	20 (Li):	Hb-236/7
			Eb-514/5
	$SiO_2-Li_2O-Na_2O$	21 (Na):	Eb/5-2238/40
	$SiO_2-Li_2SiO_3-Na_2SiO_3$	21 (Na):	Eb/5-2239
	SiO_2-MgO	27 (Mg):	Hb/B3-346/8
	$SiO_2-MgO-PbO$	47 (Pb):	Hb/C3-1177
	$SiO_2-NaCl-Na_2O$	21 (Na):	Eb/5-1981
	$SiO_2-NaF-Na_2O$	21 (Na):	Eb/5-1980/1

Hb = Main Series Volume; GD = Gmelin-Durrer; Ab = Appended Volume to Main Series; Eb = Supplement Volume to Main Series; Erg. W. = New Supplement Series Volume

O_2Si . . . . SiO_2 systems:
$SiO_2-Na_2O\cdot 4\,B_2O_3$ . . . . 21 (Na): Eb/5–1983/4
$SiO_2-Na_2CO_3-H_2O$ . . . . 21 (Na): Eb/5–1982
SiO_2-Na_2O . . . . 21 (Na): Hb–861
Eb/4–1465/7
$SiO_2-Na_2O-H_2O$ . . . . 21 (Na): Hb–873
Eb/4–1509/15
$SiO_2-Na_2O-Na_2S$ . . . . 21 (Na): Eb/5–1981
$SiO_2-Na_2O-Nb_2O_5$ . . . . 49 (Nb): Hb/B4–356
$SiO_2-Na_2O-P_2O_5$ . . . . 21 (Na): Eb/5–2013/6
SiO_2-Na_2O-PbO . . . . 47 (Pb): Hb/C3–1038/46
$SiO_2-Na_2O-PbO-TiO_2$ . . . . 47 (Pb): Hb/C4–1392/3
$SiO_2-Na_2O-ZrO_2$ . . . . 42 (Zr): Hb–407/8
$SiO_2-Na_2SO_4-Na_2SiO_3$ . . . . 21 (Na): Eb/5–1981/2
$SiO_2-Na_2SiO_3$ . . . . 21 (Na): Hb–879
$SiO_2-Na_2SiO_3-PbSiO_3$ . . . . 47 (Pb): Hb/C3–1038
$SiO_2-Nb_2O_5$ . . . . 49 (Nb): Hb/B1–401/2
$SiO_2-Nb_2O_5-TiO_2$ . . . . 49 (Nb): Hb/B4–360
$SiO_2-Nb_2O_5-Y_2O_3$ . . . . 49 (Nb): Hb/B4–359
$SiO_2-Nb_2O_5-ZnO$ . . . . 49 (Nb): Hb/B4–359
SiO_2-NiO . . . . 57 (Ni): Hb/B3–936
$SiO_2-NiO-H_2O$ . . . . 57 (Ni): Hb/B3–940/1
$SiO_2-P_2O_5$ . . . . 16 (P): Hb/C–638
$SiO_2-P_2O_5-H_2O$ . . . . 16 (P): Hb/C–639/40
$SiO_2-P_2O_5-PbO$ . . . . 47 (Pb): Hb/C3–885/6
SiO_2-PbO . . . . 47 (Pb): Hb/C3–808/10
$SiO_2-PbO-PbSO_4$ . . . . 47 (Pb): Hb/C3–856
$SiO_2-PbO-ZrO_2$ . . . . 47 (Pb): Hb/C4–1431
SiO_2-SrO . . . . 29 (Sr): Hb–210
Eb–280/1
$SiO_2-SrO-H_2O$ . . . . 29 (Sr): Eb–282
$SiO_2-Ta_2O_5$ . . . . 50 (Ta): Hb/B1–263
$SiO_2-Ta_2O_5-ZnO$ . . . . 50 (Ta): Hb/B2–349
SiO_2-TiO_2- . . . . 41 (Ti): Hb–376/7
SiO_2-ZnO . . . . 32 (Zn): Eb–989/90
SiO_2-ZrO_2 . . . . 42 (Zr): Hb–381/2
O_2Si_3 . . . . Si_3O_2 . . . . 15 (Si): Hb/B–257
O_2Sn . . . . SnO_2:
Chemical reactions . . . . 46 (Sn): Hb/C1–97/113
Colloidal forms . . . . 46 (Sn): Hb/C1–60/1
Crystallographic properties . . . . . . 46 (Sn): Hb/C1–66/71
Electrical properties . . . . 46 (Sn): Hb/C1–73/89
Electrochemical behavior . . . . 46 (Sn): Hb/C1–95/7
Formation and preparation . . . . . . 46 (Sn): Hb/C1–59/65
History . . . . 46 (Sn): Hb/A–164/7
Magnetic properties . . . . 46 (Sn): Hb/C1–73/4
Mechanical properties . . . . 46 (Sn): Hb/C1–71/3
Molecule . . . . 46 (Sn): Hb/C1–66
Optical properties . . . . 46 (Sn): Hb/C1–89/95

See the "Instructions for the user" at the beginning of the volume

O_2Sn	SnO_2:		
	Spectra	46 (Sn):	Hb/C1-89/94
	Thermal properties	46 (Sn):	Hb/C1-71/3
	Toxicity	46 (Sn):	Hb/B-412
–	$2\,SnO_2 \cdot H_2O = Sn_2O_5H_2$	46 (Sn):	Hb/C1-115, 117/8, 121
–	$SnO_2 \cdot H_2O = SnO_3H_2$	46 (Sn):	Hb/C1-115/8, 121
–	$SnO_2 \cdot 1.25\,H_2O$	46 (Sn):	Hb/C1-117
–	$SnO_2 \cdot 3\,H_2O$	46 (Sn):	Hb/C1-116
–	$SnO_2 \cdot 3.45\,H_2O$	46 (Sn):	Hb/C1-121
–	$SnO_2 \cdot xH_2O$	46 (Sn):	Hb/C1-113/27
–	$SnO_2 \cdot xH_2O$, Sols	46 (Sn):	Hb/C1-127/36
–	SnO_2 solid solutions:		
	$SnO_2-PbO-Sb_2O_5$	47 (Pb):	Hb/C4-1595
–	SnO_2 systems:		
	SnO_2-CuO	60 (Cu):	Hb/B3-1194
	SnO_2-NiO	57 (Ni):	Hb/B3-1190/1
	SnO_2-PbO	47 (Pb):	Hb/C4-1581
	SnO_2-Sn	46 (Sn):	Hb/C1-24/5
	$SnO_2-Ta_2O_5$	50 (Ta):	Hb/B2-231
	$SnO_2-ZnTa_2O_6$	50 (Ta):	Hb/B2-232
O_2Sn^{2-}	$[SnO_2]^{2-}$	46 (Sn):	Hb/C1-57
O_2Sn_2	Sn_2O_2	46 (Sn):	Hb/C1-34/5, 104
O_2Sr	SrO_2	29 (Sr):	Hb-86/7 Eb-85, 165/8
–	$SrO_2 \cdot 8\,H_2O$	29 (Sr):	Hb-238/9 Eb-168
–	$SrO_2 \cdot 9\,H_2O$	29 (Sr):	Hb-238
O_2Ta	TaO_2	50 (Ta):	Hb/A2-498, 529 Hb/B1-17, 21, 25, 28/30, 46
–	$TaO_2 \cdot xH_2O$	50 (Ta):	Hb/B1-54
–	TaO_2 solid solutions:		
	$(Ta,Nb)O_2$	50 (Ta):	Hb/B2-246
–	TaO_2 systems:		
	TaO_2-TiO_2	50 (Ta):	Hb/B2-221/2
	TaO_2-VO_2	50 (Ta):	Hb/B2-244
O_2Ta^+	$[TaO_2]^+$	50 (Ta):	Hb/A2-407 Hb/B1-59, 135
–	$[Ta_2O_4]^{2+}$	50 (Ta):	Hb/B1-178
O_2Ta^-	$[TaO_2]^-$	50 (Ta):	Hb/A2-407 Hb/B1-59
O_2Tb	TbO_2	39 (SE):	Hb/C1-103/4, 398/9
–	TbO_2 solid solutions:		
	TbO_2-PrO_2	39 (SE):	Hb/C1-404
O_2Tc	TcO_2	14 (C):	Hb/C2-168
O_2Te	TeO_2	11 (Te):	Hb-239, 275/82

Hb = Main Series Volume; GD = Gmelin-Durrer; Ab = Appended Volume to Main Series; Eb = Supplement Volume to Main Series; Erg. W. = New Supplement Series Volume

O_2Te . . . . TeO_2 glasses:
- TeO_2-Li_2O-PbO . . . . 47 (Pb): Hb/C3-992/3
- TeO_2-Na_2O-PbO . . . . 47 (Pb): Hb/C3-1034
- TeO_2-PbO . . . . 47 (Pb): Hb/C2-678/9

–. . . . . TeO_2 melts:
- $TeO_2-V_2O_5$ . . . . 48 (V): Hb/B1-320

–. . . . . TeO_2 systems:
- $TeO_2-B_2O_3-PbO$ . . . . 47 (Pb): Hb/C2-697
- $TeO_2-Bi_2O_3$ . . . . 19 (Bi): Eb-817
- TeO_2-HF . . . . 11 (Te): Hb-322
- TeO_2-HF-H_2O . . . . 11 (Te): Hb-322/3
- $TeO_2-Na_2O-H_2O$ . . . . 21 (Na): Eb/3-1205, Eb/5-1909/10
- $TeO_2-Nb_2O_5-PbO$ . . . . 49 (Nb): Hb/B4-307
- $TeO_2-PbO-SO_3$ . . . . 47 (Pb): Hb/C2-681
- $TeO_2-PbO-SeO_2$ . . . . 47 (Pb): Hb/C2-684

O_2Th . . . . ThO_2:
- Chemical reactions . . . . 44 (Th): Hb-220/6
- Crystallographic properties . . . . . . 44 (Th): Hb-205/6
- Electrical properties . . . . 44 (Th): Hb-217/20
- Formation and preparation . . . . . . 44 (Th): Hb-202/5
- Magnetic properties . . . . 44 (Th): Hb-217
- Mechanical properties . . . . 44 (Th): Hb-207/8
- Optical properties. . . . . 44 (Th): Hb-211/7
- Sols . . . . 44 (Th): Hb-238/9
- Solubility . . . . 44 (Th): Hb-225/6
- Sorption . . . . 44 (Th): Hb-208
- Thermal properties . . . . 44 (Th): Hb-209/11
- Toxicity . . . . 44 (Th): Hb-197
- Uses . . . . 44 (Th): Hb-77/81

–. . . . . $ThO_2 \cdot 0.5\ H_2O$ . . . . 44 (Th): Hb-227

–. . . . . $ThO_2 \cdot 2\ H_2O$ . . . . 44 (Th): Hb-228

–. . . . . $ThO_2 \cdot xH_2O$ . . . . 44 (Th): Hb-226/38

–. . . . . ThO_2 glasses:
- $ThO_2-P_2O_5-PbO$ . . . . 47 (Pb): Hb/C4-1471
- $ThO_2-PbO-SiO_2$ . . . . 47 (Pb): Hb/C4-1471

–. . . . . ThO_2 solid solutions:
- ThO_2-AmO_2 . . . . Erg. W. 4-79
- ThO_2-NpO_2 . . . . Erg. W. 4-49
- ThO_2-PuO_2 . . . . Erg. W. 4-58
- ThO_2-UO_2. . . . . 55 (U): Hb-239

–. . . . . ThO_2 systems:
- $ThO_2-Al_2O_3$ . . . . 44 (Th): Hb-351
- $ThO_2-Al_2O_3-BeO$ . . . . 44 (Th): Hb-351
- $ThO_2-AmO_{1.5}$ . . . . Erg. W. 4-79
- $ThO_2-BeAl_2O_4-BeO$ . . . . 44 (Th): Hb-351

O_2Th	ThO_2 systems:		
	ThO_2–BeO	44 (Th):	Hb–346
	ThO_2–BeO–MgO	44 (Th):	Hb–347
	ThO_2–CaO	44 (Th):	Hb–347
	ThO_2–CaO–ZrO_2	44 (Th):	Hb–354
	ThO_2–CeO_2	44 (Th):	Hb–352/3
	ThO_2–CrO_3–H_2O	52 (Cr):	Hb/B–907
	ThO_2–MgO	44 (Th):	Hb–346
	ThO_2–MgO–ZrO_2	44 (Th):	Hb–354
	ThO_2–NbO_2	49 (Nb):	Hb/B3–219
	ThO_2–Nb_2O_5	49 (Nb):	Hb/B3–219/20
	ThO_2–Ta_2O_5	50 (Ta):	Hb/B2–229
	ThO_2–TiO_2	44 (Th):	Hb–353/4
	ThO_2–TiO_2–ZrO_2	44 (Th):	Hb–354
	ThO_2–V_2O_5	48 (V):	Hb/B2–564
	ThO_2–ZrO_2	44 (Th):	Hb–354
O_2Ti	TiO_2:		
	Chemical reactions	41 (Ti):	Hb–252/5, 473/5
	Crystallographic properties	41 (Ti):	Hb–232/40
	Dressing and treatment of ores	41 (Ti):	Hb–81/117
	Electrical properties	41 (Ti):	Hb–247/51
	Formation	41 (Ti):	Hb–226/32, 478
	Hydrosols	41 (Ti):	Hb–263/8
	Magnetic properties	41 (Ti):	Hb–247
	Mechanical properties	41 (Ti):	Hb–240/2
	Molecule	41 (Ti):	Hb–232
	Optical properties	41 (Ti):	Hb–243/7
	Organosols	41 (Ti):	Hb–268
	Physiological action	41 (Ti):	Hb–199
	Preparation, industrial	41 (Ti):	Hb–80/117
	– of anatase	41 (Ti):	Hb–229/30
	– of brookite	41 (Ti):	Hb–230/1
	– of films	41 (Ti):	Hb–231
	– of pigments	41 (Ti):	Hb–114/6
	– of rutile	41 (Ti):	Hb–103, 111, 226/9
	– of sols	41 (Ti):	Hb–263/5, 268
	Purity testing	41 (Ti):	Hb–231
	Reduction, industrial	41 (Ti):	Hb–117/21
	Sorption	41 (Ti):	Hb–242
	Thermal properties	41 (Ti):	Hb–242/3
	Toxicity	41 (Ti):	Hb–199
	Uses	41 (Ti):	Hb–127/8, 467
–	$TiO_2 \cdot nH_2O$	41 (Ti):	Hb–256/8, 471/2
–	TiO_2 glasses:		
	TiO_2–PbO–SiO_2	47 (Pb):	Hb/C4–1392/4
	TiO_2 solid solutions:		
	$Fe_{2-x}Ti_xO_3$	59 (Fe):	D/2. Eb–430/2
	$Fe_{3-x}Ti_xO_4$	59 (Fe):	D/2. Eb–430/2

Hb = Main Series Volume; GD = Gmelin-Durrer; Ab = Appended Volume to Main Series;
Eb = Supplement Volume to Main Series; Erg. W. = New Supplement Series Volume

O_2Ti . . . . TiO_2 solid solutions:

$TiO_2-Ga_2O_3-Nb_2O_5$	49 (Nb):	Hb/B3-205
$TiO_2-In_2O_3-Nb_2O_5$	49 (Nb):	Hb/B3-205
$(Ti,Nb)O_2$	49 (Nb):	Hb/B3-179/81
$TiO_2-Nb_2O_5$	49 (Nb):	Hb/B3-187/8
$TiO_2-Nb_2O_5-Y_2O_3$	49 (Nb):	Hb/B3-209
TiO_2-NiF_2	57 (Ni):	Hb/B3-1181
$Ni_nTi_{4-n}O_{8-n}$	57 (Ni):	Hb/B3-1181
$TiO_2-PbNb_2O_6$	49 (Nb):	Hb/B3-285
$TiO_2-Pb_3Nb_4O_{13}$	49 (Nb):	Hb/B3-287

–. . . . . TiO_2 systems:

$TiO_2-Al(PO_3)_3-NaF$	41 (Ti):	Hb-464
$TiO_2-Al_2O_3$	41 (Ti):	Hb-463
$TiO_2-Al_2O_3-Cr_2O_3$	52 (Cr):	Hb/B-905
$TiO_2-Al_2O_3-SiO_2$	41 (Ti):	Hb-464
$TiO_2-Al_2TiO_5-CaTiO_3-MgTiO_3$	41 (Ti):	Hb-464
$TiO_2-B_2O_3-BaO-PbO$	47 (Pb):	Hb/C4-1412/3
$TiO_2-B_2O_3-PbO$	47 (Pb):	Hb/C4-1391
TiO_2-BaO	41 (Ti):	Hb-434
$TiO_2-BaO-CaO$	41 (Ti):	Hb-456/7
$TiO_2-BaO-MgO$	41 (Ti):	Hb-456
$TiO_2-BaO-P_2O_5-PbO$	47 (Pb):	Hb/C4-1413
$TiO_2-BaO-SrO$	41 (Ti):	Hb-457
TiO_2-BeO	41 (Ti):	Hb-416/7
$TiO_2-Bi_2O_3-PbO$	47 (Pb):	Hb/C4-1394
TiO_2-CO_2-CaO	41 (Ti):	Hb-430
$TiO_2-CO_2-CaO-K_2O$	41 (Ti):	Hb-432
$TiO_2-CO_2-CaO-Na_2O$	41 (Ti):	Hb-431
$TiO_2-CO_2-K_2O$	41 (Ti):	Hb-393/4
$TiO_2-CO_2-Na_2O$	41 (Ti):	Hb-384
$TiO_2-CaO \cdot MgO \cdot 2\,SiO_2$	41 (Ti):	Hb-433
TiO_2-CaO	41 (Ti):	Hb-424/6
$TiO_2-CaO-La_2O_3$	41 (Ti):	Hb-468
$TiO_2-CaO-Na_2O-SiO_2$	41 (Ti):	Hb-431
$TiO_2-CaO-Nb_2O_5$	49 (Nb):	Hb/B3-191/3
$TiO_2-CaO-Nb_2O_5-SiO_2$	49 (Nb):	Hb/B4-360
$TiO_2-CaO-SiO_2$	41 (Ti):	Hb-430
$TiO_2-CaO-ZrO_2$	42 (Zr):	Hb-448
$TiO_2-CaSiO_3$	41 (Ti):	Hb-430
$TiO_2-CaTiO_3-MgTiO_3$	41 (Ti):	Hb-432
$TiO_2-CoO-CuO$	60 (Cu):	Hb/B3-1264
$TiO_2-Cr_2O_3$	52 (Cr):	Hb/B-904
$TiO_2-CuO-Fe_2O_3$	60 (Cu):	Hb/B3-1327
$TiO_2-La_2O_3-Nb_2O_5-PbO$	49 (Nb):	Hb/B3-302
TiO_2-MgO	41 (Ti):	Hb-418/9
$TiO_2-MgO-Ti_2O_3$	41 (Ti):	Hb-423
$TiO_2-MgO-ZrO_2$	42 (Zr):	Hb-448
TiO_2-Na_2O	41 (Ti):	Hb-383/4
$TiO_2-Na_2O-H_2O$	41 (Ti):	Hb-384

See the "Instructions for the user" at the beginning of the volume

O_2Ti	TiO_2 systems:		
	$TiO_2-Na_2O-PbO-SiO_2$	47 (Pb):	Hb/C4–1392/3
	TiO_2-NbO_2	49 (Nb):	Hb/B3–179/81
	$TiO_2-NbO_2-Nb_2O_5$	49 (Nb):	Hb/B3–179
	$TiO_2-Nb_2O_5$	49 (Nb):	Hb/B3–181/3
	$TiO_2-Nb_2O_5-PbO$	49 (Nb):	Hb/B3–284
	$TiO_2-Nb_2O_5-SiO_2$	49 (Nb):	Hb/B4–360
	TiO_2-Ni	57 (Ni):	Hb/B1–98
	TiO_2-NiO	57 (Ni):	Hb/B3–1179
	$TiO_2-NiO-ZnO$	57 (Ni):	Hb/B3–1184/5
	TiO_2-PbO	47 (Pb):	Hb/C4–1378/9
	$TiO_2-PbO-V_2O_5$	48 (V):	Hb/B2–574
	$TiO_2-PbO-ZrO_2$	47 (Pb):	Hb/C4–1439/40
	$TiO_2-SO_3-H_2O$	41 (Ti):	Hb–345/7
	TiO_2-SiO_2	41 (Ti):	Hb–376/7
	TiO_2-TaO_2	50 (Ta):	Hb/B2–221/2
	$TiO_2-Ta_2O_5$	50 (Ta):	Hb/B2–222
	$TiO_2-Ta_2O_5-ZrO_2$	50 (Ta):	Hb/B2–229
	TiO_2-ThO_2	44 (Th):	Hb–353/4
	$TiO_2-ThO_2-ZrO_2$	44 (Th):	Hb–354
	TiO_2-TiS_2	41 (Ti):	Hb–343
	$TiO_2-VO-ZrO_2$	48 (V):	Hb/B2–564
	TiO_2-VO_2	48 (V):	Hb/B2–558/9
	TiO_2-ZnO	41 (Ti):	Hb–460
	$TiO_2-ZnTa_2O_6$	50 (Ta):	Hb/B2–226
	TiO_2-ZrO_2	42 (Zr):	Hb–446/7
O_2Ti^{2+}	$Ti(O \cdot O)^{2+}$	41 (Ti):	Hb–270/1
–	$(TiO \cdot O \cdot xH_2O)^{2+}$	41 (Ti):	Hb–471
O_2Tl	$TlO_2 = Tl_2O_4$	38 (Tl):	Hb/2–233
O_2U	UO_2	55 (U):	Hb–83/8, 100
–	UO_2, Sols	Erg. W. 8–244/5	
–	$UO_2 \cdot xH_2O$	55 (U):	Hb–99/100
–	UO_2 fuels:		
	In-pile behavior	Erg. W. 8–259/77	
	Preparation	Erg. W. 8–242/8	
–	UO_2 solid solutions:		
	UO_2-NpO_2	Erg. W. 4–49	
	UO_2-PuO_2	Erg. W. 4–60/70	
		Erg. W. 8–192	
	$UO_2-PuO_2-ZrO_2$	Erg. W. 4–71	
	UO_2-ThO_2	55 (U):	Hb–239
–	UO_2 systems:		
	UO_2-PuO_2	Erg. W. 4–59	
O_2U^{2+}	UO_2^{2+}:		
	Chemical reactions	55 (U):	Hb–257/70
	Electrochemical behavior	55 (U):	Hb–255/7
	Ionic susceptibility	55 (U):	Hb–154, 255

O_2U^{2+}	UO_2^{2+}:		
	Spectra	21 (Na):	Eb/6–354/6
		55 (U):	Hb–246/54
–	$^{239}UO_2^{2+}$, Recoil reactions	Erg. W. 8–318	
O_2V	$VO_2 = V_2O_4$:		
	Chemical reactions	48 (V):	Hb/B1–60/1
	Determination	48 (V):	Hb/A2–714
	Electrochemical behavior	48 (V):	Hb/A2–467/9, 498, 502
	Formation and preparation	48 (V):	Hb/B1–51/3
	Physical properties	48 (V):	Hb/A2–376, 380
			Hb/B1–30, 54/60
–	V_4O_8	48 (V):	Hb/B1–84
–	V_6O_{12}	48 (V):	Hb/B1–84
–	$VO_2 \cdot H_2O = VO(OH)_2$	48 (V):	Hb/A2–501, 535/6
			Hb/B1–65/70, 346
–	$V_2O_4 \cdot H_2O$	48 (V):	Hb/B1–65, 68
–	$3\ V_2O_4 \cdot 4\ H_2O$	48 (V):	Hb/B1–66
–	$V_2O_4 \cdot 7\ H_2O$	48 (V):	Hb/B1–67
–	VO_2 solid solutions:		
	VO_2–$VNbO_4$	49 (Nb):	Hb/B3–322/3
	VO_2–$VTaO_4$	50 (Ta):	Hb/B2–244
	V_2O_4–$ZrSiO_4$	48 (V):	Hb/B2–563
–	VO_2 systems:		
	VO_2–Al_2O_3	48 (V):	Hb/B2–544
	VO_2–BeO	48 (V):	Hb/B2–500
	VO_2–CaO	48 (V):	Hb/B2–511
	VO_2–CeO_2	48 (V):	Hb/B2–552
	VO_2–CoO	58 (Co):	Eb/A–873
	VO_2–HF–H_2O	48 (V):	Hb/B1–191/2
	VO_2–Li_2O–V_2O_5	48 (V):	Hb/B2–370
	VO_2–MgO	48 (V):	Hb/B2–503
	VO_2–MgO–V_2O_5	48 (V):	Hb/B2–504
	VO_2–NbO_2	49 (Nb):	Hb/B3–322/4
	VO_2–Nb_2O_5	49 (Nb):	Hb/B3–324
	VO_2–TaO_2	50 (Ta):	Hb/B2–244
	VO_2–TiO_2	48 (V):	Hb/B2–558/9
	VO_2–V_2O_5–ZnO	48 (V):	Hb/B2–534
	VO_2–ZnO	48 (V):	Hb/B2–533
O_2V^+	$[VO_2]^+$:		
	Chemical reactions	48 (V):	Hb/B1–158/60
	Detection	48 (V):	Hb/A2–647/52
	Determination	48 (V):	Hb/A2–676
	Electrochemical behavior	48 (V):	Hb/A2–468, 478, 498, 500/2, 505/6
	Properties of the ion	48 (V):	Hb/B1–157/8
	Thermodynamic data of formation	48 (V):	Hb/B1–157
	Transition reactions	48 (V):	Hb/A2–550/640

See the "Instructions for the user" at the beginning of the volume

O_2V^{3+}	$V(O_2)^{3+}$	48 (V):	Hb/B1-162/3, 166
O_2V_2	V_2O_2 *see OV ... VO*		
$O_2V_2^{4+}$	$(VO)_2^{4+}$	48 (V):	Hb/A2-601
O_2W	WO_2	54 (W):	Hb-108/15
O_2Xe	$XeO_2 \cdot 2H_2O$	Erg. W. 1-101	
O_2Xe^+	XeO_2^+	Erg. W. 1-101	
O_2Y_2	Y_2O_2	39 (SE):	Hb/C1-91, 100
O_2Zn	ZnO_2	32 (Zn):	Eb-817/8
–	$ZnO_2 \cdot 0.5\ H_2O$	32 (Zn):	Hb-146 Eb-817/8
–	$ZnO_2 \cdot H_2O$	32 (Zn):	Eb-817
O_2Zn^{2-}	$[ZnO_2]^{2-}$	32 (Zn):	Hb-112 Eb-403, 511
O_2Zr	ZrO_2:		
	Chemical reactions	42 (Zr):	Hb-238/51
	Crystallographic properties	42 (Zr):	Hb-217/22
	Electrical properties	42 (Zr):	Hb-237/8
	Electrochemical behavior	42 (Zr):	Hb-238
	Formation	42 (Zr):	Hb-215/6
	Magnetic properties	42 (Zr):	Hb-236/7
	Mechanical properties	42 (Zr):	Hb-222/6
	Optical properties	42 (Zr):	Hb-232/6
	Preparation	42 (Zr):	Hb-215/6
	–, industrial	42 (Zr):	Hb-76/7
	Solubility in glasses and enamels	42 (Zr):	Hb-248/9
	Thermal properties	42 (Zr):	Hb-226/32
	Toxicity	42 (Zr):	Hb-191
	Uses	42 (Zr):	Hb-80/4
–	$ZrO_2 \cdot xH_2O$	42 (Zr):	Hb-251/61
–	ZrO_2 solid solutions:		
	$ZrO_2-DyNbO_4$	49 (Nb):	Hb/B3-217
	$ZrO_2-ErNbO_4$	49 (Nb):	Hb/B3-217
	$ZrO_2-GdNbO_4$	49 (Nb):	Hb/B3-217
	$ZrO_2-HoNbO_4$	49 (Nb):	Hb/B3-217
	$ZrO_2-LuNbO_4$	49 (Nb):	Hb/B3-217
	$ZrO_2-PbNb_2O_6$	49 (Nb):	Hb/B3-302
	$ZrO_2-PuO_2-UO_2$	Erg. W. 4-71	
	$ZrO_2-Ta_2O_5-SiO_2$	50 (Ta):	Hb/B2-349
	$ZrO_2-TbNbO_4$	49 (Nb):	Hb/B3-217
	$ZrO_2-TmNbO_4$	49 (Nb):	Hb/B3-217
	$ZrO_2-YbNbO_4$	49 (Nb):	Hb/B3-217
–	ZrO_2 systems:		
	$ZrO_2-Al_2O_3$	42 (Zr):	Hb-443/4
	$ZrO_2-Al_2O_3-BaO$	42 (Zr):	Hb-445
	$ZrO_2-Al_2O_3-BeO$	42 (Zr):	Hb-444/5
	$ZrO_2-Al_2O_3-MgO$	42 (Zr):	Hb-445
	$ZrO_2-Al_2O_3-SiO_2$	42 (Zr):	Hb-444

Hb = Main Series Volume; GD = Gmelin-Durrer; Ab = Appended Volume to Main Series; Eb = Supplement Volume to Main Series; Erg. W. = New Supplement Series Volume

Formula	System	Volume	Reference
O_2Zr	ZrO_2 systems:		
	$ZrO_2-AmO_{1.5}$	Erg. W. 4–80	
	ZrO_2-AmO_2	Erg. W. 4–80	
	$ZrO_2-B_2O_3$	42 (Zr):	Hb–360
	ZrO_2-BaO	42 (Zr):	Hb–440
	ZrO_2-BeO	42 (Zr):	Hb–430
	$ZrO_2-BeO-CaO$	42 (Zr):	Hb–438
	$ZrO_2-BeO-Cr_2O_3$	52 (Cr):	Hb/B–906
	$ZrO_2-BeO-MgO$	42 (Zr):	Hb–434
	ZrO_2-CaCl_2	42 (Zr):	Hb–437
	ZrO_2-CaO	42 (Zr):	Hb–434/6
	$ZrO_2-CaO-MgO$	42 (Zr):	Hb–438
	$ZrO_2-CaO-ThO_2$	44 (Th):	Hb–354
	$ZrO_2-CaO-TiO_2$	42 (Zr):	Hb–448
	ZrO_2-CoO	58 (Co):	Hb/A2–473
			Eb/A–871
	$ZrO_2-Cr_2O_3$	52 (Cr):	Hb/B–905
	$ZrO_2-Cr_2O_3-MgO$	52 (Cr):	Hb/B–907
	$ZrO_2-Cr_2O_3-NiO$	57 (Ni):	Hb/B3–1217
	ZrO_2-CuO	60 (Cu):	Hb/B3–1191
	$ZrO_2-HNO_3-H_2O$	42 (Zr):	Hb–272
	$ZrO_2-HfO_2-Ta_2O_5$	50 (Ta):	Hb/B2–229
	$ZrO_2-MgO \cdot Cr_2O_3$	52 (Cr):	Hb/B–907
	ZrO_2-MgO	42 (Zr):	Hb–431/3
	$ZrO_2-MgO-ThO_2$	44 (Th):	Hb–354
	$ZrO_2-MgO-TiO_2$	42 (Zr):	Hb–448
	$ZrO_2-Na_2O-SiO_2$	42 (Zr):	Hb–407/8
	ZrO_2-NbO	49 (Nb):	Hb/B3–209/10
	ZrO_2-NbO_2	49 (Nb):	Hb/B3–209/10
	$ZrO_2-Nb_2O_5$	49 (Nb):	Hb/B3–210/1
	ZrO_2-NiO	57 (Ni):	Hb/B3–1185
	$ZrO_2-P_2O_5-PbO$	47 (Pb):	Hb/C4–1431
	ZrO_2-PbO	47 (Pb):	Hb/C4–1423
	$ZrO_2-PbO-SiO_2$	47 (Pb):	Hb/C4–1431
	$ZrO_2-PbO-TiO_2$	47 (Pb):	Hb/C4–1439/40
	ZrO_2-PuO_2	Erg. W. 4–70/1	
	$ZrO_2-SO_3-H_2O$	42 (Zr):	Hb–335/7
	ZrO_2-SiO_2	42 (Zr):	Hb–381/2
	ZrO_2-SrO	42 (Zr):	Hb–439
	$ZrO_2-Ta_2O_5$	50 (Ta):	Hb/B2–228
	$ZrO_2-Ta_2O_5-TiO_2$	50 (Ta):	Hb/B2–229
	$ZrO_2-Ta_2O_5-Y_2O_3$	50 (Ta):	Hb/B2–229
	ZrO_2-ThO_2	44 (Th):	Hb–354
	$ZrO_2-ThO_2-TiO_2$	44 (Th):	Hb–354
	ZrO_2-TiO_2	42 (Zr):	Hb–446/7
	ZrO_2-TiO_2-VO	48 (V):	Hb/B2–564
	$ZrO_2-V_2O_5$	48 (V):	Hb/B2–562
	ZrO_2-ZnO	42 (Zr):	Hb–442
$O_{2.17}V$	$VO_{2.17} = V_6O_{13}$	48 (V):	Hb/B1–17, 61/3

See the "Instructions for the user"
at the beginning of the volume

$O_{2.33}V$	$VO_{2.33} \cdot 0.65\ H_2O = V_6O_{14} \cdot 4\ H_2O$		
	$= V_3O_5(OH)_4$	48 (V):	Hb/B1–70
$O_{2.5}SrV$	$SrVO_{2.5}$	48 (V):	Hb/B2–518
–	$SrVO_{2.5}$ systems:		
	$SrVO_{2.5}-LaVO_3$	48 (V):	Hb/B2–557
	$SrVO_{2.5}-SrVO_3$	48 (V):	Hb/B2–519
$O_{2.75}SrV$	$SrVO_{2.75}$	48 (V):	Hb/B2–519
O_3	O_3		
	see O ... Ozone		
–	$^{16}O^{16}O^{18}O$	3 (O):	Hb/4–1111
–	$^{16}O^{18}O^{16}O$	3 (O):	Hb/4–1111
–	$^{16}O^{18}O^{18}O$	3 (O):	Hb/4–1111
–	$^{18}O_3$	3 (O):	Hb/4–1111
–	$O_3 \cdot H_2O = H_2O_4$	3 (O):	Hb/8–2626/7
O_3^+	O_3^+	3 (O):	Hb/4–1071, 1080
O_3^-	O_3^-	3 (O):	Hb/3–354
			Hb/4–1071
O_3^{2-}	O_3^{2-}	3 (O):	Hb/8–2625
O_3Os	OsO_3	66 (Os):	Hb–30
–	$OsO_3 \cdot H_2O$	66 (Os):	Hb–30/1
O_3OsS	$OsSO_3$	66 (Os):	Hb–60
–	$2\ OsO_3S \cdot 3\ H_2O$	66 (Os):	Hb–60
O_3Os_2	Os_2O_3	66 (Os):	Hb–25
–	$Os_2O_3 \cdot 3\ H_2O$	66 (Os):	Hb–25
O_3PRb	$RbPO_3$	24 (Rb):	Hb–234
O_3PTl	$TlPO_3$	38 (Tl):	Hb/3–431/2
–	$(TlPO_3)_n$	38 (Tl):	Hb/3–431/2
O_3PTl_2	$Tl_2PO_3 = Tl_4P_2O_6$	38 (Tl):	Hb/3–427
O_3P_2	P_2O_3	16 (P):	Hb/C–68/75
–	P_4O_6	16 (P):	Hb/C–70/1
–	$(P_2O_3)_x$	16 (P):	Hb/C–67/8
O_3Pb^{2-}	$[PbO_3]^{2-}$	47 (Pb):	Hb/B2–11
			Hb/C1–198/9
O_3PbRb_2	Rb_2PbO_3	47 (Pb):	Hb/C3–1137
O_3PbS	$PbSO_3$	47 (Pb):	Hb/C2–535/6
–	$PbSO_3 \cdot H_2O$	47 (Pb):	Hb/C2–536/7
O_3PbSSe	$PbSeSO_3$	47 (Pb):	Hb/C2–640
O_3PbS_2	PbS_2O_3	47 (Pb):	Hb/C2–595/6
–	PbS_2O_3 solutions:		
	$PbS_2O_3-H_2O$	47 (Pb):	Hb/C2–596/8
	$PbS_2O_3-Na_2S_2O_3-H_2O$	47 (Pb):	Hb/C3–1031
O_3PbSe	$PbSeO_3$	47 (Pb):	Hb/C2–632/4
–	$PbSeO_3$ systems:		
	$PbSeO_3-PbO$	47 (Pb):	Hb/C2–634/5
O_3PbSi	$PbSiO_3 = PbO \cdot SiO_2$	47 (Pb):	Hb/B2–403, 413/4
			Hb/C3–848/54
–	$PbSiO_3 \cdot xH_2O$	47 (Pb):	Hb/C3–854

Formula	Compound	System No.	Reference
O_3PbSi	$PbSiO_3$ systems:		
	$PbSiO_3-CaO \cdot TiO_2 \cdot SiO_2$	47 (Pb):	Hb/C4–1401
	$PbSiO_3-Cu_2S$	60 (Cu):	Hb/B3–1203
	$PbSiO_3-FeS-FeSiO_3-PbS$	59 (Fe):	Hb/B5–1137
	$PbSiO_3-Na_2SiO_3-SiO_2$	47 (Pb):	Hb/C3–1038
	$PbSiO_3-PbZrO_3$	47 (Pb):	Hb/C4–1431
O_3PbSn	$PbSnO_3$	47 (Pb):	Hb/C4–1581/3
–	$PbSnO_3 \cdot nH_2O$	47 (Pb):	Hb/C4–1583
–	$PbSnO_3$ solid solutions:		
	$PbSnO_3-BaSnO_3-BaTiO_3-PbTiO_3$	47 (Pb):	Hb/C4–1600/1
	$Pb(Hf, Sn, Ti)O_3$ (modified by Nb_2O_5)	49 (Nb):	Hb/B3–321
	$PbSnO_3-PbNb_2O_6$	49 (Nb):	Hb/B3–320
	$Pb(Sn, Ti, Zr)O_3$ (modified by Nb_2O_5)	49 (Nb):	Hb/B3–320/1
–	$PbSnO_3$ systems:		
	$PbSnO_3-BaSnO_3$	47 (Pb):	Hb/C4–1595/6
	$PbSnO_3-BaTiO_3$	47 (Pb):	Hb/C4–1601/2
	$PbSnO_3-PbTiO_3$	47 (Pb):	Hb/C4–1597/9
	$PbSnO_3-PbTiO_3-PbZrO_3$	47 (Pb):	Hb/C4–1605/7
	$PbSnO_3-PbZrO_3$	47 (Pb):	Hb/C4–1603/5
O_3PbSr	$SrPbO_3$	47 (Pb):	Hb/C3–1196
O_3PbTe	$3\,PbTeO_3 \cdot 2\,H_2O$	47 (Pb):	Hb/C2–679
O_3PbTh	$PbThO_3$	47 (Pb):	Hb/C4–1471
–	$PbThO_3$ solid solutions:		
	$Pb(Th, Zr)O_3$	47 (Pb):	Hb/C4–1472
O_3PbTi	$PbTiO_3$	47 (Pb):	Hb/C4–1379/90
–	$PbTiO_3$ solid solutions:		
	$(Pb, Ba)TiO_3-(Ba, Pb)Nb_2O_6$	49 (Nb):	Hb/B3–202
	$PbTiO_3-BaSnO_3-BaTiO_3$	47 (Pb):	Hb/C4–1603
	$PbTiO_3-BaSnO_3-BaTiO_3-PbSnO_3$	47 (Pb):	Hb/C4–1600/1
	$PbTiO_3-BaTiO_3$	47 (Pb):	Hb/C4–1405/12
	$(Pb, Ba)(Ti, Zr)O_3$ (modified by Nb_2O_5)	49 (Nb):	Hb/B3–318
	$PbTiO_3-BaTiO_3-CaTiO_3$	47 (Pb):	Hb/C4–1414/6
	$PbTiO_3-BaTiO_3-SrTiO_3$	47 (Pb):	Hb/C4–1416
	$PbTiO_3-Ba_5Nb_4O_{15}$	49 (Nb):	Hb/B3–297/8
	$PbTiO_3-Bi_{0.67}TiO_3-SrTiO_3$	47 (Pb):	Hb/C4–1404/5
	$(1-x)Pb(Zr_yTi_{1-y})O_3 \cdot xBiAlO_3$	47 (Pb):	Hb/C4–1462
	$PbTiO_3-CaSnO_3$	47 (Pb):	Hb/C4–1599
	$PbTiO_3-CaTiO_3$	47 (Pb):	Hb/C4–1400/1
	$(Pb, Ca)(Zr, Ti)O_3$	47 (Pb):	Hb/C4–1455
	$PbTiO_3-CdTiO_3$	47 (Pb):	Hb/C4–1417
	$PbTiO_3-K_{0.5}Bi_{0.5}TiO_3$	47 (Pb):	Hb/C4–1398/9
	$PbTiO_3-KNbO_3$	49 (Nb):	Hb/B3–290/1
	$Pb(Ti, Zr)O_3-KNbO_3$	49 (Nb):	Hb/B3–314/5
	$(1-x)Pb(Zr_yTi_{1-y})O_3 \cdot xKSbO_3$	47 (Pb):	Hb/C4–1455
	$PbTiO_3-La_{0.67}TiO_3$	47 (Pb):	Hb/C4–1417/8

See the "Instructions for the user" at the beginning of the volume

O_3PbTi	$PbTiO_3$ solid solutions:		
	$PbTiO_3-LaAlO_3$	47 (Pb):	Hb/C4-1418/9
	$(1-x)Pb(Zr_yTi_{1-y})O_3 \cdot xLaAlO_3$	47 (Pb):	Hb/C4-1462/3
	$PbTiO_3-MgTiO_3$	47 (Pb):	Hb/C4-1400
	$PbTiO_3-Na_{0.5}Bi_{0.5}TiO_3$	47 (Pb):	Hb/C4-1396/7
	$(1-x)Pb(Zr_yTi_{1-y})O_3 \cdot xNaBiO_3$	47 (Pb):	Hb/C4-1455
	$PbTiO_3-NaNbO_3$	49 (Nb):	Hb/B3-289/90
	$Pb(Ti, Zr)O_3-NaNbO_3$	49 (Nb):	Hb/B3-314
	$(1-x)Pb(Zr_yTi_{1-y})O_3 \cdot xNaSbO_3$	47 (Pb):	Hb/C4-1455
	$Pb(Hf, Ti)O_3$	47 (Pb):	Hb/C4-1467
	$PbTiO_3-PbHfO_3-PbNb_2O_6$	49 (Nb):	Hb/B3-320
	$Pb(Hf, Sn, Ti)O_3$ (modified by Nb_2O_5)	49 (Nb):	Hb/B3-321
	$Pb(Hf, Zr, Ti)O_3$	47 (Pb):	Hb/C4-1468
	$PbTiO_3-PbNb_2O_6$	49 (Nb):	Hb/B3-285/7 Hb/B4-313/4
	$Pb(Ti, Sn, Zr)O_3$ (modified by Nb_2O_5)	49 (Nb):	Hb/B3-320/1
	$Pb(Ti, Zr)O_3$	47 (Pb):	Hb/C4-1442/54
	– (modified by Nb_2O_5)	49 (Nb):	Hb/B3-309/14
	– (modified by Ta_2O_5)	50 (Ta):	Hb/B2-242/3
	$Pb(Ti, Zr)O_3-Pb_2SbNbO_6$	49 (Nb):	Hb/B3-313/4
	$Pb(Ti, Zr)O_3-Pb_3MgNb_2O_9$	49 (Nb):	Hb/B3-315/7
	$(Pb, Sr) (Zr, Ti)O_3$	47 (Pb):	Hb/C4-1455/8
	– (modified by Nb_2O_5)	49 (Nb):	Hb/B3-317/8
	$PbTiO_3-Pb_2ScNbO_6$	49 (Nb):	Hb/B3-300/1
	$PbTiO_3-Pb_2YbNbO_6$	49 (Nb):	Hb/B3-302
	$PbTiO_3-Pb_3MgNb_2O_9$	49 (Nb):	Hb/B3-291/2
	$PbTiO_3-Pb_3ZnNb_2O_9$	49 (Nb):	Hb/B3-298/9
	$PbTiO_3-Pb_5Nb_4O_{15}$	49 (Nb):	Hb/B3-285
	$PbTiO_3-SrSnO_3$	47 (Pb):	Hb/C4-1600
	$PbTiO_3-SrTiO_3$	47 (Pb):	Hb/C4-1401/4
	$PbTiO_3-SrZrO_3$	47 (Pb):	Hb/C4-1458
	$PbTiO_3-Sr_5Nb_4O_{15}$	49 (Nb):	Hb/B3-294/5
	$PbTiO_3-ZnBi_2Nb_2O_9$	49 (Nb):	Hb/B3-299/300
	$PbTiO_3-ZnTiO_3$	47 (Pb):	Hb/C4-1416/7
–	$PbTiO_3$ systems:		
	$PbTiO_3-B_2O_3-PbO-PbZrO_3$	47 (Pb):	Hb/C4-1454
	$PbTiO_3-BaSnO_3$	47 (Pb):	Hb/C4-1602/3
	$PbTiO_3-BaTiO_3-BaZrO_3-PbZrO_3$	47 (Pb):	Hb/C4-1459/60
	$PbTiO_3-BaTiO_3-Na_2SiO_3$	47 (Pb):	Hb/C4-1413/4
	$PbTiO_3-BaZrO_3$	47 (Pb):	Hb/C4-1461/2
	$PbTiO_3-Ba_3(PO_4)_2$	47 (Pb):	Hb/C4-1413
	$PbTiO_3-K_2CO_3$	47 (Pb):	Hb/C4-1398
	$PbTiO_3-K_2F_2$	47 (Pb):	Hb/C4-1398
	$PbTiO_3-K_2SiO_3$	47 (Pb):	Hb/C4-1398
	$PbTiO_3-K_2SiO_3-Na_2SiO_3$	47 (Pb):	Hb/C4-1399/400
	$PbTiO_3-K_2TiO_3$	47 (Pb):	Hb/C4-1397/8
	$(PbTiO_3)_2-K_4P_2O_7$	47 (Pb):	Hb/C4-1398

Hb = Main Series Volume; GD = Gmelin-Durrer; Ab = Appended Volume to Main Series; Eb = Supplement Volume to Main Series; Erg. W. = New Supplement Series Volume

O_3PbTi	$PbTiO_3$ systems:		
	$PbTiO_3-LaNbO_4$	49 (Nb):	Hb/B3-302
	$PbTiO_3-Na_2CO_3$	47 (Pb):	Hb/C4-1396
	$PbTiO_3-Na_2F_2$	47 (Pb):	Hb/C4-1396
	$PbTiO_3-Na_2SiO_3$	47 (Pb):	Hb/C4-1396
	$(PbTiO_3)_2-Na_4P_2O_7$	47 (Pb):	Hb/C4-1396
	$PbTiO_3-PbNb_2O_6-PbZrO_3$	49 (Nb):	Hb/B3-308/9
	$PbTiO_3-PbSnO_3$	47 (Pb):	Hb/C4-1597/9
	$PbTiO_3-PbSnO_3-PbZrO_3$	47 (Pb):	Hb/C4-1605/7
	$PbTiO_3-PbTa_2O_6$	50 (Ta):	Hb/B2-242
	$PbTiO_3-PbZrO_3$	47 (Pb):	Hb/C4-1440/2
	$PbTiO_3-PbZrO_3-SrTiO_3-SrZrO_3$	47 (Pb):	Hb/C4-1455
	$PbTiO_3-Pb_2ScTaO_6$	50 (Ta):	Hb/B2-243
O_3PbZr	$PbZrO_3$	47 (Pb):	Hb/C4-1423/30
–	$PbZrO_3$ solid solutions:		
	$PbZrO_3-BaTiO_3$	47 (Pb):	Hb/C4-1460
	$(Pb, Ba)(Zr, Ti)O_3$ (modified by Nb_2O_5)	49 (Nb):	Hb/B3-318
	$(1-x)Pb(Zr_yTi_{1-y})O_3 \cdot xBiAlO_3$	47 (Pb):	Hb/C4-1462
	$(Pb, Ca)(Zr, Ti)O_3$	47 (Pb):	Hb/C4-1455
	$PbZrO_3-CdZrO_3$	47 (Pb):	Hb/C4-1439
	$K_xPb_{1-2x}Bi_xZrO_3$	47 (Pb):	Hb/C4-1433
	$Pb(Zr, Ti)O_3-KNbO_3$	49 (Nb):	Hb/B3-314/5
	$(1-x)Pb(Zr_yTi_{1-y})O_3 \cdot xKSbO_3$	47 (Pb):	Hb/C4-1455
	$(1-x)Pb(Zr_yTi_{1-y})O_3 \cdot xLaAlO_3$	47 (Pb):	Hb/C4-1462/3
	$(1-x)Pb(Zr_yTi_{1-y})O_3 \cdot xNaBiO_3$	47 (Pb):	Hb/C4-1455
	$Na_xPb_{1-2x}Bi_xZrO_3$	47 (Pb):	Hb/C4-1432/3
	$PbZrO_3-NaNbO_3$	49 (Nb):	Hb/B3-304/5
	$Pb(Zr, Ti)O_3-NaNbO_3$	49 (Nb):	Hb/B3-314
	$(1-x)Pb(Zr_yTi_{1-y})O_3 \cdot xNaSbO_3$	47 (Pb):	Hb/C4-1455
	$PbZrO_3-PbCeO_3$	47 (Pb):	Hb/C4-1439
	$Pb(Hf, Zr)O_3$	47 (Pb):	Hb/C4-1467/8
	$Pb(Hf, Zr, Ti)O_3$	47 (Pb):	Hb/C4-1468
	$PbZrO_3-PbNb_2O_6$	49 (Nb):	Hb/B3-232, 302/4
	$Pb(Zr, Sn, Ti)O_3$ (modified by Nb_2O_5)	49 (Nb):	Hb/B3-320/1
	$Pb(Th, Zr)O_3$	47 (Pb):	Hb/C4-1472
	$Pb(Zr, Ti)O_3$	47 (Pb):	Hb/C4-1442/54
	– (modified by Nb_2O_5)	49 (Nb):	Hb/B3-309/14
	– (modified by Ta_2O_5)	50 (Ta):	Hb/B2-242/3
	$Pb(Zr,Ti)O_3-Pb_2SbNbO_6$	49 (Nb):	Hb/B3-313/4
	$Pb(Zr,Ti)O_3-Pb_3MgNb_2O_9$	49 (Nb):	Hb/B3-315/7
	$Pb(Zr,Ti)O_3-Pb_3ZnNb_2O_9$	49 (Nb):	Hb/B3-314
	$(Pb,Sr)(Zr,Ti)O_3$	47 (Pb):	Hb/C4-1455/8
	– (modified by Nb_2O_5)	49 (Nb):	Hb/B3-317/8
	$PbZrO_3-Pb_2ScNbO_6$	49 (Nb):	Hb/B3-307/8
	$PbZrO_3-Pb_3MgNb_2O_9$	49 (Nb):	Hb/B3-305
	$PbZrO_3-SrZrO_3$	47 (Pb):	Hb/C4-1435/7
	$PbZrO_3-ZnZrO_3$	47 (Pb):	Hb/C4-1438/9

See the "Instructions for the user" at the beginning of the volume

O_3PbZr . . . . $PbZrO_3$ systems:
$PbZrO_3-B_2O_3-PbO-PbTiO_3$ . . . . . . 47 (Pb): Hb/C4-1454
$PbZrO_3-BaNb_2O_6-BaZrO_3-PbNb_2O_6$ . . . . 49 (Nb): Hb/B3-305/7
$PbZrO_3-BaTiO_3-BaZrO_3-PbTiO_3$ . . . 47 (Pb): Hb/C4-1459/60
$PbZrO_3-BaZrO_3$. . . . . 47 (Pb): Hb/C4-1437/8
$PbZrO_3-CaZrO_3$. . . . . 47 (Pb): Hb/C4-1433/4
$PbZrO_3-K_2B_2O_4$ . . . . 47 (Pb): Hb/C4-1432
$PbZrO_3-K_2CO_3$ . . . . 47 (Pb): Hb/C4-1432
$PbZrO_3-K_2Cl_2$ . . . . 47 (Pb): Hb/C4-1432
$PbZrO_3-K_2F_2$ . . . . 47 (Pb): Hb/C4-1432
$PbZrO_3-K_2SO_4$ . . . . 47 (Pb): Hb/C4-1432
$PbZrO_3-K_2TiO_3$ . . . . 47 (Pb): Hb/C4-1432
$PbZrO_3-Li_2B_2O_4$. . . . . 47 (Pb): Hb/C4-1432
$PbZrO_3-Li_2Cl_2$. . . . . 47 (Pb): Hb/C4-1432
$PbZrO_3-Li_2F_2$ . . . . 47 (Pb): Hb/C4-1432
$PbZrO_3-Li_2SO_4$ . . . . 47 (Pb): Hb/C4-1432
$PbZrO_3-Na_2B_2O_4$ . . . . 47 (Pb): Hb/C4-1432
$PbZrO_3-Na_2CO_3$ . . . . 47 (Pb): Hb/C4-1432
$PbZrO_3-Na_2Cl_2$ . . . . 47 (Pb): Hb/C4-1432
$PbZrO_3-Na_2F_2$ . . . . 47 (Pb): Hb/C4-1432
$PbZrO_3-Na_2SO_4$ . . . . 47 (Pb): Hb/C4-1432
$PbZrO_3-Na_2SiO_3$ . . . . 47 (Pb): Hb/C4-1432
$PbZrO_3-Na_2TiO_3$ . . . . 47 (Pb): Hb/C4-1432
$PbZrO_3-Na_4P_2O_7$ . . . . 47 (Pb): Hb/C4-1432
$PbZrO_3-Pb(BO_2)_2$. . . . . 47 (Pb): Hb/C4-1431
$PbZrO_3-PbF_2$ . . . . 47 (Pb): Hb/C4-1431
$PbZrO_3-PbNb_2O_6-PbTiO_3$ . . . . 49 (Nb): Hb/B3-308/9
$PbZrO_3-Pb(PO_3)_2$. . . . . 47 (Pb): Hb/C4-1431
$PbZrO_3-PbSiO_3$. . . . . 47 (Pb): Hb/C4-1431
$PbZrO_3-PbSnO_3$ . . . . 47 (Pb): Hb/C4-1603/5
$PbZrO_3-PbSnO_3-PbTiO_3$ . . . . 47 (Pb): Hb/C4-1605/7
$PbZrO_3-PbTa_2O_6$ . . . . 50 (Ta): Hb/B2-242
$PbZrO_3-PbTiO_3$. . . . . 47 (Pb): Hb/C4-1440/2
$PbZrO_3-PbTiO_3-SrTiO_3-SrZrO_3$ . . . 47 (Pb): Hb/C4-1455
$PbZrO_3-Pb_2ScTaO_6$. . . . . 50 (Ta): Hb/B2-243
$PbZrO_3-Pb_3(PO_4)_2$ . . . . 47 (Pb): Hb/C4-1431
$PbZrO_3-SrZrO_3$ . . . . 47 (Pb): Hb/C4-1434

O_3Pb_2 . . . . Pb_2O_3. 47 (Pb): Hb/A1-111, Hb/C1-139/44
–. . . . . $Pb_2O_3 \cdot 3\,H_2O$ . . . . 47 (Pb): Hb/C1-144
–. . . . . Pb_2O_3 systems:
$Pb_2O_3-Pb_3O_4$ . . . . 47 (Pb): Hb/C1-54/5

O_3Pd_2 . . . . $Pd_2O_3 \cdot xH_2O$ . . . . 65 (Pd): Hb/2-265/6
O_3PmSc. . . . . $PmScO_3 = Sc_2O_3 \cdot Pm_2O_3$. . . . . 39 (SE): Hb/C1-113, 124, 313
O_3Pm_2 . . . . Pm_2O_3:
Chemical reactions . . . . 39 (SE): Hb/C1-313
Crystallographic properties . . . . . . 39 (SE): Hb/C1-104/6, 119/21, 312

Hb = Main Series Volume; GD = Gmelin-Durrer; Ab = Appended Volume to Main Series; Eb = Supplement Volume to Main Series; Erg. W. = New Supplement Series Volume

O_3Pm_2 . . . . Pm_2O_3:
Formation and preparation . . . . . . 39 (SE): Hb/C1–132, 312
Magnetic properties . . . . 39 (SE): Hb/C1–142/3, 312/3
Mechanical properties . . . . 39 (SE): Hb/C1–127/8
Optical properties. . . . . 39 (SE): Hb/C1–150/1
Thermal properties . . . . 39 (SE): Hb/C1–130, 139, 312

–. . . . . $Pm_2O_3 \cdot 3\,H_2O = Pm(OH)_3$ . . . . 39 (SE): Hb/C2–9, 17, 56/7
–. . . . . Pm_2O_3 solid solutions:
Pm_2O_3–Sm_2O_3 . . . . 39 (SE): Hb/C1–322

O_3Po . . . . PoO_3 . . . . 12 (Po): Hb–133
O_3Po^{2-} . . . . $[PoO_3]^{2-}$ . . . . 12 (Po): Hb–102
O_3Po_2 . . . . Po_2O_3. 12 (Po): Hb–133
O_3PrSc . . . . $PrScO_3 = Sc_2O_3 \cdot Pr_2O_3$ . . . . 39 (SE): Hb/C1–113, 124, 127
O_3PrV . . . . $PrVO_3$. 48 (V): Hb/B2–551/2
O_3Pr_2 . . . . Pr_2O_3:
Catalytic properties . . . . 39 (SE): Hb/C1–275
Chemical reactions . . . . 39 (SE): Hb/C1–156/75, 273/5
Crystallographic properties . . . . . . 39 (SE): Hb/C1–104/6, 113/23, 270/2
Electrical properties . . . . 39 (SE): Hb/C1–144/6, 149, 273
Electrochemical behavior . . . . 39 (SE): Hb/C1–155
Formation and preparation . . . . . . 39 (SE): Hb/C1–106, 131/3, 269/70
Magnetic properties . . . . 39 (SE): Hb/C1–142/3, 272/3
Mechanical properties . . . . 39 (SE): Hb/C1–127/8, 272
Optical properties. . . . . 39 (SE): Hb/C1–151, 273
Thermal properties . . . . 39 (SE): Hb/C1–130/41, 272

–. . . . . $Pr_2O_3 \cdot H_2O = PrO(OH)$ . . . . 39 (SE): Hb/C2–15, 93, 95, 102
–. . . . . $Pr_2O_3 \cdot 2\,H_2O$ . . . . 39 (SE): Hb/C2–15
–. . . . . $Pr_2O_3 \cdot 3\,H_2O$
see H_3O_3Pr ... $Pr(OH)_3$

–. . . . . Pr_2O_3 glasses:
Pr_2O_3–PbO–SiO_2 . . . . 47 (Pb): Hb/C4–1376

–. . . . . Pr_2O_3 solid solutions:
Pr_2O_3–Nd_2O_3 . . . . 39 (SE): Hb/C1–308
Pr_2O_3–Tb_2O_3 . . . . 39 (SE): Hb/C1–400

–. . . . . Pr_2O_3 systems:
Pr_2O_3–H_2O . . . . 39 (SE): Hb/C2–50, 102
Pr_2O_3–Lu_2O_3 . . . . 39 (SE): Hb/C1–436
Pr_2O_3–Nb_2O_5 . . . . 49 (Nb): Hb/B3–155
Pr_2O_3–Nb_2O_5–H_2O . . . . 49 (Nb): Hb/B3–155
$PrO_{1.5}$–NpO_2 . . . . Erg. W. 4–46/8
Pr_2O_3–$Pr(NO_3)_3$–$Pr_2(CO_3)_3$–H_2O. . . . 39 (SE): Hb/C2–50
Pr_2O_3–$Pr_2(CO_3)_3$–H_2O . . . . 39 (SE): Hb/C2–50
Pr_2O_3–Y_2O_3 . . . . 39 (SE): Hb/C1–294
Pr_2O_3–Yb_2O_3 . . . . 39 (SE): Hb/C1–433

O_3Pt. . . . . $PtO_3 \cdot H_2O$ . . . . 68 (Pt): Hb/C1–53/4

See the "Instructions for the user"
at the beginning of the volume

Formula	Compound	Volume	Reference
O_3Pt_2	$Pt_2O_3 \cdot 2\,H_2O$	68 (Pt):	Hb/C1–26, 53
$O_3Pt_2Sn_3$	$Pt_2Sn_3O_3$	68 (Pt):	Hb/C3–310
O_3Pu	$PuO_3 \cdot 0.8\,H_2O$	Erg. W. 4–29	
–	$PuO_3 \cdot H_2O$	Erg. W. 4–29	
–	$PuO_3 \cdot 2\,H_2O$	Erg. W. 4–29	
O_3PuSc	$PuScO_3$	Erg. W. 4–56	
O_3PuSr	$SrPuO_3$	Erg. W. 4–53/5	
O_3PuV	$PuVO_3$	Erg. W. 4–56, 71	
O_3Pu_2	Pu_2O_3	Erg. W. 4–12/4	
		Erg. W. 8–192/3, 372	
O_3RbReS	$RbReO_3S$	70 (Re):	Hb–144
O_3RbTa	$RbTaO_3$	50 (Ta):	Hb/B2–170/1
–	$RbTaO_3$ solid solutions:		
	$Rb(Ta,Nb)O_3$	50 (Ta):	Hb/B2–260
O_3RbV	$RbVO_3 = Rb_2O \cdot V_2O_5$	48 (V):	Hb/A2–368
			Hb/B2–486/7
–	$RbVO_3$ systems:		
	$RbVO_3$–$RbCl$	48 (V):	Hb/B2–489
	$RbVO_3$–RbF	48 (V):	Hb/B2–488
	$RbVO_3$–$Rb_3[VO_3F_2]$	48 (V):	Hb/B2–488
O_3Rb_2	Rb_2O_3	24 (Rb):	Hb–111
O_3Rb_2S	Rb_2SO_3	24 (Rb):	Hb–207
O_3Rb_2SSe	Rb_2SeSO_3	24 (Rb):	Hb–218
$O_3Rb_2S_2$	$Rb_2S_2O_3$	24 (Rb):	Hb–214/5
–	$Rb_2S_2O_3 \cdot 2\,H_2O$	24 (Rb):	Hb–215
O_3Rb_2Se	$Rb_2SeO_3 \cdot H_2O$	24 (Rb):	Hb–216
O_3Rb_2Si	Rb_2SiO_3	24 (Rb):	Hb–232
O_3Rb_2Tb	Rb_2TbO_3	39 (SE):	Hb/C2–141
O_3Rb_2Ti	Rb_2TiO_3	41 (Ti):	Hb–411
O_3Re	ReO_3	70 (Re):	Hb–90/3
O_3ReSTl	$TlReO_3S$	70 (Re):	Hb–148/9
O_3Re_2	$Re_2O_3 \cdot xH_2O$	70 (Re):	Hb–96/7
O_3Rh_2	Rh_2O_3	64 (Rh):	Hb–46/7
–	$Rh_2O_3 \cdot H_2O$	64 (Rh):	Hb–47
–	$Rh_2O_3 \cdot 3\,H_2O$	64 (Rh):	Hb–48
–	$Rh_2O_3 \cdot 5\,H_2O$	64 (Rh):	Hb–47
O_3Rn	RnO_3	Erg. W. 1–130	
O_3Ru	RuO_3	63 (Ru):	Eb–232/3, 240
O_3Ru^+	RuO_3^+	63 (Ru):	Eb–233
O_3Ru^-	RuO_3^-	63 (Ru):	Eb–233
O_3Ru^{2-}	RuO_3^{2-}	63 (Ru):	Eb–249
O_3RuSr	$SrRuO_3$	63 (Ru):	Eb–254/5
–	$SrRuO_3$ solid solutions:		
	$(Ba_{1-x}Sr_x)RuO_3$	63 (Ru):	Eb–256/8
O_3Ru_2	Ru_2O_3	63 (Ru):	Hb–25
O_3S	SO_3:		
	Absorption by H_2SO_4	9 (S):	Hb/A2–408/11

Hb = Main Series Volume; GD = Gmelin-Durrer; Ab = Appended Volume to Main Series; Eb = Supplement Volume to Main Series; Erg. W. = New Supplement Series Volume

Formula	Entry	Volume	Reference
O_3S	SO_3:		
	Behavior as solvent	9 (S):	Hb/B1–366/7
	Chemical reactions	9 (S):	Hb/B1–353/66
	– with metal compounds	9 (S):	Hb/B1–360/5
	– with metals	9 (S):	Hb/B1–356
	– with nonmetal compounds	9 (S):	Hb/B1–356/60
	– with nonmetals	9 (S):	Hb/B1–354/5
	– with organic substances	9 (S):	Hb/B1–365/6
	Colloidal forms	9 (S):	Hb/B1–334
	Crystallographic properties	9 (S):	Hb/B1–341/3
	Decomposition	9 (S):	Hb/B1–353/4
	Electrical properties	9 (S):	Hb/B1–353
	Formation	9 (S):	Hb/B1–323/31, 335/7
	History	9 (S):	Hb/A1–59/60
	Magnetic properties	9 (S):	Hb/B1–352/3
	Mechanical properties	9 (S):	Hb/B1–343/4, 349/50
	Molecule	9 (S):	Hb/B1–337/41
	Optical properties	9 (S):	Hb/B1–350/2
	Preparation	9 (S):	Hb/B1–332/5
	–, industrial	9 (S):	Hb/A2–320/426
	–, –, contact process	9 (S):	Hb/A2–364/408
	–, –, further treatment	9 (S):	Hb/A2–408/22
	–, –, process control	9 (S):	Hb/A2–422/5
	–, –, purification of gases	9 (S):	Hb/A2–351/64
	–, –, theoretical	9 (S):	Hb/A2–324/51
	–, –, waste gas utilization	9 (S):	Hb/A2–425/6
	– in pure state	9 (S):	Hb/B1–332/3
	– of modifications	9 (S):	Hb/B1–333/4
	– of special forms	9 (S):	Hb/B1–334/5
	Spectra	9 (S):	Hb/B1–351/2
	Thermal properties	9 (S):	Hb/B1–344/9
–	S_2O_6	9 (S):	Hb/B1–338, 351/2
–	S_3O_9	9 (S):	Hb/B1–337/8
–	$(SO_3)_m$	9 (S):	Hb/B1–337
–	$11\ SO_3 \cdot 2\ H_2O = 2\ H_2SO_4 \cdot 9\ SO_3$	9 (S):	Hb/B2–626
–	$4\ SO_3 \cdot H_2O = H_2SO_4 \cdot 3\ SO_3$	9 (S):	Hb/B2–625/6
–	$3\ SO_3 \cdot H_2O = H_2SO_4 \cdot 2\ SO_3$	9 (S):	Hb/B2–625
–	$2\ SO_3 \cdot H_2O = H_2SO_4 \cdot SO_3 = H_2S_2O_7$	9 (S):	Hb/B2–624/5, 795/8
–	$4\ SO_3 \cdot 3\ H_2O$	9 (S):	Hb/B2–622, 624
–	$7\ SO_3 \cdot 6\ H_2O$	9 (S):	Hb/B2–622
–	$SO_3 \cdot H_2O$ *see H_2O_4S ... H_2SO_4*		
–	$SO_3 \cdot 2\ H_2O = H_2SO_4 \cdot H_2O$	9 (S):	Hb/B2–614/5, 621/2
–	$SO_3 \cdot 3\ H_2O = H_2SO_4 \cdot 2\ H_2O$	9 (S):	Hb/B2–614/5, 621
–	$SO_3 \cdot 4\ H_2O$	9 (S):	Hb/B2–614/5, 620/1
–	$SO_3 \cdot 5\ H_2O$	9 (S):	Hb/B2–614/5, 620
–	$SO_3 \cdot 7\ H_2O$	9 (S):	Hb/B2–614/5, 620
–	$SO_3 \cdot 9\ H_2O$	9 (S):	Hb/B2–620

See the "Instructions for the user" at the beginning of the volume

O_3S . . . . SO_3 solutions:

SO_3–HCN	9 (S):	Hb/B1–366
SO_3–HNO_3–H_2SO_4	9 (S):	Hb/B3–1678/80
SO_3–H_2O *see H_2O_4S ... H_2SO_4*		
SO_3–H_2SO_4:		
Chemical reactions	9 (S):	Hb/B2–796/8
Electrochemical behavior	9 (S):	Hb/B2–795/6
Physical properties	9 (S):	Hb/B2–787/95
Thermodynamic data of formation and solution	9 (S):	Hb/B2–784/7
SO_3–N_2O	9 (S):	Hb/B1–366
SO_3–organic solvents	9 (S):	Hb/B1–366
SO_3–SO_2	9 (S):	Hb/B1–366

–. . . . . SO_3 systems:

SO_3–B_2O_3–H_2O	13 (B):	Eb–210
SO_3–B_2O_3–PbO	47 (Pb):	Hb/C2–697
SO_3–Bi_2O_3–H_2O	19 (Bi):	Eb–747
SO_3–CaO	28 (Ca):	Hb/B3–658, 675/9
SO_3–CaO–SiO_2	28 (Ca):	Hb/B3–1107/8
SO_3–CrO_3–H_2O	52 (Cr):	Hb/B–331/2
SO_3–CrO_3–K_2O–H_2O	52 (Cr):	Hb/B–646/7
SO_3–CuO–Fe_2O_3–H_2O	60 (Cu):	Hb/B3–1297
SO_3–Fe_2O_3–H_2O	59 (Fe):	Hb/B2–446/7
SO_3–HCl	9 (S):	Hb/B3–1842
SO_3–HCl–H_2O	9 (S):	Hb/B3–1830
SO_3–HNO_3	9 (S):	Hb/B3–1667/9, 1678/9
SO_3–HNO_3–H_2SO_4	9 (S):	Hb/B3–1667/9, 1680
SO_3–HSO_3Cl	9 (S):	Hb/B3–1842/4
SO_3–H_2O	9 (S):	Hb/B2–613/9
SO_3–H_2S–HgO–H_2O	34 (Hg):	Hb/B3–1021/2
SO_3–H_2SO_4–K_2SO_4	22 (K):	Hb/3–744
SO_3–H_2SO_4–$(NH_4)_2SO_4$	23 (NH_4):	Hb/2–293
SO_3–H_2SeO_4	10 (Se):	Hb/B–185
SO_3–HgO–H_2O	34 (Hg):	Hb/B3–1021
SO_3–K_2O–P_2O_5–H_2O	22 (K):	Hb/5–1002
SO_3–NH_3–P_2O_5–H_2O	23 (NH_4):	Hb/2–430/1
SO_3–$(NH_4)_2SO_4$–Nb_2O_5–H_2O	49 (Nb):	Hb/B4–303
SO_3–$(NH_4)_2SO_4$–Ta_2O_5–H_2O	50 (Ta):	Hb/B2–316
SO_3–N_2O_3–H_2O	9 (S):	Hb/B3–1650
SO_3–N_2O_5–H_2O	9 (S):	Hb/B3–1665/96
SO_3–Na_2O	21 (Na):	Eb/3–1146
SO_3–Na_2O–V_2O_5	48 (V):	Hb/B2–405
SO_3–Na_2O–V_2O_5–H_2O	48 (V):	Hb/B2–406
SO_3–Nb_2O_5–H_2O	49 (Nb):	Hb/B1–269
SO_3–NiO–UO_3–H_2O	57 (Ni):	Hb/B3–1235
SO_3–NiO–UO_3–D_2O	57 (Ni):	Hb/B3–1235

O_3S	SO_3 systems:		
	SO_3-POCl_3	9 (S):	Hb/B1–360
		16 (P):	Hb/C–597
	$SO_3-PbO-TeO_2$	47 (Pb):	Hb/C2–681
	$SO_3-SO_2Cl_2$	9 (S):	Hb/B3–1820
	$SO_3-Ta_2O_5-H_2O$	50 (Ta):	Hb/B1–176
	$SO_3-TiO_2-H_2O$	41 (Ti):	Hb–345/7
	$SO_3-Tl_2O_3-H_2O$	38 (Tl):	Hb/3–364/5
	$SO_3-ZrO_2-H_2O$	42 (Zr):	Hb–335/7
O_3S^{2-}	$[SO_3]^{2-}$		
	see H_2O_3S ... H_2SO_3		
O_3SSe	$SSeO_3$	10 (Se):	Hb/B–176/7, 181
–	$SeSO_3$	10 (Se):	Hb/B–176/7
–	$(SeSO_3)_2$	10 (Se):	Hb/B–179
–	$SeSO_3$ solutions:		
	$SeSO_3-H_2SO_4$	10 (Se):	Hb/B–177/81
O_3SSe^{2-}	$[SeSO_3]^{2-}$	10 (Se):	Hb/B–181/5
O_3SSr	$SrSO_3$	29 (Sr):	Hb–171/2
			Eb–249
–	$SrSO_3 \cdot 0.5\ H_2O$	29 (Sr):	Hb–172
O_3STe	$TeSO_3$	11 (Te):	Hb–355/7
–	$TeSO_3$ solutions:		
	$TeSO_3-H_2SO_4$	11 (Te):	Hb–357
	$TeSO_3-H_2SO_4-H_2O$	11 (Te):	Hb–357/9
O_3STl_2	Tl_2SO_3	38 (Tl):	Hb/3–349
O_3SV^{3-}	$[VO_3S]^{3-}$	48 (V):	Hb/B1–281
O_3SZn	$ZnSO_3 \cdot 2\ H_2O$	32 (Zn):	Hb–219/20
			Eb–935
–	$ZnSO_3 \cdot 2.5\ H_2O$	32 (Zn):	Hb–219/20
			Eb–935
–	$ZnSO_3$ systems:		
	$ZnSO_3-Na_2SO_3-H_2O$	32 (Zn):	Eb–1006
	$ZnSO_3-SO_2-H_2O$	32 (Zn):	Eb–935
O_3S_2	S_2O_3	9 (S):	Hb/B1–179/81
$O_3S_2^{2-}$	$[S_2O_3]^{2-}$		
	see $H_2O_3S_2$ ... $H_2S_2O_3$ solutions		
O_3S_2Sr	SrS_2O_3	29 (Sr):	Hb–181
			Eb–258/9
–	$SrS_2O_3 \cdot H_2O$	29 (Sr):	Hb–182
			Eb–258/9
–	$SrS_2O_3 \cdot 3\ H_2O$	29 (Sr):	Eb–258
–	$SrS_2O_3 \cdot 5\ H_2O$	29 (Sr):	Hb–181/2
			Eb–258
–	SrS_2O_3 systems:		
	$SrS_2O_3-H_2O$	29 (Sr):	Eb–258/9
$O_3S_2Tl^{-}$	$[Tl_2(S_2O_3)_2]^{2-}$	38 (Tl):	Hb/3–368
$O_3S_2Tl_2$	$Tl_2S_2O_3$	38 (Tl):	Hb/3–367/8

See the "Instructions for the user" at the beginning of the volume

Formula	Compound	Volume	Reference
O_3S_2Zn	ZnS_2O_3	32 (Zn):	Hb–241/2 Eb–965/6
$O_3S_4Sb_4$	$Sb_4O_3S_4$	18 (Sb):	Hb/B3–541
$O_3S_6Sb_6$	$2\,Sb_2S_3 \cdot Sb_2O_3 = Sb_2OS_2$	18 (Sb):	Hb/A3–305 Hb/B3–540
$O_3S_{15}Sb_{12}$	$5\,Sb_2S_3 \cdot Sb_2O_3 = Sb_4OS_5$	18 (Sb):	Hb/B3–540
O_3Sb^{3-}	SbO_3^{3-}	18 (Sb):	Hb/B2–368/9
O_3Sb_2	Sb_2O_3:		
	Chemical reactions	18 (Sb):	Hb/B2–359/65
	Formation and preparation	18 (Sb):	Hb/B2–349/52
	History	18 (Sb):	Hb/A1–23/4, 29/30, 36/9
	Physical properties	18 (Sb):	Hb/B2–352/9
	Toxicity	18 (Sb):	Hb/B2–276
	Uses	18 (Sb):	Hb/A3–305
–	$Sb_2O_3 \cdot xH_2O$	18 (Sb):	Hb/B2–365/7
–	Sb_2O_3 glasses:		
	$Sb_2O_3-PbO-SiO_2$	47 (Pb):	Hb/C3–941
–	Sb_2O_3 solid solutions:		
	$Sb_2O_3-PbNb_2O_6$	49 (Nb):	Hb/B3–241
	$Sb_2O_3-Pb_2Nb_2O_7$	49 (Nb):	Hb/B3–241
–	Sb_2O_3 systems:		
	$Sb_2O_3-Ag_2O-Sb_2O_5$	61 (Ag):	Hb/B4–89
	$Sb_2O_3-Bi_2O_3$	19 (Bi):	Eb–858
	Sb_2O_3-CaO	28 (Ca):	Hb/B3–1239
	$Sb_2O_3-HCl-SbCl_3-H_2O$	18 (Sb):	Hb/B2–462/5
	$Sb_2O_3-NaOH-H_2O$	21 (Na):	Eb/4–1710
	$Sb_2O_3-Na_2O-P_2O_5$	21 (Na):	Eb/5–2026
	$Sb_2O_3-(NaPO_3)_x$	21 (Na):	Eb/5–2026
	Sb_2O_3-PbO	47 (Pb):	Hb/C3–927/8
	$Sb_2O_3-PbO-Sb_2S_3$	47 (Pb):	Hb/C3–939
	$Sb_2O_3-Sb_2S_3$	18 (Sb):	Hb/B3–539/40
O_3ScSm	$SmScO_3 = Sc_2O_3 \cdot Sm_2O_3$	39 (SE):	Hb/C1–113, 124, 321/2
O_3ScY	$YScO_3 = Sc_2O_3 \cdot Y_2O_3$	39 (SE):	Hb/C1–113, 124/5, 127, 205
O_3Sc_2	Sc_2O_3:		
	Chemical reactions	39 (SE):	Hb/C1–157, 162/4, 168/70, 187/8
	Crystallographic properties	39 (SE):	Hb/C1–104/6, 121/2, 183
	Diffusion	39 (SE):	Hb/C1–128
	Electrical properties	39 (SE):	Hb/C1–144/50, 185
	Electrochemical behavior	39 (SE):	Hb/C1–155
	Formation	39 (SE):	Hb/C1–106/10, 131/3, 181/3
	Mechanical properties	39 (SE):	Hb/C1–127, 184
	Optical properties	39 (SE):	Hb/C1–150/1, 154/5, 186/7
	Preparation	39 (SE):	Hb/C1–106/12, 181/3

Formula	Compound	Vol.	Reference
O_3Sc_2	Sc_2O_3:		
	Preparation of special forms	39 (SE):	Hb/C1–106/12, 182
	Spectra	39 (SE):	Hb/C1–154/5, 186
	Thermal properties	39 (SE):	Hb/C1–129/41, 184/5
–	Sc_2O_{3-x}	39 (SE):	Hb/C1–177/8
–	$Sc_2O_3 \cdot H_2O$ = $ScO(OH)$	39 (SE):	Hb/C2–23/4, 93, 98/9
–	$Sc_2O_3 \cdot 3\,H_2O$ *see H_3O_3Sc ... $Sc(OH)_3$*		
–	$Sc_2O_3 \cdot nH_2O$ = $Sc(OH)_3 \cdot nH_2O$	39 (SE):	Hb/C2–19/26
–	Sc_2O_3 systems:		
	$ScO_{1.5}$–AmO_2	Erg. W. 4–79	
	Sc_2O_3–CeO_2	39 (SE):	Hb/C1–262
	Sc_2O_3–Er_2O_3	39 (SE):	Hb/C1–420
	Sc_2O_3–HNO_3–H_2O	39 (SE):	Hb/C2–25
	Sc_2O_3–Ho_2O_3	39 (SE):	Hb/C1–413/4
	Sc_2O_3–La_2O_3	39 (SE):	Hb/C1–217/8
	Sc_2O_3–Lu_2O_3	39 (SE):	Hb/C1–435
	Sc_2O_3–N_2O_5–H_2O	39 (SE):	Hb/C2–225/6
	Sc_2O_3–Nb_2O_5	49 (Nb):	Hb/B3–136/7
	Sc_2O_3–Nd_2O_3	39 (SE):	Hb/C1–306/7
	$ScO_{1.5}$–NpO_2	Erg. W. 4–45	
	Sc_2O_3–Ta_2O_5	50 (Ta):	Hb/B2–204
	Sc_2O_3–Y_2O_3	39 (SE):	Hb/C1–205
	Sc_2O_3–Yb_2O_3	39 (SE):	Hb/C1–430
O_3Se	SeO_3	10 (Se):	Hb/B–43/4
–	SeO_3 systems:		
	SeO_3–CrO_3–H_2O	52 (Cr):	Hb/B–341
	SeO_3–Nb_2O_5–H_2O	49 (Nb):	Hb/B1–81
	SeO_3–Tl_2O_3–H_2O	38 (Tl):	Hb/3–376
O_3Se^{2-}	$[SeO_3]^{2-}$ *see H_2O_3Se ... H_2SeO_3*		
O_3SeSr	$SrSeO_3$	29 (Sr):	Hb–186 Eb–261
–	$SrSeO_3 \cdot 3.5\,H_2O$	29 (Sr):	Hb–186
O_3SeTe	$TeSeO_3$	11 (Te):	Hb–362/3
O_3SeTl_2	Tl_2SeO_3	38 (Tl):	Hb/3–373
O_3SeZn	$ZnSeO_3$	32 (Zn):	Hb–246 Eb–968
–	$ZnSeO_3 \cdot H_2O$	32 (Zn):	Hb–247
–	$ZnSeO_3 \cdot 2\,H_2O$	32 (Zn):	Hb–247
O_3Se_2	Se_2O_3	10 (Se):	Hb/B–25
O_3Si^{2-}	$[SiO_3]^{2-}$	15 (Si):	Hb/B–421/2
O_3SiSr	$SrO \cdot SiO_2$ = $SrSiO_3$	29 (Sr):	Hb–210/1 Eb–282
–	$SrO \cdot SiO_2 \cdot 0.6\,H_2O$	29 (Sr):	Eb–282
–	$SrSiO_3$ systems:		
	$SrSiO_3$–$CaSiO_3$	29 (Sr):	Hb–237

See the "Instructions for the user"
at the beginning of the volume

O_3SiSr	$SrSiO_3$ systems:		
	$SrSiO_3-Li_2SiO_3$	29 (Sr):	Hb-223
	$SrSiO_3-Na_2SiO_3$	29 (Sr):	Hb-226
O_3SiTl_2	Tl_2SiO_3	38 (Tl):	Hb/3-424
O_3SiZn	$ZnSiO_3$	32 (Zn):	Hb-273/4 Eb-989
–	$ZnSiO_3$ systems:		
	$ZnSiO_3-Li_2SiO_3$	32 (Zn):	Hb-294
O_3Si_2	Si_2O_3	15 (Si):	Hb/B-257
O_3SmV	$SmVO_3$	48 (V):	Hb/B2-551/2
O_3Sm_2	Sm_2O_3:		
	Catalytic properties	39 (SE):	Hb/C1-321
	Chemical reactions	39 (SE):	Hb/C1-156/76, 319/21
	Crystallographic properties	39 (SE):	Hb/C1-104/6, 115/23, 316/7
	Diffusion	39 (SE):	Hb/C1-317
	Electrical properties	39 (SE):	Hb/C1-144/50, 319
	Electrochemical behavior	39 (SE):	Hb/C1-155
	Formation	39 (SE):	Hb/C1-106/10, 131/3, 315
	Magnetic properties	39 (SE):	Hb/C1-142/3, 318
	Mechanical properties	39 (SE):	Hb/C1-127/8, 317
	Optical properties	39 (SE):	Hb/C1-150/5, 319
	Preparation	39 (SE):	Hb/C1-106/12, 315
	– of special forms	39 (SE):	Hb/C1-110/2, 315
	Spectra	39 (SE):	Hb/C1-153/5, 319
	Thermal properties	39 (SE):	Hb/C1-129/41, 317
–	$Sm_2O_3 \cdot H_2O = SmO(OH)$	39 (SE):	Hb/C2-14, 15, 93/7, 103
–	$Sm_2O_3 \cdot 3\,H_2O$ *see H_3O_3Sm ... $Sm(OH)_3$*		
–	$Sm_2O_3 \cdot nH_2O$	39 (SE):	Hb/C2-15
–	Sm_2O_3 solid solutions:		
	$Sm_2O_3-Gd_2O_3$	39 (SE):	Hb/C1-386
	$Sm_2O_3-La_2O_3$	39 (SE):	Hb/C1-322
	$Sm_2O_3-Nd_2O_3$	39 (SE):	Hb/C1-322
	$Sm_2O_3-Pm_2O_3$	39 (SE):	Hb/C1-322
–	Sm_2O_3 systems:		
	$Sm_2O_3-BaO-Nb_2O_5$	49 (Nb):	Hb/B3-162
	$Sm_2O_3-CeO_2$	39 (SE):	Hb/C1-323
	$Sm_2O_3-Dy_2O_3$	39 (SE):	Hb/C1-409
	$Sm_2O_3-Er_2O_3$	39 (SE):	Hb/C1-421
	$Sm_2O_3-Eu_2O_3$	39 (SE):	Hb/C1-377
	$Sm_2O_3-Ho_2O_3$	39 (SE):	Hb/C1-414
	$Sm_2O_3-La_2O_3-Lu_2O_3$	39 (SE):	Hb/C1-436
	$Sm_2O_3-Lu_2O_3$	39 (SE):	Hb/C1-436
	$Sm_2O_3-Nb_2O_5$	49 (Nb):	Hb/B3-160

O_3Sm_2	Sm_2O_3 systems:		
	$SmO_{1.5}$–NpO_2	Erg. W. 4–46/8	
	Sm_2O_3–PrO_2	39 (SE):	Hb/C1–295
	Sm_2O_3–PrO_x	39 (SE):	Hb/C1–294/5
	Sm_2O_3–Sm	39 (SE):	Hb/C1–313/4
	Sm_2O_3–Ta_2O_5	50 (Ta):	Hb/B2–214
	Sm_2O_3–Tm_2O_3	39 (SE):	Hb/C1–425
	Sm_2O_3–Yb_2O_3	39 (SE):	Hb/C1–433
O_3Sn	2 $SnO_3 \cdot H_2O$	46 (Sn):	Hb/C1–139
–	$SnO_3 \cdot 2\,H_2O$	46 (Sn):	Hb/C1–139
O_3SnSr	$SrSnO_3$ solid solutions:		
	$SrSnO_3$–$PbTiO_3$	47 (Pb):	Hb/C4–1600
	$SrSnO_3$–$SrNb_2O_6$	49 (Nb):	Hb/B3–223
O_3SnTa	$SnTaO_3$	50 (Ta):	Hb/B2–231
O_3Sn_2	Sn_2O_3	46 (Sn):	Hb/C1–57/9
O_3Sn_3	Sn_3O_3	46 (Sn):	Hb/C1–34/5, 104
O_3SrTe	$SrTeO_3$	29 (Sr):	Hb–187
			Eb–262
O_3SrTh	$SrThO_3$	44 (Th):	Hb–348
O_3SrTi	$SrTiO_3$	41 (Ti):	Hb–433/4
–	$SrTiO_3$ solid solutions:		
	$SrTiO_3$–$BaTiO_3$–$NaNbO_3$	49 (Nb):	Hb/B3–199
	$SrTiO_3$–$BaTiO_3$–$PbTiO_3$	47 (Pb):	Hb/C4–1416
	$SrTiO_3$–$Bi_{0.67}TiO_3$–$PbTiO_3$	47 (Pb):	Hb/C4–1404/5
	$SrTiO_3$–$Bi_2Ti_3O_9$–$PbNb_2O_6$	49 (Nb):	Hb/B3–295/6
	$SrTiO_3$–$NaNbO_3$	49 (Nb):	Hb/B3–195
	$SrTiO_3$–$PbNb_2O_6$	49 (Nb):	Hb/B3–292/4
	$SrTiO_3$–$PbTiO_3$	47 (Pb):	Hb/C4–1401/4
	(Pb,Sr) (Zr,Ti)O_3	47 (Pb):	Hb/C4–1455/8
	– (modified by Nb_2O_5)	49 (Nb):	Hb/B3–317/8
	$SrTiO_3$–$SrNb_2O_6$	49 (Nb):	Hb/B3–194
–	$SrTiO_3$ systems:		
	$SrTiO_3$–$BaTiO_3$	41 (Ti):	Hb–457/60
	$SrTiO_3$–$PbTiO_3$–$PbZrO_3$–$SrZrO_3$	47 (Pb):	Hb/C4–1455
O_3SrV	$SrVO_3$	48 (V):	Hb/B2–519
–	$SrVO_3$ solid solutions:		
	$Sr_xLa_{1-x}VO_3$	48 (V):	Hb/B2–557
–	$SrVO_3$ systems:		
	$SrVO_3$–$SrVO_{2.5}$	48 (V):	Hb/B2–519
O_3SrZr	$SrZrO_3$	42 (Zr):	Hb–439
–	$SrZrO_3$ solid solutions:		
	$SrZrO_3$–$PbTiO_3$	47 (Pb):	Hb/C4–1458
	(Pb,Sr) (Zr,Ti)O_3	47 (Pb):	Hb/C4–1455/8
	– (modified by Nb_2O_5)	49 (Nb):	Hb/B3–317/8
	$SrZrO_3$–$PbZrO_3$	47 (Pb):	Hb/C4–1435/7
	$SrZrO_3$–$SrNb_2O_6$	49 (Nb):	Hb/B3–214

See the "Instructions for the user"
at the beginning of the volume

Formula	Compound	Volume	Reference
O_3SrZr	$SrZrO_3$ systems:		
	$SrZrO_3-PbTiO_3-PbZrO_3-SrTiO_3$	47 (Pb):	Hb/C4–1455
	$SrZrO_3-PbZrO_3$	47 (Pb):	Hb/C4–1434
	$SrZrO_3-SrTa_2O_6$	50 (Ta):	Hb/B2–228/9
O_3Ta	TaO_3	50 (Ta):	Hb/A1–8 Hb/B1–30
O_3Ta^-	$[TaO_3]^-$	50 (Ta):	Hb/A2–407 Hb/B1–57, 59/60
O_3Ta_2	Ta_2O_3	50 (Ta):	Hb/A1–8 Hb/A2–316 Hb/B1–30
$O_3Ta_3^-$	$[Ta_3O_3]^-$	50 (Ta):	Hb/B1–59
O_3Tb_2	Tb_2O_3:		
	Chemical reactions	39 (SE):	Hb/C1–156/75, 391
	Crystallographic properties	39 (SE):	Hb/C1–104/6, 115/23, 388/9
	Electrical properties	39 (SE):	Hb/C1–144/50, 390
	Formation and preparation	39 (SE):	Hb/C1–109, 111, 131/3, 387/8
	Magnetic properties	39 (SE):	Hb/C1–142/3, 389/90
	Mechanical properties	39 (SE):	Hb/C1–127/8
	Optical properties	39 (SE):	Hb/C1–150/2, 390
	Thermal properties	39 (SE):	Hb/C1–130/9
–	$Tb_2O_3 \cdot H_2O = TbO(OH)$	39 (SE):	Hb/C2–14, 15, 93/7, 103/4
–	$Tb_2O_3 \cdot 3\,H_2O$ *see H_3O_3Tb ... $Tb(OH)_3$*		
–	$Tb_2O_3 \cdot nH_2O$	39 (SE):	Hb/C2–65/6, 68
–	Tb_2O_3 solid solutions:		
	$Tb_2O_3-Pr_2O_3$	39 (SE):	Hb/C1–400
	$Tb_2O_3-Y_2O_3$	39 (SE):	Hb/C1–400
–	Tb_2O_3 systems:		
	$Tb_2O_3-Er_2O_3$	39 (SE):	Hb/C1–421
	$TbO_{1.5}-NpO_2$	Erg. W. 4–46/8	
	$Tb_2O_3-Tb(NO_3)_3-H_2O$	39 (SE):	Hb/C2–272
O_3Te	TeO_3	11 (Te):	Hb–282/4
–	$TeO_3 \cdot 3\,H_2O$ *see H_6O_6Te ... H_6TeO_6*		
–	TeO_3 systems:		
	$TeO_3-NaOH-H_2O$	21 (Na):	Eb/5–1910
O_3Te^{2-}	$[TeO_3]^{2-}$	11 (Te):	Hb–286/7
O_3Ti	$TiO_3 \cdot 2\,H_2O = Ti(OH)_3(OOH)$	41 (Ti):	Hb–268/71
–	$TiO_3 \cdot 3\,H_2O$	41 (Ti):	Hb–269
O_3Ti^{2-}	$[TiO_3]^{2-}$	41 (Ti):	Hb–468
O_3TiV	$TiVO_3$	48 (V):	Hb/B2–558
O_3TiZn	$ZnTiO_3 = Zn_2TiO_4 \cdot TiO_2$	41 (Ti):	Hb–460, 461/2

Hb = Main Series Volume; GD = Gmelin-Durrer; Ab = Appended Volume to Main Series;
Eb = Supplement Volume to Main Series; Erg. W. = New Supplement Series Volume

O_3TiZn . . . . $ZnTiO_3$ solid solutions:
$ZnTiO_3$–$BaTiO_3$–$NiZrO_3$ . . . . 57 (Ni): Hb/B3–1186
$ZnTiO_3$–$PbTiO_3$ . . . . 47 (Pb): Hb/C4–1416/7

O_3Ti_2 . . . . Ti_2O_3 . . . . 41 (Ti): Hb–221/4, 478
–. . . . . Ti_2O_3 systems:
Ti_2O_3–Al_2O_3. . . . . 41 (Ti): Hb–463
Ti_2O_3–Fe–FeO–FeO · TiO_2–Fe_2Ti . . . 59 (Fe): Hb/A9–1873
Ti_2O_3–MgO–TiO_2. . . . . 41 (Ti): Hb–423
Ti_2O_3–TiC . . . . 41 (Ti): Hb–368
Ti_2O_3–V_2O_3 . . . . 48 (V): Hb/B2–558

O_3TlV . . . . $TlVO_3$ = Tl_2O · V_2O_5 . . . . 48 (V): Hb/B2–548
O_3Tl_2 . . . . Tl_2O_3 . . . . 38 (Tl): Hb/2–222/9
–. . . . . Tl_2O_3 · H_2O = TlO(OH) . . . . 38 (Tl): Hb/2–232
–. . . . . Tl_2O_3 · 3 H_2O = $Tl(OH)_3$ . . . . 38 (Tl): Hb/2–232/3
–. . . . . Tl_2O_3 systems:
Tl_2O_3–Nb_2O_5 . . . . 49 (Nb): Hb/B3–127
Tl_2O_3–SO_3–H_2O . . . . 38 (Tl): Hb/3–364/5
Tl_2O_3–SeO_3–H_2O . . . . 38 (Tl): Hb/3–376
Tl_2O_3–Ta_2O_5. . . . . 50 (Ta): Hb/B2–201

O_3Tm_2 . . . . Tm_2O_3:
Chemical reactions . . . . 39 (SE): Hb/C1–156/75, 424/5
Crystallographic properties . . . . . . 39 (SE): Hb/C1–104/6, 115/23, 422/3
Diffusion. . . . . 39 (SE): Hb/C1–128
Electrical properties . . . . 39 (SE): Hb/C1–145/8
Formation . . . . 39 (SE): Hb/C1–106/10, 131/3, 422
Mechanical properties . . . . 39 (SE): Hb/C1–127/8, 423/4
Optical properties. . . . . 39 (SE): Hb/C1–150/4, 424
Preparation . . . . 39 (SE): Hb/C1–106/12, 422
– of special forms. . . . . 39 (SE): Hb/C1–110/2, 422
Spectra . . . . 39 (SE): Hb/C1–154/5
Thermal properties . . . . 39 (SE): Hb/C1–129/41, 424

–. . . . . Tm_2O_3 · H_2O = TmO(OH) . . . . 39 (SE): Hb/C2–14, 15, 93/7, 104

–. . . . . Tm_2O_3 · 3 H_2O
see H_3O_3Tm ... $Tm(OH)_3$

–. . . . . Tm_2O_3 · nH_2O . . . . 39 (SE): Hb/C2–75
–. . . . . Tm_2O_3 solid solutions:
Tm_2O_3–Yb_2O_3. . . . . 39 (SE): Hb/C1–433/4

–. . . . . Tm_2O_3 systems:
Tm_2O_3–Dy_2O_3. . . . . 39 (SE): Hb/C1–425
Tm_2O_3–Er_2O_3 . . . . 39 (SE): Hb/C1–425
Tm_2O_3–Eu_2O_3. . . . . 39 (SE): Hb/C1–425
Tm_2O_3–Gd_2O_3 . . . . 39 (SE): Hb/C1–425
Tm_2O_3–Ho_2O_3. . . . . 39 (SE): Hb/C1–425
Tm_2O_3–La_2O_3 . . . . 39 (SE): Hb/C1–425

See the "Instructions for the user"
at the beginning of the volume

O_3Tm_2 . . . . Tm_2O_3 systems:
Tm_2O_3–Lu_2O_3 . . . . 39 (SE): Hb/C1–436
Tm_2O_3–Nd_2O_3 . . . . 39 (SE): Hb/C1–425
$TmO_{1.5}$–NpO_2 . . . . Erg. W. 4–46/8
$TmO_{1.5}$–PuO_2 . . . . Erg. W. 4–57
Tm_2O_3–Sm_2O_3 . . . . 39 (SE): Hb/C1–425

O_3U . . . . UO_3 . . . . 55 (U): Hb–95/9
–. . . . . $UO_3 \cdot 0.5\ H_2O = H_2U_2O_7$ . . . . 55 (U): Hb–104
–. . . . . $UO_3 \cdot H_2O$ . . . . 55 (U): Hb–102/4
–. . . . . $UO_3 \cdot 1.5\ H_2O$ . . . . 55 (U): Hb–101
–. . . . . $UO_3 \cdot 2\ H_2O$ . . . . 55 (U): Hb–101/2
–. . . . . UO_3 systems:
UO_3–NiO–SO_3–H_2O . . . . 57 (Ni): Hb/B3–1235
UO_3–NiO–SO_3–D_2O . . . . 57 (Ni): Hb/B3–1235

O_3V^+ . . . . $[VO_3]^+ = [V(O)(O)_2]^+$ . . . . 48 (V): Hb/A2–588/9, 638/40
Hb/B1–162/3, 165/6

O_3V^- . . . . $[VO_3]^-$:
Chemical reactions . . . . 48 (V): Hb/B1–132/45, 149
Detection . . . . 48 (V): Hb/A2–647/52, 655
Electrochemical behavior . . . . 48 (V): Hb/A2–474/6, 493/4, 506/9
Spectra . . . . 48 (V): Hb/A2–368/9
Hb/B1–149
Transition reactions . . . . 48 (V): Hb/A2–550/640

O_3V^{2-} . . . . $[VO_3]^{2-}$ . . . . 48 (V): Hb/A2–555/6
O_3VY . . . . YVO_3 . . . . 48 (V): Hb/B2–551/2
O_3V_2 . . . . V_2O_3:
Chemical reactions . . . . 48 (V): Hb/B1–45/9
Crystallographic properties . . . . . . 48 (V): Hb/B1–36/7
Determination . . . . 48 (V): Hb/A2–714
Electrical properties . . . . 48 (V): Hb/B1–30, 43/5
Electrochemical behavior . . . . 48 (V): Hb/A1–306
Hb/A2–465/8, 498/9
Formation and preparation . . . . . . 48 (V): Hb/B1–32/6
History . . . . 48 (V): Hb/A1–7
Magnetic properties . . . . 48 (V): Hb/B1–16/7, 40/3
Mechanical properties . . . . 48 (V): Hb/B1–37/9
Spectra . . . . 48 (V): Hb/A2–376/7, 379/80, 387
Hb/B1–40
Thermal properties . . . . 48 (V): Hb/B1–37/40
Toxicity . . . . 48 (V): Hb/A2–725/8

–. . . . . V_2O_3 systems:
V_2O_3–Al_2O_3 . . . . 48 (V): Hb/B2–543/4
V_2O_3–C–CO . . . . 48 (V): Hb/B1–47
V_2O_3–Cr_2O_3 . . . . 52 (Cr): Hb/B–939
V_2O_3–Ti_2O_3 . . . . 48 (V): Hb/B2–558
V_2O_3–V_2O_5 . . . . 48 (V): Hb/B1–13

$O_3V_2^+$ . . . . $[V_2O_3]^+$ . . . . 48 (V): Hb/B1–84
$O_3V_2^{2+}$ . . . . $[(VO)_2O]^{2+}$ . . . . 48 (V): Hb/A2–601
O_3W . . . . WO_3:
Chemical reactions . . . . 54 (W): Hb–108/10, 123/7
Colloidal forms . . . . 54 (W): Hb–132/7
Formation and preparation . . . . . . 54 (W): Hb–117/21
Physical properties . . . . 54 (W): Hb–121/2
Preparation . . . . 54 (W): Hb–19/23

–. . . . . $WO_3 \cdot H_2O = H_2WO_4$ . . . . 54 (W): Hb–127/32
–. . . . . $WO_3 \cdot 2\,H_2O$ . . . . 54 (W): Hb–127/31
–. . . . . $WO_3 \cdot xH_2O$ . . . . 54 (W): Hb–127/8
–. . . . . WO_3 systems:
$WO_3–Ag_2WO_4$ . . . . 61 (Ag): Hb/B4–349
$WO_3–K_2WO_4$ . . . . 54 (W): Hb–246/7
$WO_3–Li_2WO_4$ . . . . 54 (W): Hb–209/10
$WO_3–NaPO_3$ . . . . 54 (W): Hb–241
$WO_3–Na_2WO_4$ . . . . 54 (W): Hb–219/20
$WO_3–PbO$ . . . . 54 (W): Hb–317

O_3W_2 . . . . W_2O_3 . . . . 54 (W): Hb–111
O_3W_4 . . . . W_4O_3 . . . . 54 (W): Hb–111
O_3Xe . . . . XeO_3 . . . . Erg. W. 1–101/12
O_3Xe^+ . . . . XeO_3^+ . . . . Erg. W. 1–101
O_3Y_2 . . . . Y_2O_3:
Catalytic properties . . . . 39 (SE): Hb/C1–204
Chemical reactions . . . . 39 (SE): Hb/C1–157/76, 201/4
Crystallographic properties . . . . . . 39 (SE): Hb/C1–104/6, 115/7, 120/3, 192/3
Diffusion . . . . 39 (SE): Hb/C1–128, 193
Electrical properties . . . . 39 (SE): Hb/C1–144/50, 195/9
Electrochemical behavior . . . . 39 (SE): Hb/C1–155
Formation . . . . 39 (SE): Hb/C1–106/9, 131/3, 189/90
Magnetic properties . . . . 39 (SE): Hb/C1–142/3, 195
Mechanical properties . . . . 39 (SE): Hb/C1–127/8, 193/4
Optical properties . . . . 39 (SE): Hb/C1–150/5, 199/201
Preparation . . . . 39 (SE): Hb/C1–106/12, 189/92
– of special forms . . . . 39 (SE): Hb/C1–110/2, 190/2
Spectra . . . . 39 (SE): Hb/C1–154/5
Thermal properties . . . . 39 (SE): Hb/C1–129/41, 194/5

–. . . . . Y_2O_{3-x} . . . . 39 (SE): Hb/C1–177/9
–. . . . . $Y_2O_3 \cdot H_2O = YO(OH)$ . . . . 39 (SE): Hb/C2–15, 94/7, 99/100
–. . . . . $Y_2O_3 \cdot 2\,H_2O$ . . . . 39 (SE): Hb/C2–15
–. . . . . $Y_2O_3 \cdot 3\,H_2O$
see H_3O_3Y ... $Y(OH)_3$

–. . . . . $Y_2O_3 \cdot 4.2\,H_2O$ . . . . 39 (SE): Hb/C2–15
–. . . . . Y_2O_3 solid solutions:
$Y_2O_3–CeO_2$ . . . . 39 (SE): Hb/C1–263/5

See the "Instructions for the user"
at the beginning of the volume

O_3Y_2 . . . . Y_2O_3 solid solutions:
Y_2O_3–Dy_2O_3 . . . . 39 (SE): Hb/C1–409
Y_2O_3–Ho_2O_3 . . . . 39 (SE): Hb/C1–414
Y_2O_3–La_2O_3 . . . . 39 (SE): Hb/C1–218/9
Y_2O_3–Nb_2O_5–TiO_2 . . . . 49 (Nb): Hb/B3–209
Y_2O_3–Nd_2O_3 . . . . 39 (SE): Hb/C1–307/8
Y_2O_3–TbO_x . . . . 39 (SE): Hb/C1–400/1
Y_2O_3–Tb_2O_3 . . . . 39 (SE): Hb/C1–400

–. . . . . Y_2O_3 systems:
Y_2O_3–Al_2O_3–B_2O_3–PbO . . . . 47 (Pb): Hb/C4–1375/6
Y_2O_3–CeO_2 . . . . 39 (SE): Hb/C1–262/4
Y_2O_3–CuO . . . . 60 (Cu): Hb/B3–1186
Y_2O_3–Eu_2O_3 . . . . 39 (SE): Hb/C1–377
Y_2O_3–La_2O_3 . . . . 39 (SE): Hb/C1–218/9
Y_2O_3–Nb_2O_5 . . . . 49 (Nb): Hb/B3–138/9
Y_2O_3–Nb_2O_5–SiO_2 . . . . 49 (Nb): Hb/B4–359
Y_2O_3–Nd_2O_3 . . . . 39 (SE): Hb/C1–306/8
Y_2O_3–PrO_x . . . . 39 (SE): Hb/C1–294/5
Y_2O_3–Pr_2O_3 . . . . 39 (SE): Hb/C1–294
$YO_{1.5}$–PuO_2 . . . . Erg. W. 4–57
Y_2O_3–Sc_2O_3 . . . . 39 (SE): Hb/C1–205
Y_2O_3–Ta . . . . 50 (Ta): Hb/B2–205
Y_2O_3–Ta_2O_5 . . . . 50 (Ta): Hb/B2–205
Y_2O_3–Ta_2O_5–ZrO_2 . . . . 50 (Ta): Hb/B2–229

O_3Yb_2 . . . . Yb_2O_3:
Chemical reactions . . . . 39 (SE): Hb/C1–156/75, 429/30
Crystallographic properties . . . . . . 39 (SE): Hb/C1–104/6, 115/23, 427
Electrical properties . . . . 39 (SE): Hb/C1–144/50, 429
Formation . . . . 39 (SE): Hb/C1–106/10, 131/3, 426/7
Magnetic properties . . . . 39 (SE): Hb/C1–142/3, 428/9
Mechanical properties . . . . 39 (SE): Hb/C1–127/8, 428
Optical properties . . . . 39 (SE): Hb/C1–150/5, 429
Preparation . . . . 39 (SE): Hb/C1–106/12, 426/7
– of special forms . . . . 39 (SE): Hb/C1–110/2
Spectra . . . . 39 (SE): Hb/C1–154/5, 429
Thermal properties . . . . 39 (SE): Hb/C1–129/41, 428

–. . . . . $Yb_2O_3 \cdot H_2O$ = YbO(OH) . . . . 39 (SE): Hb/C2–14, 93/7, 104/5
–. . . . . $Yb_2O_3 \cdot nH_2O$ . . . . 39 (SE): Hb/C2–76
–. . . . . Yb_2O_3 solid solutions:
Yb_2O_3–Dy_2O_3 . . . . 39 (SE): Hb/C1–433/4
Yb_2O_3–Er_2O_3 . . . . 39 (SE): Hb/C1–433/4
Yb_2O_3–Ho_2O_3 . . . . 39 (SE): Hb/C1–433/4
Yb_2O_3–Tm_2O_3 . . . . 39 (SE): Hb/C1–433/4

–. . . . . Yb_2O_3 systems:
Yb_2O_3–CeO_2 . . . . 39 (SE): Hb/C1–434
Yb_2O_3–Eu_2O_3 . . . . 39 (SE): Hb/C1–433

Formula	Compound	Volume	Reference
O_3Yb_2	Yb_2O_3 systems:		
	$Yb_2O_3-Gd_2O_3$	39 (SE):	Hb/C1–433
	$Yb_2O_3-La_2O_3$	39 (SE):	Hb/C1–430/2
	$Yb_2O_3-Lu_2O_3$	39 (SE):	Hb/C1–436
	$Yb_2O_3-Nd_2O_3$	39 (SE):	Hb/C1–433
	$YbO_{1.5}-NpO_2$	Erg. W. 4–46/8	
	$Yb_2O_3-Pr_2O_3$	39 (SE):	Hb/C1–433
	$Yb_2O_3-Sc_2O_3$	39 (SE):	Hb/C1–430
	$Yb_2O_3-Sm_2O_3$	39 (SE):	Hb/C1–433
O_3ZnZr	$ZnZrO_3$	42 (Zr):	Hb–442
–	$ZnZrO_3$ solid solutions:		
	$ZnZrO_3-PbZrO_3$	47 (Pb):	Hb/C4–1438/9
O_3Zn_2	Zn_2O_3	32 (Zn):	Hb–145
–	$Zn_2O_3 \cdot H_2O$	32 (Zn):	Eb–817
O_3Zr	$ZrO_3 \cdot xH_2O$	42 (Zr):	Hb–261/3
O_3Zr_2	Zr_2O_3	42 (Zr):	Hb–215
$O_{3.5}S_{0.5}Zr$	$ZrO_2 \cdot 0.5\ SO_3$	42 (Zr):	Hb–340
–	$ZrO_2 \cdot 0.5\ SO_3 \cdot xH_2O$	42 (Zr):	Hb–351
$O_{3.5}U$	$UO_{3.5} \cdot 2\ H_2O$	55 (U):	Hb–105
–	$UO_{3.5} \cdot 3\ H_2O$	55 (U):	Hb–105
O_4	O_4	3 (O):	Hb/3–352/4, 362, 367, 575/8
–	$(O_2)_2$	3 (O):	Hb/3–362
O_4^+	$(O_4)^+$	3 (O):	Hb/4–1071
O_4^-	O_4^-	3 (O):	Hb/3–354
O_4^{2-}	O_4^{2-}	3 (O):	Hb/8–2627
O_4Os	OsO_4:		
	Chemical reactions	66 (Os):	Hb–41/2
	Formation	66 (Os):	Hb–31/33, 36
	Physical properties	66 (Os):	Hb–36/41
	Physiological action	66 (Os):	Hb–42
	Preparation	66 (Os):	Hb–33/6
	Solubility	66 (Os):	Hb–43
–	$(OsO_4)_4$	66 (Os):	Hb–48
–	OsO_4 solutions:		
	OsO_4-CCl_4	66 (Os):	Hb–48
	OsO_4-H_2O	66 (Os):	Hb–43/8
	$OsO_4-H_2SO_4$	66 (Os):	Hb–48
	OsO_4–organic solvents	66 (Os):	Hb–48/9
	OsO_4-POCl_3	66 (Os):	Hb–48
O_4P^{3-}	$[PO_4]^{3-}$	16 (P):	Hb/B–343/8, 364/77 Hb/C–170/2
	see also H_3O_4P ... H_3PO_4 solutions		
$O_4PPb_{0.5}Th_{0.5}$	$Pb_{0.5}Th_{0.5}PO_4$	47 (Pb):	Hb/C4–1471
O_4PPu	$PuPO_4$	Erg. W. 4–252	
–	$PuPO_4 \cdot 0.5\ H_2O$	Erg. W. 4–251/2	
O_4PRb_3	$Rb_3PO_4 \cdot 4\ H_2O$	24 (Rb):	Hb–233

See the "Instructions for the user"
at the beginning of the volume

O_4PRh	$RhPO_4$	64 (Rh):	Hb–72
O_4PSb	$5\,SbPO_4 \cdot 2\,H_2O$	18 (Sb):	Hb/B3–561
–	$SbPO_4 \cdot 2\,H_2O$	18 (Sb):	Hb/B3–560/1
O_4PTl	$TlPO_4 \cdot 2\,H_2O$	38 (Tl):	Hb/3–430
O_4PTl_3	Tl_3PO_4	38 (Tl):	Hb/3–427/8
O_4P_2	P_2O_4	16 (P):	Hb/C–75/7
–	P_2O_4 systems:		
	P_2O_4–Na_2O–H_2O	21 (Na):	Hb–890
$O_4P_2Pb_3S_4$	$Pb_3(PO_2S_2)_2 \cdot 4\,H_2O$	47 (Pb):	Hb/C3–884
$O_4P_2S_4Zn_3$	$Zn_3P_2S_4O_4 \cdot H_2O$	32 (Zn):	Eb–997
$O_4P_4S_3$	$P_4S_3O_4$	16 (P):	Hb/C–582
$O_4P_{10}Zn_5$	$Zn_5P_{10}O_4$	32 (Zn):	Hb–276
$O_4Pb_{0.5}Th_{0.5}V$	$Th_{0.5}Pb_{0.5}VO_4$	48 (V):	Hb/B2–574
O_4Pb^{4-}	$[PbO_4]^{4-}$	47 (Pb):	Hb/C1–198/9
O_4PbS	$PbSO_4$:		
	Chemical reactions	47 (Pb):	Hb/C2–560/9
	Colloidal forms	47 (Pb):	Hb/C2–546
	Electrochemical behavior	47 (Pb):	Hb/C2–554/60
	Formation	47 (Pb):	Hb/C2–537/42, 544/5
	History	47 (Pb):	Hb/A1–140/3
	Physical properties	47 (Pb):	Hb/C2–546/54
	Preparation	47 (Pb):	Hb/C2–537/46
	–, industrial	47 (Pb):	Hb/C2–542/4
	Solubility	47 (Pb):	Hb/C2–569/80
			Hb/C3–1136
	Special forms	47 (Pb):	Hb/C2–545/6
–	$PbSO_4$ solid solutions:		
	$PbSO_4$–$BaSO_4$	30 (Ba):	Hb–267
			Eb–424
		47 (Pb):	Hb/C3–1206
	$PbSO_4$–$RaSO_4$	47 (Pb):	Hb/C3–1212
	$PbSO_4$–$SrSO_4$	47 (Pb):	Hb/C3–1198
–	$PbSO_4$ solutions:		
	$PbSO_4$–acids	47 (Pb):	Hb/B2–283/4
			Hb/C2–570/3
	$PbSO_4$–H_2O	47 (Pb):	Hb/C2–569/70
	$PbSO_4$–organic solvents	47 (Pb):	Hb/C2–578/80
	$PbSO_4$–salt solutions	47 (Pb):	Hb/C2–573
			Hb/C3–1136
–	$PbSO_4$ systems:		
	$PbSO_4$–$AgCl$–Ag_2SO_4–$PbCl_2$	61 (Ag):	Hb/B4–267
	$PbSO_4$–$Al(NO_3)_3$–H_2O	47 (Pb):	Hb/C4–1295
	$PbSO_4$–$BaSO_4$–H_2O	47 (Pb):	Hb/C3–1207
	$PbSO_4$–$Ca(NO_3)_2$–H_2O	47 (Pb):	Hb/C3–1190
	$PbSO_4$–$CsCl$–Cs_2SO_4–$PbCl_2$	47 (Pb):	Hb/C3–1156
	$PbSO_4$–Cs_2SO_4	47 (Pb):	Hb/C3–1155
	$PbSO_4$–Cs_2SO_4–K_2SO_4	47 (Pb):	Hb/C3–1158
	$PbSO_4$–Cs_2SO_4–Li_2SO_4	47 (Pb):	Hb/C3–1157

Formula	Compound	Volume	Reference
O_4PbS	$PbSO_4$ systems:		
	$PbSO_4-Cs_2SO_4-Rb_2SO_4$	47 (Pb):	Hb/C3–1158
	$PbSO_4-H_2SO_4-H_2O$	47 (Pb):	Hb/C2–572
	$PbSO_4-KCl-K_2SO_4-PbCl_2$	47 (Pb):	Hb/C3–1096/7
	$PbSO_4-KF-K_2SO_4-PbF_2$	47 (Pb):	Hb/C3–1096
	$PbSO_4-K_2SO_4$	47 (Pb):	Hb/C3–1093/4
	$PbSO_4-K_2SO_4-H_2O$	47 (Pb):	Hb/C3–1094/5
	$PbSO_4-K_2SO_4-Li_2SO_4$	47 (Pb):	Hb/C3–1115
	$PbSO_4-K_2SO_4-Na_2SO_4$	47 (Pb):	Hb/C3–1119
	$PbSO_4-K_2SO_4-Rb_2SO_4$	47 (Pb):	Hb/C3–1146
	$PbSO_4-K_2SO_4-Tl_2SO_4$	47 (Pb):	Hb/C4–1357
	$PbSO_4-LiCl-Li_2SO_4-PbCl_2$	47 (Pb):	Hb/C3–992
	$PbSO_4-LiF-Li_2SO_4-PbF_2$	47 (Pb):	Hb/C3–992
	$PbSO_4-Li_2SO_4$	47 (Pb):	Hb/C3–992
	$PbSO_4-Li_2SO_4-H_2O$	47 (Pb):	Hb/C3–992
	$PbSO_4-Li_2SO_4-Mg(NO_3)_2-H_2O$	47 (Pb):	Hb/C3–1178
	$PbSO_4-Li_2SO_4-Na_2SO_4$	47 (Pb):	Hb/C3–1051
	$PbSO_4-Li_2SO_4-Rb_2SO_4$	47 (Pb):	Hb/C3–1145
	$PbSO_4-Li_2SO_4-Tl_2SO_4$	47 (Pb):	Hb/C4–1355
	$PbSO_4-NaCl-H_2O$	47 (Pb):	Hb/C3–1033
	$PbSO_4-NaCl-Na_2SO_4-PbCl_2$	47 (Pb):	Hb/C3–1033
	$PbSO_4-NaF-Na_2SO_4-PbF_2$	47 (Pb):	Hb/C3–1032
	$PbSO_4-Na_2SO_4$	47 (Pb):	Hb/C3–1030
	$PbSO_4-Na_2SO_4-H_2O$	47 (Pb):	Hb/C3–1030/1
	$PbSO_4-Na_2SO_4-Tl_2SO_4$	47 (Pb):	Hb/C4–1356
	$PbSO_4-PbCl_2$	47 (Pb):	Hb/C2–606
	$PbSO_4-PbCrO_4$	52 (Cr):	Hb/B–933/4
	$PbSO_4-PbF_2$	47 (Pb):	Hb/C2–605
	$PbSO_4-PbMoO_4$	53 (Mo):	Hb–310/1
	$PbSO_4-PbO$	47 (Pb):	Hb/C2–582/3
	$PbSO_4-PbO-SiO_2$	47 (Pb):	Hb/C3–856
	$PbSO_4-PbWO_4$	54 (W):	Hb–321
	$PbSO_4-Rb_2SO_4$	47 (Pb):	Hb/C3–1143
	$PbSO_4-Tl_2SO_4$	47 (Pb):	Hb/C4–1353
O_4PbS_2	$PbS_2O_4 \cdot xH_2O$	47 (Pb):	Hb/C2–535
O_4PbSb_2	$Pb(SbO_2)_2 = PbO \cdot Sb_2O_3$	47 (Pb):	Hb/C3–930
O_4PbSe	$PbSeO_4$	47 (Pb):	Hb/C2–636/8
$O_4PbSiZn$	$PbZnSiO_4$	47 (Pb):	Hb/C4–1229
–	$PbZnSiO_4$ solid solutions:		
	$(Pb,Ca)_4Zn_4Si_4O_{16}$	47 (Pb):	Hb/C4–1231
O_4PbSr_2	Sr_2PbO_4	47 (Pb):	Hb/C3–1195
O_4PbU	$PbUO_4$	55 (U):	Hb–239
O_4PbW	$PbWO_4$	54 (W):	Hb–317/9
–	$PbWO_4$ systems:		
	$PbWO_4-Bi_2(WO_4)_3$	54 (W):	Hb–321
	$PbWO_4-PbCrO_4$	54 (W):	Hb–321
	$PbWO_4-PbMoO_4$	54 (W):	Hb–321
	$PbWO_4-PbSO_4$	54 (W):	Hb–321

See the "Instructions for the user"
at the beginning of the volume

O_4Pb_2S . . . . Pb_2SO_4 . . . . 47 (Pb): Hb/C2–537
$O_4Pb_2S^{2+}$ . . . . $[Pb_2SO_4]^{2+}$ . . . . 47 (Pb): Hb/C2–580
$O_4Pb_2S_2$. . . . . $PbS_2O_3 \cdot PbO \cdot 2\ H_2O$ . . . . 47 (Pb): Hb/C2–600
–. . . . . $PbS_2O_3 \cdot PbO \cdot xH_2O$. . . . . 47 (Pb): Hb/C2–598
O_4Pb_2Se . . . . $PbSeO_3 \cdot PbO$ . . . . 47 (Pb): Hb/C2–635
O_4Pb_2Si. . . . . $Pb_2SiO_4 = 2\ PbO \cdot SiO_2$ . . . . 47 (Pb): Hb/B2–403, 413/4
Hb/C3–842/7
–. . . . . $2\ PbO \cdot SiO_2 \cdot H_2O$ . . . . 47 (Pb): Hb/C3–838
–. . . . . Pb_2SiO_4 solid solutions:
$(Pb,Ba)_2SiO_4$. . . . . 47 (Pb): Hb/C3–1210
Pb_2SiO_4–$KPbAsO_4$ . . . . 47 (Pb): Hb/C3–1112
Pb_2SiO_4–$NaPbAsO_4$ . . . . 47 (Pb): Hb/C3–1050/1
Pb_2SiO_4–$NaPbPO_4$ . . . . 47 (Pb): Hb/C3–1050
Pb_2SiO_4–$Pb_2MgSi_2O_7$ . . . . 47 (Pb): Hb/C3–1178
O_4Pb_2Sn . . . . Pb_2SnO_4 . . . . 47 (Pb): Hb/C4–1581/2
O_4Pb_2Ti. . . . . Pb_2TiO_4. . . . . 47 (Pb): Hb/C4–1379
O_4Pb_2Zr. . . . . Pb_2ZrO_4. . . . . 47 (Pb): Hb/C4–1423
O_4Pb_3 . . . . Pb_3O_4:
Adulterating. . . . . 47 (Pb): Hb/C1–127
Aftertreatment . . . . 47 (Pb): Hb/C1–126
Chemical reactions . . . . 47 (Pb): Hb/C1–135/9
Electrochemical behavior . . . . 47 (Pb): Hb/C1–133/5
Extending . . . . 47 (Pb): Hb/C1–127
Formation . . . . 47 (Pb): Hb/C1–122/3
History. . . . . 47 (Pb): Hb/A1–111/23
Performance tests . . . . 47 (Pb): Hb/C1–127
Physical properties . . . . 47 (Pb): Hb/C1–129/33
Hb/C4–1745
Preparation . . . . 47 (Pb): Hb/C1–122/9
–, industrial . . . . 47 (Pb): Hb/C1–124/9
Purification . . . . 47 (Pb): Hb/C1–126
Purity testing . . . . 47 (Pb): Hb/C1–128/9
Standards . . . . 47 (Pb): Hb/C1–127
–. . . . . Pb_3O_4 systems:
Pb_3O_4–KNO_3–$Pb(NO_3)_2$ . . . . 47 (Pb): Hb/C3–1064
Pb_3O_4–PbO . . . . 47 (Pb): Hb/C1–54
Pb_3O_4–PbO_2. . . . . 47 (Pb): Hb/C1–55/6
Pb_3O_4–Pb_2O_3 . . . . 47 (Pb): Hb/C1–54/5
Pb_3O_4–$Pb_{12}O_{19}$ . . . . 47 (Pb): Hb/C1–54/5
O_4PdS. . . . . $PdSO_4 \cdot xH_2O$ . . . . 65 (Pd): Hb/2–295
O_4PdSe . . . . $PdSeO_4$. . . . . 65 (Pd): Hb/2–296/7
O_4PmTa. . . . . $PmTaO_4$ . . . . 50 (Ta): Hb/B2–202/3, 213
O_4PrTa . . . . $PrTaO_4$ . . . . 50 (Ta): Hb/B2–202/3, 211
–. . . . . $PrTaO_4$ solid solutions:
$(Pr,Nd)TaO_4$ . . . . 50 (Ta): Hb/B2–213
O_4PrV . . . . $PrVO_4$. 48 (V): Hb/B2–553/5
O_4Pr_2 . . . . $Pr_2O_4 \cdot nH_2O$ . . . . 39 (SE): Hb/C2–106/8, 112

Hb = Main Series Volume; GD = Gmelin-Durrer; Ab = Appended Volume to Main Series; Eb = Supplement Volume to Main Series; Erg. W. = New Supplement Series Volume

O_4Pt^{2-}	$[PtO_4]^{2-}$	68 (Pt):	Hb/B4–265
$O_4Pt_2Sn_3$	$Pt_2Sn_3O_4$	68 (Pt):	Hb/C3–310
O_4Pt_3	$Pt_3O_4 \cdot 9\,H_2O$	68 (Pt):	Hb/C1–26, 52
O_4Pu	$PuO_4 \cdot nH_2O$	Erg. W. 4–28	
O_4PuRb_2	Rb_2PuO_4	Erg. W. 4–52/3	
O_4PuSi	$PuSiO_4$	Erg. W. 4–56, 70	
O_4PuSr	$SrPuO_4$	Erg. W. 4–53/5	
O_4PuTa	$PuTaO_4$	Erg. W. 4–71	
$O_4Pu_2^{4+}$	$[Pu_2O_4]^{4+}$	Erg. W. 4–28	
$O_4Pu_2S_3U_2$	$U_2Pu_2O_4S_3$	Erg. W. 4–173/4	
$O_4Pu_4S_3$	$Pu_4O_4S_3$	Erg. W. 4–172/3	
O_4RaS	$RaSO_4$	31 (Ra):	Hb–44/5, 59, 62/3
–	$RaSO_4$ solid solutions:		
	$RaSO_4$–$BaSO_4$	30 (Ba):	Eb–424
	$RaSO_4$–$CsClO_4$	25 (Cs):	Hb/2–169
	$RaSO_4$–$PbSO_4$	47 (Pb):	Hb/C3–1212
	$RaSO_4$–$RbClO_4$	24 (Rb):	Hb–160
O_4RbRe	$RbReO_4$	70 (Re):	Hb–143
O_4RbS^-	$RbSO_4^-$	24 (Rb):	Hb–213
O_4Rb_2	Rb_2O_4	24 (Rb):	Hb–111/2
O_4Rb_2Ru	Rb_2RuO_4	63 (Ru):	Hb–116
O_4Rb_2S	Rb_2SO_4	24 (Rb):	Hb–207/11
–	Rb_2SO_4 solutions:		
	Rb_2SO_4–H_2O	24 (Rb):	Hb–211/3
	Rb_2SO_4–NH_3–H_2O	24 (Rb):	Hb–213
	Rb_2SO_4–organic solvents	24 (Rb):	Hb–213
–	Rb_2SO_4 systems:		
	Rb_2SO_4–Ag_2SO_4	61 (Ag):	Hb/B3–106
	Rb_2SO_4–Ag_2SO_4–H_2O	61 (Ag):	Hb/B3–106
	Rb_2SO_4–$CaSO_4$	28 (Ca):	Hb/B3–1335
	Rb_2SO_4–$CaSO_4$–H_2O	28 (Ca):	Hb/B3–1335/6
	Rb_2SO_4–$CdSO_4$–H_2O	33 (Cd):	Eb–702/3
	Rb_2SO_4–$CoCl_2$–$CoSO_4$–$RbCl$	58 (Co):	Eb/A–831
	Rb_2SO_4–$CoSO_4$–H_2O	58 (Co):	Eb/A–829/30
	Rb_2SO_4–Cs_2SO_4–$PbSO_4$	47 (Pb):	Hb/C3–1158
	Rb_2SO_4–KCl–K_2SO_4–$RbCl$	24 (Rb):	Hb–249
	Rb_2SO_4–K_2SO_4	24 (Rb):	Hb–249
	Rb_2SO_4–K_2SO_4–$PbSO_4$	47 (Pb):	Hb/C3–1146
	Rb_2SO_4–Li_2SO_4–$PbSO_4$	47 (Pb):	Hb/C3–1145
	Rb_2SO_4–$NiSO_4$–H_2O	57 (Ni):	Hb/B3–1096
	Rb_2SO_4–$PbSO_4$	47 (Pb):	Hb/C3–1143
	Rb_2SO_4–$RbCl$	24 (Rb):	Hb–213
$O_4Rb_2S_2$	$Rb_2S_2O_4$	24 (Rb):	Hb–206
O_4Rb_2Se	Rb_2SeO_4	24 (Rb):	Hb–216/7
O_4Rb_2Te	$Rb_2TeO_4 \cdot 3\,H_2O = Rb_2H_4[TeO_6] \cdot H_2O$	24 (Rb):	Hb–219
O_4Rb_2U	Rb_2UO_4	55 (U):	Hb–226
O_4Rb_2W	Rb_2WO_4	54 (W):	Hb–282

See the "Instructions for the user"
at the beginning of the volume

O_4Rb_2W	Rb_2WO_4 systems:		
	$Rb_2WO_4-AgI-Ag_2WO_4-RbI$	61 (Ag):	Hb/B4-353/4
	$Rb_2WO_4-Ag_2WO_4$	61 (Ag):	Hb/B4-351/3
O_4Rb_3V	$Rb_3VO_4 = 3\,Rb_2O \cdot V_2O_5$	48 (V):	Hb/B2-486
O_4Re^-	$[ReO_4]^-$	70 (Re):	Hb-61, 99
O_4ReTl	$TlReO_4$	70 (Re):	Hb-147/8
O_4Ru	RuO_4	63 (Ru):	Hb-27/31 Eb-232/3, 240/7
O_4Ru^+	$[RuO_4]^+$	63 (Ru):	Eb-233
O_4Ru^-	$[RuO_4]^-$	63 (Ru):	Hb-91 Eb-233, 251/2
O_4Ru^{2-}	$[RuO_4]^{2-}$	63 (Ru):	Hb-90 Eb-233, 249/51
O_4RuS	$RuSO_4$	63 (Ru):	Eb-316
O_4RuSr_2	Sr_2RuO_4	63 (Ru):	Eb-255
O_4S	SO_4	9 (S):	Hb/B1-371/2
O_4S^{2-}	$[SO_4]^{2-}$ *see H_2O_4S ... H_2SO_4* *H_2O_4S ... H_2SO_4 solutions*		
O_4SSb_2	Sb_2O_4S	18 (Sb):	Hb/B3-541
O_4SSn	$SnSO_4$ solutions:		
	$SnSO_4-H_2O$	46 (Sn):	Hb/B-306/24, 413
O_4SSn^{2+}	$[SnSO_4]^{2+}$	46 (Sn):	Hb/B-393
O_4SSr	$SrSO_4$:		
	Chemical reactions	29 (Sr):	Hb-176/8 Eb-255/6
	Electrochemical behavior	29 (Sr):	Eb-254/5
	Formation	29 (Sr):	Hb-172/4 Eb-249/50
	Physical properties	29 (Sr):	Hb-174/6 Eb-250/4
	Preparation	29 (Sr):	Hb-172/4
	–, industrial	29 (Sr):	Hb-16 Eb-87
	Solubility	29 (Sr):	Hb-178/80 Eb-256/7
	Uses	29 (Sr):	Eb-87
–	$SrSO_4 \cdot 2\,H_2O$	29 (Sr):	Eb-256
–	$SrSO_4 \cdot xH_2O$	29 (Sr):	Hb-178
–	$SrSO_4$ solid solutions:		
	$SrSO_4-BaSO_4$	30 (Ba):	Hb-267 Eb-424, 533/4
	$SrSO_4-PbSO_4$	47 (Pb):	Hb/C3-1198
–	$SrSO_4$ solutions:		
	$SrSO_4-C_2H_5OH-H_2O$	29 (Sr):	Eb-257
	$SrSO_4-H_2O$	29 (Sr):	Hb-180/1 Eb-256/7
	$SrSO_4-H_2SO_4$	29 (Sr):	Hb-181

O_4SSr	$SrSO_4$ systems:		
	$SrSO_4-BaSO_4$	30 (Ba):	Hb–388
	$SrSO_4-BaSO_4-CaSO_4$	30 (Ba):	Hb–390
	$SrSO_4-CaF_2$	29 (Sr):	Eb–304
	$SrSO_4-CaSO_4$	29 (Sr):	Hb–237
	$SrSO_4-K_2SO_4$	29 (Sr):	Hb–229/30
	$SrSO_4-K_2SO_4-H_2O$	29 (Sr):	Hb–230/1
	$SrSO_4-LiF-Li_2SO_4-SrF_2$	29 (Sr):	Eb–299
	$SrSO_4-Li_2SO_4$	29 (Sr):	Hb–223
	$SrSO_4-(NH_4)_2SO_4-H_2O$	29 (Sr):	Hb–233
	$SrSO_4-NaCl$	29 (Sr):	Hb–224
	$SrSO_4-NaCl-Na_2CO_3$	29 (Sr):	Hb–225
	$SrSO_4-Na_2CO_3-Na_2SO_4$		
	$-SrCO_3-H_2O$	29 (Sr):	Hb–225
	$SrSO_4-Na_2SO_4$	29 (Sr):	Hb–224
	$SrSO_4-Na_2SO_4-H_2O$	29 (Sr):	Hb–224
	$SrSO_4-S-SO_2-SrO-SrS$	29 (Sr):	Eb–257
	$SrSO_4-S-SO_2-SrS$	29 (Sr):	Eb–257
O_4STl^-	$[TlSO_4]^-$	38 (Tl):	Hb/3–358
O_4STl_2	Tl_2SO_4	38 (Tl):	Hb/1–23, 29 Hb/3–350/60
–.	Tl_2SO_4 systems:		
	$Tl_2SO_4-AgBr-Ag_2SO_4-TlBr$	61 (Ag):	Hb/B4–226/7
	$Tl_2SO_4-AgCl-Ag_2SO_4-TlCl$	61 (Ag):	Hb/B4–226
	$Tl_2SO_4-AgI-Ag_2SO_4-TlI$	61 (Ag):	Hb/B4–227
	$Tl_2SO_4-AgNO_3-Ag_2SO_4-TlNO_3$	61 (Ag):	Hb/B4–226
	$Tl_2SO_4-Ag_2SO_4$	61 (Ag):	Hb/B4–225
	$Tl_2SO_4-Am_2(SO_4)_3-H_2O$	Erg. W. 4–186	
	$Tl_2SO_4-CdSO_4-H_2O$	38 (Tl):	Hb/3–486
	$Tl_2SO_4-CoSO_4-H_2O$	58 (Co):	Eb/A–868
	$Tl_2SO_4-CuSO_4-H_2O$	60 (Cu):	Hb/B3–1184
	$Tl_2SO_4-CuSO_4-K_2SO_4-H_2O$	60 (Cu):	Hb/B3–1186
	$Tl_2SO_4-CuSO_4-(NH_4)_2SO_4-H_2O$ . . .	60 (Cu):	Hb/B3–1186
	$Tl_2SO_4-FeSO_4-H_2O$	59 (Fe):	Hb/B5–1123/4
	$Tl_2SO_4-H_2SO_4-H_2O$	38 (Tl):	Hb/3–360
	$Tl_2SO_4-H_2SO_4-Tl_2(SO_4)_3-H_2O$	38 (Tl):	Hb/3–362
	$Tl_2SO_2-HgBr_2-HgSO_4-TlBr$	38 (Tl):	Hb/3–490
	$Tl_2SO_4-HgCl_2-HgSO_4-TlCl$	38 (Tl):	Hb/3–489/90
	$Tl_2SO_4-HgI_2-HgSO_4-TlI$	38 (Tl):	Hb/3–490
	$Tl_2SO_4-K_2SO_4-H_2O$	38 (Tl):	Hb/3–458
	$Tl_2SO_4-K_2SO_4-PbSO_4$	47 (Pb):	Hb/C4–1357
	$Tl_2SO_4-Li_2SO_4-PbSO_4$	47 (Pb):	Hb/C4–1355
	$Tl_2SO_4-MgSO_4-H_2O$	38 (Tl):	Hb/3–474
	$Tl_2SO_4-NaBr-Na_2SO_4-TlBr$	38 (Tl):	Hb/3–444
	$Tl_2SO_4-NaCl-NaNO_3-Na_2SO_4$		
	$-TlCl-TlNO_3$	38 (Tl):	Hb/3–444
	$Tl_2SO_4-NaCl-Na_2SO_4-TlCl$	38 (Tl):	Hb/3–444
	$Tl_2SO_4-Na_2SO_4-PbSO_4$	47 (Pb):	Hb/C4–1356
	$Tl_2SO_4-NiSO_4-H_2O$	57 (Ni):	Hb/B3–1166/7

See the "Instructions for the user"
at the beginning of the volume

Formula	Compound	Gmelin Vol.	Reference
O_4STl_2	Tl_2SO_4 systems:		
	$Tl_2SO_4-PbSO_4$	47 (Pb):	Hb/C4–1353
	Tl_2SO_4-TlCl	38 (Tl):	Hb/3–360
	$Tl_2SO_4-ZnSO_4-H_2O$	38 (Tl):	Hb/3–480
O_4SU	$UOSO_3 \cdot 2\,H_2O = U(OH)_2SO_3 \cdot H_2O$	55 (U):	Hb–145
O_4SV	VSO_4		
	see also V^{2+} ... V^{2+}		
–	$VSO_4 \cdot 6\,H_2O$	48 (V):	Hb/B1–281/2
–	$VSO_4 \cdot 7\,H_2O = [V(H_2O)_6]SO_4 \cdot H_2O$	48 (V):	Hb/A2–337, 359/60 Hb/B1–281/2
–	VSO_4 solid solutions:		
	$(V,Fe)SO_4 \cdot 7\,H_2O$	59 (Fe):	Hb/B5–1138
	$(V,Mg)SO_4 \cdot 7\,H_2O$	48 (V):	Hb/B2–508
–	VSO_4 solutions:		
	VSO_4-H_2O	48 (V):	Hb/A2–502/3, 511, 553 Hb/B1–282/4
O_4SZn	$ZnSO_4$	32 (Zn):	Hb–220/2 Eb–936/41
–	$ZnSO_4 \cdot H_2O$	32 (Zn):	Hb–227 Eb–946/7
–	$ZnSO_4 \cdot 2\,H_2O$	32 (Zn):	Hb–227 Eb–946
–	$ZnSO_4 \cdot 3\,H_2O$	32 (Zn):	Hb–227 Eb–946
–	$ZnSO_4 \cdot 4\,H_2O$	32 (Zn):	Eb–946
–	$ZnSO_4 \cdot 5\,H_2O$	32 (Zn):	Hb–227
–	$ZnSO_4 \cdot 6\,H_2O$	32 (Zn):	Hb–226/7 Eb–946
–	$ZnSO_4 \cdot 7\,H_2O = [Zn(H_2O)_6]\,[SO_4 \cdot H_2O]$ $= Zn(OH)HSO_4 \cdot 6\,H_2O$	32 (Zn):	Hb–222/6 Eb–944/6
–	$ZnSO_4 \cdot 13\,H_2O$	32 (Zn):	Hb–222
–	$ZnSO_4 \cdot 20\,H_2O$	32 (Zn):	Hb–222
–	$ZnSO_4$ solid solutions:		
	$(Zn,Cu)SO_4 \cdot H_2O$	60 (Cu):	Hb/B3–1149
	$(Zn,Cu)SO_4 \cdot 5\,H_2O$	60 (Cu):	Hb/B3–1148/9
	$(Zn,Cu)SO_4 \cdot 7\,H_2O$	60 (Cu):	Hb/B3–1148/9
	$(Zn,Cu,Fe)SO_4$	60 (Cu):	Hb/B3–1326
	$ZnSO_4 \cdot H_2O-FeSO_4 \cdot H_2O$	59 (Fe):	Hb/B5–1102
	$ZnSO_4 \cdot 7\,H_2O-FeSO_4 \cdot 7\,H_2O$	59 (Fe):	Hb/B5–1102
	$(Zn,Mg)SO_4 \cdot 7\,H_2O$	32 (Zn):	Hb–324 Eb–1024
	$(Ni,Zn,Mg)SO_4 \cdot 7\,H_2O$	57 (Ni):	Hb/B3–1140
	$ZnSO_4-(NH_4)_2PO_3F$	32 (Zn):	Eb–1020
–	$ZnSO_4$ solutions:		
	$ZnSO_4-HF$	32 (Zn):	Eb–958
	$ZnSO_4-H_2O$	32 (Zn):	Hb–228/39 Eb–231/3, 947/58

O_4SZn	$ZnSO_4$ solutions:		
	$ZnSO_4-H_2SO_4$	32 (Zn):	Eb-958
	$ZnSO_4$-organic solvents	32 (Zn):	Hb-239/40 Eb-958/9
	$ZnSO_4$-salt solutions	32 (Zn):	Eb-958
–	$ZnSO_4$ systems:		
	$ZnSO_4-AgNO_3$	61 (Ag):	Hb/B4-141
	$ZnSO_4-BeSO_4-H_2O$	32 (Zn):	Eb-1022/3
	$ZnSO_4-(CH_2OH)_2CHOH-H_2O$	32 (Zn):	Eb-959
	$ZnSO_4-(CH_3)_3COH-H_2O$	32 (Zn):	Eb-958
	$ZnSO_4-CuSO_4-FeSO_4-H_2SO_4-H_2O$	60 (Cu):	Hb/B3-1326
	$ZnSO_4-CuSO_4-H_2O$	60 (Cu):	Hb/B3-1148/50
	$ZnSO_4-H_2O$	32 (Zn):	Hb-225/6 Eb-941/3
	$ZnSO_4-H_2SO_4-H_2O$	32 (Zn):	Eb-962/3
	$ZnSO_4-KCl-K_2SO_4-NaCl$ $-Na_2SO_4-ZnCl_2$	32 (Zn):	Eb-1016
	$ZnSO_4-K_2SO_4-H_2O$	32 (Zn):	Eb-1011/2
	$ZnSO_4-MgSO_4-H_2O$	32 (Zn):	Eb-1024
	$ZnSO_4-NH_3-H_2O$	32 (Zn):	Eb-959/60
	$ZnSO_4-(NH_4)_2SO_4-H_2O$	32 (Zn):	Eb-1018/9
	$ZnSO_4-Na_2SO_4-H_2O$	32 (Zn):	Hb-297/8 Eb-1006/7
	$ZnSO_4-NiSO_4$	57 (Ni):	Hb/B3-1136
	$ZnSO_4-Tl_2SO_4-H_2O$	38 (Tl):	Hb/3-480
	$ZnSO_4-ZnO-H_2O$	32 (Zn):	Eb-963/4
O_4SZn_2	$ZnSO_3 \cdot ZnO \cdot 2\,H_2O$	32 (Zn):	Hb-220
$O_4S_2^{2-}$	$[S_2O_4]^{2-}$	9 (S):	Hb/B2-390/1
O_4S_2Sr	SrS_2O_4	29 (Sr):	Hb-171
–	$SrS_2O_4 \cdot 4.5\,H_2O$	29 (Sr):	Eb-249
$O_4S_2Th_3$	$Th_3O_4S_2$	44 (Th):	Hb-280
O_4S_2Zn	ZnS_2O_4	32 (Zn):	Hb-242/3 Eb-934/5
O_4Sb^{3-}	$[SbO_4]^{3-}$	18 (Sb):	Hb/B2-388/9
O_4SbTa	$SbTaO_4$	50 (Ta):	Hb/B1-273/4
–	$SbTaO_4$ systems:		
	$SbTaO_4-SbNbO_4$	50 (Ta):	Hb/B2-248
O_4SbV	$SbVO_4$	48 (V):	Hb/B1-367
O_4Sb_2	Sb_2O_4	18 (Sb):	Hb/A1-23/4, 29/30 Hb/B2-371/5
–	$Sb_2O_4 \cdot H_2O = H_2Sb_2O_5$	18 (Sb):	Hb/B2-375
–	$Sb_2O_4 \cdot xH_2O$	18 (Sb):	Hb/B2-375
–	Sb_2O_4 systems:		
	Sb_2O_4-PbO	47 (Pb):	Hb/C3-928/9
O_4Sb_2Zn	$ZnSb_2O_4$	32 (Zn):	Eb-999
O_4ScTa	$ScTaO_4$	50 (Ta):	Hb/B2-202/3, 205
O_4ScV	$ScVO_4$	48 (V):	Hb/B2-553/5
O_4Sc_2	$Sc_2O_4 \cdot nH_2O$	39 (SE):	Hb/C2-106

See the "Instructions for the user" at the beginning of the volume

Formula	Compound	Volume	Reference
O_4Sc_3	Sc_3O_4	39 (SE):	Hb/C1–181
O_4Se^{2-}	$[SeO_4]^{2-}$ *see H_2O_4Se ... H_2SeO_4*		
O_4SeSr	$SrSeO_4$	29 (Sr):	Hb–186/7 Eb–261
O_4SeTi	$TiO_2 \cdot SeO_2 \cdot H_2O$	41 (Ti):	Hb–353
O_4SeTl_2	Tl_2SeO_4	38 (Tl):	Hb/3–373/5
O_4SeV	$VOSeO_3 \cdot 2\,H_2O$	48 (V):	Hb/B1–313
–	$VOSeO_3 \cdot 3\,H_2O$	48 (V):	Hb/B1–313
O_4SeZn	$ZnSeO_4$	32 (Zn):	Eb–968
–	$ZnSeO_4 \cdot H_2O$	32 (Zn):	Eb–968
–	$ZnSeO_4 \cdot 2\,H_2O$	32 (Zn):	Hb–247
–	$ZnSeO_4 \cdot 5\,H_2O$	32 (Zn):	Hb–247 Eb–968
–	$ZnSeO_4 \cdot 6\,H_2O$	32 (Zn):	Hb–247 Eb–968
–	$ZnSeO_4 \cdot 7\,H_2O$	32 (Zn):	Hb–247
–	$ZnSeO_4$ systems:		
	$ZnSeO_4$–H_2O	32 (Zn):	Eb–968
O_4SeZr	$ZrO \cdot SeO_3 \cdot 2\,H_2O$		
	$= ZrO_2 \cdot SeO_2 \cdot 2\,H_2O$	42 (Zr):	Hb–355
O_4Se_3	Se_3O_4	10 (Se):	Hb/B–25
O_4Si	$Si(O_2)_2$	15 (Si):	Hb/B–582
O_4Si^{4-}	$[SiO_4]^{4-}$	15 (Si):	Hb/B–421/2
O_4SiSr_2	$2\,SrO \cdot SiO_2 = Sr_2SiO_4$	29 (Sr):	Hb–211 Eb–281/2
–	$2\,SrO \cdot SiO_2$ systems:		
	$2\,SrO \cdot SiO_2$–$2\,BaO \cdot SiO_2$	30 (Ba):	Eb–535
O_4SiTh	$ThSiO_4$	44 (Th):	Hb–313
–	$ThSiO_4 \cdot xH_2O$	44 (Th):	Hb–313
O_4SiTl_4	Tl_4SiO_4	38 (Tl):	Hb/3–424
O_4SiZn_2	Zn_2SiO_4	32 (Zn):	Hb–272/3 Eb–989/92
–	$Zn_2SiO_4 \cdot H_2O$	32 (Zn):	Hb–273 Eb–992/3
–	Zn_2SiO_4 systems:		
	Zn_2SiO_4–Ni_2SiO_4	57 (Ni):	Hb/B3–1138
	Zn_2SiO_4–Zn_2GeO_4	45 (Ge):	Eb–574
O_4SiZr	$ZrSiO_4$	42 (Zr):	Hb–81/3, 382/9
–	$ZrSiO_4$ solid solutions:		
	$ZrSiO_4$–$NbSiO_4$	49 (Nb):	Hb/B4–360
	$(Ta,Zr)SiO_4$	50 (Ta):	Hb/B2–349
	$ZrSiO_4$–V_2O_4	48 (V):	Hb/B2–563
–	$ZrSiO_4$ systems:		
	$ZrSiO_4$–Li_4SiO_4	42 (Zr):	Hb–401
O_4Si_3	Si_3O_4	15 (Si):	Hb/B–257

Hb = Main Series Volume; GD = Gmelin-Durrer; Ab = Appended Volume to Main Series;
Eb = Supplement Volume to Main Series; Erg. W. = New Supplement Series Volume

Formula	Compound	Gmelin System No.	Reference
O_4SmTa	$SmTaO_4$	50 (Ta):	Hb/B2–202/3, 214
O_4SmV	$SmVO_4$	48 (V):	Hb/B2–553/5
O_4Sm_2	$Sm_2O_4 \cdot nH_2O$	39 (SE):	Hb/C2–106/7
O_4SnW	$SnWO_4 \cdot 6\,H_2O$	54 (W):	Hb–316
O_4SnZn_2	Zn_2SnO_4 solid solutions:		
	$Ni_xZn_{2-x}SnO_4$	57 (Ni):	Hb/B3–1193
–	Zn_2SnO_4 systems:		
	$Zn_2SnO_4–Co_2SnO_4$	58 (Co):	Eb/A–872
O_4Sn_3	Sn_3O_4	46 (Sn):	Hb/C1–58/9
O_4Sn_4	Sn_4O_4	46 (Sn):	Hb/C1–34/5, 104
$O_4Sr_{0.5}Th_{0.5}V$	$Sr_{0.5}Th_{0.5}VO_4$	48 (V):	Hb/B2–566
O_4SrTe	$SrTeO_4$	29 (Sr):	Hb–187
			Eb–262
O_4SrU	$SrUO_4$	29 (Sr):	Hb–74
		55 (U):	Hb–232
O_4SrV_2	SrV_2O_4	48 (V):	Hb/B2–518
O_4SrW	$SrWO_4$	54 (W):	Hb–295/6
O_4Sr_2Ti	Sr_2TiO_4	41 (Ti):	Hb–433
O_4Sr_2V	Sr_2VO_4	48 (V):	Hb/B2–519
O_4Sr_2Zr	Sr_2ZrO_4	42 (Zr):	Hb–439
O_4TaTb	$TbTaO_4$	50 (Ta):	Hb/B2–202/3, 217
O_4TaTi	$TiTaO_4$	50 (Ta):	Hb/B2–224
O_4TaTl	$TlTaO_4$	50 (Ta):	Hb/B2–201
O_4TaTm	$TmTaO_4$	50 (Ta):	Hb/B2–202/3, 220
O_4TaV	$VTaO_4$	50 (Ta):	Hb/B2–245
–	$VTaO_4$ solid solutions:		
	$VTaO_4–VO_2$	50 (Ta):	Hb/B2–244
O_4TaY	$YTaO_4$	50 (Ta):	Hb/B2–202/3, 206
–	$YTaO_4$ solid solutions:		
	$Y(Ta,Nb)O_4$	50 (Ta):	Hb/B2–264
–	$YTaO_4$ systems:		
	$YTaO_4–B_2O_3–BaO$	50 (Ta):	Hb/B2–187, 188
	$YTaO_4–BaO$	50 (Ta):	Hb/B2–187
O_4TaYb	$YbTaO_4$	50 (Ta):	Hb/B2–202/3, 220
$O_4Ta_2^{2+}$	$[Ta_2O_4]^{2+}$	50 (Ta):	Hb/B1–178
O_4TbV	$TbVO_4$	48 (V):	Hb/B2–553/5
O_4Te^{2-}	$[TeO_4]^{2-}$	11 (Te):	Hb–292, 301
O_4TeTl_2	Tl_2TeO_4	38 (Tl):	Hb/3–377
–	$Tl_2TeO_4 \cdot 0.5\,H_2O$	38 (Tl):	Hb/3–377
O_4TeZn	$ZnTeO_4$	32 (Zn):	Eb–934, 970
O_4TeZr	$ZrO_2 \cdot TeO_2 \cdot H_2O$		
	$= Zr(TeO_3)_2 \cdot Zr(OH)_4$	42 (Zr):	Hb–355/6
O_4ThZr	$ZrO_2 \cdot ThO_2$	44 (Th):	Hb–354
O_4TiV	$TiVO_4 = Ti_2O_3 \cdot V_2O_5$	48 (V):	Hb/B2–559/60
O_4TiZn_2	Zn_2TiO_4	41 (Ti):	Hb–460/1
–	Zn_2TiO_4 solid solutions:		
	$Zn_2TiO_4–LiZnNbO_4$	49 (Nb):	Hb/B3–204

See the "Instructions for the user" at the beginning of the volume

Formula	Compound	Volume	Reference
O_4TiZn_2	Zn_2TiO_4 systems:		
	$Zn_2TiO_4-Co_2TiO_4$	58 (Co):	Eb/A–870
O_4TiZr	$ZrTiO_4$ solid solutions:		
	$ZrTiO_4-PbNb_2O_6$	49 (Nb):	Hb/B3–308
	$ZrTiO_4-Zr_6Ta_2O_{17}$	50 (Ta):	Hb/B2–229
O_4Ti_3	Ti_3O_4	41 (Ti):	Hb–211
O_4TlV	$TlVO_4 = Tl_2O_3 \cdot V_2O_5$	48 (V):	Hb/B2–548
O_4Tl_2	$Tl_2O_4 = TlO_2$	38 (Tl):	Hb/2–233
O_4Tl_2W	Tl_2WO_4	54 (W):	Hb–314
O_4Tl_3V	$Tl_3VO_4 = 3\ Tl_2O \cdot V_2O_5$	48 (V):	Hb/B2–547
O_4Tl_4	$Tl_4O_4 = Tl_2O \cdot Tl_2O_3$	38 (Tl):	Hb/2–221
O_4TmV	$TmVO_4$	48 (V):	Hb/B2–553/5
O_4U	$UO_4 \cdot 2\ H_2O$	55 (U):	Hb–105/7
–	$UO_4 \cdot 3\ H_2O$	55 (U):	Hb–106
–	$UO_4 \cdot 4.5\ H_2O$	55 (U):	Hb–105
O_4UV	$VO_2 \cdot UO_2$	48 (V):	Hb/B1–91
O_4UZn	$ZnUO_4$	55 (U):	Hb–236
O_4V^-	$[VO_4]^-$	48 (V):	Hb/A2–469, 639
			Hb/B1–165
O_4V^{3-}	$[VO_4]^{3-}$	48 (V):	Hb/A2–391, 468, 478, 498, 550/640
			Hb/B1–132/47
–	$[VO_4(H_2O)_6]^{3-}$	48 (V):	Hb/B1–146
O_4VY	YVO_4	48 (V):	Hb/A2–343
			Hb/B2–553/5
O_4VYb	$YbVO_4$	48 (V):	Hb/B2–553/5
O_4VZn_2	Zn_2VO_4	48 (V):	Hb/B2–533
O_4V_2	V_2O_4		
	see O_2V ... VO_2		
O_4V_2Zn	ZnV_2O_4	48 (V):	Hb/B2–532/3
–	ZnV_2O_4 solid solutions:		
	$(Mg,Zn)\ (V,Al)_2O_4$	48 (V):	Hb/B2–547
O_4W^{2-}	$[WO_4]^{2-}$	54 (W):	Hb–144
O_4WZn	$ZnWO_4$	54 (W):	Hb–304/6
–	$ZnWO_4 \cdot H_2O$	54 (W):	Hb–306
O_4Xe	XeO_4	Erg. W. 1–112/3	
O_4Xe^+	$[XeO_4]^+$	Erg. W. 1–101	
O_4Xe^{2-}	$[XeO_4]^{2-}$	Erg. W. 1–109	
O_4Zn_2Zr	Zn_2ZrO_4	42 (Zr):	Hb–442
$O_{4.5}SZn_{1.5}$	$ZnSO_4 \cdot 0.5\ ZnO$	32 (Zn):	Hb–240
	see also $O_9S_2Zn_3$ ... $ZnO \cdot 2\ ZnSO_4$		
$O_{4.9}Pr_2$	$Pr_2O_{4.9} \cdot 5\ H_2O$	39 (SE):	Hb/C2–112
$O_{4.98}Rb_{2.98}S$	$PbSO_3 \cdot 1.98\ PbO \cdot 0.61\ H_2O$	47 (Pb):	Hb/C2–537
O_5^+	O_5^+	3 (O):	Hb/3–354
O_5Os_2	Os_2O_5	66 (Os):	Hb–30
$O_5Os_3S_7$	$Os_3O_5S_7 \cdot 2\ H_2O$	66 (Os):	Hb–60
O_5P^{3-}	$[PO_5]^{3-}$	16 (P):	Hb/C–300

Hb = Main Series Volume; GD = Gmelin-Durrer; Ab = Appended Volume to Main Series; Eb = Supplement Volume to Main Series; Erg. W. = New Supplement Series Volume

O_5PTa	$TaOPO_4$		
	see $O_{10}P_2Ta_2$ ... $Ta_2O_5 \cdot P_2O_5$		
O_5PV	$VOPO_4$		
	see $O_{10}P_2V_2$ ... $V_2O_5 \cdot P_2O_5$		
O_5P_2	P_2O_5:		
	Chemical reactions	16 (P):	Hb/C–87/93
	Formation	16 (P):	Hb/C–79/80, 82
	Handling	16 (P):	Hb/B–74/6
	Modifications	16 (P):	Hb/C–77/9
	Physical properties	16 (P):	Hb/C–83/7
	Preparation	16 (P):	Hb/C–80/2
	–, industrial		
	see also P ... Phosphorus		
	–, –, oxidation of phosphorus	16 (P):	Hb/B–50/8
	– of special forms	16 (P):	Hb/B–70/2
	Purity testing	16 (P):	Hb/B–73
	Shipment	16 (P):	Hb/B–76
	Standards	16 (P):	Hb/B–72/3
	Toxicity	16 (P):	Hb/B–93
–	P_4O_{10}	16 (P):	Hb/C–77/9, 82/3
–	P_2O_5 glasses:		
	$P_2O_5-Ag_2O-MoO_3$	61 (Ag):	Hb/B4–343
	$P_2O_5-B_2O_3-Na_2O$	21 (Na):	Eb/5–2013
	$P_2O_5-Bi_2O_3$	19 (Bi):	Eb–847
	$P_2O_5-Bi_2O_3-PbO$	47 (Pb):	Hb/C3–978
	P_2O_5-CdO	33 (Cd):	Eb–662
	$P_2O_5-Li_2O$	20 (Li):	Eb–521
	$P_2O_5-Na_2O-SiO_2$	21 (Na):	Eb/5–2016
	P_2O_5-PbO	47 (Pb):	Hb/C3–864/5
	$P_2O_5-PbO-ThO_2$	47 (Pb):	Hb/C4–1471
–	P_2O_5 melts:		
	$P_2O_5-BaO-Na_2O-V_2O_5$	48 (V):	Hb/B2–532
	$P_2O_5-BaO-V_2O_5$	48 (V):	Hb/B2–531
–	P_2O_5 systems:		
	$P_2O_5-Ag_2O-H_2O$	61 (Ag):	Hb/B4–7
	$P_2O_5-As_2O_5-Li_2O-Na_2O$	21 (Na):	Eb/5–2244
	$P_2O_5-As_2O_5-Na_2O$	21 (Na):	Eb/5–2021
	$P_2O_5-As_2O_5-Na_2O-H_2O$	21 (Na):	Eb/5–2021
	$P_2O_5-B_2O_3-H_2O$	16 (P):	Hb/C–613/4
	$P_2O_5-B_2O_3-K_2O$	13 (B):	Eb–135
	$P_2O_5-B_2O_3-SiO_2$	16 (P):	Hb/C–642
	$P_2O_5-BaO-H_2O$	30 (Ba):	Hb–339
	$P_2O_5-BaO-PbO-TiO_2$	47 (Pb):	Hb/C4–1413
	$P_2O_5-Bi_2O_3$	19 (Bi):	Eb–847
	$P_2O_5-Bi_2O_3-Na_2O$	21 (Na):	Eb/5–2026
	P_2O_5-CaO	28 (Ca):	Hb/B3–1122/6
	$P_2O_5-CaO-HF-H_2O$	28 (Ca):	Hb/B3–1199

See the "Instructions for the user" at the beginning of the volume

O_5P_2 . . . . P_2O_5 systems:

$P_2O_5-CaO-H_2O$ . . . . 28 (Ca): Hb/B3–1126/39
$P_2O_5-CaO-H_2SiF_6-N_2O_5-H_2O$ . . . . 28 (Ca): Hb/B3–1222
$P_2O_5-CaO-MgO-H_2O$ . . . . 28 (Ca): Hb/B3–1379
$P_2O_5-CaO-N_2O_5-H_2O$ . . . . 28 (Ca): Hb/B1–243
Hb/B3–1189/92
$P_2O_5-CaO-SiO_2$ . . . . 28 (Ca): Hb/B1–204/6
Hb/B3–1215/9
$P_2O_5-Ca_2P_2O_7$ . . . . 28 (Ca): Hb/B3–1123/5
$P_2O_5-[Co(H_2O)(NH_3)_5]_2O_3-H_2O$ . . . . 58 (Co): Eb/B2–388
$P_2O_5-[Co(NH_3)_6]_2O_3$ . . . . 58 (Co): Eb/B2–341
$P_2O_5-Cr_2O_3-H_2O$ . . . . 52 (Cr): Hb/B–440
$P_2O_5-FeO-H_2O$ . . . . 59 (Fe): Hb/B4–771
$P_2O_5-Fe_2O_3-H_2O$ . . . . 59 (Fe): Hb/B4–773
$P_2O_5-HF-H_2O$ . . . . 16 (P): Hb/C–404
$P_2O_5-H_2O$ . . . . 16 (P): Hb/C–155/60
$P_2O_5-K_2O-H_2O$ . . . . 22 (K): Hb/5–984/8
$P_2O_5-K_2O-NH_3-H_2O$ . . . . 23 (NH_4): Hb/2–517/8
$P_2O_5-K_2O-SO_3-H_2O$ . . . . 22 (K): Hb/5–1002
$P_2O_5-Li_2O$ . . . . 20 (Li): Eb–521
$P_2O_5-Li_2O-H_2O$ . . . . 20 (Li): Eb–521/2
$P_2O_5-NH_3-H_2O$ . . . . 23 (NH_4): Hb/2–418/9
$P_2O_5-NH_3-SO_3-H_2O$ . . . . 23 (NH_4): Hb/2–430/1
$P_2O_5-N_2O_5-H_2O$ . . . . 16 (P): Hb/C–332/5
$P_2O_5-Na_2O$ . . . . 21 (Na): Eb/4–1561/2
$P_2O_5-Na_2O-H_2O$ . . . . 21 (Na): Eb/4–1562/7
$P_2O_5-Na_2O-Sb_2O_3$ . . . . 21 (Na): Eb/5–2026
$P_2O_5-Na_2O-SiO_2$ . . . . 21 (Na): Eb/5–2013/6
$P_2O_5-Nb_2O_5-PbO$ . . . . 49 (Nb): Hb/B4–361
$P_2O_5-Nb_2O_5-ZnO$ . . . . 49 (Nb): Hb/B4–361
$P_2O_5-POCl_3-PSCl_3-P_2S_5$ . . . . 16 (P): Hb/C–597/8
P_2O_5-PbO . . . . 47 (Pb): Hb/C3–863/5
$P_2O_5-PbO-H_2O$ . . . . 47 (Pb): Hb/C3–870
$P_2O_5-PbO-SiO_2$ . . . . 47 (Pb): Hb/C3–885/6
$P_2O_5-PbO-V_2O_5$ . . . . 48 (V): Hb/B2–572
$P_2O_5-PbO-ZrO_2$ . . . . 47 (Pb): Hb/C4–1431
$P_2O_5-SiO_2$ . . . . 16 (P): Hb/C–638
$P_2O_5-SiO_2-H_2O$ . . . . 16 (P): Hb/C–639/40
$P_2O_5-SrO-H_2O$ . . . . 29 (Sr): Hb–213
$P_2O_5-Ta_2O_5-ZnO$ . . . . 50 (Ta): Hb/B2–349
$P_2O_5-V_2O_5$ . . . . 48 (V): Hb/B1–361
$P_2O_5-V_2O_5-H_2O$ . . . . 48 (V): Hb/B1–362
$P_2O_5-ZnO-H_2O$ . . . . 32 (Zn): Hb–277

$O_5P_{14}Zn_7$ . . . . $Zn_7P_{14}O_5$ . . . . 32 (Zn): Hb–276
O_5Pa_2 . . . . Pa_2O_5 . . . . 51 (Pa): Hb–67
– . . . . Pa_2O_5 systems:

$PaO_{2.5}-PuO_2$ . . . . Erg. W. 4–58

O_5PbS . . . . $PbOSO_4 \cdot H_2O$ . . . . 47 (Pb): Hb/C2–594

Hb = Main Series Volume; GD = Gmelin-Durrer; Ab = Appended Volume to Main Series;
Eb = Supplement Volume to Main Series; Erg. W. = New Supplement Series Volume

O_5PbSb_2	$PbSb_2O_5 = PbO \cdot Sb_2O_4$	47 (Pb):	Hb/C3–930
O_5PbSe_2	$PbSe_2O_5$	47 (Pb):	Hb/C2–636
$O_5PbSiZr$	$PbZrSiO_5$	42 (Zr):	Hb–438
		47 (Pb):	Hb/C4–1431
O_5PbSi_2	$PbSi_2O_5$	47 (Pb):	Hb/C3–854
O_5Pb_2	Pb_2O_5	47 (Pb):	Hb/C1–196
O_5Pb_2S	$PbSO_4 \cdot PbO$	47 (Pb):	Hb/C2–584/8
$O_5Pb_2S_2V_2$	$Pb_2V_2O_5S_2$	48 (V):	Hb/B2–572
O_5Pb_2Se	$PbSeO_4 \cdot PbO$	47 (Pb):	Hb/C2–638/9
O_5Pb_2W	$2\,PbO \cdot WO_3$	54 (W):	Hb–317
O_5Pb_3Se	$PbSeO_3 \cdot 2\,PbO$	47 (Pb):	Hb/C2–635
O_5Pb_3Si	Pb_3SiO_5	47 (Pb):	Hb/C3–842
O_5Pr_2	$Pr_2O_5 \cdot nH_2O$	39 (SE):	Hb/C2–106/8, 112
O_5Pu^{3-}	$[PuO_5]^{3-}$	Erg. W. 8–76	
O_5PuRb_3	Rb_3PuO_5	Erg. W. 4–52/3	
$O_5Rb_2S_2$	$Rb_2S_2O_5$	24 (Rb):	Hb–207
$O_5Rb_2S_3$	$Rb_2S_2O_3 \cdot SO_2$	24 (Rb):	Hb–215
$O_5Rb_2Ti_2$	$Rb_2Ti_2O_5$	41 (Ti):	Hb–411
–	$Rb_2Ti_2O_5$ systems:		
	$Rb_2Ti_2O_5$–RbF	41 (Ti):	Hb–412
O_5Re_2	$Re_2O_5 \cdot xH_2O$	70 (Re):	Hb–97
O_5RuS	$RuOSO_4$	63 (Ru):	Eb–314/5
O_5Ru_2	Ru_2O_5	63 (Ru):	Hb–27
–	$Ru_2O_5 \cdot 2\,H_2O$	63 (Ru):	Hb–27
			Eb–240
O_5SSb_2	Sb_2O_5S	18 (Sb):	Hb/B3–541
O_5SSe	$SeO_2 \cdot SO_3$	10 (Se):	Hb/B–181
O_5STh	$ThOSO_4 \cdot H_2O$	44 (Th):	Hb–293
–	$ThOSO_4 \cdot 2\,H_2O$	44 (Th):	Hb–293
–	$ThOSO_4 \cdot 3\,H_2O$	44 (Th):	Hb–292
–	$ThOSO_4 \cdot 5\,H_2O$	44 (Th):	Hb–292
O_5STi	$TiOSO_4 = TiO_2 \cdot SO_3$	41 (Ti):	Hb–347, 348
–	$TiOSO_4 \cdot H_2O$	41 (Ti):	Hb–346/7, 349
–	$TiOSO_4 \cdot 2\,H_2O$	41 (Ti):	Hb–345/7, 348/9
–	$TiOSO_4 \cdot 5\,H_2O$	41 (Ti):	Hb–348
O_5SU	$UOSO_4 \cdot 2\,H_2O$	55 (U):	Hb–151
–	$UO_2SO_3 \cdot 2.75\,H_2O$		
	$= 4\,UO_3 \cdot 4\,SO_2 \cdot 11\,H_2O$	55 (U):	Hb–146
–	$UO_2SO_3 \cdot 4\,H_2O = UO_3 \cdot SO_2 \cdot 4\,H_2O$	55 (U):	Hb–145/6
O_5SV	$VOSO_4$	48 (V):	Hb/B1–92, 288/91
	see also OV^{2+} ... VO^{2+}		
–	$VOSO_4 \cdot H_2O$	48 (V):	Hb/B1–295/6
–	$VOSO_4 \cdot 1.5\,H_2O$	48 (V):	Hb/B1–291, 295, 299
–	$VOSO_4 \cdot 2\,H_2O$	48 (V):	Hb/B1–72, 295
–	$VOSO_4 \cdot 2.5\,H_2O$	48 (V):	Hb/B1–295
–	$VOSO_4 \cdot 3\,H_2O$	48 (V):	Hb/A2–376, 621
			Hb/B1–291, 294/5, 299
–	$VOSO_4 \cdot 3.5\,H_2O$	48 (V):	Hb/B1–293/4

See the "Instructions for the user" at the beginning of the volume

Formula	Compound	Volume	Reference
O_5SV	$VOSO_4 \cdot 4\,H_2O = [VO(H_2O)_4SO_4]$	48 (V):	Hb/B1–293, 296
–	$VOSO_4 \cdot 5\,H_2O$	48 (V):	Hb/B1–71/2, 292/3, 299
–	$VOSO_4 \cdot 6\,H_2O$	48 (V):	Hb/B1–291/2, 299
–	$VOSO_4 \cdot 6.5\,H_2O$	48 (V):	Hb/B1–291/2
–	$VOSO_4$ solutions:		
	$VOSO_4–H_2O$	48 (V):	Hb/A2–480/1 Hb/B1–296/8
	$VOSO_4–H_2SO_4–H_2O$	48 (V):	Hb/A2–476, 510/1 Hb/B1–301/2
	$VOSO_4–LiCl–H_2O$	48 (V):	Hb/B2–376
	$VOSO_4$–organic solvents	48 (V):	Hb/B1–298
–	$VOSO_4$ systems:		
	$VOSO_4–C_4H_6O_6–H_2O$	48 (V):	Hb/B1–350
	$VOSO_4–H_2C_2O_4–H_2O$	48 (V):	Hb/B1–348
	$VOSO_4–H_2O$	48 (V):	Hb/B1–291
	$VOSO_4–H_2SO_4–H_2O$	48 (V):	Hb/B1–298/300
	$VOSO_4–NaC_4H_5O_6–H_2O$	48 (V):	Hb/B1–350
	$VOSO_4–NaKC_4H_4O_6–H_2O$	48 (V):	Hb/B1–350
	$VOSO_4–Na_2C_2O_4–H_2O$	48 (V):	Hb/B1–348
O_5SV^+	$[VO(H_2O)SO_4]^+$	48 (V):	Hb/B1–305
O_5SZn_2	$ZnSO_4 \cdot ZnO$	32 (Zn):	Hb–240
–	$ZnSO_4 \cdot ZnO \cdot H_2O$	32 (Zn):	Hb–241
–	$ZnSO_4 \cdot ZnO \cdot xH_2O$	32 (Zn):	Hb–240
O_5SZr	$ZrOSO_4 \cdot 4\,H_2O$	42 (Zr):	Hb–346
–	$ZrOSO_4 \cdot xH_2O$	42 (Zr):	Hb–345/6
$O_5S_2^{2-}$	$[S_2O_5]^{2-}$	9 (S):	Hb/B2–613
O_5S_2U	$UO_2S_2O_3$	55 (U):	Hb–156
O_5Sb_2	Sb_2O_5	18 (Sb):	Hb/B2–276, 376/80
–	$Sb_2O_5 \cdot H_2O = HSbO_3$	18 (Sb):	Hb/B2–381
–	$3\,Sb_2O_5 \cdot 5\,H_2O = H_5Sb_3O_{10}$	18 (Sb):	Hb/B2–384/5
–	$Sb_2O_5 \cdot 2\,H_2O = H_4Sb_2O_7$	18 (Sb):	Hb/B2–381
–	$Sb_2O_5 \cdot 3\,H_2O = H_3SbO_4$	18 (Sb):	Hb/B2–381
–	$Sb_2O_5 \cdot xH_2O$	18 (Sb):	Hb/B2–380/7
–	Sb_2O_5 solid solutions:		
	$Sb_2O_5–PbO–SnO_2$	47 (Pb):	Hb/C4–1595
–	Sb_2O_5 systems:		
	$Sb_2O_5–Ag_2O–Sb_2O_3$	61 (Ag):	Hb/B4–89
	$Sb_2O_5–Na_2O–H_2O$	21 (Na):	Eb/4–1711/2
	$Sb_2O_5–PbO$	47 (Pb):	Hb/C3–929
O_5SeTi	$TiO_2 \cdot SeO_3 \cdot H_2O$	41 (Ti):	Hb–353
O_5SeU	UO_2SeO_3	55 (U):	Hb–157
–	$UO_2SeO_3 \cdot 2\,H_2O$	55 (U):	Hb–158
$O_5SiSrTi$	$SrTiSiO_5$	41 (Ti):	Hb–434
$O_5SiSrZr$	$SrZrSiO_5$	42 (Zr):	Hb–438
O_5SiSr_3	$3\,SrO \cdot SiO_2$	29 (Sr):	Eb–281
$O_5SiZnZr$	$ZnZrSiO_5$	42 (Zr):	Hb–438
O_5Si_2	Si_2O_5	15 (Si):	Hb/B–257

Hb = Main Series Volume; GD = Gmelin-Durrer; Ab = Appended Volume to Main Series; Eb = Supplement Volume to Main Series; Erg. W. = New Supplement Series Volume

$O_5Si_2^{2-}$	$[Si_2O_5]^{2-}$	15 (Si):	Hb/B–421/2
O_5Sm_2	$Sm_2O_5 \cdot nH_2O$	39 (SE):	Hb/C2–106/7, 113
O_5SrTa_2	$SrTa_2O_5$	50 (Ta):	Hb/B2–181
–	$SrTa_2O_5$ solid solutions:		
	$(Sr,Ba)Ta_2O_5$	50 (Ta):	Hb/B2–192
O_5SrV_2	SrV_2O_5	48 (V):	Hb/B2–519
$O_5Sr_2V_2$	$Sr_2V_2O_5$ *see $O_{2.5}SrV$ ... $SrVO_{2.5}$*		
O_5TaV	$VTaO_5$	50 (Ta):	Hb/B2–245
O_5Ta_2	Ta_2O_5:		
	Chemical reactions	50 (Ta):	Hb/B1–45/53
	Electrochemical behavior	50 (Ta):	Hb/A2–412, 490/3 Hb/B1–45
	Formation and preparation	50 (Ta):	Hb/B1–17, 31/4
	History	50 (Ta):	Hb/A1–8
	Physical properties	50 (Ta):	Hb/A2–322/3 Hb/B1–34/45
	Purity testing	49 (Nb):	Hb/A–352/3
	Toxicity	50 (Ta):	Hb/A1–276
	Uses	50 (Ta):	Hb/A1–276
–	$Ta_2O_5 \cdot xH_2O$	50 (Ta):	Hb/A1–247 Hb/B1–54/8, 136
–	Ta_2O_5 solid solutions:		
	$Ta_2O_5-BaTiO_3$	50 (Ta):	Hb/B2–224/6
	$Ta_2O_5-K_2O-Nb_2O_5-PbO$	50 (Ta):	Hb/B2–265
	$Ta_2O_5-Na_2O-Nb_2O_5-PbO$	50 (Ta):	Hb/B2–265
	$Ta_2O_5-SiO_2-ZrO_2$	50 (Ta):	Hb/B2–349
–	Ta_2O_5 systems:		
	$Ta_2O_5-Ag_2O$	61 (Ag):	Hb/B4–285
	$Ta_2O_5-Al_2O_3$	50 (Ta):	Hb/B2–197
	Ta_2O_5-BaO	50 (Ta):	Hb/B2–186
	$Ta_2O_5-Bi_2O_3$	50 (Ta):	Hb/B1–275
	Ta_2O_5-CaO	50 (Ta):	Hb/B2–176/7
	Ta_2O_5-CdO	50 (Ta):	Hb/B2–195
	$Ta_2O_5-CeO_2$	50 (Ta):	Hb/B2–210
	$Ta_2O_5-Ce_2O_3$	50 (Ta):	Hb/B2–210
	Ta_2O_5-CuO	60 (Cu):	Hb/B3–1214
	$Ta_2O_5-Dy_2O_3$	50 (Ta):	Hb/B2–217
	$Ta_2O_5-Eu_2O_3$	50 (Ta):	Hb/B2–215
	$Ta_2O_5-Ga_2O_3$	50 (Ta):	Hb/B2–199
	$Ta_2O_5-Gd_2O_3$	50 (Ta):	Hb/B2–216
	$Ta_2O_5-GeO_2-K_2O$	50 (Ta):	Hb/B2–230
	$Ta_2O_5-GeO_2-PbO$	50 (Ta):	Hb/B2–243
	$Ta_2O_5-HF-H_2O$	50 (Ta):	Hb/B1–88
	$Ta_2O_5-HfO_2-ZrO_2$	50 (Ta):	Hb/B2–229
	$Ta_2O_5-In_2O_3$	50 (Ta):	Hb/B2–200
	Ta_2O_5-KCl	50 (Ta):	Hb/B2–287

Formula	Compound	Gmelin System No.	Reference
O_5Ta_2	Ta_2O_5 systems:		
	$Ta_2O_5-KCl-K_2TaF_7$	50 (Ta):	Hb/B2-289/90
	Ta_2O_5-KF	50 (Ta):	Hb/B2-275
	$Ta_2O_5-KF-K_2TaF_7$	50 (Ta):	Hb/B2-275/6
	$Ta_2O_5-K_2CO_3$	50 (Ta):	Hb/B2-151/2
	$Ta_2O_5-K_2O$	50 (Ta):	Hb/B2-151/2
	$Ta_2O_5-K_2O-C_2H_5OH-H_2O$	50 (Ta):	Hb/B2-165
	$Ta_2O_5-K_2O-H_2O$	50 (Ta):	Hb/B2-163/5
	$Ta_2O_5-K_2O-SiO_2$	50 (Ta):	Hb/B2-348
	$Ta_2O_5-K_2TaF_7$	50 (Ta):	Hb/B2-275
	$Ta_2O_5-La_2O_3$	50 (Ta):	Hb/B2-207/8
	$Ta_2O_5-La_2O_3-Li_2O$	50 (Ta):	Hb/B2-209
	$Ta_2O_5-Li_2O$	50 (Ta):	Hb/B2-126
	Ta_2O_5-MgO	50 (Ta):	Hb/B2-173
	$Ta_2O_5-(NH_4)_2SO_4-SO_3-H_2O$	50 (Ta):	Hb/B2-316
	$Ta_2O_5-Na_2O$	50 (Ta):	Hb/B2-141
	$Ta_2O_5-Na_2O-H_2O$	50 (Ta):	Hb/B2-147
	$Ta_2O_5-Nb_2O_5$	50 (Ta):	Hb/B2-246/8
	$Ta_2O_5-Nd_2O_3$	50 (Ta):	Hb/B2-212
	$Ta_2O_5-P_2O_5-ZnO$	50 (Ta):	Hb/B2-349
	Ta_2O_5-PbO	50 (Ta):	Hb/B2-232/3
	$Ta_2O_5-SO_3-H_2O$	50 (Ta):	Hb/B1-176
	$Ta_2O_5-Sc_2O_3$	50 (Ta):	Hb/B2-204
	$Ta_2O_5-SiO_2$	50 (Ta):	Hb/B1-263
	$Ta_2O_5-SiO_2-ZnO$	50 (Ta):	Hb/B2-349
	$Ta_2O_5-Sm_2O_3$	50 (Ta):	Hb/B2-214
	Ta_2O_5-SnO	50 (Ta):	Hb/B2-231
	$Ta_2O_5-SnO_2$	50 (Ta):	Hb/B2-231
	Ta_2O_5-SrO	50 (Ta):	Hb/B2-181
	$Ta_2O_5-ThO_2$	50 (Ta):	Hb/B2-229
	$Ta_2O_5-TiO_2$	50 (Ta):	Hb/B2-222
	$Ta_2O_5-TiO_2-ZrO_2$	50 (Ta):	Hb/B2-229
	$Ta_2O_5-Tl_2O_3$	50 (Ta):	Hb/B2-201
	$Ta_2O_5-V_2O_5$	50 (Ta):	Hb/B2-244
	$Ta_2O_5-Y_2O_3$	50 (Ta):	Hb/B2-205
	$Ta_2O_5-Y_2O_3-ZrO_2$	50 (Ta):	Hb/B2-229
	Ta_2O_5-ZnO	50 (Ta):	Hb/B2-193
	$Ta_2O_5-ZrO_2$	50 (Ta):	Hb/B2-228
$O_5Ta_2^-$	$[Ta_2O_5]^-$	50 (Ta):	Hb/B1-59
O_5TeTh	$ThOTeO_4 \cdot 4\,H_2O$	44 (Th):	Hb-295
–	$ThOTeO_4 \cdot 8\,H_2O$	44 (Th):	Hb-295
O_5TiV_2	V_2TiO_5	48 (V):	Hb/B2-558
O_5Ti_2	$Ti_2O_5 \cdot xH_2O$	41 (Ti):	Hb-269
O_5Ti_2V	Ti_2VO_5	48 (V):	Hb/B2-558
O_5Ti_3	Ti_3O_5	41 (Ti):	Hb-224/6
–	Ti_3O_5 systems:		
	$Ti_3O_5-V_3O_5$	48 (V):	Hb/B2-558
O_5Tl_3	$Tl_3O_5 = Tl_2O_3 \cdot TlO_2$	38 (Tl):	Hb/2-233

Hb = Main Series Volume; GD = Gmelin-Durrer; Ab = Appended Volume to Main Series;
Eb = Supplement Volume to Main Series; Erg. W. = New Supplement Series Volume

Formula	Compound	Volume	Reference
O_5UV	UVO_5	48 (V):	Hb/B1–91
O_5U_2	U_2O_5	55 (U):	Hb–88/9
O_5V^-	$[V(O)(O_2)_2]^-$	48 (V):	Hb/B1–162/3, 165
O_5V_2	V_2O_5:		
	Catalytic properties	48 (V):	Hb/B1–84
	Chemical reactions	48 (V):	Hb/B1–84/95
	Determination	48 (V):	Hb/A2–714
	Diffusion of gases	48 (V):	Hb/B1–95/6
	Electrochemical behavior	48 (V):	Hb/A2–468/9, 488/9, 496/8 Hb/B1–84
	Formation	48 (V):	Hb/B1–73
	Gels	48 (V):	Hb/B1–130/1
	History	48 (V):	Hb/A1–6/10
	Physical properties	48 (V):	Hb/A2–376/80, 387 Hb/B1–30, 73/83
	Preparation	48 (V):	Hb/A1–280/5
	Purification	48 (V):	Hb/A1–283/4
	Purity testing	48 (V):	Hb/A1–284/5
	Sols	48 (V):	Hb/B1–98/130
	Solubility	48 (V):	Hb/B1–97
	Sorption of gases	48 (V):	Hb/B1–95/7
	Special forms	48 (V):	Hb/A1–270, 284 Hb/B1–72, 98/131
	Toxicity	48 (V):	Hb/A2–725/8
–	V_4O_{10}	48 (V):	Hb/B1–84
–	$V_2O_5 \cdot 0.5\ H_2O$	48 (V):	Hb/B1–98
–	$V_2O_5 \cdot H_2O$	48 (V):	Hb/B1–97/8
–	$V_2O_5 \cdot 2\ H_2O$	48 (V):	Hb/B1–97
–	$V_2O_5 \cdot nH_2O$	48 (V):	Hb/B1–97
–	V_2O_5 melts:		
	$V_2O_5-As_2O_3$	48 (V):	Hb/B1–366
	$V_2O_5-B_2O_3-ZnO$	48 (V):	Hb/B2–536
	$V_2O_5-BaO-Na_2O-P_2O_5$	48 (V):	Hb/B2–532
	$V_2O_5-BaO-P_2O_5$	48 (V):	Hb/B2–531
	$V_2O_5-Cs_2O$	48 (V):	Hb/B2–492
	$V_2O_5-Li_2SO_4$	48 (V):	Hb/B2–376
	$V_2O_5-Na_2SO_4$	48 (V):	Hb/B2–405/6
	$V_2O_5-Na_2S_2O_7$	48 (V):	Hb/B2–406
	$V_2O_5-Rb_2O$	48 (V):	Hb/B2–486
	$V_2O_5-TeO_2$	48 (V):	Hb/B1–320
–	V_2O_5 solutions:		
	$V_2O_5-H_2O$	48 (V):	Hb/B1–132/45
–	V_2O_5 systems:		
	$V_2O_5-Ag_2O$	61 (Ag):	Hb/B4–277
	V_2O_5-Al	48 (V):	Hb/B2–544/5
	$V_2O_5-Al_2O_3$	48 (V):	Hb/B2–545
	$V_2O_5-B_2O_3$	48 (V):	Hb/B1–325

See the "Instructions for the user"
at the beginning of the volume

O_5V_2	V_2O_5 systems:		
	V_2O_5-BaO	48 (V):	Hb/B2–525
	V_2O_5-BeO	48 (V):	Hb/B2–500
	$V_2O_5-Bi_2O_3$	48 (V):	Hb/B1–368
	$V_2O_5-Bi_2O_3-PbO$	48 (V):	Hb/B2–573
	V_2O_5-CaO	48 (V):	Hb/B2–511/2
	V_2O_5-CoO	58 (Co):	Hb/A2–478 Eb/A–873
	$V_2O_5-Cr_2O_3$	52 (Cr):	Hb/B–939
	$V_2O_5-Cr_2O_3-NiO$	57 (Ni):	Hb/B3–1217
	$V_2O_5-Cs_2O$	48 (V):	Hb/B2–492
	$V_2O_5-Cs_2SO_4$	48 (V):	Hb/B2–499
	V_2O_5-CuO	60 (Cu):	Hb/B3–1209
	$V_2O_5-CuO-Fe_2O_3$	60 (Cu):	Hb/B3–1330
	$V_2O_5-HF-H_2O$	48 (V):	Hb/B1–194
	$V_2O_5-H_2SO_4-H_2O$	48 (V):	Hb/B1–303/5
	$V_2O_5-KPO_3$	48 (V):	Hb/B2–444
	$V_2O_5-KVO_3$	48 (V):	Hb/B2–412
	$V_2O_5-K_2O$	48 (V):	Hb/B2–412/3
	$V_2O_5-K_2SO_4$	48 (V):	Hb/B2–438/9
	$V_2O_5-K_2S_2O_7$	48 (V):	Hb/B2–439
	$V_2O_5-Li_2O$	48 (V):	Hb/B2–371
	$V_2O_5-Li_2O-VO_2$	48 (V):	Hb/B2–370
	V_2O_5-MgO	48 (V):	Hb/B2–504/5
	$V_2O_5-MgO-VO_2$	48 (V):	Hb/B2–504
	$V_2O_5-NH_3-H_2O$	48 (V):	Hb/B2–449/50
	$V_2O_5-Na_2O$	48 (V):	Hb/B2–380/1
	$V_2O_5-Na_2O-H_2O$	48 (V):	Hb/B2–382
	$V_2O_5-Na_2O-SO_3$	48 (V):	Hb/B2–405
	$V_2O_5-Na_2O-SO_3-H_2O$	48 (V):	Hb/B2–406
	$V_2O_5-Nb_2O_5$	49 (Nb):	Hb/B3–324/6
	$V_2O_5-Nb_2O_5-ZnO$	49 (Nb):	Hb/B3–329
	V_2O_5-NiO	57 (Ni):	Hb/B3–1198/9
	$V_2O_5-P_2O_5$	48 (V):	Hb/B1–361
	$V_2O_5-P_2O_5-H_2O$	48 (V):	Hb/B1–362
	$V_2O_5-P_2O_5-PbO$	48 (V):	Hb/B2–572
	V_2O_5-PbO	48 (V):	Hb/B2–567/8
	$V_2O_5-PbO-TiO_2$	48 (V):	Hb/B2–574
	$V_2O_5-Ta_2O_5$	50 (Ta):	Hb/B2–244
	$V_2O_5-ThO_2$	48 (V):	Hb/B2–564
	V_2O_5-VO	48 (V):	Hb/B1–13
	$V_2O_5-VO_2-ZnO$	48 (V):	Hb/B2–534
	$V_2O_5-V_2O_3$	48 (V):	Hb/B1–13
	V_2O_5-ZnO	48 (V):	Hb/B2–534
	$V_2O_5-ZrO_2$	48 (V):	Hb/B2–562
$O_5V_2^{2-}$	$[V_2O_5]^{2-}$	48 (V):	Hb/A2–477
$O_5V_2Zn_x$	$Zn_xV_2O_5$	48 (V):	Hb/B2–534
O_5V_3	$V_3O_5 = VO_{1.67}$	48 (V):	Hb/B1–13, 17, 49/50, 86

Hb = Main Series Volume; GD = Gmelin-Durrer; Ab = Appended Volume to Main Series; Eb = Supplement Volume to Main Series; Erg. W. = New Supplement Series Volume

O_5V_3	V_3O_5 systems:		
	$V_3O_5-Ti_3O_5$	48 (V):	Hb/B2–558
O_5W_2	W_2O_5	54 (W):	Hb–107/11, 115/7
O_5Zn_3	Zn_3O_5	32 (Zn):	Hb–145
–	$Zn_3O_5 \cdot H_2O$	32 (Zn):	Hb–146
			Eb–817
O_5Zr_2	$Zr_2O_5 \cdot xH_2O$	42 (Zr):	Hb–261
O_6	O_6	3 (O):	Hb/4–1041
O_6^+	$(2\ O_3)^+$	3 (O):	Hb/4–1071
O_6P_2	P_2O_6	16 (P):	Hb/C–93/4
O_6P_2Pb	$[Pb(PO_3)_2]_x$	47 (Pb):	Hb/C3–875/6
–	$Pb(PO_3)_2$ systems:		
	$Pb(PO_3)_2-BaTiO_3$	47 (Pb):	Hb/C4–1413
	$Pb(PO_3)_2-PbZrO_3$	47 (Pb):	Hb/C4–1431
$O_6P_2Pb_2$	$Pb_2P_2O_6$	47 (Pb):	Hb/C3–863
$O_6P_2Pb_3S_2$	$Pb_3(PO_3S)_2$	47 (Pb):	Hb/C3–884
$O_6P_2Rb_4$	$Rb_4P_2O_6 \cdot 10\ H_2O$	24 (Rb):	Hb–233
O_6P_2Sr	$Sr(PO_3)_2$	29 (Sr):	Hb–216
			Eb–286
–	$Sr(PO_3)_2 \cdot 2\ H_2O$	29 (Sr):	Hb–216
$O_6P_2Sr_2$	$Sr_2P_2O_6$	29 (Sr):	Eb–283
–	$Sr_2P_2O_6 \cdot 2\ H_2O$	29 (Sr):	Eb–283
O_6P_2Th	$ThP_2O_6 \cdot 11\ H_2O$	44 (Th):	Hb–316
$O_6P_2Tl_4$	$Tl_4P_2O_6 = Tl_2PO_3$	38 (Tl):	Hb/3–427
O_6P_2U	UP_2O_6	55 (U):	Hb–180
O_6P_2Zn	$Zn(PO_3)_2$	32 (Zn):	Hb–279
–	$Zn_2(PO_3)_4$	32 (Zn):	Hb–280
–	$Zn_3(PO_3)_6$	32 (Zn):	Eb–997
–	$Zn(PO_3)_2 \cdot 4\ H_2O$	32 (Zn):	Hb–279/80
–	$Zn_2(PO_3)_4 \cdot 10\ H_2O$	32 (Zn):	Hb–280
–	$Zn_3(PO_3)_6 \cdot 9\ H_2O$	32 (Zn):	Hb–280
$O_6P_2Zn_2$	$Zn_2P_2O_6 \cdot 2\ H_2O$	32 (Zn):	Hb–277
O_6P_2Zr	$ZrP_2O_6 \cdot H_2O$	42 (Zr):	Hb–392
O_6P_4	P_4O_6	16 (P):	Hb/C–70/1
$O_6P_4S_4$	$P_4S_4O_6$	16 (P):	Hb/C–582/3
$O_6P_4Se_4$	$P_4O_6Se_4$	16 (P):	Hb/C–75
O_6PbS_2	PbS_2O_6	47 (Pb):	Hb/B2–284/6
–	$PbS_2O_6 \cdot H_2O$	47 (Pb):	Hb/C2–598
–	$PbS_2O_6 \cdot 4\ H_2O$	47 (Pb):	Hb/C2–598/9
–	PbS_2O_6 solid solutions:		
	$(Pb,Ba)S_2O_6 \cdot 4\ H_2O$	30 (Ba):	Hb–284
		47 (Pb):	Hb/C3–1207
–	PbS_2O_6 systems:		
	$PbS_2O_6-SrS_2O_6-H_2O$	47 (Pb):	Hb/C3–1198
O_6PbS_3	PbS_3O_6	47 (Pb):	Hb/C2–600
O_6PbS_4	PbS_4O_6	47 (Pb):	Hb/B2–286
–	$PbS_4O_6 \cdot 2\ H_2O$	47 (Pb):	Hb/C2–600

See the "Instructions for the user"
at the beginning of the volume

$O_6PbS_4^{2-}$	$[Pb(S_2O_3)_2]^{2-}$	47 (Pb):	Hb/B1–365 Hb/C2–597
O_6PbS_5	PbS_5O_6	47 (Pb):	Hb/C2–600
O_6PbSb_2	$Pb(SbO_3)_2$	47 (Pb):	Hb/C3–932/3
–	$Pb(SbO_3)_2 \cdot nH_2O$	47 (Pb):	Hb/C3–933
O_6PbTa_2	$PbTa_2O_6$	50 (Ta):	Hb/B2–233/6
–	$PbTa_2O_6$ solid solutions:		
	$(Pb,Ba,Ca)Ta_2O_6$	50 (Ta):	Hb/B2–239
	$(Pb,Ba,Sr)Ta_2O_6$	50 (Ta):	Hb/B2–239
	$Pb(Ta,Nb)_2O_6$	50 (Ta):	Hb/B2–265/6
	$(Pb,Sr)Ta_2O_6$	50 (Ta):	Hb/B2–239
–	$PbTa_2O_6$ systems:		
	$PbTa_2O_6$–$PbTiO_3$	50 (Ta):	Hb/B2–242
	$PbTa_2O_6$–$PbZrO_3$	50 (Ta):	Hb/B2–242
O_6PbV_2	$PbO \cdot V_2O_5 = Pb(VO_3)_2$	48 (V):	Hb/B2–570
O_6Pb_2PrTa	Pb_2PrTaO_6	50 (Ta):	Hb/B2–240
O_6Pb_2SbTa	Pb_2SbTaO_6	50 (Ta):	Hb/B2–238
O_6Pb_2ScTa	Pb_2ScTaO_6	50 (Ta):	Hb/B2–240
–	Pb_2ScTaO_6 systems:		
	Pb_2ScTaO_6–$PbHfO_3$	50 (Ta):	Hb/B2–243
	Pb_2ScTaO_6–$PbTiO_3$	50 (Ta):	Hb/B2–243
	Pb_2ScTaO_6–$PbZrO_3$	50 (Ta):	Hb/B2–243
O_6Pb_2SmTa	Pb_2SmTaO_6	50 (Ta):	Hb/B2–240
O_6Pb_2TaY	Pb_2YTaO_6	50 (Ta):	Hb/B2–240
O_6Pb_2TaYb	Pb_2YbTaO_6	50 (Ta):	Hb/B2–240/1
O_6Pb_3S	$PbSO_4 \cdot 2\ PbO$	47 (Pb):	Hb/C2–588/9
$O_6Pb_3Sb_2$	$Pb_3(SbO_3)_2 = 3\ PbO \cdot Sb_2O_3$	47 (Pb):	Hb/C3–930
O_6Pb_4S	$PbSO_3 \cdot 3\ PbO \cdot 0.6\ H_2O$	47 (Pb):	Hb/C2–537
O_6Pb_4Si	Pb_4SiO_6	47 (Pb):	Hb/C3–840/2
O_6Pd_5	Pd_5O_6	65 (Pd):	Hb/2–259
$O_6Pm_2Sc_2$	$Sc_2O_3 \cdot Pm_2O_3 = PmScO_3$	39 (SE):	Hb/C1–113, 124, 313
O_6PrSr_2Ta	Sr_2PrTaO_6	50 (Ta):	Hb/B2–212
$O_6Pr_2Sc_2$	$Sc_2O_3 \cdot Pr_2O_3 = PrScO_3$	39 (SE):	Hb/C1–113, 124, 127
O_6PtS_2	$Pt(SO_3)_2$	68 (Pt):	Hb/C1–109
$O_6PtS_2^{2-}$	$[Pt(SO_3)_2]^{2-}$	68 (Pt):	Hb/C2–158
O_6PuSe_2	$Pu(SeO_3)_2$	Erg. W. 4–191	
O_6PuSr_3	Sr_3PuO_6	Erg. W. 4–54/5	
$O_6Rb_2S_2$	$Rb_2S_2O_6$	24 (Rb):	Hb–215/6
$O_6Rb_2S_2Se$	$Rb_2SeS_2O_6$	24 (Rb):	Hb–218
$O_6Rb_2S_3$	$Rb_2S_3O_6$	24 (Rb):	Hb–216
$O_6Rb_2S_4$	$Rb_2S_4O_6$	24 (Rb):	Hb–216
O_6Rb_2U	$Rb_2UO_6 \cdot 3\ H_2O$	55 (U):	Hb–226
$O_6Rb_2V_2$	$Rb_2O \cdot V_2O_5$ *see O_3RbV ... $RbVO_3$*		
O_6Rb_4Xe	Rb_4XeO_6	Erg. W. 1–116	
O_6RuS_2	RuS_2O_6	63 (Ru):	Hb–77
O_6SSb_2	$Sb_2O_3 \cdot SO_3 = (SbO)_2SO_4$	18 (Sb):	Hb/B3–545
–	$Sb_2O_3 \cdot SO_3 \cdot H_2O$	18 (Sb):	Hb/B3–545

Hb = Main Series Volume; GD = Gmelin-Durrer; Ab = Appended Volume to Main Series; Eb = Supplement Volume to Main Series; Erg. W. = New Supplement Series Volume

Formula	Compound	Volume	Reference
O_6STh	$ThO_2SO_4 \cdot 3\,H_2O$	44 (Th):	Hb–293
O_6SU	UO_2SO_4	55 (U):	Hb–151/2, 246/54
–	$UO_2SO_4 \cdot H_2O$	55 (U):	Hb–154
–	$UO_2SO_4 \cdot 2\,H_2O$	55 (U):	Hb–152
–	$UO_2SO_4 \cdot 3\,H_2O$	55 (U):	Hb–153/4
–	$UO_2SO_4 \cdot 10\,H_2O$	55 (U):	Hb–153
–	UO_2SO_4 solutions:		
	$UO_2SO_4-H_2O$	55 (U):	Hb–154/5, 246/57
–	UO_2SO_4 systems:		
	$UO_2SO_4-Ag_2SO_4-H_2O$	61 (Ag):	Hb/B4–358
	$UO_2SO_4-CuSO_4-H_2SO_4-H_2O$	60 (Cu):	Hb/B3–1241
	$UO_2SO_4-CuSO_4-H_2SO_4-D_2O$	60 (Cu):	Hb/B3–1241
	$UO_2SO_4-CuSO_4-H_2SO_4$ $-NiSO_4-H_2O$	60 (Cu):	Hb/B3–1262
	$UO_2SO_4-H_2SO_4-H_2O$	55 (U):	Hb–152/3
	$UO_2SO_4-K_2SO_4-H_2O$	55 (U):	Hb–205
	$UO_2SO_4-(NH_4)_2SO_4-H_2O$	55 (U):	Hb–216
	$UO_2SO_4-Na_2SO_4-H_2O$	55 (U):	Hb–193
O_6SV_2	$2\,VO_2 \cdot SO_2 \cdot nH_2O$	48 (V):	Hb/B1–281
O_6SW	$WO_3 \cdot SO_3$	54 (W):	Hb–186
O_6S_2	S_2O_6	9 (S):	Hb/B1–338, 351/2
$O_6S_2^{2-}$	$[S_2O_6]^{2-}$ see *$H_2O_6S_2$ ... $H_2S_2O_6$*		
O_6S_2Se	$(SeSO_3)SO_3$	10 (Se):	Hb/B–179
$O_6S_2Se^{2-}$	$[SeS_2O_6]^{2-}$	10 (Se):	Hb/B–186/9
$O_6S_2Se_2$	$(SeSO_3)_2$	10 (Se):	Hb/B–179
$O_6S_2Se_2^{2-}$	$[Se_2S_2O_6]^{2-}$	10 (Se):	Hb/B–187, 190/1
$O_6S_2Se_3^{2-}$	$[Se_3S_2O_6]^{2-}$	10 (Se):	Hb/B–187, 190
O_6S_2Sr	SrS_2O_6	29 (Sr):	Hb–183/4 Eb–259
–	$SrS_2O_6 \cdot H_2O$	29 (Sr):	Eb–259
–	$SrS_2O_6 \cdot 4\,H_2O$	29 (Sr):	Hb–183/4 Eb–259
–	SrS_2O_6 solid solutions:		
	$SrS_2O_6-K_2S_2O_6-Tl_2S_2O_6$	38 (Tl):	Hb/3–477
	$SrS_2O_6-Tl_2S_2O_6$	38 (Tl):	Hb/3–477
–	SrS_2O_6 systems:		
	$SrS_2O_6-H_2O$	29 (Sr):	Eb–259
	$SrS_2O_6-(NH_4)_2S_2O_6-H_2O$	29 (Sr):	Hb–234
	$SrS_2O_6-Na_2S_2O_6-H_2O$	29 (Sr):	Hb–224
	$SrS_2O_6-PbS_2O_6-H_2O$	47 (Pb):	Hb/C3–1198
O_6S_2Th	$Th(SO_3)_2 \cdot H_2O$	44 (Th):	Hb–283
–	$Th(SO_3)_2 \cdot 4\,H_2O$	44 (Th):	Hb–283
$O_6S_2Tl_2$	$Tl_2S_2O_6$	38 (Tl):	Hb/3–368
–	$Tl_2S_2O_6$ solid solutions:		
	$Tl_2S_2O_6 \cdot 2\,H_2O-BaS_2O_6 \cdot 2\,H_2O$	30 (Ba):	Hb–284
	$Tl_2S_2O_6-K_2S_2O_6-SrS_2O_6$	38 (Tl):	Hb/3–477
	$Tl_2S_2O_6-SrS_2O_6$	38 (Tl):	Hb/3–477

See the "Instructions for the user" at the beginning of the volume

Formula	Compound	Volume	Reference
$O_6S_2Tl_2$	$Tl_2S_2O_6$ systems:		
	$Tl_2S_2O_6-K_2S_2O_6-H_2O$	38 (Tl):	Hb/3–459
$O_6S_2Tl_2Zn$	$Tl_2[Zn(SO_3)_2] = Tl_2SO_3 \cdot ZnSO_3$	38 (Tl):	Hb/3–479
O_6S_2U	$U(SO_3)_2$	55 (U):	Hb–145
O_6S_2Zn	$ZnS_2O_6 \cdot 6\,H_2O$	32 (Zn):	Hb–243 Eb–966
–	$ZnS_2O_6 \cdot nH_2O$	32 (Zn):	Eb–966
O_6S_2Zr	$Zr(SO_3)_2 \cdot 7\,H_2O$	42 (Zr):	Hb–333
$O_6S_3^{2-}$	$[S_3O_6]^{2-}$	9 (S):	Hb/B2–856, 979/92, 1031/2, 1041
$O_6S_3Se^{2-}$	$[SeS_3O_6]^{2-}$	10 (Se):	Hb/B–189/91
$O_6S_3Tl_2$	$Tl_2S_3O_6$	38 (Tl):	Hb/3–369
O_6S_3Zn	ZnS_3O_6	32 (Zn):	Hb–243 Eb–966
$O_6S_4^{2-}$	$[S_4O_6]^{2-}$	9 (S):	Hb/B2–992/1011, 1032
$O_6S_4Se^{2-}$	$[SeS_4O_6]^{2-}$	10 (Se):	Hb/B–190/1
O_6S_4Sr	$SrS_4O_6 \cdot 2\,H_2O$	29 (Sr):	Hb–184
–	$SrS_4O_6 \cdot 6\,H_2O$	29 (Sr):	Hb–184
O_6S_4Th	$Th(S_2O_3)_2$	44 (Th):	Hb–293
$O_6S_4Tl_2^{2-}$	$[Tl_2(S_2O_3)_2]^{2-}$	38 (Tl):	Hb/3–368
O_6S_4Zn	ZnS_4O_6	32 (Zn):	Hb–243 Eb–966
O_6S_4Zr	$Zr(S_2O_3)_2$	42 (Zr):	Hb–353
$O_6S_5^{2-}$	$[S_5O_6]^{2-}$	9 (S):	Hb/B2–1011/24, 1032/3, 1041/2
O_6S_5Zn	ZnS_5O_6	32 (Zn):	Hb–243 Eb–966
$O_6S_6^{2-}$	$[S_6O_6]^{2-}$	9 (S):	Hb/B2–1024/8, 1033, 1036/46
O_6S_6Sr	$SrS_6O_6 \cdot 2\,H_2O$	29 (Sr):	Eb–260
O_6Sb_2Sr	$Sr(SbO_3)_2$	29 (Sr):	Hb–220 Eb–294/5
–	– (doped with Nb_2O_5)	49 (Nb):	Hb/B3–31
–	$Sr(SbO_3)_2 \cdot 6\,H_2O$	29 (Sr):	Hb–220
O_6Sb_2Zn	$ZnSb_2O_6 = Zn(SbO_3)_2$	32 (Zn):	Hb–290/1 Eb–999/1000
–	$Zn(SbO_3)_2 \cdot nH_2O$	32 (Zn):	Hb–290/1
$O_6ScTaTi$	$ScTiTaO_6$	50 (Ta):	Hb/B2–227
$O_6Sc_2Sm_2$	$Sc_2O_3 \cdot Sm_2O_3 = SmScO_3$	39 (SE):	Hb/C1–113, 124, 321/2
$O_6Sc_2Y_2$	$Sc_2O_3 \cdot Y_2O_3 = YScO_3$	39 (SE):	Hb/C1–113, 124/5, 127, 205
O_6SeTi_2	$2\,TiO_2 \cdot SeO_2 \cdot H_2O$	41 (Ti):	Hb–353
O_6SeU	$UO_2SeO_4 \cdot xH_2O$	55 (U):	Hb–158
O_6Se_2Th	$Th(SeO_3)_2 \cdot H_2O$	44 (Th):	Hb–294
–	$Th(SeO_3)_2 \cdot 8\,H_2O$	44 (Th):	Hb–294
O_6Se_2U	$U(SeO_3)_2$	55 (U):	Hb–157
O_6Se_2Zr	$Zr(SeO_3)_2$	42 (Zr):	Hb–354
–	$Zr(SeO_3)_2 \cdot H_2O$	42 (Zr):	Hb–354
O_6SiZn_4	$ZnSiO_3 \cdot 3\,ZnO$	32 (Zn):	Eb–993

Hb = Main Series Volume; GD = Gmelin-Durrer; Ab = Appended Volume to Main Series; Eb = Supplement Volume to Main Series; Erg. W. = New Supplement Series Volume

$O_6Si_2^{4-}$	$[Si_2O_6]^{4-}$	15 (Si):	Hb/B-421/2
O_6Si_2Th	$ThSi_2O_6$	44 (Th):	Hb-313
O_6SmSr_2Ta	Sr_2SmTaO_6	50 (Ta):	Hb/B2-214
$O_6SmTaTi$	$SmTiTaO_6$	50 (Ta):	Hb/B2-227/8
$O_6Sn_{0.4}Sr_{1.2}Ta_{1.6}$	$Sr_{1.2}(Ta_{0.8}Sn_{0.2})_2O_6$		
	$= 3\ SrO \cdot SnO_2 \cdot 2\ Ta_2O_5$	50 (Ta):	Hb/B2-232
O_6SnTa_2	$SnTa_2O_6$	50 (Ta):	Hb/B2-231
O_6Sn_5	Sn_5O_6	46 (Sn):	Hb/C1-58/9
O_6SrTa_2	$SrTa_2O_6$	50 (Ta):	Hb/B2-181, 182
–	$SrTa_2O_6$ solid solutions:		
	$(Sr,Ba,Pb)Ta_2O_6$	50 (Ta):	Hb/B2-239
	$(Sr,Pb)Ta_2O_6$	50 (Ta):	Hb/B2-239
–	$SrTa_2O_6$ systems:		
	$SrTa_2O_6-SrZrO_3$	50 (Ta):	Hb/B2-228/9
O_6SrV_2	$Sr(VO_3)_2 = SrO \cdot V_2O_5$	48 (V):	Hb/B2-520
–	$Sr(VO_3)_2 \cdot 4\ H_2O$	48 (V):	Hb/B2-520/1
$O_6Sr_{1.2}Ta_{1.6}Zr_{0.4}$	$Sr_{1.2}(Zr_{0.2}Ta_{0.8})_2O_6$		
	$= 3\ SrO \cdot ZrO_2 \cdot 2\ Ta_2O_5$	50 (Ta):	Hb/B2-228/9
O_6Sr_2TaTb	Sr_2TbTaO_6	50 (Ta):	Hb/B2-217
O_6Sr_2TaTm	Sr_2TmTaO_6	50 (Ta):	Hb/B2-220
O_6Sr_2TaY	Sr_2YTaO_6	50 (Ta):	Hb/B2-207
O_6Sr_2TaYb	Sr_2YbTaO_6	50 (Ta):	Hb/B2-221
$O_6TaTbTi$	$TbTiTaO_6$ solid solutions:		
	$TbTi(Ta,Nb)O_6$	50 (Ta):	Hb/B2-264
O_6TaTiY	$YTiTaO_6$	50 (Ta):	Hb/B2-227
$O_6TaTiYb$	$YbTiTaO_6$	50 (Ta):	Hb/B2-227/8
$O_6Ta_2^-$	$[Ta_2O_6]^-$	50 (Ta):	Hb/B1-59
O_6Ta_2V	VTa_2O_6	50 (Ta):	Hb/B2-244
O_6Ta_2Zn	$ZnTa_2O_6$	50 (Ta):	Hb/B2-193/4, 349
–	$ZnTa_2O_6$ solid solutions:		
	$ZnTa_2O_6-BaTiO_3$	50 (Ta):	Hb/B2-226
–	$ZnTa_2O_6$ systems:		
	$ZnTa_2O_6-GeO_2$	50 (Ta):	Hb/B2-230
	$ZnTa_2O_6-SnO_2$	50 (Ta):	Hb/B2-232
	$ZnTa_2O_6-TiO_2$	50 (Ta):	Hb/B2-226
O_6TeZn_3	Zn_3TeO_6	32 (Zn):	Hb-248
O_6ThTi_2	$ThO_2 \cdot 2\ TiO_2$	44 (Th):	Hb-353/4
$O_6Ti_2Zn_2$	$Zn_2TiO_4 \cdot TiO_2 = ZnTiO_3$	41 (Ti):	Hb-460, 461/2
$O_6Tl_2V_2$	$Tl_2O \cdot V_2O_5 = TlVO_3$	48 (V):	Hb/B2-548
O_6UW	$UO_3 \cdot WO_3 \cdot 2\ H_2O$	55 (U):	Hb-244
O_6V^{3-}	$[V(O)_2(O_2)_2]^{3-}$	48 (V):	Hb/A2-588/9 Hb/B1-162/3, 165 Hb/B2-424
O_6V_2Zn	$ZnO \cdot V_2O_5 = Zn(VO_3)_2$	48 (V):	Hb/B2-534, 535
–	$Zn(VO_3)_2 \cdot 2\ H_2O$	48 (V):	Hb/B2-535
$O_6V_3^+$	$[V_3O_6]^+$	48 (V):	Hb/B1-84
O_6Xe^{4-}	$[XeO_6]^{4-}$	Erg. W. 1-109	

See the "Instructions for the user" at the beginning of the volume

O_6Y_2	$Y_2(O_2)_3$	39 (SE):	Hb/C1–204/5
$O_{6.3}Ta_2Tl_2$	$Tl_2Ta_2O_{6.3}$	50 (Ta):	Hb/B2–201
O_7^+	O_7^+	3 (O):	Hb/3–354
O_7^-	O_7^-	3 (O):	Hb/3–354
$O_7P_2^{4-}$	$[P_2O_7]^{4-}$	16 (P):	Hb/B–348/9, 377/9
			Hb/C–224, 228/39
O_7P_2Pb	PbP_2O_7	47 (Pb):	Hb/C3–877/8
$O_7P_2Pb^{2-}$	$[PbP_2O_7]^{2-}$	16 (P):	Hb/C–238/9
		47 (Pb):	Hb/B1–367
			Hb/C3–874
$O_7P_2Pb_2$	$Pb_2P_2O_7$	47 (Pb):	Hb/C3–873
–	$Pb_2P_2O_7$ solutions:		
	$Pb_2P_2O_7–H_2O$	47 (Pb):	Hb/B2–371/80
			Hb/C3–874
O_7P_2Pt	PtP_2O_7	68 (Pt):	Hb/C1–137
O_7P_2Pu	PuP_2O_7	Erg. W. 4–252/3	
$O_7P_2Rb_4$	$Rb_4P_2O_7$	24 (Rb):	Hb–234
O_7P_2Si	$P_2O_5 \cdot SiO_2$	16 (P):	Hb/C–638/9
–	$P_2O_5 \cdot SiO_2 \cdot H_2O$	16 (P):	Hb/C–640
–	$P_2O5 \cdot SiO_2 \cdot 2\ H_2O$	16 (P):	Hb/C–640
$O_7P_2Sn^{2-}$	$[SnP_2O_7]^{2-}$	16 (P):	Hb/C–239
		46 (Sn):	Hb/B–387/8
$O_7P_2Sn_2$	$Sn_2P_2O_7$	46 (Sn):	Hb/B–329/31
$O_7P_2Sr^{2-}$	$[SrP_2O_7]^{2-}$	16 (P):	Hb/C–238
$O_7P_2Sr_2$	$Sr_2P_2O_7$	29 (Sr):	Hb–215
			Eb–285/6
–	$Sr_2P_2O_7 \cdot H_2O$	29 (Sr):	Hb–215
–	$Sr_2P_2O_7 \cdot 2.5\ H_2O$	29 (Sr):	Hb–215
O_7P_2Th	$ThP_2O_7 \cdot 2\ H_2O$	44 (Th):	Hb–318/9
O_7P_2Ti	$TiO_2 \cdot P_2O_5$	41 (Ti):	Hb–380
$O_7P_2Tl^{3-}$	$[TlP_2O_7]^{3-}$	16 (P):	Hb/C–238
$O_7P_2Tl_4$	$Tl_4P_2O_7$	38 (Tl):	Hb/3–430/1
–	$Tl_4P_2O_7 \cdot 2\ H_2O$	38 (Tl):	Hb/3–431
O_7P_2U	UP_2O_7	55 (U):	Hb–182
–	$UP_2O_7 \cdot 3\ H_2O$	55 (U):	Hb–181
$O_7P_2Zn^{2-}$	$[ZnP_2O_7]^{2-}$	16 (P):	Hb/C–238, 239
$O_7P_2Zn_2$	$Zn_2P_2O_7$	32 (Zn):	Hb–279
			Eb–996
–	$Zn_2P_2O_7 \cdot 5\ H_2O$	32 (Zn):	Eb–996
O_7P_2Zr	$ZrP_2O_7 = ZrO_2 \cdot P_2O_5$	42 (Zr):	Hb–396/8
–	$ZrO_2 \cdot P_2O_5 \cdot 1.5\ H_2O$	42 (Zr):	Hb–393
–	$ZrO_2 \cdot P_2O_5 \cdot 2\ H_2O$	42 (Zr):	Hb–393
–	$ZrO_2 \cdot P_2O_5 \cdot nH_2O$	42 (Zr):	Hb–393
O_7PbS_2	PbS_2O_7	47 (Pb):	Hb/C2–600/1
O_7PbTi_3	$PbTi_3O_7$	47 (Pb):	Hb/C4–1390
O_7PbU_2	$PbO \cdot 2\ UO_3$	55 (U):	Hb–239/40
–	$PbO \cdot 2\ UO_3 \cdot 2.5\ H_2O$	55 (U):	Hb–240
O_7PbW_2	$PbO \cdot 2\ WO_3 \cdot 2\ H_2O$	54 (W):	Hb–319

$O_7Pb_2Pr_2$	$Pr_2Pb_2O_7$	47 (Pb):	Hb/C4–1373
$O_7Pb_2Rb_2Si_2$	$Rb_2Pb_2Si_2O_7$	47 (Pb):	Hb/C3–1145
$O_7Pb_2Ru_2$	$Pb_2Ru_2O_7$	63 (Ru):	Eb–259
$O_7Pb_2Sb_2$	$Pb_2Sb_2O_7$	47 (Pb):	Hb/C3–932
–	$Pb_2Sb_2O_7 \cdot nH_2O$	47 (Pb):	Hb/C3–932
$O_7Pb_2Se_2$	$Pb_2Se_2O_7$	47 (Pb):	Hb/C2–636
$O_7Pb_2Si_2Zn$	$Pb_2ZnSi_2O_7$	47 (Pb):	Hb/C4–1229
$O_7Pb_2Sm_2$	$Sm_2Pb_2O_7$	47 (Pb):	Hb/C4–1373
$O_7Pb_2Ta_2$	$Pb_2Ta_2O_7$	50 (Ta):	Hb/B2–232/3, 237/8
$O_7Pb_2Tb_2$	$Tb_2Pb_2O_7$	47 (Pb):	Hb/C4–1373
$O_7Pb_2V_2$	$2\ PbO \cdot V_2O_5 = Pb_2V_2O_7$	48 (V):	Hb/B2–567/8, 569
–	$2\ PbO \cdot V_2O_5 \cdot xH_2O = Pb_2V_2O_7 \cdot xH_2O$	48 (V):	Hb/B2–569/70
$O_7Pb_2Y_2$	$Y_2Pb_2O_7$	47 (Pb):	Hb/C4–1373
$O_7Pb_3Sb_2$	$Pb_3Sb_2O_7 = 3\ PbO \cdot Sb_2O_4$	47 (Pb):	Hb/C3–930
$O_7Pb_3Si_2$	$Pb_3Si_2O_7$	47 (Pb):	Hb/C3–847/8
–	$Pb_3Si_2O_7$ solid solutions: $Pb_9(Si_2O_7)_3$–$Pb_5(SiO_4)_2SO_4$	47 (Pb):	Hb/C3–857
O_7Pb_4S	$PbSO_4 \cdot 3\ PbO \cdot H_2O$	47 (Pb):	Hb/C2–589/91
O_7Pb_5Se	$PbSeO_3 \cdot 4\ PbO$	47 (Pb):	Hb/C2–635/6
$O_7Pr_2Ru_2$	$Pr_2Ru_2O_7$	63 (Ru):	Eb–258
$O_7Pu_2Zr_2$	$Pu_2Zr_2O_7$	Erg. W. 4–70	
$O_7Rb_2S_2$	$Rb_2S_2O_7 = Rb_2SO_4 \cdot SO_3$	24 (Rb):	Hb–214
$O_7Rb_4V_2$	$2\ Rb_2O \cdot V_2O_5 = Rb_4V_2O_7$	48 (V):	Hb/B2–486
O_7Re_2	Re_2O_7	70 (Re):	Hb–93/5
$O_7Ru_2Sm_2$	$Sm_2Ru_2O_7$	63 (Ru):	Eb–258
$O_7Ru_2Tb_2$	$Tb_2Ru_2O_7$	63 (Ru):	Eb–258
$O_7Ru_2Tm_2$	$Tm_2Ru_2O_7$	63 (Ru):	Eb–258
$O_7Ru_2Y_2$	$Y_2Ru_2O_7$	63 (Ru):	Eb–258
$O_7Ru_2Yb_2$	$Yb_2Ru_2O_7$	63 (Ru):	Eb–258
O_7SSe^{2-}	$[SeSO_7]^{2-}$	10 (Se):	Hb/B–185
O_7STe_2	$2\ TeO_2 \cdot SO_3$	11 (Te):	Hb–359/60
O_7STi_2	$2\ TiO_2 \cdot SO_3$	41 (Ti):	Hb–349, 351
–	$2\ TiO_2 \cdot SO_3 \cdot xH_2O$	41 (Ti):	Hb–349
O_7SZn_4	$ZnSO_4 \cdot 3\ ZnO \cdot H_2O$	32 (Zn):	Hb–241
–	$ZnSO_4 \cdot 3\ ZnO \cdot 2\ H_2O$	32 (Zn):	Hb–241
–	$ZnSO_4 \cdot 3\ ZnO \cdot 3\ H_2O$	32 (Zn):	Hb–241
–	$ZnSO_4 \cdot 3\ ZnO \cdot 5\ H_2O$	32 (Zn):	Hb–241
–	$ZnSO_4 \cdot 3\ ZnO \cdot 6\ H_2O$	32 (Zn):	Hb–241
–	$ZnSO_4 \cdot 3\ ZnO \cdot 7\ H_2O$	32 (Zn):	Hb–241
–	$ZnSO_4 \cdot 3\ ZnO \cdot 8\ H_2O$	32 (Zn):	Hb–241
–	$ZnSO_4 \cdot 3\ ZnO \cdot xH_2O$	32 (Zn):	Eb–956
O_7SZr_2	$ZrOSO_4 \cdot ZrO_2 = Zr(SO_4)_2 \cdot 3\ ZrO_2$	42 (Zr):	Hb–350
–	$ZrOSO_4 \cdot ZrO_2 \cdot xH_2O$	42 (Zr):	Hb–350/1
$O_7S_{1.5}V_2$	$2\ VO_2 \cdot 1.5\ SO_2 \cdot nH_2O$	48 (V):	Hb/B1–281
O_7S_2	S_2O_7	9 (S):	Hb/B1–367/70
–	$(S_2O_7)_x$	9 (S):	Hb/B1–367/70
O_7S_2Sr	SrS_2O_7	29 (Sr):	Hb–181
$O_7S_2Tl_2$	$Tl_2S_2O_7 = Tl_2SO_4 \cdot SO_3$	38 (Tl):	Hb/3–367

See the "Instructions for the user" at the beginning of the volume

O_7S_2V — $VOS_2O_6 \cdot xH_2O$ — 48 (V): Hb/B1–307
O_7S_2Zn — ZnS_2O_7 — 32 (Zn): Hb–241
$O_7S_2Zn_3$ — $ZnO \cdot 2\,ZnSO_3 \cdot 3\,H_2O$ — 32 (Zn): Eb–935
$O_7Sb_2Se_2$ — $Sb_2O_3 \cdot 2\,SeO_2 \cdot 0.5\,H_2O$ — 18 (Sb): Hb/B3–553
$O_7Sb_2Sr_2$ — $Sr_2Sb_2O_7$ — 29 (Sr): Eb–294
O_7Sb_2Zr — $ZrO_2 \cdot Sb_2O_5 \cdot 7.5\,H_2O$ — 42 (Zr): Hb–400
O_7Sb_4 — Sb_4O_7 — 18 (Sb): Hb/B2–359/60
O_7Sc_3Ta — Sc_3TaO_7 — 50 (Ta): Hb/B2–205
O_7SeTe_2 — $2\,TeO_2 \cdot SeO_3$ — 11 (Te): Hb–363
O_7SeTi_2 — $2\,TiO_2 \cdot SeO_3 \cdot H_2O$ — 41 (Ti): Hb–353
$O_7Si_2Tl_6$ — $3\,Tl_2O \cdot 2\,SiO_2 \cdot H_2O$ — 38 (Tl): Hb/3–424
$O_7Sm_2Ti_2$ — $Sm_2Ti_2O_7$ — 41 (Ti): Hb–468
O_7Sm_3Ta — Sm_3TaO_7 — 50 (Ta): Hb/B2–214
O_7Sn_2 — $Sn_2O_7 \cdot 3\,H_2O$ — 46 (Sn): Hb/C1–139
– — $Sn_2O_7 \cdot 5\,H_2O$ — 46 (Sn): Hb/C1–139
$O_7Sn_2Sr_3$ — $Sr_3Sn_2O_7$ — 29 (Sr): Eb–160
$O_7Sn_2Ta_2$ — $Sn_2Ta_2O_7$ — 50 (Ta): Hb/B2–231/2
– — $Sn_2Ta_2O_7$ solid solutions:
  $Sn_2(Ta,Nb)_2O_7$ — 50 (Ta): Hb/B2–264
O_7SrU_2 — $SrO \cdot 2\,UO_3$ — 55 (U): Hb–232
O_7SrW_2 — $SrO \cdot 2\,WO_3 \cdot 3\,H_2O$ — 54 (W): Hb–296
$O_7Sr_2Ta_2$ — $Sr_2Ta_2O_7$ — 50 (Ta): Hb/B2–181, 183
– — $Sr_2Ta_2O_7$ solid solutions:
  $(Sr,Ba)_2Ta_2O_7$ — 50 (Ta): Hb/B2–192/3
  $(Sr,Ca)_2Ta_2O_7$ — 50 (Ta): Hb/B2–185
  $Sr_2Ta_2O_7{-}NaNbO_3$ — 50 (Ta): Hb/B2–262
  $Sr_2(Ta,Nb)_2O_7$ — 50 (Ta): Hb/B2–262
$O_7Sr_2V_2$ — $2\,SrO \cdot V_2O_5 = Sr_2V_2O_7$ — 48 (V): Hb/B2–519/20
– — $Sr_2V_2O_7 \cdot 10\,H_2O$ — 48 (V): Hb/B2–520
O_7TaY_3 — Y_3TaO_7 — 50 (Ta): Hb/B2–205, 206
$O_7Ta_2^-$ — $[Ta_2O_7]^-$ — 50 (Ta): Hb/B1–59
O_7Ta_2Ti — $TiTa_2O_7$ — 50 (Ta): Hb/B2–221/4
– — $TiTa_2O_7$ solid solutions:
  $Ti(Ta,Nb)_2O_7$ — 50 (Ta): Hb/B2–264
$O_7Ta_2Tl_2$ — $Tl_2Ta_2O_7$ — 50 (Ta): Hb/B2–201
$O_7Ta_2Zn_2$ — $Zn_2Ta_2O_7$ — 50 (Ta): Hb/B2–194
O_7Tb_4 — $Tb_4O_7 = TbO_{1.75}$ — 39 (SE): Hb/C1–130, 156, 397/8
– — $Tb_4O_7 \cdot nH_2O$ — 39 (SE): Hb/C2–2
O_7Tc_2 — Tc_2O_7 — 14 (C): Hb/C2–168
O_7TeZn_4 — $4\,ZnO \cdot TeO_3$ — 32 (Zn): Eb–970
– — $ZnTeO_4 \cdot 3\,ZnO \cdot 4\,H_2O$ — 32 (Zn): Eb–970
O_7Te_3 — $Te_3O_7 = 2\,TeO_2 \cdot TeO_3$ — 11 (Te): Hb–282
O_7ThV_2 — $ThO_2 \cdot V_2O_5 = ThV_2O_7$ — 48 (V): Hb/B2–565
– — $ThV_2O_7 \cdot H_2O$ — 48 (V): Hb/B2–565
O_7Th_2 — $Th_2O_7 \cdot 4\,H_2O$ — 44 (Th): Hb–240
$O_7Ti_2Y_2$ — $Y_2Ti_2O_7$ — 41 (Ti): Hb–468
$O_7Ti_2Zn_3$ — $3\,ZnO \cdot 2\,TiO_2$ — 41 (Ti): Hb–462

Hb = Main Series Volume; GD = Gmelin-Durrer; Eb = Supplement Volume to Main Series; Ab = Appended Volume to Main Series; Erg. W. = New Supplement Series Volume

Formula	Compound	Volume	Reference
O_7Ti_3Zn	$ZnO \cdot 3\,TiO_2$	41 (Ti):	Hb–462
$O_7Tl_2W_2$	$Tl_2O \cdot 2\,WO_3 \cdot H_2O$	54 (W):	Hb–314
$O_7Tl_4V_2$	$2\,Tl_2O \cdot V_2O_5 = Tl_4V_2O_7$	48 (V):	Hb/B2–547
O_7U_3	U_3O_7 systems:		
	$U_3O_7-PuO_2$	Erg. W. 4–60	
O_7V^{3-}	$[V(O)(O_2)_3]^{3-}$	48 (V):	Hb/B1–162/3, 164
$O_7V_2{}^{4-}$	$[V_2O_7]^{4-}$	48 (V):	Hb/A2–498 Hb/B1–132/45, 146/8
$O_7V_2Zn_2$	$2\,ZnO \cdot V_2O_5 = Zn_2V_2O_7$	48 (V):	Hb/B2–534, 535
O_7V_2Zr	$ZrO_2 \cdot V_2O_5 = ZrV_2O_7$	48 (V):	Hb/B2–562, 563
O_7V_3	$V_3O_7 \cdot 3\,H_2O$	48 (V):	Hb/B1–70
O_7V_4	$V_4O_7 = VO_{1.75}$	48 (V):	Hb/B1–13, 30, 50/1
O_7W_2Zn	$ZnO \cdot 2\,WO_3 \cdot 3\,H_2O$	54 (W):	Hb–306
O_7Zn_4	Zn_4O_7	32 (Zn):	Hb–145
–	$Zn_4O_7 \cdot 2\,H_2O$	32 (Zn):	Hb–146
$O_{7.5}SZn_{4.5}$	$ZnSO_4 \cdot 3.5\,ZnO \cdot 3\,H_2O$	32 (Zn):	Eb–963/4
–	$ZnSO_4 \cdot 3.5\,ZnO \cdot 7\,H_2O$	32 (Zn):	Hb–240
–	$ZnSO_4 \cdot 3.5\,ZnO \cdot 12\,H_2O$	32 (Zn):	Hb–240
$O_8P_2{}^{4-}$	$[P_2O_8]^{4-}$	16 (P):	Hb/C–300/2
$O_8P_2PbZn_2$	$Zn_2Pb(PO_4)_2$	47 (Pb):	Hb/C1–117
$O_8P_2Pb_2$	$Pb_2P_2O_8$	47 (Pb):	Hb/C3–879
$O_8P_2Pb_3$	$Pb_3(PO_4)_2$	47 (Pb):	Hb/C3–865/9
–	$Pb_3(PO_4)_2$ systems:		
	$Pb_3(PO_4)_2-BaTiO_3$	47 (Pb):	Hb/C4–1413
	$Pb_3(PO_4)_2-PbCl_2$	47 (Pb):	Hb/C3–881
	$Pb_3(PO_4)_2-PbF_2$	47 (Pb):	Hb/C3–880
	$Pb_3(PO_4)_2-PbZrO_3$	47 (Pb):	Hb/C4–1431
$O_8P_2Rb_4$	$Rb_4P_2O_8$	24 (Rb):	Hb–234
O_8P_2SrTh	$SrO \cdot ThO_2 \cdot P_2O_5 = Sr[Th(PO_4)_2]$	44 (Th):	Hb–348
O_8P_2SrU	$SrU(PO_4)_2$	55 (U):	Hb–232
$O_8P_2Sr_3$	$Sr_3(PO_4)_2$	29 (Sr):	Hb–212/3 Eb–283/4
–	$Sr_3(PO_4)_2$ systems:		
	$Sr_3(PO_4)_2-SrCl_2$	29 (Sr):	Hb–216
	$Sr_3(PO_4)_2-SrF_2$	29 (Sr):	Hb–216
O_8P_2U	$UO_2(PO_3)_2$	55 (U):	Hb–184
$O_8P_2Zn_2$	$Zn_2P_2O_8 \cdot 4\,H_2O$	32 (Zn):	Eb–997
$O_8P_2Zn_3$	$Zn_3(PO_4)_2$	32 (Zn):	Hb–277/8 Eb–995
–	$Zn_3(PO_4)_2 \cdot 4\,H_2O$	32 (Zn):	Hb–277/8 Eb–995/6
$O_8PbRb_2S_2$	$Rb_2Pb(SO_4)_2$	47 (Pb):	Hb/C3–1144
$O_8PbRb_2Se_2$	$Rb_2Pb(SeO_4)_2$	47 (Pb):	Hb/C3–1144
O_8PbS_2	$Pb(SO_4)_2$	47 (Pb):	Hb/C2–580/1
–	$PbS_2O_8 \cdot 3\,H_2O$	47 (Pb):	Hb/C2–601
$O_8PbS_2Tl_2$	$Tl_2Pb(SO_4)_2$	47 (Pb):	Hb/C4–1353
$O_8PbSe_2Tl_2$	$Tl_2Pb(SeO_4)_2$	47 (Pb):	Hb/C4–1353

See the "Instructions for the user"
at the beginning of the volume

O_8PbV_2 . . . . $Pb(VO_4)_2$ . . . . 48 (V): Hb/B2–570/1
$O_8Pb_3S_2{}^{2+}$ . . . . . . $[Pb_3(SO_4)_2]^{2+}$ . . . . 47 (Pb): Hb/C2–580
$O_8Pb_3Sb_2$ . . . . $Pb_3(SbO_4)_2$ . . . . 47 (Pb): Hb/C3–931
$O_8Pb_3Ta_2$ . . . . $Pb_3Ta_2O_8$ . . . . 50 (Ta): Hb/B2–232/3, 238
$O_8Pb_3V_2$. . . . . $3\ PbO \cdot V_2O_5 = Pb_3(VO_4)_2$ . . . . 48 (V): Hb/B2–568/9
–. . . . . $Pb_3(VO_4)_2$ solid solutions:
$(Pb,Pr)_3(VO_4)_2$ . . . . 48 (V): Hb/B2–569
–. . . . . $Pb_3(VO_4)_2$ systems:
$Pb_3(VO_4)_2–LiCl–Li_3VO_4–PbCl_2$. . . . . 48 (V): Hb/B2–573
$Pb_3(VO_4)_2–PbCl_2$ . . . . 48 (V): Hb/B2–571
O_8Pb_5 . . . . Pb_5O_8. 47 (Pb): Hb/C1–144/7
O_8Pb_5S . . . . $PbSO_4 \cdot 4\ PbO$ . . . . 47 (Pb): Hb/C2–591/3
$O_8Pb_5Sb_2$ . . . . $Pb_5(SbO_4)_2 = 5\ PbO \cdot Sb_2O_3$ . . . . 47 (Pb): Hb/C3–929
O_8Pb_5Se . . . . $PbSeO_4 \cdot 4\ PbO$ . . . . 47 (Pb): Hb/C2–638/9
O_8Pb_6Si . . . . $6\ PbO \cdot SiO_2 \cdot 4\ H_2O$ . . . . 47 (Pb): Hb/C3–839
$O_8Pr_3V_2$ . . . . $Pr_3(VO_4)_2$ solid solutions:
$(Pb,Pr)_3(VO_4)_2$ . . . . 48 (V): Hb/B2–569
O_8PtS_2 . . . . $Pt(SO_4)_2$ . . . . 68 (Pt): Hb/C1–111
–. . . . . $Pt(SO_4)_2 \cdot 4\ H_2O$
$= H_2[Pt(SO_4)_2(OH)_2] \cdot 2\ H_2O$ . . . . . . 68 (Pt): Hb/C1–111/2
O_8PuRbS_2 . . . . . . $RbPu(SO_4)_2 \cdot 4\ H_2O$ . . . . Erg. W. 4–176/7
O_8PuS_2 . . . . $Pu(SO_4)_2$ . . . . Erg. W. 4–178/9, 184
–. . . . . $Pu(SO_4)_2 \cdot 0.33\ H_2O$ . . . . Erg. W. 4–181/2
–. . . . . $Pu(SO_4)_2 \cdot 4\ H_2O$ . . . . Erg. W. 4–179/82
–. . . . . $Pu(SO_4)_2 \cdot xH_2O$ . . . . Erg. W. 4–179
O_8PuS_2Tl . . . . $TlPu(SO_4)_2 \cdot 4\ H_2O$ . . . . Erg. W. 4–176, 178
O_8RbS_2Sb . . . . . . $Rb[Sb(SO_4)_2]$ . . . . 24 (Rb): Hb–239
O_8RbS_2Ti . . . . $RbTi(SO_4)_2 \cdot 12\ H_2O$ . . . . 41 (Ti): Hb–413
O_8RbS_2Tl . . . . $RbTl(SO_4)_2 = Rb_2SO_4 \cdot Tl_2(SO_4)_3$ . . . . . 38 (Tl): Hb/3–469
–. . . . . $RbTl(SO_4)_2 \cdot 4\ H_2O$. . . . . 38 (Tl): Hb/3–469
O_8RbS_2V . . . . $RbV(SO_4)_2 \cdot 6\ H_2O$ . . . . 48 (V): Hb/B2–491
–. . . . . $RbV(SO_4)_2 \cdot 12\ H_2O$ . . . . 48 (V): Hb/B2–489/91
$O_8Rb_2S_2$ . . . . $Rb_2S_2O_8$ . . . . 24 (Rb): Hb–214
$O_8Rb_2S_2V$. . . . . . . $Rb_2V(SO_4)_2 \cdot 6\ H_2O$. . . . . 48 (V): Hb/B2–489
$O_8Rb_2S_2Zn$ . . . . . . $Rb_2Zn(SO_4)_2 \cdot 6\ H_2O$
$= [Zn(H_2O)_6]\ (SO_4Rb)_2$ . . . . 32 (Zn): Hb–111, 318
Eb–1020
$O_8Rb_2Se_2Zn$ . . . . . $ZnSeO_4 \cdot Rb_2SeO_4 \cdot 6\ H_2O$. . . . . 32 (Zn): Hb–319
O_8Rb_2W. . . . . $Rb_2WO_8 \cdot 3\ H_2O$ . . . . 54 (W): Hb–283
O_8Rb_3Ta . . . . Rb_3TaO_8 . . . . 50 (Ta): Hb/B2–171
O_8Rb_4U . . . . $Rb_4UO_8 \cdot 8\ H_2O$ . . . . 55 (U): Hb–226
$O_8Rb_6V_2$. . . . . $3\ Rb_2O \cdot V_2O_5 = Rb_3VO_4$ . . . . 48 (V): Hb/B2–486
O_8Re_2 . . . . Re_2O_8. 70 (Re): Hb–95/6
O_8Re_2Sr. . . . . $Sr(ReO_4)_2 \cdot 2\ H_2O$ . . . . 70 (Re): Hb–146
O_8Re_2Zn . . . . $Zn(ReO_4)_2 \cdot xH_2O$. . . . . 70 (Re): Hb–147
O_8RuS_2 . . . . $Ru(SO_4)_2$ . . . . 63 (Ru): Hb–76
O_8STa_2 . . . . $Ta_2O_4SO_4 = (TaO_2)_2SO_4$ . . . . 50 (Ta): Hb/B1–177
–. . . . . $Ta_2O_4SO_4 \cdot 3\ H_2O$ . . . . 50 (Ta): Hb/B1–177

Hb = Main Series Volume; GD = Gmelin-Durrer; Ab = Appended Volume to Main Series;
Eb = Supplement Volume to Main Series; Erg. W. = New Supplement Series Volume

O_8STa_2 . . . . $Ta_2O_5SO_3 \cdot nH_2O$ . . . . 50 (Ta): Hb/B1–176
O_8SV_2 . . . . $(VO_2)_2SO_4$ . . . . 48 (V): Hb/A2–505, 581, 637/8
–. . . . . $V_2O_5 \cdot SO_3$ . . . . 48 (V): Hb/B1–302
O_8SZn_5 . . . . $ZnSO_4 \cdot 4\,ZnO$ . . . . 32 (Zn): Hb–240
–. . . . . $ZnSO_4 \cdot 4\,ZnO \cdot 4\,H_2O$ . . . . 32 (Zn): Hb–241
$O_8S_2^{2-}$ . . . . $[S_2O_8]^{2-}$
see $H_2O_8S_2$ ... $H_2S_2O_8$ solutions

O_8S_2SbTl . . . . $Tl[Sb(SO_4)_2]$ . . . . 38 (Tl): Hb/3–437
O_8S_2Sr . . . . $SrS_2O_8 \cdot H_2O$ . . . . 29 (Sr): Hb–181
–. . . . . $SrS_2O_8 \cdot 4\,H_2O$ . . . . 29 (Sr): Hb–181
O_8S_2Th . . . . $Th(SO_4)_2$ . . . . 44 (Th): Hb–283/4
–. . . . . $Th(SO_4)_2 \cdot 2\,H_2O$ . . . . 44 (Th): Hb–289
–. . . . . $Th(SO_4)_2 \cdot 4\,H_2O$ . . . . 44 (Th): Hb–289, 291
–. . . . . $Th(SO_4)_2 \cdot 6\,H_2O$ . . . . 44 (Th): Hb–288
–. . . . . $Th(SO_4)_2 \cdot 8\,H_2O$ . . . . 44 (Th): Hb–287/8, 291
–. . . . . $Th(SO_4)_2 \cdot 9\,H_2O$ . . . . 44 (Th): Hb–286/7
–. . . . . $Th(SO_4)_2$ solid solutions:
$Th(SO_4)_2 \cdot 4\,H_2O–Ce(SO_4)_2 \cdot 4\,H_2O$ . . 44 (Th): Hb–353
$Th(SO_4)_2 \cdot 8\,H_2O–U(SO_4)_2 \cdot 8\,H_2O$ . . . 55 (U): Hb–239

–. . . . . $Th(SO_4)_2$ solutions:
$Th(SO_4)_2–H_2O$. . . . . 44 (Th): Hb–290
$Th(SO_4)_2$–organic solvents . . . . . . 44 (Th): Hb–291
$Th(SO_4)_2$–salt solutions . . . . 44 (Th): Hb–291

–. . . . . $Th(SO_4)_2$ systems:
$Th(SO_4)_2–H_2O$. . . . . 44 (Th): Hb–284/6
$Th(SO_4)_2–H_2SO_4–H_2O$ . . . . 44 (Th): Hb–286
$Th(SO_4)_2–K_2SO_4–H_2O$ . . . . 44 (Th): Hb–333
$Th(SO_4)_2–Li_2SO_4–H_2O$ . . . . 44 (Th): Hb–322
$Th(SO_4)_2–(NH_4)_2SO_4–H_2O$ . . . . 44 (Th): Hb–338/9
$Th(SO_4)_2–Na_2SO_4–H_2O$ . . . . 44 (Th): Hb–325

O_8S_2Ti . . . . $Ti(SO_4)_2$. . . . . 41 (Ti): Hb–347, 349/50
–. . . . . $Ti(SO_4)_2 \cdot 3\,H_2O$ . . . . 41 (Ti): Hb–348
–. . . . . $Ti(SO_4)_2 \cdot 4\,H_2O$ . . . . 41 (Ti): Hb–347
–. . . . . $Ti(SO_4)_2 \cdot 9\,H_2O$ . . . . 41 (Ti): Hb–347
$O_8S_2Ti^{2-}$ . . . . $[Ti(SO_4)_2]^{2-}$ . . . . 41 (Ti): Hb–468
O_8S_2TlV. . . . . $TlV(SO_4)_2 \cdot 12\,H_2O$ . . . . 48 (V): Hb/B2–550/1
$O_8S_2Tl_2$ . . . . $Tl_2S_2O_8$ . . . . 38 (Tl): Hb/3–367
$O_8S_2Tl_2Zn$ . . . . . . $Tl_2Zn(SO_4)_2 \cdot 6\,H_2O$
$= Tl_2SO_4 \cdot ZnSO_4 \cdot 6\,H_2O$. . . . . 38 (Tl): Hb/3–480/1
O_8S_2U. . . . . $U(SO_4)_2$. . . . . 55 (U): Hb–146, 245
–. . . . . $U(SO_4)_2 \cdot 2\,H_2O$ . . . . 55 (U): Hb–150
–. . . . . $U(SO_4)_2 \cdot 4\,H_2O$ . . . . 55 (U): Hb–149/50
–. . . . . $U(SO_4)_2 \cdot 8\,H_2O$ . . . . 55 (U): Hb–148/9
–. . . . . $U(SO_4)_2 \cdot 9\,H_2O$ . . . . 55 (U): Hb–147/8
–. . . . . $U(SO_4)_2$ solid solutions:
$U(SO_4)_2 \cdot 8\,H_2O–Th(SO_4)_2 \cdot 8\,H_2O$ . . . 55 (U): Hb–239
$U(SO_4)_2 \cdot 4\,H_2O–Zr(SO_4)_2 \cdot 4\,H_2O$ . . . 55 (U): Hb–239

See the "Instructions for the user" at the beginning of the volume

Formula	Compound	Gmelin Vol.	Reference
O_8S_2U	$U(SO_4)_2$ solutions:		
	$U(SO_4)_2-H_2O$	55 (U):	Hb–150/1, 245/6
O_8S_2Zn	ZnS_2O_8	32 (Zn):	Hb–241
			Eb–965
$O_8S_2Zn^{2-}$	$[Zn_2(SO_4)_4]^{4-}$	32 (Zn):	Eb–1007
O_8S_2Zr	$Zr(SO_4)_2 = ZrOSO_4 \cdot SO_3$	42 (Zr):	Hb–334/5
–	$Zr(SO_4)_2 \cdot H_2O$	42 (Zr):	Hb–338, 340
–	$Zr(SO_4)_2 \cdot 4\ H_2O$	42 (Zr):	Hb–334, 337/8
–	$Zr(SO_4)_2$ solid solutions:		
	$Zr(SO_4)_2 \cdot 4\ H_2O-U(SO_4)_2 \cdot 4\ H_2O$	55 (U):	Hb–239
–	$Zr(SO_4)_2$ solutions:		
	$Zr(SO_4)_2-CH_3OH$	42 (Zr):	Hb–342
	$Zr(SO_4)_2-H_2O$	42 (Zr):	Hb–339/42
O_8S_3	$SO(S_2O_7) = 2\ SO_3 \cdot SO_2$	9 (S):	Hb/B1–323, 358
O_8Sb_5	Sb_5O_8	18 (Sb):	Hb/B2–359
O_8Se_2Th	$Th(SeO_4)_2 \cdot 9\ H_2O$	44 (Th):	Hb–294
$O_8Se_2Tl_2Zn$	$Tl_2Zn(SeO_4)_2 \cdot 6\ H_2O$	38 (Tl):	Hb/3–481
O_8Se_2Zr	$Zr(SeO_4)_2 \cdot 4\ H_2O$	42 (Zr):	Hb–355
$O_8Si_3^{4-}$	$[Si_3O_8]^{4-}$	15 (Si):	Hb/B–421/2
$O_8Sn_3Ta_2$	$Sn_3Ta_2O_8$	50 (Ta):	Hb/B2–231/2
O_8SrV_2	$Sr(VO_4)_2$	48 (V):	Hb/B2–522
O_8SrW_2	$SrO \cdot 2\ WO_3 \cdot O \cdot 6\ H_2O$	54 (W):	Hb–297
$O_8Sr_2Ti_3$	$2\ SrO \cdot 3\ TiO_2$	41 (Ti):	Hb–434
$O_8Sr_2Zr_3$	$2\ SrO \cdot 3\ ZrO_2$	42 (Zr):	Hb–439/40
$O_8Sr_3V_2$	$3\ SrO \cdot V_2O_5 = Sr_3(VO_4)_2$	48 (V):	Hb/B1–93
			Hb/B2–519
$O_8Ta_2^-$	$[Ta_2O_8]^-$	50 (Ta):	Hb/B1–59
$O_8Ta_2Zn_3$	$Zn_3Ta_2O_8$	50 (Ta):	Hb/B2–193, 194, 349
–	$Zn_3Ta_2O_8$ solid solutions:		
	$Zn_3(Ta,Nb)_2O_8$	50 (Ta):	Hb/B2–263
O_8ThTi_2Zr	$ThO_2 \cdot ZrO_2 \cdot 2\ TiO_2$	44 (Th):	Hb–354
$O_8Ti_2V_2$	$Ti_2O_3 \cdot V_2O_5 = TiVO_4$	48 (V):	Hb/B2–559/60
$O_8Tl_2V_2$	$Tl_2O_3 \cdot V_2O_5 = TlVO_4$	48 (V):	Hb/B2–548
$O_8Tl_6V_2$	$3\ Tl_2O \cdot V_2O_5 = Tl_3VO_4$	48 (V):	Hb/B2–547
O_8UV_2	$UO_3 \cdot V_2O_5$	48 (V):	Hb/B1–33
–	$UO_3 \cdot V_2O_5 \cdot H_2O$	55 (U):	Hb–241
O_8UV_3	$3\ VO_2 \cdot UO_2$	48 (V):	Hb/B1–91
O_8U_3	U_3O_8	55 (U):	Hb–89/94
–	$U_3O_8 \cdot xH_2O$	55 (U):	Hb–100/1
–	U_3O_8 solid solutions:		
	$(U,Pu)_3O_{8-x}$	Erg. W. 4–59	
O_8V^{3-}	$[V(O_2)_4]^{3-}$	48 (V):	Hb/B1–162/3, 164
			Hb/B2–424
O_8V_2	$[V(O_2)]\ [V(O)_2(O_2)_2]$	48 (V):	Hb/B1–163, 166
$O_8V_2Zn_3$	$3\ ZnO \cdot V_2O_5 = Zn_3(VO_4)_2$	48 (V):	Hb/B2–534, 535
O_8V_4	V_4O_8	48 (V):	Hb/B1–84
O_8W_2Th	$Th(WO_4)_2$	54 (W):	Hb–316

Hb = Main Series Volume; GD = Gmelin-Durrer; Ab = Appended Volume to Main Series;
Eb = Supplement Volume to Main Series; Erg. W. = New Supplement Series Volume

O_8W_2Zr	$Zr(WO_4)_2$	54 (W):	Hb-315
O_8Zn_5	Zn_5O_8	32 (Zn):	Hb-145
$O_9P_2Pb_2$	$2\ PbO_2 \cdot P_2O_5 \cdot nH_2O$	47 (Pb):	Hb/C3-877
$O_9P_2Pb_2Ti$	$Pb_2TiP_2O_9$	47 (Pb):	Hb/C4-1394
$O_9P_2Pb_2Zr$	$Pb_2ZrP_2O_9$	47 (Pb):	Hb/C4-1432
$O_9P_2Pb_4$	$Pb_3(PO_4)_2 \cdot PbO = 4\ PbO \cdot P_2O_5$	47 (Pb):	Hb/C3-869/70, 881
$O_9P_2Si_2$	$P_2O_5 \cdot 2\ SiO_2$	16 (P):	Hb/C-639
$O_9P_2Ti_2$	$2\ TiO_2 \cdot P_2O_5$	41 (Ti):	Hb-380
–	$2\ TiO_2 \cdot P_2O_5 \cdot H_2O$	41 (Ti):	Hb-380
–	$2\ TiO_2 \cdot P_2O_5 \cdot 3\ H_2O$	41 (Ti):	Hb-380
$O_9P_2U_2$	$2\ UO_2 \cdot P_2O_5 = UP_2O_7 \cdot UO_2$	55 (U):	Hb-183
$O_9P_2Zn_4$	$4\ ZnO \cdot P_2O_5 \cdot H_2O = Zn_3(PO_4)_2 \cdot Zn(OH)_2$	32 (Zn):	Hb-279 Eb-996
$O_9P_2Zr_2$	$2\ ZrO_2 \cdot P_2O_5 = (ZrO)_2P_2O_7$	42 (Zr):	Hb-398
$O_9P_3^{3-}$	$[P_3O_9]^{3-}$	16 (P):	Hb/C-280/7
$O_9P_3Sr^-$	$[SrP_3O_9]^-$	16 (P):	Hb/C-287
O_9P_3Ti	$Ti(PO_3)_3$	41 (Ti):	Hb-379
$O_9PbRb_4S_6$	$Rb_4Pb(S_2O_3)_3$	47 (Pb):	Hb/C3-1144
$O_9PbS_6^{4-}$	$[Pb(S_2O_3)_3]^{4-}$	47 (Pb):	Hb/B1-365 Hb/C2-597
$O_9Pb_2S_6^{2-}$	$[Pb_2(S_2O_3)_3]^{2-}$	47 (Pb):	Hb/C2-598
$O_9Pb_3Te_2$	$3\ PbO \cdot 2\ TeO_3 \cdot 2\ H_2O$	47 (Pb):	Hb/C2-679
$O_9Pb_3U_2$	$3\ PbO \cdot 2\ UO_3$	55 (U):	Hb-239
O_9Pb_4SSi	$2\ PbO \cdot PbSiO_3 \cdot PbSO_4$	47 (Pb):	Hb/C3-856
O_9PrTa_3	$PrTa_3O_9$	50 (Ta):	Hb/B2-201/2, 211
O_9PrV_3	$Pr(VO_3)_3 \cdot 4\ H_2O$	48 (V):	Hb/B2-555/6
$O_9Pu_2S_3$	$Pu_2(SO_3)_3 \cdot nH_2O$	Erg. W. 4-175	
$O_9Rb_2SV_2$	$V_2O_5 \cdot Rb_2SO_4$	48 (V):	Hb/B2-491
$O_9Rb_3S_6Sb$	$Rb_3[Sb(S_2O_3)_3]$	24 (Rb):	Hb-239
$O_9Rh_2S_3$	$Rh_2(SO_3)_3 \cdot 6\ H_2O$	64 (Rh):	Hb-66
O_9RuS_2	$RuO(SO_4)_2 = RuO_2S_2O_7$	63 (Ru):	Eb-316
$O_9Ru_2S_3$	$Ru_2(SO_3)_3$	63 (Ru):	Hb-73/4
O_9Ru_4	Ru_4O_9	63 (Ru):	Hb-27
–	$Ru_4O_9 \cdot xH_2O$	63 (Ru):	Hb-27
O_9SSb_4	$2\ Sb_2O_3 \cdot SO_3$	18 (Sb):	Hb/B3-545
–	$2\ Sb_2O_3 \cdot SO_3 \cdot nH_2O$	18 (Sb):	Hb/B3-543, 545
O_9STi_3	$3\ TiO_2 \cdot SO_3$	41 (Ti):	Hb-349
O_9SZn_6	$ZnSO_4 \cdot 5\ ZnO$	32 (Zn):	Hb-240
–	$ZnSO_4 \cdot 5\ ZnO \cdot 10\ H_2O$	32 (Zn):	Hb-240
O_9SZr_3	$3\ ZrO_2 \cdot SO_3 \cdot xH_2O$	42 (Zr):	Hb-352
$O_9S_2Sb_2$	$Sb_2O_3 \cdot 2\ SO_3 = Sb(SbO)\ (SO_4)_2$	18 (Sb):	Hb/B3-544
–	$Sb_2O_3 \cdot 2\ SO_3 \cdot H_2O = Sb(OH)SO_4$	18 (Sb):	Hb/B3-544
–	$Sb_2O_3 \cdot 2\ SO_3 \cdot 3\ H_2O$	18 (Sb):	Hb/B3-544
$O_9S_2Ti^{2-}$	$[TiO(SO_4)_2]^{2-}$	41 (Ti):	Hb-468
$O_9S_2Tl_2$	$Tl_2O(SO_4)_2 \cdot 3\ H_2O$	38 (Tl):	Hb/3-365
–	$Tl_2O(SO_4)_2 \cdot 5\ H_2O$	38 (Tl):	Hb/3-365
O_9S_2U	$UO_2S_2O_7$	55 (U):	Hb-156

See the "Instructions for the user"
at the beginning of the volume

$O_9S_2V^-$	$[VO(SO_4)_2]^-$	48 (V):	Hb/B1–305
$O_9S_2V^{2-}$	$[VO(SO_4)_2]^{2-}$	48 (V):	Hb/A2–536/7 Hb/B1–298
$O_9S_2Zn_3$	$ZnO \cdot 2\ ZnSO_4$ *see also* $O_{4.5}SZn_{1.5}$ ... $ZnSO_4 \cdot 0.5\ ZnO$	32 (Zn):	Eb–931
$O_9S_2Zn_5$	$2\ ZnSO_3 \cdot 3\ ZnO \cdot 3\ H_2O$	32 (Zn):	Hb–220
$O_9S_2Zr^{2-}$	$[ZrO(SO_4)_2]^{2-}$	42 (Zr):	Hb–340
O_9S_3	S_3O_9	9 (S):	Hb/B1–337/8
$O_9S_3Sb_2$	$Sb_2(SO_3)_3$	18 (Sb):	Hb/B3–541
$O_9S_6SbTl_3$	$Tl_3[Sb(S_2O_3)_3]$	38 (Tl):	Hb/3–437
$O_9S_6Zn^{4-}$	$[Zn(S_2O_3)_3]^{4-}$	32 (Zn):	Eb–966
O_9SeU^{6-}	$[UO_5(SeO_4)]^{6-}$	55 (U):	Hb–158
$O_9Se_3Tl_2$	$Tl_2(SeO_3)_3$	38 (Tl):	Hb/3–373
O_9SmTa_3	$SmTa_3O_9$	50 (Ta):	Hb/B2–201/2, 214
O_9SmV_3	$Sm(VO_3)_3 \cdot 4\ H_2O$	48 (V):	Hb/B2–555/6
$O_9Sr_3Ta_2Zn$	$ZnSr_3Ta_2O_9$	50 (Ta):	Hb/B2–195
$O_9Sr_4Ta_2$	$Sr_4Ta_2O_9$	50 (Ta):	Hb/B2–181, 183
$O_9Ta_2^-$	$[Ta_2O_9]^-$	50 (Ta):	Hb/B1–59
$O_9Ta_2Th_2$	$Th_2Ta_2O_9$	50 (Ta):	Hb/B2–229, 230
$O_9Ta_2Zr_2$	$(ZrO)_2Ta_2O_7$	50 (Ta):	Hb/B2–228
O_9Ta_3Tb	$TbTa_3O_9$	50 (Ta):	Hb/B2–201/2, 217
O_9Ta_3Y	YTa_3O_9	50 (Ta):	Hb/B2–201/2, 205/6
O_9Ta_3Yb	$YbTa_3O_9$	50 (Ta):	Hb/B2–201/2, 220
O_9Ta_{32}	$Ta_{32}O_9$	50 (Ta):	Hb/B1–24/5
O_9Tl_{10}	$3\ Tl_2O \cdot 2\ Tl_2O_3$	38 (Tl):	Hb/2–221
$O_9V_3^{3-}$	$[V_3O_9]^{3-}$	48 (V):	Hb/B1–132/45, 149/51
O_9V_3Y	$Y(VO_3)_3 \cdot 4\ H_2O$	48 (V):	Hb/B2–555/6
O_9V_4	V_4O_9	48 (V):	Hb/B1–14/5, 18
$O_9V_4^+$	$[V_4O_9]^+$	48 (V):	Hb/B1–84
$O_9V_4^{2-}$	$[V_4O_9]^{2-}$	48 (V):	Hb/A2–477, 498 Hb/B1–160
O_9V_5	$V_5O_9 = VO_{1.8}$	48 (V):	Hb/B1–13, 17, 50/1
O_9W_5	W_5O_9	54 (W):	Hb–111
O_9Y_4	$Y_4O_9 \cdot 9\ H_2O = Y(OH)_3 \cdot 0.75\ H_2O_2$	39 (SE):	Hb/C2–106, 110
O_9Zn_5	Zn_5O_9	32 (Zn):	Hb–145
–	$4\ ZnO_2 \cdot ZnO \cdot 2\ H_2O$	32 (Zn):	Eb–817
$O_{10}P_2STh$	$Th(SO_4)\ (PO_3)_2$	44 (Th):	Hb–320
$O_{10}P_2Ta_2$	$Ta_2O_5 \cdot P_2O_5 = TaOPO_4$	50 (Ta):	Hb/B1–269 Hb/B2–349
–	$Ta_2O_5 \cdot P_2O_5 \cdot 5\ H_2O$	50 (Ta):	Hb/B1–269
–	$Ta_2O_5 \cdot P_2O_5 \cdot 9\ H_2O$	50 (Ta):	Hb/B1–268, 269
–	$Ta_2O_5 \cdot P_2O_5 \cdot xH_2O$	50 (Ta):	Hb/B1–268
$O_{10}P_2V_2$	$V_2O_5 \cdot P_2O_5 = VPO_5$	48 (V):	Hb/B1–361/2
–	$V_2O_5 \cdot P_2O_5 \cdot 2\ H_2O$	48 (V):	Hb/B1–363
–	$V_2O_5 \cdot P_2O_5 \cdot 4\ H_2O$	48 (V):	Hb/B1–362/3
–	$V_2O_5 \cdot P_2O_5 \cdot nH_2O$ $= (VO_2)H_2PO_4 \cdot nH_2O$	48 (V):	Hb/B1–362

Hb = Main Series Volume; GD = Gmelin-Durrer; Ab = Appended Volume to Main Series; Eb = Supplement Volume to Main Series; Erg. W. = New Supplement Series Volume

Formula	Compound	System	Reference
$O_{10}P_3^{5-}$	$[P_3O_{10}]^{5-}$	16 (P):	Hb/B–380 Hb/C–239/41
$O_{10}P_3Pb^{3-}$	$[PbP_3O_{10}]^{3-}$	16 (P): 47 (Pb):	Hb/C–248 Hb/B1–367
$O_{10}P_3Sr^{3-}$	$[SrP_3O_{10}]^{3-}$	16 (P):	Hb/C–247
$O_{10}P_3Zn^{3-}$	$[ZnP_3O_{10}]^{3-}$	16 (P):	Hb/C–247
$O_{10}P_4$	P_4O_{10}	16 (P):	Hb/C–77/9, 82/3
$O_{10}P_6S_5$	$P_6S_5O_{10}$	16 (P):	Hb/C–583
$O_{10}PbU_3$	$PbO \cdot 3\,UO_3$	55 (U):	Hb–240
$O_{10}PbW_3$	$PbO \cdot 3\,WO_3 \cdot 2\,H_2O$	54 (W):	Hb–319
$O_{10}Pb_3SSi_2$	$2\,PbSiO_3 \cdot PbSO_4$	47 (Pb):	Hb/C3–856
$O_{10}Pb_3Sb_4$	$Pb_3Sb_4O_{10} = 3\,PbO \cdot Sb_2O_3 \cdot Sb_2O_4$	47 (Pb):	Hb/C3–930
$O_{10}Pb_3U_2$	$Pb_3U_2O_{10} = 2\,PbO \cdot UO_4 \cdot PbUO_4$	55 (U):	Hb–240
$O_{10}Pb_6$	$Pb_6O_{10} = PbO_{1.66}$	47 (Pb):	Hb/C1–144/7
$O_{10}Pt_2Sn_6$	$Pt_2O \cdot 3\,Sn_2O_3 = Pt_2Sn_6O_{10}$	68 (Pt):	Hb/C3–310
$O_{10}Rb_2S_2U$	$Rb_2SO_4 \cdot UO_2SO_4 \cdot 2\,H_2O$ $= Rb_2UO_2(SO_4)_2 \cdot 2\,H_2O$	55 (U):	Hb–227
$O_{10}RuS_4$	$Ru(S_2O_5)_2$	63 (Ru):	Hb–74
$O_{10}STh_2$	$Th_2O_7SO_3$	44 (Th):	Hb–293
$O_{10}SZr_2$	$Zr_2O_6SO_4 \cdot 8\,H_2O$	42 (Zr):	Hb–353
$O_{10}S_2Ti^{2-}$	$[TiO \cdot O(SO_4)_2]^{2-}$	41 (Ti):	Hb–270, 471/2
$O_{10}S_2Tl_2U$	$Tl_2SO_4 \cdot UO_2SO_4 \cdot 3\,H_2O$ $= Tl_2UO_2(SO_4)_2 \cdot 3\,H_2O$	55 (U):	Hb–238
$O_{10}S_2V_3$	$3\,VO_2 \cdot 2\,SO_2 \cdot 4.5\,H_2O$	48 (V):	Hb/B1–281
$O_{10}S_2Zr^{2-}$	$[ZrO_2(SO_4)_2]^{2-}$	42 (Zr):	Hb–263
$O_{10}S_3Zr_2$	$2\,ZrO_2 \cdot 3\,SO_2$	42 (Zr):	Hb–344
$O_{10}Sb_3^{5-}$	$[Sb_3O_{10}]^{5-}$	18 (Sb):	Hb/B2–390
$O_{10}Se_3U_2$	$U(SeO_3)_2 \cdot UO(SeO_3)$	55 (U):	Hb–157
$O_{10}Sn_4^{4-}$	$[Sn_4O_{10}]^{4-}$	46 (Sn):	Hb/C1–138
$O_{10}SrW_3$	$SrO \cdot 3\,WO_3 \cdot 5\,H_2O$	54 (W):	Hb–297
$O_{10}Sr_5Ta_2$	$Sr_5Ta_2O_{10}$	50 (Ta):	Hb/B2–183
–	$Sr_5Ta_2O_{10}$ solid solutions: $(Sr,Ba)_5Ta_2O_{10}$ $(Sr,Ba,Ca)_5Ta_2O_{10}$	 50 (Ta): 50 (Ta):	 Hb/B2–193, 199 Hb/B2–193
$O_{10}UV_3$	UV_3O_{10}	48 (V):	Hb/B1–91
$O_{10}U_3$	$U_3O_{10} \cdot 2\,H_2O$	55 (U):	Hb–100
$O_{10}V_3^{5-}$	$[V_3O_{10}]^{5-}$	48 (V):	Hb/B1–150
$O_{10}V_3Zn_4$	$4\,ZnO \cdot 3\,VO_2$	48 (V):	Hb/B2–533
$O_{10}V_4$	V_4O_{10}	48 (V):	Hb/B1–84
$O_{10}W_3Zn$	$2\,ZnO \cdot 6\,WO_3 \cdot 10\,H_2O$	54 (W):	Hb–306
$O_{11}P_2Pu_2$	$(PuO_2)_2P_2O_7$	Erg. W. 4–254	
$O_{11}P_2Sb_4$	$2\,Sb_2O_3 \cdot P_2O_5$	18 (Sb):	Hb/B3–561
$O_{11}P_2Ti_2$	$2\,TiO_3 \cdot P_2O_5 \cdot 3\,H_2O$	41 (Ti):	Hb–380
$O_{11}P_2Ti_3$	$3\,TiO_2 \cdot P_2O_5 \cdot 6\,H_2O$	41 (Ti):	Hb–380
$O_{11}P_2Tl_4$	$2\,Tl_2O_3 \cdot P_2O_5 \cdot 5\,H_2O$	38 (Tl):	Hb/3–430
$O_{11}P_2U_2$	$(UO_2)_2P_2O_7$	55 (U):	Hb–183
–	$(UO_2)_2P_2O_7 \cdot 4\,H_2O$	55 (U):	Hb–183
–	$(UO_2)_2P_2O_7 \cdot 5\,H_2O$	55 (U):	Hb–183

See the "Instructions for the user" at the beginning of the volume

Formula	Compound	Volume	Reference
$O_{11}P_2U_2$	$(UO_2)_2P_2O_7 \cdot xH_2O$	55 (U):	Hb–183
$O_{11}P_2V_3$	$(VO)_3(PO_4)_2 \cdot xH_2O$	48 (V):	Hb/B1–361
$O_{11}P_4$	P_4O_{11}	16 (P):	Hb/C–94
$O_{11}P_4Pb$	$[PbP_4O_{11}]_n$ solid solutions:		
	$[(Pb,Sr)P_4O_{11}]_n$	47 (Pb):	Hb/C3–1199
$O_{11}P_4Sr$	$[SrP_4O_{11}]_n$ solid solutions:		
	$[(Sr,Pb)P_4O_{11}]_n$	47 (Pb):	Hb/C3–1199
$O_{11}PbTa_4$	$PbTa_4O_{11}$	50 (Ta):	Hb/B2–232, 233
$O_{11}PbV_4$	$PbO \cdot 2\,V_2O_5 = PbV_4O_{11}$	48 (V):	Hb/B2–570
$O_{11}Pb_3Si_4$	$3\,PbO \cdot 4\,SiO_2 \cdot H_2O$	47 (Pb):	Hb/C3–838
$O_{11}Pb_7$	Pb_7O_{11}	47 (Pb):	Hb/C1–144/7
$O_{11}PdSn_5$	$PdO \cdot 5\,SnO_2 \cdot 4\,H_2O$	65 (Pd):	Hb/2–339
$O_{11}Pd_8S$	$PdSO_4 \cdot 7\,PdO \cdot 6\,H_2O$	65 (Pd):	Hb/2–295
–	$PdSO_4 \cdot 7\,PdO \cdot 10\,H_2O$	65 (Pd):	Hb/2–295
$O_{11}Pr_6$	$Pr_6O_{11} = PrO_{1.833}$:		
	Chemical reactions	39 (SE):	Hb/C1–156, 166/7, 170, 288/90
	Formation and preparation	39 (SE):	Hb/C1–106/7, 286/7
	Physical properties	39 (SE):	Hb/C1–130, 144/5, 152/5, 286/8
–	$Pr_6O_{11} \cdot nH_2O$	39 (SE):	Hb/C1–289 Hb/C2–2, 15
–	Pr_6O_{11} systems:		
	$Pr_6O_{11}–H_2O$	39 (SE):	Hb/C2–50
$O_{11}PtSn_5$	$PtO \cdot 5\,SnO_2 \cdot 4\,H_2O$	68 (Pt):	Hb/C3–310
$O_{11}Rb_2S_2Zr_2$	$Rb_2Zr_2O_3(SO_4)_2 \cdot 15\,H_2O$	42 (Zr):	Hb–427
$O_{11}Rb_2V_4$	$Rb_2O \cdot 2\,V_2O_5$	48 (V):	Hb/B2–487
$O_{11}S_2Ta_2$	$Ta_2O_3(SO_4)_2$	50 (Ta):	Hb/B2–146
$O_{11}S_2V_2$	$V_2O_5 \cdot 2\,SO_3$	48 (V):	Hb/B1–302/3
–	$V_2O_5 \cdot 2\,SO_3 \cdot H_2O$	48 (V):	Hb/B1–302
–	$V_2O_5 \cdot 2\,SO_3 \cdot 3\,H_2O$	48 (V):	Hb/B1–304
–	$V_2O_5 \cdot 2\,SO_3 \cdot 8\,H_2O$	48 (V):	Hb/B1–304
$O_{11}S_2V_3Zn$	$ZnSO_3 \cdot VOSO_3 \cdot 2\,VO_2$	48 (V):	Hb/B2–536
$O_{11}S_2Zr_2^{2-}$	$[Zr_2O_3(SO_4)_2]^{2-}$	42 (Zr):	Hb–340
$O_{11}S_3$	S_3O_{11}	9 (S):	Hb/B1–370/1
$O_{11}Sb_2Se_4$	$Sb_2O_3 \cdot 4\,SeO_2$	18 (Sb):	Hb/B3–553
$O_{11}SnTa_4$	$SnTa_4O_{11}$	50 (Ta):	Hb/B2–231
$O_{11}SrTa_4$	$SrTa_4O_{11}$	50 (Ta):	Hb/B2–182
$O_{11}SrV_4$	$SrO \cdot 2\,V_2O_5 = SrV_4O_{11}$	48 (V):	Hb/B2–521
$O_{11}Sr_6Ta_2$	$Sr_6Ta_2O_{11}$	50 (Ta):	Hb/B2–181, 183
$O_{11}Tb_6$	$Tb_6O_{11} = TbO_{1.833}$	39 (SE):	Hb/C1–387
$O_{11}TiV_9$	TiV_9O_{11}	48 (V):	Hb/B2–558
$O_{11}UW_3$	$UO_2 \cdot 3\,WO_3 \cdot 6\,H_2O$	55 (U):	Hb–244
$O_{11}U_2V_2$	$2\,UO_3 \cdot V_2O_5$	55 (U):	Hb–241
$O_{11}V_2^{4-}$	$[V_2(O)_3(O_2)_4]^{4-}$	48 (V):	Hb/B2–424
$O_{11}V_2Zn_6$	$6\,ZnO \cdot V_2O_5 \cdot xH_2O$	48 (V):	Hb/B2–535

$O_{11}V_4^{2-}$ . . . . $[V_4O_{11}]^{2-}$ . . . . 48 (V): Hb/B1–144
Hb/B2–397
$O_{11}V_4Zn$ . . . . $ZnO \cdot 2\,V_2O_5 = ZnV_4O_{11}$ . . . . 48 (V): Hb/B2–536
$O_{11}V_6$ . . . . $V_6O_{11} = VO_{1.83}$ . . . . 48 (V): Hb/B1–13, 30, 50/1
$O_{11}W_4$ . . . . W_4O_{11} . . . . 54 (W): Hb–107/11, 115/7
O_{12} . . . . O_{12} . . . . 3 (O): Hb/4–1116
$O_{12}PPb_5SiV$ . . . . . $Pb_5SiO_4PO_4VO_4$ . . . . 48 (V): Hb/B2–573
$O_{12}P_2Pb_3SSr$ . . . . . $Pb_3SrSO_4(PO_4)_2$ . . . . 47 (Pb): Hb/C3–1199
$O_{12}P_2Pb_3SZn$ . . . . . $Pb_3ZnSO_4(PO_4)_2$ . . . . 47 (Pb): Hb/C4–1229
$O_{12}P_2Pb_4S$ . . . . . . $Pb_3(PO_4)_2 \cdot PbSO_4$ . . . . 47 (Pb): Hb/C3–884
$O_{12}P_2Pb_5Si$ . . . . . . $Pb_5(PO_4)_2SiO_4$ . . . . 47 (Pb): Hb/C3–886/7
$O_{12}P_2SZr_2$ . . . . $2\,ZrO_2 \cdot P_2O_5 \cdot SO_3$ . . . . 42 (Zr): Hb–398/9
$O_{12}P_3Pb_4Rb$ . . . . . $RbPb_4(PO_4)_3$ . . . . 47 (Pb): Hb/C3–1145
$O_{12}P_3Pb_4Tl$ . . . . . . $TlPb_4(PO_4)_3$ . . . . 47 (Pb): Hb/C4–1355
$O_{12}P_4^{4-}$ . . . . $[P_4O_{12}]^{4-}$ . . . . 16 (P): Hb/C–287/93
$O_{12}P_4Pb_2$ . . . . $Pb_2P_4O_{12}$ . . . . 47 (Pb): Hb/C3–877
–. . . . . $Pb_2P_4O_{12} \cdot 4\,H_2O$ . . . . 47 (Pb): Hb/C3–877
$O_{12}P_4Pu$ . . . . $Pu(PO_3)_4$ . . . . Erg. W. 4–252/3
$O_{12}P_4Si$ . . . . $2\,P_2O_5 \cdot SiO_2 \cdot 2\,H_2O$
$= Si(PO_3)_4 \cdot 2\,H_2O$ . . . . 16 (P): Hb/C–640
–. . . . . $2\,P_2O_5 \cdot SiO_2 \cdot 4\,H_2O = H_8[Si(PO_4)_4]$ . . . 16 (P): Hb/C–640
$O_{12}P_4Sr^{2-}$ . . . . . . $[SrP_4O_{12}]^{2-}$ . . . . 16 (P): Hb/C–293
$O_{12}P_4Sr_2$ . . . . $Sr_2(PO_3)_4$ . . . . 29 (Sr): Hb–216
–. . . . . $Sr_2(PO_3)_4 \cdot 8\,H_2O$ . . . . 29 (Sr): Hb–216
$O_{12}P_4Th$ . . . . $Th(PO_3)_4$ . . . . 44 (Th): Hb–319
$O_{12}P_4U$ . . . . $U(PO_3)_4$ . . . . 55 (U): Hb–183
–. . . . . $U(PO_3)_4 \cdot 6\,H_2O$ . . . . 55 (U): Hb–184
$O_{12}P_4Zn_2$ . . . . $Zn_2(PO_3)_4$ . . . . 32 (Zn): Hb–280
–. . . . . $Zn_2(PO_3)_4 \cdot 10\,H_2O$ . . . . 32 (Zn): Hb–280
$O_{12}PbRb_2S_3$ . . . . . . $Rb_2Pb(SO_4)_3$ . . . . 47 (Pb): Hb/C2–582
$O_{12}PbS_3Sn$ . . . . . . $PbSn(SO_4)_3 \cdot 3\,H_2O$ . . . . 47 (Pb): Hb/C4–1586
$O_{12}PbS_8^{6-}$ . . . . . . $[Pb(S_2O_3)_4]^{6-}$ . . . . 47 (Pb): Hb/B1–365
Hb/C2–597
$O_{12}Pb_2Rb_2S_3$ . . . . . $Rb_2Pb_2(SO_4)_3$ . . . . 47 (Pb): Hb/C3–1143
$O_{12}Pb_2S_3Tl_2$ . . . . . $Tl_2Pb_2(SO_4)_3$ . . . . 47 (Pb): Hb/C4–1353
$O_{12}Pb_3S_8^{2-}$ . . . . . . $[Pb_3(S_2O_3)_4]^{2-}$ . . . . 47 (Pb): Hb/C2–598
$O_{12}Pb_4TlV_3$ . . . . . . $TlPb_4(VO_4)_3$ . . . . 48 (V): Hb/B2–574
$O_{12}Pb_5SSi_2$ . . . . . . $Pb_5(SiO_4)_2SO_4$ . . . . 47 (Pb): Hb/C3–856
–. . . . . $Pb_5(SiO_4)_2SO_4$ solid solutions:
$Pb_5(SiO_4)_2SO_4–Pb_9(Si_2O_7)_3$ . . . . 47 (Pb): Hb/C3–857
$O_{12}Pb_5Sb_4$ . . . . . . $Pb_5Sb_4O_{12} = 5\,PbO \cdot Sb_2O_3 \cdot Sb_2O_4$ . . . 47 (Pb): Hb/C3–930
$O_{12}Pb_5SiV_2$ . . . . . . $Pb_5SiO_4(VO_4)_2$ . . . . 48 (V): Hb/B2–572
$O_{12}Pr_6$ . . . . $Pr_6O_{12} \cdot nH_2O$ . . . . 39 (SE): Hb/C2–106
$O_{12}Pr_7$ . . . . $Pr_7O_{12} = PrO_{1.714}$ . . . . 39 (SE): Hb/C1–284/5
$O_{12}PtS_4^{6-}$ . . . . $[Pt(SO_3)_4]^{6-}$ . . . . 68 (Pt): Hb/C2–158
$O_{12}PuS_3Tl_3$ . . . . . . $Tl_3Pu(SO_4)_3$ . . . . Erg. W. 4–178
$O_{12}PuTa_4$ . . . . $PuO_2 \cdot 2\,Ta_2O_5$ . . . . Erg. W. 4–56, 72
$O_{12}Pu_2S_3$ . . . . $Pu_2(SO_4)_3$ . . . . Erg. W. 4–175/6

*See the "Instructions for the user"
at the beginning of the volume*

Formula	Compound	Volume	Page
$O_{12}Pu_2S_3$	$Pu_2(SO_4)_3 \cdot 5\,H_2O$	Erg. W. 4–176	
–	$Pu_2(SO_4)_3 \cdot 7\,H_2O$	Erg. W. 4–175/6	
–	$Pu_2(SO_4)_3 \cdot xH_2O$	Erg. W. 4–179	
–	$Pu_2(SO_4)_3$ systems:		
	$Pu_2(SO_4)_3-K_2SO_4-H_2O$	Erg. W. 4–177	
$O_{12}Rb_2S_3Th$	$Rb_2SO_4 \cdot Th(SO_4)_2 \cdot 2\,H_2O$	44 (Th):	Hb–344/5
$O_{12}Rb_2W_4$	$Rb_2W_4O_{12}$	54 (W):	Hb–283
$O_{12}Rb_4S_3V_2$	$2\,Rb_2SO_3 \cdot VOSO_3 \cdot VO_2 \cdot H_2O$	48 (V):	Hb/B2–489
$O_{12}Rh_2S_3$	$Rh_2(SO_4)_3$	64 (Rh):	Hb–67
–	$Rh_2(SO_4)_3 \cdot 4\,H_2O$	64 (Rh):	Hb–68
–	$Rh_2(SO_4)_3 \cdot 6\,H_2O$	64 (Rh):	Hb–68
–	$Rh_2(SO_4)_3 \cdot 9\,H_2O$	64 (Rh):	Hb–68
–	$Rh_2(SO_4)_3 \cdot 12\,H_2O$	64 (Rh):	Hb–68
–	$Rh_2(SO_4)_3 \cdot 15\,H_2O$	64 (Rh):	Hb–67/8
–	$Rh_2(SO_4)_3$ systems:		
	$Rh_2(SO_4)_3-H_2O$	64 (Rh):	Hb–67
$O_{12}Ru_2S_3$	$Ru_2(SO_4)_3$	63 (Ru):	Hb–74
$O_{12}SSb_6$	$3\,Sb_2O_3 \cdot SO_3$	18 (Sb):	Hb/B3–545
–	$3\,Sb_2O_3 \cdot SO_3 \cdot 1.67\,H_2O$	18 (Sb):	Hb/B3–545
$O_{12}SU_3$	$3\,UO_3 \cdot SO_3 \cdot 2\,H_2O$	55 (U):	Hb–156
$O_{12}S_2V_2^{2-}$	$[V_2O_4(SO_4)_2]^{2-}$	48 (V):	Hb/B2–440
$O_{12}S_2Zr_3$	$Zr(SO_4)_2 \cdot 2\,ZrO_2$	42 (Zr):	Hb–349
–	$3\,ZrO_2 \cdot 2\,SO_3 \cdot 6\,H_2O$	42 (Zr):	Hb–349
–	$2\,ZrOSO_4 \cdot ZrO_2 \cdot 8\,H_2O$	42 (Zr):	Hb–349
–	$2\,ZrOSO_4 \cdot ZrO_2 \cdot 8.5\,H_2O$	42 (Zr):	Hb–349
–	$3\,ZrO_2 \cdot 2\,SO_3 \cdot 12\,H_2O$	42 (Zr):	Hb–349
–	$3\,ZrO_2 \cdot 2\,SO_3 \cdot xH_2O$	42 (Zr):	Hb–349
$O_{12}S_3Sb_2$	$Sb_2(SO_4)_3$	18 (Sb):	Hb/B3–542/3
$O_{12}S_3SrTi$	$SrTi(SO_4)_3$	41 (Ti):	Hb–434
$O_{12}S_3ThTl_2$	$Tl_2SO_4 \cdot Th(SO_4)_2 \cdot 4\,H_2O$	44 (Th):	Hb–351
$O_{12}S_3Ti_2$	$Ti_2(SO_4)_3$	41 (Ti):	Hb–343/5, 469/70, 478/9
–	$Ti_2(SO_4)_3 \cdot 8\,H_2O$	41 (Ti):	Hb–345
$O_{12}S_3Tl_2$	$Tl_2(SO_4)_3$	38 (Tl):	Hb/3–365, 366/7
–	$Tl_2(SO_4)_3 \cdot H_2O$	38 (Tl):	Hb/3–365
–	$Tl_2(SO_4)_3 \cdot 7\,H_2O$	38 (Tl):	Hb/3–365
–	$Tl_2(SO_4)_3$ systems:		
	$Tl_2(SO_4)_3-H_2SO_4-Tl_2SO_4-H_2O$	38 (Tl):	Hb/3–362
$O_{12}S_3Tl_2Zr_2$	$Tl_2SO_4 \cdot 2\,Zr(SO_4) \cdot 8\,H_2O$	42 (Zr):	Hb–446
$O_{12}S_3Tl_4$	$Tl_4(SO_4)_3$ *see* $O_{24}S_6Tl_8$ ... $3\,Tl_2SO_4 \cdot Tl_2(SO_4)_3$		
$O_{12}S_3Tl_4V_2$	$2\,Tl_2SO_3 \cdot VOSO_3 \cdot VO_2 \cdot 4\,H_2O$	48 (V):	Hb/B2–550
$O_{12}S_3V_2$	$V_2(SO_4)_3$	48 (V):	Hb/A2–380 Hb/B1–284
–	$V_2(SO_4)_3 \cdot 3\,H_2O$	48 (V):	Hb/B1–285/6
–	$V_2(SO_4)_3 \cdot 4\,H_2O$	48 (V):	Hb/B1–285
–	$V_2(SO_4)_3 \cdot nH_2O$	48 (V):	Hb/B1–285

Hb = Main Series Volume; GD = Gmelin-Durrer; Ab = Appended Volume to Main Series;
Eb = Supplement Volume to Main Series; Erg. W. = New Supplement Series Volume

$O_{12}S_3V_2$ . . . . $V_2(SO_4)_3$ solutions:
$V_2(SO_4)_3-H_2O$ . . . . 48 (V): Hb/A2–335, 476
Hb/B1–287/8

$O_{12}S_3ZnZr$ . . . . . . $ZnO \cdot ZrO_2 \cdot (SO_3)_3 \cdot 6\ H_2O$ . . . . 42 (Zr): Hb–442
$O_{12}S_4Th$. . . . . $Th(S_2O_6)_2 \cdot 4\ H_2O$ . . . . 44 (Th): Hb–293
$O_{12}Se_2U^{6-}$ . . . . . . $[UO_4(SeO_4)_2]^{6-}$ . . . . 55 (U): Hb–158
$O_{12}SnTa_4$ . . . . $SnTa_4O_{12}$ . . . . 50 (Ta): Hb/B2–231, 232
$O_{12}Ta_2Ti_3Zn$ . . . . . $ZnTa_2O_6 \cdot 3\ TiO_2$ . . . . 50 (Ta): Hb/B2–226
$O_{12}Ta_4Th$ . . . . $ThTa_4O_{12}$ . . . . 50 (Ta): Hb/B2–229/30
–. . . . . $ThTa_4O_{12}$ solid solutions:
$ThTa_4O_{12}-CaTa_2O_6$ . . . . 50 (Ta): Hb/B2–230
$Th(Ta,Nb)_4O_{12}$. . . . . 50 (Ta): Hb/B2–266

$O_{12}Ta_4U$ . . . . $U(TaO_3)_4$ . . . . 50 (Ta): Hb/B1–51
$O_{12}Tb_7$ . . . . $Tb_7O_{12} = TbO_{1.714}$. . . . . 39 (SE): Hb/C1–387, 394/5
$O_{12}ThV_4$. . . . . $ThO_2 \cdot 2\ V_2O_5 = Th(VO_3)_4$. . . . . 48 (V): Hb/B2–565/6
$O_{12}TiV_9$ . . . . TiV_9O_{12} . . . . 48 (V): Hb/B2–558
$O_{12}Ti_2V_8$ . . . . $Ti_2V_8O_{12}$ . . . . 48 (V): Hb/B2–558
$O_{12}Ti_3V_7$ . . . . $Ti_3V_7O_{12}$ . . . . 48 (V): Hb/B2–558
$O_{12}Ti_7$ . . . . Ti_7O_{12}. . . . . 41 (Ti): Hb–224/6, 473
$O_{12}Ti_9V$ . . . . Ti_9VO_{12} . . . . 48 (V): Hb/B2–558
$O_{12}UW_3$. . . . . $UO_3 \cdot 3\ WO_3 \cdot 5\ H_2O$ . . . . 55 (U): Hb–244
$O_{12}U_3$ . . . . $2\ UO_3 \cdot UO_6 \cdot 6\ H_2O$ . . . . 55 (U): Hb–106
$O_{12}V_4^{4-}$. . . . . $[V_4O_{12}]^{4-}$ . . . . 48 (V): Hb/A2–498
Hb/B1–132/45, 151/2
$O_{12}V_6$ . . . . V_6O_{12} . . . . 48 (V): Hb/B1–84
$O_{12}Zn_7$ . . . . $10\ ZnO_2 \cdot 4\ ZnO \cdot 5\ H_2O$ . . . . 32 (Zn): Hb–146
$O_{13}P_2Pb_8$ . . . . $Pb_3(PO_4)_2 \cdot 5\ PbO = 8\ PbO \cdot P_2O_5$. . . . . 47 (Pb): Hb/C3–869
$O_{13}P_2Rb_4Se_2$ . . . . . $2\ Rb_2O \cdot P_2O_5 \cdot 2\ SeO_3 \cdot 3\ H_2O$ . . . . . . 24 (Rb): Hb–234
$O_{13}P_4^{6-}$. . . . . $[P_4O_{13}]^{6-}$ . . . . 16 (P): Hb/C–249/51
$O_{13}P_4Pb_3$ . . . . $Pb_3P_4O_{13}$ . . . . 47 (Pb): Hb/C3–875
$O_{13}P_4Rh_2$ . . . . $Rh_2O_3 \cdot 2\ P_2O_5 \cdot 6\ H_2O$ . . . . 64 (Rh): Hb–72
$O_{13}P_4S$ . . . . $P_4O_{10} \cdot SO_3$. . . . . 16 (P): Hb/C–583
$O_{13}P_4Zn_3$ . . . . $Zn_3P_4O_{13}$ . . . . 32 (Zn): Eb–996
$O_{13}PbW_4$ . . . . $PbO \cdot 4\ WO_3 \cdot 7\ H_2O$ . . . . 54 (W): Hb–319/20
$O_{13}Pb_3Sb_4$ . . . . . . $Pb_3Sb_4O_{13} = 3\ PbO \cdot 2\ Sb_2O_5$. . . . . . . 47 (Pb): Hb/C3–932
$O_{13}Pb_3Ta_4$ . . . . . . $Pb_3Ta_4O_{13}$ . . . . 50 (Ta): Hb/B2–232, 235, 237
$O_{13}Pb_3V_4$ . . . . $3\ PbO \cdot 2\ V_2O_5 \cdot 2\ H_2O$
$= Pb_3V_4O_{13} \cdot 2\ H_2O$ . . . . 48 (V): Hb/B2–570
$O_{13}Pb_4SSi_3$ . . . . . . $3\ PbSiO_3 \cdot PbSO_4$ . . . . 47 (Pb): Hb/C3–856
$O_{13}Pb_6SSi_2$ . . . . . . $3\ PbO \cdot 2\ PbSiO_3 \cdot PbSO_4$ . . . . 47 (Pb): Hb/C3–856
$O_{13}Pb_8V_2$ . . . . $8\ PbO \cdot V_2O_5 = 5\ PbO \cdot Pb_3(VO_4)_2$ . . . . 48 (V): Hb/B2–568
$O_{13}Pr_6$. . . . . $Pr_6O_{13} \cdot nH_2O$. . . . . 39 (SE): Hb/C2–106
$O_{13}Pu_2S_3$ . . . . $Pu_2O(SO_4)_3 \cdot 8\ H_2O$ . . . . Erg. W. 4–182/3
$O_{13}Rb_2W_4$. . . . . . . $Rb_2O \cdot 4\ WO_3 \cdot 8\ H_2O$
$= Rb_6H_4[H_2(W_2O_7)_6] \cdot 21\ H_2O$ . . . . . 54 (W): Hb–282
$O_{13}STi_5$ . . . . $5\ TiO_2 \cdot SO_3 \cdot 5\ H_2O$ . . . . 41 (Ti): Hb–349
$O_{13}SV_4$ . . . . $2\ V_2O_5 \cdot SO_3$ . . . . 48 (V): Hb/B1–302
$O_{13}S_3Ti_2$ . . . . $2\ TiO_2 \cdot 3\ SO_3 \cdot 3\ H_2O$ . . . . 41 (Ti): Hb–348

See the "Instructions for the user"
at the beginning of the volume

Formula	Compound	Volume	Reference
$O_{13}S_3Ti_2$	$Ti_2O(SO_4)_3 \cdot 5\,H_2O$	41 (Ti):	Hb–346
$O_{13}S_3V^{3-}$	$[VO(SO_4)_3]^{3-}$	48 (V):	Hb/B1–305
$O_{13}S_3Zr_2$	$2\,ZrO_2 \cdot 3\,SO_3 \cdot 5\,H_2O$		
	$= H_4[Zr_4(SO_4)_6(OH)_8] \cdot 4\,H_2O$	42 (Zr):	Hb–344
$O_{13}Sb_6$	$Sb_6O_{13} = Sb_2O_3 \cdot 2\,Sb_2O_5$		
	$= 2\,Sb_2O_4 \cdot Sb_2O_5$	18 (Sb):	Hb/B2–347, 370
$O_{13}SrW_4$	$SrO \cdot 4\,WO_3 \cdot 8\,H_2O$		
	$= Sr_3H_4[H_2(W_2O_7)_6] \cdot 21\,H_2O$	54 (W):	Hb–297
$O_{13}Sr_3V_4$	$3\,SrO \cdot 2\,V_2O_5 \cdot 2\,H_2O$	48 (V):	Hb/B2–520
$O_{13}Ta_4Tl_6$	$Tl_6Ta_4O_{13} \cdot 4\,H_2O$	50 (Ta):	Hb/B2–201
$O_{13}Tl_2W_4$	$Tl_2O \cdot 4\,WO_3 \cdot 3\,H_2O$	54 (W):	Hb–314
$O_{13}Tl_6V_4$	$3\,Tl_2O \cdot 2\,V_2O_5 = Tl_6V_4O_{13}$	48 (V):	Hb/B2–389, 547/8
$O_{13}UV_4$	$UO_3 \cdot 2\,V_2O_5 \cdot xH_2O$	55 (U):	Hb–241
$O_{13}U_4Zn$	$ZnO \cdot 4\,UO_3$	55 (U):	Hb–236
$O_{13}V_4^{6-}$	$[V_4O_{13}]^{6-}$	48 (V):	Hb/B1–144, 148, 152
			Hb/B2–389
$O_{13}V_6$	$V_6O_{13} = VO_{2.17}$	48 (V):	Hb/B1–17, 61/3
$O_{13}V_7$	$V_7O_{13} = VO_{1.86}$	48 (V):	Hb/B1–13, 30, 50/1
$O_{13}W_4Zn$	$ZnO \cdot 4\,WO_3 \cdot 8\,H_2O$		
	$= Zn_3H_4[H_2(W_2O_7)_6] \cdot 21\,H_2O$	54 (W):	Hb–306/7
$O_{14}P_2S_3$	$P_2O_5 \cdot 3\,SO_3$	16 (P):	Hb/C–583
$O_{14}P_2U_3$	$(UO_2)_3(PO_4)_2$	55 (U):	Hb–181
$O_{14}P_4Pb^{6-}$	$[Pb(P_2O_7)_2]^{6-}$	16 (P):	Hb/C–239
		47 (Pb):	Hb/B1–367
			Hb/C3–874
$O_{14}P_4Tl^{7-}$	$[Tl(P_2O_7)_2]^{7-}$	16 (P):	Hb/C–238
$O_{14}P_4V_2$	$V_2O_4 \cdot 2\,P_2O_5 \cdot 6\,H_2O$	48 (V):	Hb/B1–361
$O_{14}P_4Zn^{6-}$	$[Zn(P_2O_7)_2]^{6-}$	16 (P):	Hb/C–238, 239
$O_{14}P_7$	P_7O_{14}	16 (P):	Hb/C–76
$O_{14}Pb_5Si_4Zn$	$Pb_5Zn(Si_2O_7)_2$	47 (Pb):	Hb/C4–1229
$O_{14}Pb_7SSi_2$	$4\,PbO \cdot 2\,PbSiO_3 \cdot PbSO_4$	47 (Pb):	Hb/C3–856/7
$O_{14}Pb_7Sb_4$	$Pb_7Sb_4O_{14} = 7\,PbO \cdot Sb_2O_3 \cdot Sb_2O_4$	47 (Pb):	Hb/C3–930
$O_{14}Pr_6$	$Pr_6O_{14} \cdot nH_2O$	39 (SE):	Hb/C2–106
$O_{14}S_2Zr_4$	$Zr(SO_4)_2 \cdot 3\,ZrO_2 = ZrOSO_4 \cdot ZrO_2$	42 (Zr):	Hb–350
$O_{14}S_3Ta_2$	$Ta_2O_2(SO_4)_3 = Ta_2O_5(SO_3)_3$		
	$= (TaO_2)_2S_3O_{10}$	50 (Ta):	Hb/B1–177
$O_{14}S_3V_2$	$V_2O_5 \cdot 3\,SO_3$	48 (V):	Hb/B1–302/3
–	$V_2O_5 \cdot 3\,SO_3 \cdot 3\,H_2O$	48 (V):	Hb/B1–302
$O_{14}S_3V_4$	$2\,VO_2 \cdot 1.5\,SO_2 \cdot nH_2O$	48 (V):	Hb/B1–281
$O_{14}S_4Ti$	$Ti(S_2O_7)_2 = TiO_2 \cdot 4\,SO_3$	41 (Ti):	Hb–346
–	$TiO_2 \cdot 4\,SO_3 \cdot 2\,H_2O$	41 (Ti):	Hb–346
$O_{14}Se_3Zr_4$	$3\,ZrOSeO_3 \cdot ZrO_2 \cdot 18\,H_2O$		
	$= 4\,ZrO_2 \cdot 3\,SeO_2 \cdot 18\,H_2O$	42 (Zr):	Hb–354/5
$O_{14}Ta_2Ti_4Zn$	$ZnTa_2O_6 \cdot 4\,TiO_2$	50 (Ta):	Hb/B2–226
$O_{14}Ta_2W_3$	$3\,WO_3 \cdot Ta_2O_5$	50 (Ta):	Hb/B1–51
$O_{14}Ti_5Zn_4$	$4\,ZnO \cdot 5\,TiO_2$	41 (Ti):	Hb–462
$O_{14}V_5^{3-}$	$[V_5O_{14}]^{3-}$	48 (V):	Hb/B1–140/1
$O_{14}V_6$	V_6O_{14}	48 (V):	Hb/B1–84

Hb = Main Series Volume; GD = Gmelin-Durrer; Ab = Appended Volume to Main Series; Eb = Supplement Volume to Main Series; Erg. W. = New Supplement Series Volume

$O_{14}V_6$ . . . . $V_6O_{14} \cdot 4\,H_2O = VO_{2.33} \cdot 0.65\,H_2O$
$= V_3O_5(OH)_4$ . . . . 48 (V): Hb/B1–70
–. . . . . $V_6O_{14} \cdot 6\,H_2O = V_6O_{13}(OH)_2 \cdot 5\,H_2O$ . . . 48 (V): Hb/B1–70
$O_{14}W_5$. . . . . W_5O_{14} . . . . 54 (W): Hb–117
$O_{14.5}Se_{3.5}V_2$. . . . . . $2\,VO_2 \cdot 3.5\,SeO_3 \cdot H_2O$ . . . . 48 (V): Hb/B1–314
$O_{15}P_2Ta_4$ . . . . $2\,Ta_2O_5 \cdot P_2O_5 \cdot 6\,H_2O$
$= (TaO_2)_2HPO_4 \cdot 2.5\,H_2O$ . . . . 50 (Ta): Hb/B1–268/9
$O_{15}P_2V_4$ . . . . $2\,V_2O_5 \cdot P_2O_5$ . . . . 48 (V): Hb/B1–362
$O_{15}P_4Pb_5$ . . . . $5\,PbO \cdot 2\,P_2O_5$ . . . . 47 (Pb): Hb/C3–870
$O_{15}P_4Zn_5$ . . . . $10\,ZnO \cdot 4\,P_2O_5 \cdot 10\,H_2O$. . . . . 32 (Zn): Hb–278
$O_{15}Pb_2S_{10}{}^{6-}$ . . . . . $[Pb_2(S_2O_3)_5]^{6-}$ . . . . 47 (Pb): Hb/C2–598
$O_{15}Pb_5Ta_4$ . . . . . . $Pb_5Ta_4O_{15}$ . . . . 50 (Ta): Hb/B2–232/3, 237/8
$O_{15}Pr_6$. . . . . $Pr_6O_{15} \cdot nH_2O$. . . . . 39 (SE): Hb/C2–106, 108
$O_{15}Rb_2S_3V_4$. . . . . . $Rb_2SO_3 \cdot 2\,VOSO_3 \cdot 2\,VO_2 \cdot 4\,H_2O$. . . . . 48 (V): Hb/B2–489
$O_{15}Rb_4W_4$. . . . . . . $2\,Rb_2O \cdot 4\,WO_3 \cdot O \cdot 3\,H_2O$. . . . . 54 (W): Hb–283
$O_{15}Rh_4S_3$ . . . . $Rh_2(SO_4)_3 \cdot Rh_2O_3$ . . . . 64 (Rh): Hb–67
$O_{15}SU_4$ . . . . $4\,UO_3 \cdot SO_3 \cdot 7\,H_2O$ . . . . 55 (U): Hb–156
$O_{15}S_4$ . . . . S_4O_{15} . . . . 9 (S): Hb/B1–367
$O_{15}S_4Sb_2$ . . . . $Sb_2O_3 \cdot 4\,SO_3 = Sb_2(SO_4)_3 \cdot SO_3$ . . . . . 18 (Sb): Hb/B3–544
$O_{15}S_5Sb_2$ . . . . $Sb_2(SO_3)_5$. . . . . 18 (Sb): Hb/B2–449
$O_{15}Se_3U^{6-}$ . . . . . . $[UO_3(SeO_4)_3]^{6-}$ . . . . 55 (U): Hb–158
$O_{15}SnSr_3Ta_4$ . . . . . $3\,SrO \cdot SnO_2 \cdot 2\,Ta_2O_5$
$= Sr_{1.2}(Ta_{0.8}Sn_{0.2})_2O_6$. . . . . 50 (Ta): Hb/B2–232
$O_{15}Sr_3Ta_4Zr$ . . . . . $3\,SrO \cdot ZrO_2 \cdot 2\,Ta_2O_5$
$= Sr_{1.2}(Zr_{0.2}Ta_{0.8})_2O_6$ . . . . 50 (Ta): Hb/B2–228/9
$O_{15}Sr_5Ta_4$. . . . . . . $Sr_5Ta_4O_{15}$. . . . . 50 (Ta): Hb/B2–181, 183
–. . . . . $Sr_5Ta_4O_{15}$ solid solutions:
$(Sr,Ba)_5Ta_4O_{15}$. . . . . 50 (Ta): Hb/B2–193
$O_{15}V_8$ . . . . $V_8O_{15} = VO_{1.87}$. . . . . 48 (V): Hb/B1–13, 17, 50/1
$O_{16}Os_4$ . . . . $(OsO_4)_4$ . . . . 66 (Os): Hb–48
$O_{16}P_4Pu_3$ . . . . $Pu_3(PO_4)_4 \cdot nH_2O$ . . . . Erg. W. 4–253
$O_{16}P_4Si_3$ . . . . $2\,P_2O_5 \cdot 3\,SiO_2 = Si_3(PO_4)_4$ . . . . 16 (P): Hb/C–639
$O_{16}P_4Th_3$ . . . . $Th_3(PO_4)_4$. . . . . 44 (Th): Hb–316
–. . . . . $Th_3(PO_4)_4 \cdot xH_2O$ . . . . 44 (Th): Hb–317/8
–. . . . . $Th_3(PO_4)_4$ systems:
$Th_3(PO_4)_4–H_3PO_4–H_2O$. . . . . 44 (Th): Hb–318
$O_{16}P_4U_3$. . . . . $U_3(PO_4)_4$ . . . . 55 (U): Hb–181
$O_{16}P_4Zr_3$ . . . . $3\,ZrO_2 \cdot 2\,P_2O_5 \cdot nH_2O$ . . . . 42 (Zr): Hb–393
$O_{16}P_6Pb$. . . . . $PbO \cdot 3\,P_2O_5$ . . . . 47 (Pb): Hb/C3–877
$O_{16}P_6Pt$ . . . . $PtO \cdot 3\,P_2O_5$ . . . . 68 (Pt): Hb/C1–137
$O_{16}P_8$ . . . . P_8O_{16} . . . . 16 (P): Hb/C–76
$O_{16}Pr_6$. . . . . $Pr_6O_{16} \cdot nH_2O$. . . . . 39 (SE): Hb/C2–106
$O_{16}Pr_9$. . . . . $Pr_9O_{16} = PrO_{1.778}$ . . . . 39 (SE): Hb/C1–285
$O_{16}Pt_6Rb_6S_4$ . . . . . $Rb_6Pt_6(SO_4)_4 \cdot 17\,H_2O$ . . . . 68 (Pt): Hb/C2–255
$O_{16}PuRb_4S_4$. . . . . . $Rb_4Pu(SO_4)_4$ . . . . Erg. W. 4–183/4
–. . . . . $Rb_4Pu(SO_4)_4 \cdot 2\,H_2O$ . . . . Erg. W. 4–183/4
$O_{16}PuS_4Tl_5$ . . . . . . $Tl_5Pu(SO_4)_4$. . . . . Erg. W. 4–176, 178
$O_{16}Rb_2Rh_2S_4$ . . . . . $Rb_2SO_4 \cdot Rh_2(SO_4)_3 \cdot 24\,H_2O$ . . . . 64 (Rh): Hb–95

See the "Instructions for the user"
at the beginning of the volume

Formula	Compound	Volume	Reference
$O_{16}Rb_2S_4Tl_2$	$Rb_2SO_4 \cdot Tl_2(SO_4)_3 = RbTl(SO_4)_2$	38 (Tl):	Hb/3–469
–	$Rb_2SO_4 \cdot Tl_2(SO_4)_3 \cdot 8\ H_2O$	38 (Tl):	Hb/3–469
$O_{16}Rb_2V_6$	$Rb_2O \cdot 3\ V_2O_5 = Rb_2V_6O_{16}$	48 (V):	Hb/B2–487
$O_{16}Rb_2W_5$	$Rb_2O \cdot 5\ WO_3$	54 (W):	Hb–282/3
$O_{16}Rb_7Ta_5$	$Rb_7Ta_5O_{16} \cdot 11\ H_2O$	50 (Ta):	Hb/B2–171
$O_{16}Rh_2S_4Tl_2$	$Tl_2SO_4 \cdot Rh_2(SO_4)_3 \cdot 24\ H_2O$	64 (Rh):	Hb–99
$O_{16}S_2Zr_5$	$5\ ZrO_2 \cdot 2\ SO_3 \cdot xH_2O$	42 (Zr):	Hb–351/2
$O_{16}S_3V_2$	$V_2(O_2)_2(SO_4)_3$	48 (V):	Hb/B2–393
$O_{16}S_4Sb_2Sr$	$SrSO_4 \cdot Sb_2(SO_4)_3 \cdot 6\ H_2O$	29 (Sr):	Hb–221
$O_{16}S_4Tl_4$	$Tl_2SO_4 \cdot Tl_2(SO_4)_3$	38 (Tl):	Hb/3–363
$O_{16}S_4Tl_4Zr$	$2\ Tl_2SO_4 \cdot Zr(SO_4)_2 \cdot 4\ H_2O$	42 (Zr):	Hb–446
$O_{16}S_4Zn_2{}^{4-}$	$[Zn_2(SO_4)_4]^{4-}$	32 (Zn):	Eb–1007
$O_{16}ScTaTe_6$	$ScTa(Te_3O_8)_2$	50 (Ta):	Hb/B2–319
$O_{16}Se_4Tl_4$	$Tl_2SeO_4 \cdot Tl_2(SeO_4)_3$	38 (Tl):	Hb/3–375
$O_{16}Si_{10}$	$Si_{10}O_{16}$	15 (Si):	Hb/B–591
$O_{16}Sn_5Ta_2Zn$	$ZnTa_2O_6 \cdot 5\ SnO_2$	50 (Ta):	Hb/B2–232
$O_{16}SrV_6$	$SrO \cdot 3\ V_2O_5$	48 (V):	Hb/B2–522
–	$SrO \cdot 3\ V_2O_5 \cdot 13.5\ H_2O$	48 (V):	Hb/B2–522
$O_{16}Ta_5{}^{7-}$	$[Ta_5O_{16}]^{7-}$	50 (Ta):	Hb/B1–59/60
$O_{16}Th_3V_4$	$3\ ThO_2 \cdot 2\ V_2O_5 = Th_3(VO_4)_4$	48 (V):	Hb/B2–565
–	$Th_3(VO_4)_4$ solid solutions:		
	$Ce_{4(1-x)}Th_{3x}(VO_4)_4$	48 (V):	Hb/B2–566
	$Na_{1.33}Th_{3-0.33x}(VO_4)_4$	48 (V):	Hb/B2–566
$O_{16}V_4Zr_3$	$3\ ZrO_2 \cdot 2\ V_2O_5 \cdot 9\ H_2O$ $= (ZrO)_3V_4O_{13} \cdot 9\ H_2O$	48 (V):	Hb/B2–562/3
$O_{16}V_6Zn_{1+x}$	$Zn_{1+y}(V_3O_8)_2$	48 (V):	Hb/B2–534
$O_{17}P_2Sb_8$	$4\ Sb_2O_3 \cdot P_2O_5$	18 (Sb):	Hb/B3–561
$O_{17}P_2Zr_6$	$6\ ZrO_2 \cdot P_2O_5$	42 (Zr):	Hb–393
$O_{17}P_6Pb_2$	$2\ PbO \cdot 3\ P_2O_5$	47 (Pb):	Hb/C3–877
$O_{17}P_6Sr_2$	$2\ SrO \cdot 3\ P_2O_5 \cdot xH_2O$	29 (Sr):	Hb–215
$O_{17}Pb_2V_6$	$2\ PbO \cdot 3\ V_2O_5 = Pb_2V_6O_{17}$	48 (V):	Hb/B2–570
$O_{17}Pb_{10}SSi_2$	$10\ PbO \cdot 2\ SiO_2 \cdot SO_3$	47 (Pb):	Hb/C3–856
$O_{17}Pb_{10}TiV_2$	$10\ PbO \cdot V_2O_5 \cdot TiO_2$	48 (V):	Hb/B2–574
$O_{17}Pb_{12}$	$Pb_{12}O_{17} = PbO_{1.41}$	47 (Pb):	Hb/C1–139/44
$O_{17}Pr_6$	$Pr_6O_{17} \cdot nH_2O$	39 (SE):	Hb/C2–106
$O_{17}Pt_8S$	$Pt_8O_{13}SO_4 \cdot 16\ H_2O$	68 (Pt):	Hb/C1–111
$O_{17}S_2U_6$	$6\ UO_2 \cdot S_2O_5 \cdot 10\ H_2O$	55 (U):	Hb–156
$O_{17}S_3Ti_4$	$4\ TiO_2 \cdot 3\ SO_3$	41 (Ti):	Hb–349
$O_{17}S_3Zr_4$	$4\ ZrO_2 \cdot 3\ SO_3 = (ZrOSO_4)_3 \cdot ZrO_2$	42 (Zr):	Hb–346/7
–	$4\ ZrO_2 \cdot 3\ SO_3 \cdot 5\ H_2O$ $= Zr_4(SO_4)_3(OH)_{10}$	42 (Zr):	Hb–340, 348
–	$4\ ZrO_2 \cdot 3\ SO_3 \cdot 7\ H_2O$	42 (Zr):	Hb–348
–	$4\ ZrO_2 \cdot 3\ SO_3 \cdot 8\ H_2O$	42 (Zr):	Hb–348
–	$4\ ZrO_2 \cdot 3\ SO_3 \cdot 9\ H_2O$	42 (Zr):	Hb–348
–	$4\ ZrO_2 \cdot 3\ SO_3 \cdot 15\ H_2O$	42 (Zr):	Hb–347/8
$O_{17}S_4V_2$	$V_2O_5 \cdot 4\ SO_3$	48 (V):	Hb/B1–303
–	$V_2O_5 \cdot 4\ SO_3 \cdot 4\ H_2O$	48 (V):	Hb/B1–304
$O_{17}Si_7U$	$UO_3 \cdot 7\ SiO_2$	55 (U):	Hb–179

Hb = Main Series Volume; GD = Gmelin-Durrer; Ab = Appended Volume to Main Series;
Eb = Supplement Volume to Main Series; Erg. W. = New Supplement Series Volume

Formula	Compound	Volume	Reference
$O_{17}Sr_2V_6$	$2\,SrO \cdot 3\,V_2O_5$	48 (V):	Hb/B2-521
$O_{17}Sr_2W_5$	$4\,SrO \cdot 10\,WO_3 \cdot 26\,H_2O$	54 (W):	Hb-297
$O_{17}Ta_2W_4$	$4\,WO_3 \cdot Ta_2O_5$	50 (Ta):	Hb/B1-51
$O_{17}Ta_2Zr_6$	$Zr_6Ta_2O_{17} = Ta_2O_5 \cdot 6\,ZrO_2$	50 (Ta):	Hb/B2-49, 228
–	$Zr_6Ta_2O_{17}$ solid solutions:		
	$Zr_6Ta_2O_{17}-ZrTiO_4$	50 (Ta):	Hb/B2-229
$O_{17}Ta_6Tl_4$	$Tl_4Ta_6O_{17} \cdot 6\,H_2O$	50 (Ta):	Hb/B2-201
$O_{17}V_2W_4$	$V_2O_5 \cdot 4\,WO_3 \cdot 10.25\,H_2O$		
	$= 4\,V_2O_5 \cdot 16\,WO_3 \cdot 41\,H_2O$	54 (W):	Hb-385/6
$O_{17}V_4Zn_7$	$7\,ZnO \cdot 2\,V_2O_5$	48 (V):	Hb/B2-535
$O_{17}V_6^{4-}$	$[V_6O_{17}]^{4-}$	48 (V):	Hb/B1-141, 153
$O_{17}V_6Zn_2$	$2\,ZnO \cdot 3\,V_2O_5 \cdot 15\,H_2O$		
	$= Zn_2V_6O_{17} \cdot 15\,H_2O$	48 (V):	Hb/B2-535/6
$O_{17}W_5Zn_2$	$4\,ZnO \cdot 10\,WO_3 \cdot 18\,H_2O$	54 (W):	Hb-306
$O_{17}Zn_{10}$	$Zn_{10}O_{17}$	32 (Zn):	Hb-145
$O_{18}P_6Pb_3$	$Pb_3[(PO_3)_3]_2 \cdot 3\,H_2O$	47 (Pb):	Hb/C3-876/7
$O_{18}P_6S_2$	$3\,P_2O_4 \cdot 2\,SO_3$	16 (P):	Hb/C-583
$O_{18}P_6Sr_3$	$Sr_3(PO_3)_6$	29 (Sr):	Hb-216
$O_{18}P_6Zn_3$	$Zn_3(PO_3)_6$	32 (Zn):	Eb-997
–	$Zn_3(PO_3)_6 \cdot 9\,H_2O$	32 (Zn):	Hb-280
$O_{18}PbS_{12}^{10-}$	$[Pb(S_2O_3)_6]^{10-}$	47 (Pb):	Hb/C2-598
$O_{18}Pr_6$	$Pr_6O_{18} \cdot nH_2O$	39 (SE):	Hb/C2-106
$O_{18}Pr_{10}$	$Pr_{10}O_{18} = PrO_{1.80}$	39 (SE):	Hb/C1-285
$O_{18}STa_{10}$	$5\,Ta_2O_3 \cdot SO_3 \cdot 8\,H_2O$	50 (Ta):	Hb/B1-177
$O_{18}S_4Ti_3$	$3\,TiO_2 \cdot 4\,SO_3 \cdot 15\,H_2O$	41 (Ti):	Hb-350
$O_{18}S_4V_2Zn$	$Zn(VO)_2(SO_4)_4$	48 (V):	Hb/B2-536
$O_{18}S_4Zr_3$	$3\,ZrO_2 \cdot 4\,SO_3 \cdot 15\,H_2O$	42 (Zr):	Hb-345
$O_{18}Sb_{12}W_5$	$Sb_2O_3 \cdot 5\,WO_3 \cdot 4\,H_2O$	54 (W):	Hb-207
$O_{18}Se_3U_2^{6-}$	$[UO_2(UO_4)\,(SeO_4)_3]^{6-}$	55 (U):	Hb-158
$O_{18}Si_2Zr_7$	$7\,ZrO_2 \cdot 2\,SiO_2$	42 (Zr):	Hb-389
$O_{19}P_4Tl_6$	$Tl_6P_4O_{19} \cdot 12\,H_2O$	38 (Tl):	Hb/3-430
$O_{19}P_6Si_2$	$3\,P_2O_5 \cdot 2\,SiO_2 \cdot H_2O$	16 (P):	Hb/C-640
$O_{19}P_6V_2$	$V_2O_4 \cdot 3\,P_2O_5 \cdot 10\,H_2O$	48 (V):	Hb/B1-361
$O_{19}Pb_4Ta_6$	$Pb_4Ta_6O_{19} \cdot 16.5\,H_2O$	50 (Ta):	Hb/B2-237
$O_{19}Pb_4U_5$	$4\,PbO \cdot 5\,UO_3 \cdot 4\,H_2O$	55 (U):	Hb-239
$O_{19}Pb_8SSi_4$	$3\,PbO \cdot 4\,PbSiO_3 \cdot PbSO_4$	47 (Pb):	Hb/C3-856
–	$8\,PbO \cdot 4\,SiO_2 \cdot SO_3$	47 (Pb):	Hb/C3-857
$O_{19}Pb_{12}$	$Pb_{12}O_{19}$	47 (Pb):	Hb/C1-144/7
–	$Pb_{12}O_{19}$ systems:		
	$Pb_{12}O_{19}-PbO_2$	47 (Pb):	Hb/C1-55
	$Pb_{12}O_{19}-Pb_3O_4$	47 (Pb):	Hb/C1-54/5
$O_{19}PrTa_7$	$PrTa_7O_{19}$	50 (Ta):	Hb/B2-201
$O_{19}Pr_6$	$Pr_6O_{19} \cdot nH_2O$	39 (SE):	Hb/C2-106
$O_{19}Rb_8Ta_6$	$Rb_8Ta_6O_{19} \cdot 14\,H_2O$	50 (Ta):	Hb/B2-171
$O_{19}S_2U_7$	$7\,UO_2 \cdot S_2O_5 \cdot 8\,H_2O$	55 (U):	Hb-156
$O_{19}S_3Zr_5$	$5\,ZrO_2 \cdot 3\,SO_3 \cdot 13\,H_2O$	42 (Zr):	Hb-350
$O_{19}S_5Ti_2$	$2\,TiO_2 \cdot 5\,SO_3 \cdot 5\,H_2O$	41 (Ti):	Hb-347
$O_{19}Se_5V_2$	$2\,VO_2 \cdot 5\,SeO_3 \cdot 10\,H_2O$	48 (V):	Hb/B1-314

See the "Instructions for the user" at the beginning of the volume

$O_{19}SmTa_7$	$SmTa_7O_{19}$	50 (Ta):	Hb/B2–201, 214
$O_{19}Sr_4Ta_6$	$Sr_4Ta_6O_{19} \cdot 16.5\ H_2O$	50 (Ta):	Hb/B2–182/3
$O_{19}Ta_6^{8-}$	$[Ta_6O_{19}]^{8-}$	50 (Ta):	Hb/B1–59/61 Hb/B2–167
$O_{20}P_2Ta_6$	$3\ Ta_2O_5 \cdot P_2O_5$	50 (Ta):	Hb/B1–268
$O_{20}P_5Ta_3$	$Ta_3(PO_4)_5$	50 (Ta):	Hb/B1–269
$O_{20}P_6Pb_5$	$Pb_5(P_3O_{10})_2$	47 (Pb):	Hb/C3–874/5
$O_{20}P_6Sr_5$	$Sr_5(P_3O_{10})_2$	29 (Sr):	Eb–286
$O_{20}P_6V_2$	$V_2O_5 \cdot 3\ P_2O_5$	48 (V):	Hb/B1–362
$O_{20}P_6Zn_5$	$Zn_5(P_3O_{10})_2$	32 (Zn):	Hb–279
$O_{20}Pr_6$	$Pr_6O_{20} \cdot nH_2O$	39 (SE):	Hb/C2–106
$O_{20}Pr_{11}$	$Pr_{11}O_{20} = PrO_{1.818}$	39 (SE):	Hb/C1–285
$O_{20}S_2Ti_7$	$7\ TiO_2 \cdot 2\ SO_3 \cdot xH_2O$	41 (Ti):	Hb–349
$O_{20}S_5Ta_2$	$Ta_2(SO_4)_5 = (TaO_2)_2S_5O_{16}$	50 (Ta):	Hb/B1–176/7
$O_{20}S_5Tl_6$	$2\ Tl_2SO_4 \cdot Tl_2(SO_4)_3$	38 (Tl):	Hb/3–362
$O_{20}Tb_{11}$	$Tb_{11}O_{20} = TbO_{1.818}$	39 (SE):	Hb/C1–387, 395/6
$O_{20}U_7$	U_7O_{20}	55 (U):	Hb–82, 94
$O_{20}W_6Zn_2$	$2\ ZnO \cdot 6\ WO_3 \cdot 10\ H_2O$	54 (W):	Hb–306
$O_{20}Zn_{12}$	$8\ ZnO_2 \cdot 4\ ZnO \cdot 7\ H_2O$	32 (Zn):	Hb–146
$O_{21}P_6Sb_4$	$Sb_4(P_2O_7)_3 \cdot 2\ H_2O$	18 (Sb):	Hb/B3–561
$O_{21}P_6V_4$	$V_4(P_2O_7)_3 \cdot 30\ H_2O$	48 (V):	Hb/B1–360
$O_{21}Pb_4Rb_2Si_8$	$Rb_2Pb_4Si_8O_{21}$	47 (Pb):	Hb/C3–1145
$O_{21}Pb_4Si_8Tl_2$	$Tl_2Pb_4Si_8O_{21}$	47 (Pb):	Hb/C4–1354
$O_{21}Pb_8Si_6Sr$	$Pb_8Sr(Si_2O_7)_3$	47 (Pb):	Hb/C3–1199
$O_{21}Pb_8Si_6Zn$	$Pb_8Zn(Si_2O_7)_3$	47 (Pb):	Hb/C4–1229
$O_{21}Pb_{8.5}Si_6Sr_{0.5}$	$Pb_{8.5}Sr_{0.5}(Si_2O_7)_3$	47 (Pb):	Hb/C3–1199
$O_{21}Pr_6$	$Pr_6O_{21} \cdot nH_2O$	39 (SE):	Hb/C2–106
$O_{21}S_2Sb_{10}$	$5\ Sb_2O_3 \cdot 2\ SO_3 \cdot 7\ H_2O$	18 (Sb):	Hb/B3–543, 545
$O_{21}S_2U_8$	$8\ UO_2 \cdot S_2O_5 \cdot 21\ H_2O$	55 (U):	Hb–156
$O_{21}S_4Tl_2V_6$	$Tl_2SO_3 \cdot 3\ VOSO_3 \cdot 3\ VO_2 \cdot 8\ H_2O$	48 (V):	Hb/B2–550
$O_{21}Sb_2W_6$	$Sb_2O_3 \cdot 6\ WO_3 \cdot 8\ H_2O$	54 (W):	Hb–207
$O_{21}SrV_8$	$SrO \cdot 4\ V_2O_5 \cdot 11\ H_2O$	48 (V):	Hb/B2–522
$O_{21}U_2V_6$	$2\ UO_3 \cdot 3\ V_2O_5$	48 (V):	Hb/B1–33
$O_{21}W_6^{6-}$	$[W_6O_{21}]^{6-}$	54 (W):	Hb–144
$O_{22}Pr_6$	$Pr_6O_{22} \cdot nH_2O$	39 (SE):	Hb/C2–106
$O_{22}S_4U_5$	$4\ UOSO_4 \cdot UO_2 \cdot 8\ H_2O$	55 (U):	Hb–151
$O_{22}S_7Tl_8$	$3\ Tl_2S_2O_6 \cdot Tl_2SO_4$	38 (Tl):	Hb/3–368/9
$O_{23}P_4SrU_4$	$SrO \cdot 4\ UO_3 \cdot 2\ P_2O_5 \cdot 21\ H_2O$ $= Sr(UO_2)_4H_2(PO_4)_4 \cdot 20\ H_2O$	55 (U):	Hb–232
$O_{23}Pr_6$	$Pr_6O_{23} \cdot nH_2O$	39 (SE):	Hb/C2–106
$O_{23}S_3Zr_7$	$7\ ZrO_2 \cdot 3\ SO_3 \cdot 14\ H_2O$	42 (Zr):	Hb–351
$O_{23}S_5Ti_4$	$4\ TiO_2 \cdot 5\ SO_3 \cdot 12.5\ H_2O$	41 (Ti):	Hb–350
$O_{23}S_5U_4$	$3\ UOSO_4 \cdot U(SO_4)_2 \cdot xH_2O$	55 (U):	Hb–151
$O_{23}Sr_3V_8$	$3\ SrO \cdot 4\ V_2O_5 \cdot 14\ H_2O$	48 (V):	Hb/B2–521
$O_{24}PbS_{16}^{14-}$	$[Pb(S_2O_3)_8]^{14-}$	47 (Pb):	Hb/C2–598
$O_{24}Pb_3W_7$	$3\ PbO\ \ 7\ WO_3 \cdot 10\ H_2O$	54 (W):	Hb–319
$O_{24}Pr_6$	$Pr_6O_{24} \cdot nH_2O$	39 (SE):	Hb/C2–106
$O_{24}Rb_4V_{10}$	$Rb_4[V_{10}O_{24}] \cdot xH_2O$ $= 2\ Rb_2O \cdot 3\ V_2O_4 \cdot 2\ V_2O_5 \cdot xH_2O$	48 (V):	Hb/B2–486

Hb = Main Series Volume; GD = Gmelin-Durrer; Ab = Appended Volume to Main Series;
Eb = Supplement Volume to Main Series; Erg. W. = New Supplement Series Volume

Formula	Compound	Gmelin	Reference
$O_{24}S_5Sb_6$	$3\,Sb_2O_3 \cdot 5\,SO_3 \cdot H_2O$	18 (Sb):	Hb/B3-544/5
$O_{24}S_6Tl_8$	$3\,Tl_2SO_4 \cdot Tl_2(SO_4)_3$	38 (Tl):	Hb/3-362
–	$3\,Tl_2SO_4 \cdot Tl_2(SO_4)_3$ solid solutions: $(NH_4, Tl)_3Tl(SO_4)_3$	38 (Tl):	Hb/3-463
$O_{24}Sr_3W_7$	$3\,SrO \cdot 7\,WO_3 \cdot 16\,H_2O$	54 (W):	Hb-297
$O_{24}V_{10}{}^{4-}$	$[V_{10}O_{24}]^{4-}$	48 (V):	Hb/B1-161
$O_{24}Zn_{14}$	$10\,ZnO_2 \cdot 4\,ZnO \cdot 5\,H_2O$	32 (Zn):	Hb-146
$O_{25}P_2Pb_{16}Si_2$	$16\,PbO \cdot P_2O_5 \cdot 2\,SiO_2$	47 (Pb):	Hb/C3-886
$O_{25}P_6Pb_{10}$	$3\,Pb_3(PO_4)_2 \cdot PbO = 10\,PbO \cdot 3\,P_2O_5$	47 (Pb):	Hb/C3-870
–	$3\,Pb_3(PO_4)_2 \cdot PbO$ systems: $3\,Pb_3(PO_4)_2 \cdot PbO{-}Pb_5F(PO_4)_3$	47 (Pb):	Hb/C3-881
$O_{25}P_6V_4$	$2\,V_2O_5 \cdot 3\,P_2O_5$	48 (V):	Hb/B1-363
–	$2\,V_2O_5 \cdot 3\,P_2O_5 \cdot 9\,H_2O$	48 (V):	Hb/B1-363
–	$2\,V_2O_5 \cdot 3\,P_2O_5 \cdot 13\,H_2O$ $= (VO)_2(HPO_4)_3 \cdot 5\,H_2O$ $= [VO_2H_2PO_4]_2H_3PO_4 \cdot 3\,H_2O$	48 (V):	Hb/B1-362
$O_{25}P_6Zr_5$	$5\,ZrO_2 \cdot 3\,P_2O_5 \cdot 9\,H_2O$	42 (Zr):	Hb-393
$O_{25}Pb_{10}Sb_6$	$3\,Pb_3(SbO_4)_2 \cdot PbO$	47 (Pb):	Hb/C3-934
$O_{25}Rb_2S_8$	$Rb_2S_2O_7 \cdot 6\,SO_3 = Rb_2SO_4 \cdot 7\,SO_3$	24 (Rb):	Hb-214
$O_{25}Rb_2W_8$	$Rb_2O \cdot 8\,WO_3$	54 (W):	Hb-283
$O_{25}Rb_4V_{10}$	$Rb_4[V_{10}O_{25}] \cdot xH_2O$ $= 2\,Rb_2O \cdot 2\,V_2O_4 \cdot 3\,V_2O_5 \cdot xH_2O$	48 (V):	Hb/B2-486
$O_{25}S_8Tl_2$	$Tl_2S_2O_7 \cdot 6\,SO_3 = Tl_2SO_4 \cdot 7\,SO_3$	38 (Tl):	Hb/3-367
$O_{25}Sb_6U_5$	$5\,UO_2 \cdot 3\,Sb_2O_5 \cdot 15\,H_2O$	55 (U):	Hb-185
$O_{25}Sr_{10}V_6$	$10\,SrO \cdot 3\,V_2O_5$	48 (V):	Hb/B2-519
$O_{25}Ta_9V$	VTa_9O_{25}	50 (Ta):	Hb/B2-244, 245
$O_{25}V_{10}{}^{4-}$	$[V_{10}O_{25}]^{4-}$	48 (V):	Hb/B1-161
$O_{26}Pb_8SSi_6$	$Pb_8(SiO_4)_2(Si_2O_7)_2SO_4$	47 (Pb):	Hb/C3-856
$O_{26}V_{10}{}^{4-}$	$[V_{10}O_{26}]^{4-}$	48 (V):	Hb/B1-161
$O_{26}V_{10}Zn$	$ZnO \cdot 5\,V_2O_5 = ZnV_{10}O_{26}$	48 (V):	Hb/B2-536
$O_{26.19}P_{5.98}V_{5.12}$	$3.12\,VO_2 \cdot V_2O_5 \cdot 2.99\,P_2O_5 \cdot 21.35\,H_2O$	48 (V):	Hb/B1-363
$O_{27}P_4Sr_2U_5$	$2\,SrO \cdot 5\,UO_3 \cdot 2\,P_2O_5 \cdot 24\,H_2O$	55 (U):	Hb-232
$O_{27}P_6Rh_8$	$4\,Rh_2O_3 \cdot 3\,P_2O_5 \cdot 32\,H_2O$	64 (Rh):	Hb-72
$O_{27}P_6Tl_8$	$Tl_8P_6O_{27} \cdot 13\,H_2O$	38 (Tl):	Hb/3-430
$O_{27}P_6U_4$	$4\,UO_3 \cdot 3\,P_2O_5$	55 (U):	Hb-184
$O_{27}S_2Sb_{14}$	$7\,Sb_2O_3 \cdot 2\,SO_3$	18 (Sb):	Hb/B3-545
–	$7\,Sb_2O_3 \cdot 2\,SO_3 \cdot nH_2O$	18 (Sb):	Hb/B3-545
$O_{27}S_5Ti_6$	$6\,TiO_2 \cdot 5\,SO_3 \cdot 4\,H_2O$	41 (Ti):	Hb-347
$O_{27}S_5Zr_6$	$(ZrOSO_4)_5 \cdot ZrO_2 = 5\,Zr(SO_4)_2 \cdot 7\,ZrO_2$	42 (Zr):	Hb-346
–	$(ZrOSO_4)_5 \cdot ZrO_2 \cdot 12\,H_2O$	42 (Zr):	Hb-346
$O_{27}S_7$	S_7O_{27}	9 (S):	Hb/B1-367
$O_{27}S_8Sb_2$	$Sb_2O_3 \cdot 8\,SO_3$	18 (Sb):	Hb/B3-544
$O_{27}Sb_6W_4$	$3\,Sb_2O_5 \cdot 4\,WO_3 \cdot 11\,H_2O$	54 (W):	Hb-380
$O_{27}V_{10}{}^{4-}$	$[V_{10}O_{27}]^{4-}$	48 (V):	Hb/B1-152, 155
$O_{27}V_{10}Zn_2$	$2\,ZnO \cdot 5\,V_2O_5 = Zn_2V_{10}O_{27}$	48 (V):	Hb/B2-536
$O_{28}Rb_6V_{10}$	$3\,Rb_2O \cdot 5\,V_2O_5 \cdot 8.5\,H_2O$ $= Rb_6V_{10}O_{28} \cdot 8.5\,H_2O$	48 (V):	Hb/B2-487

See the "Instructions for the user" at the beginning of the volume

$O_{28}Sr_3V_{10}$ . . . . $3\,SrO \cdot 5\,V_2O_5 \cdot 24\,H_2O$
= $Sr_3V_{10}O_{28} \cdot 24\,H_2O$ . . . . 48 (V): Hb/B2–521
–. . . . . $3\,SrO \cdot 5\,V_2O_5 \cdot xH_2O$ . . . . 48 (V): Hb/B2–520
$O_{28}V_{10}{}^{6-}$ . . . . $[V_{10}O_{28}]^{6-}$ . . . . 48 (V): Hb/A2–498
Hb/B1–132/45, 153/7
$O_{28}V_{10}Zn_3$. . . . . . . $3\,ZnO \cdot 5\,V_2O_5 \cdot 8\,H_2O$
= $Zn_3V_{10}O_{28} \cdot 8\,H_2O$ . . . . 48 (V): Hb/B2–536
–. . . . . $3\,ZnO \cdot 5\,V_2O_5 \cdot 16\,H_2O$ . . . . 48 (V): Hb/B2–536
$O_{29}S_5Zr_7$ . . . . $7\,ZrO_2 \cdot 5\,SO_3 \cdot 30\,H_2O$ . . . . 42 (Zr): Hb–348/9
$O_{29}S_7Th_4$ . . . . $(ThOSO_4 \cdot 2\,H_2O) \cdot (3\,Th(SO_4)_2 \cdot 2\,H_2O)$. . 44 (Th): Hb–292
$O_{30}P_8Zn_{10}$. . . . . . . $10\,ZnO \cdot 4\,P_2O_5 \cdot 10\,H_2O$. . . . . 32 (Zn): Hb–278
$O_{30}P_8Zr_5$ . . . . $5\,ZrO_2 \cdot 4\,P_2O_5 \cdot 8\,H_2O$. . . . . 42 (Zr): Hb–393
$O_{30}P_{10}Sr_5$ . . . . $Sr_5(PO_3)_{10} \cdot 13\,H_2O$ . . . . 29 (Sr): Hb–216
$O_{30}S_9Sb_2$ . . . . $Sb_2O_3 \cdot 9\,SO_3$. . . . . 18 (Sb): Hb/B3–544
$O_{31}S_5Zr_8$ . . . . $8\,ZrO_2 \cdot 5\,SO_3 \cdot 19\,H_2O$
= $(Zr_4(SO_4)_2(OH)_{11})_2SO_4 \cdot 8\,H_2O$ . . . . 42 (Zr): Hb–340, 350
$O_{31}Tl_{12}V_{10}$ . . . . . . $6\,Tl_2O \cdot 5\,V_2O_5 = Tl_{12}V_{10}O_{31}$ . . . . 48 (V): Hb/B2–548
$O_{31}W_7Zr_5$ . . . . $5\,ZrO_2 \cdot 7\,WO_3 \cdot 21\,H_2O$ . . . . 54 (W): Hb–315
$O_{32}Pb_5U_9$ . . . . $5\,PbO \cdot 9\,UO_3 \cdot 10\,H_2O$. . . . . 55 (U): Hb–240
$O_{32}PtW_{10}$ . . . . $PtO_2 \cdot 10\,WO_3 \cdot xH_2O$ . . . . 68 (Pt): Hb/C3–321
$O_{32}S_6Zr_7$ . . . . $7\,ZrO_2 \cdot 6\,SO_3 \cdot xH_2O$ . . . . 42 (Zr): Hb–346
$O_{32}S_8Tl_{12}$ . . . . $5\,Tl_2SO_4 \cdot Tl_2(SO_4)_3$ . . . . 38 (Tl): Hb/3–363
$O_{32}Se_8Tl_{12}$ . . . . . . $5\,Tl_2SeO_4 \cdot Tl_2(SeO_4)_3$ . . . . 38 (Tl): Hb/3–375
$O_{32}SiW_{10}$ . . . . $SiO_2 \cdot 10\,WO_3 \cdot 7\,H_2O$
= $H_8[SiO(W_2O_7)_5] \cdot 3\,H_2O$ . . . . 54 (W): Hb–345
$O_{32}Ta_{12}Zr$. . . . . . . $ZrO_2 \cdot 6\,Ta_2O_5$ . . . . 50 (Ta): Hb/B2–349
$O_{32}TiW_{10}$ . . . . $4\,H_2O \cdot TiO_2 \cdot 10\,WO_3 \cdot xH_2O$ . . . . 54 (W): Hb–383
$O_{33}S_7Zr_6$ . . . . $6\,ZrO_2 \cdot 7\,SO_3 \cdot 19\,H_2O$ . . . . 42 (Zr): Hb–345
$O_{33}V_{12}{}^{6-}$ . . . . $[V_{12}O_{33}]^{6-}$ . . . . 48 (V): Hb/B2–397
$O_{34}Sr_4W_{10}$ . . . . . . $4\,SrO \cdot 10\,WO_3 \cdot 26\,H_2O$ . . . . 54 (W): Hb–297
$O_{34}V_{10}Yb_6$ . . . . . . $3\,Yb_2O_3 \cdot 5\,V_2O_5$ . . . . 48 (V): Hb/B2–556
$O_{34}W_{10}Zn_4$ . . . . . . $4\,ZnO \cdot 10\,WO_3 \cdot 18\,H_2O$. . . . . 54 (W): Hb–306
$O_{35}Ta_{12}U_2$. . . . . . . $U_2O_5 \cdot 6\,Ta_2O_5$ . . . . 50 (Ta): Hb/B1–51
$O_{35}Ta_{12}V_2$. . . . . . . $V_2Ta_{12}O_{35}$ . . . . 50 (Ta): Hb/B2–244
$O_{35}V_2W_{10}$ . . . . $V_2O_5 \cdot 10\,WO_3 \cdot 22\,H_2O$ . . . . 54 (W): Hb–385
$O_{36}PV_{12}{}^{7-}$ . . . . . . $[P(V_3O_9)_4]^{7-}$ . . . . 48 (V): Hb/B1–362
Hb/B2–444
$O_{37}W_9Zr_5$ . . . . $5\,ZrO_2 \cdot 9\,WO_3 \cdot 33\,H_2O$ . . . . 54 (W): Hb–315
$O_{38}SiW_{12}$ . . . . $SiO_2 \cdot 12\,WO_3 \cdot 8\,H_2O$
= $H_8[Si(W_2O_7)_6] \cdot 4\,H_2O$ . . . . 54 (W): Hb–343/4
–. . . . . $SiO_2 \cdot 12\,WO_3 \cdot 9\,H_2O$ . . . . 54 (W): Hb–343/4
–. . . . . $SiO_2 \cdot 12\,WO_3 \cdot 19\,H_2O$ . . . . 54 (W): Hb–343
–. . . . . $SiO_2 \cdot 12\,WO_3 \cdot 24\,H_2O$ . . . . 54 (W): Hb–344/5
–. . . . . $SiO_2 \cdot 12\,WO_3 \cdot 26\,H_2O$ . . . . 54 (W): Hb–343
–. . . . . $SiO_2 \cdot 12\,WO_3 \cdot 32\,H_2O$ . . . . 54 (W): Hb–340/3
$O_{38}TiW_{12}$ . . . . $4\,H_2O \cdot TiO_2 \cdot 12\,WO_3 \cdot xH_2O$ . . . . 54 (W): Hb–383
$O_{39}Rb_2SiW_{12}$ . . . . . $Rb_2O \cdot SiO_2 \cdot 12\,WO_3 \cdot 6\,H_2O$
= $Rb_2H_6[Si(W_2O_7)_6] \cdot 3\,H_2O$ . . . . . . 54 (W): Hb–349

$O_{39}SiTl_2W_{12}$	$Tl_2O \cdot SiO_2 \cdot 12\ WO_3 \cdot 10\ H_2O$		
	$= Tl_2H_6[Si(W_2O_7)_6] \cdot 7\ H_2O$	54 (W):	Hb–353
$O_{39}Sr_4V_{14}$	$4\ SrO \cdot 7\ V_2O_5 \cdot 30\ H_2O$	48 (V):	Hb/B2–521
$O_{40}Pb_2SiW_{12}$	$2\ PbO \cdot SiO_2 \cdot 12\ WO_3 \cdot 21\ H_2O$		
	$= Pb_2H_4[Si(W_2O_7)_6] \cdot 19\ H_2O$	54 (W):	Hb–353
$O_{40}Rb_2S_{10}Ti_6$	$Rb_2SO_4 \cdot 3\ Ti_2(SO_4)_3 \cdot 24\ H_2O$	41 (Ti):	Hb–413
$O_{40}S_{10}Tl_{16}$	$7\ Tl_2SO_4 \cdot Tl_2(SO_4)_3 \cdot 6\ H_2O$	38 (Tl):	Hb/3–362
$O_{40}SiSr_2W_{12}$	$2\ SrO \cdot SiO_2 \cdot 12\ WO_3 \cdot 16\ H_2O$		
	$= Sr_2H_4[Si(W_2O_7)_6] \cdot 14\ H_2O$	54 (W):	Hb–351
–	$2\ SrO \cdot SiO_2 \cdot 12\ WO_3 \cdot 17\ H_2O$	54 (W):	Hb–351
–	$2\ SrO \cdot SiO_2 \cdot 12\ WO_3 \cdot 23\ H_2O$	54 (W):	Hb–350
–	$2\ SrO \cdot SiO_2 \cdot 12\ WO_3 \cdot 27\ H_2O$	54 (W):	Hb–350
$O_{40}SiThW_{12}$	$ThO_2 \cdot SiO_2 \cdot 12\ WO_3 \cdot 27\ H_2O$		
	$= ThH_4[Si(W_2O_7)_6] \cdot 25\ H_2O$	54 (W):	Hb–353
–	$ThO_2 \cdot SiO_2 \cdot 12\ WO_3 \cdot 30\ H_2O$	54 (W):	Hb–353
$O_{40}SiW_{12}Zn_2$	$2\ ZnO \cdot SiO_2 \cdot 12\ WO_3 \cdot 18\ H_2O$		
	$= Zn_2H_4[Si(W_2O_7)_6] \cdot 16\ H_2O$	54 (W):	Hb–352
–	$2\ ZnO \cdot SiO_2 \cdot 12\ WO_3 \cdot 27\ H_2O$	54 (W):	Hb–351/2
–	$2\ ZnO \cdot SiO_2 \cdot 12\ WO_3 \cdot 29\ H_2O$	54 (W):	Hb–351
$O_{41}P_2W_{12}$	$P_2O_5 \cdot 12\ WO_3 \cdot 42\ H_2O$	54 (W):	Hb–374
$O_{41}Rb_{10}W_{12}$	$5\ Rb_2O \cdot 12\ WO_3 \cdot 18\ H_2O$	54 (W):	Hb–282
$O_{41}SiUW_{12}$	$3\ UO_3 \cdot 3(SiO_2 \cdot 12\ WO_3) \cdot 88\ H_2O$	55 (U):	Hb–244
$O_{41}Sr_6V_{14}$	$6\ SrO \cdot 7\ V_2O_5$	48 (V):	Hb/B2–521
$O_{41}Tl_{10}W_{12}$	$5\ Tl_2O \cdot 12\ WO_3 \cdot 9\ H_2O$	54 (W):	Hb–314
$O_{41}W_{12}Zn_5$	$5\ ZnO \cdot 12\ WO_3 \cdot 35\ H_2O$	54 (W):	Hb–306
$O_{42}P_2Tl_2W_{12}$	$Tl_2O \cdot P_2O_5 \cdot 12\ WO_3 \cdot 4\ H_2O$	54 (W):	Hb–375
$O_{42}Pb_4SiW_{12}$	$4\ PbO \cdot SiO_2 \cdot 12\ WO_3 \cdot 20\ H_2O$		
	$= Pb_4[Si(W_2O_7)_6] \cdot 20\ H_2O$	54 (W):	Hb–353
$O_{42}Rb_8SiW_{12}$	$4\ Rb_2O \cdot SiO_2 \cdot 12\ WO_3$		
	$= Rb_8[Si(W_2O_7)_6]$	54 (W):	Hb–349
$O_{43}P_2Pb_2W_{12}$	$2\ PbO \cdot P_2O_5 \cdot 12\ WO_3 \cdot 6\ H_2O$	54 (W):	Hb–375
$O_{43}P_2Sr_2W_{12}$	$2\ SrO \cdot P_2O_5 \cdot 12\ WO_3 \cdot 17\ H_2O$	54 (W):	Hb–375
$O_{43}P_2W_{12}Zn_2$	$2\ ZnO \cdot P_2O_5 \cdot 12\ WO_3 \cdot 7\ H_2O$	54 (W):	Hb–375
$O_{44}Rb_{10}W_{12}$	$5\ Rb_2O \cdot 12\ WO_3 \cdot 3\ O \cdot 12\ H_2O$	54 (W):	Hb–283
$O_{44}S_{11}Th_2Tl_{14}$	$7\ Tl_2SO_4 \cdot 2\ Th(SO_4)_2$	44 (Th):	Hb–352
$O_{44}S_{11}Tl_{10}$	$2\ Tl_2SO_4 \cdot 3\ Tl_2(SO_4)_3 \cdot 22\ H_2O$	38 (Tl):	Hb/3–362
$O_{44}S_{11}Tl_{14}Zr_2$	$7\ Tl_2SO_4 \cdot 2\ Zr(SO_4)_2$	42 (Zr):	Hb–446
$O_{44}SiU_2W_{12}$	$6\ UO_3 \cdot 3(SiO_2 \cdot 12\ WO_3) \cdot 91\ H_2O$	55 (U):	Hb–244
$O_{47}S_{12}Tl_{10}$	$2\ Tl_2O \cdot 3\ Tl_2O_3 \cdot 12\ SO_3 \cdot 25\ H_2O$	38 (Tl):	Hb/3–362
$O_{50}P_2Ta_{18}$	$9\ Ta_2O_5 \cdot P_2O_5$	50 (Ta):	Hb/B1–268
$O_{50}Ta_{17}V_3$	$V_3Ta_{17}O_{50}$	50 (Ta):	Hb/B2–244, 245
$O_{51}Pr_4V_{18}$	$Pr_4(V_6O_{17})_3 \cdot 48\ H_2O$	48 (V):	Hb/B2–555/6
$O_{51}STh_{24}$	$24\ ThO_2 \cdot SO_3 \cdot 12\ H_2O$	44 (Th):	Hb–293
$O_{51}Sm_4V_{18}$	$Sm_4(V_6O_{17})_3 \cdot 48\ H_2O$	48 (V):	Hb/B2–555/6
$O_{51}V_{18}Y_4$	$Y_4(V_6O_{17})_3 \cdot 48\ H_2O$	48 (V):	Hb/B2–555/6
$O_{53}P_2W_{16}$	$P_2O_5 \cdot 16\ WO_3 \cdot 69\ H_2O$	54 (W):	Hb–374
$O_{54}S_{10}Zr_{12}$	$5\ Zr(SO_4)_2 \cdot 7\ ZrO_2 = (ZrOSO_4)_5 \cdot ZrO_2$	42 (Zr):	Hb–346
$O_{55}P_2Sr_{10}V_{16}$	$10\ SrO \cdot 8\ V_2O_5 \cdot P_2O_5 \cdot xH_2O$	48 (V):	Hb/B2–523

See the "Instructions for the user" at the beginning of the volume

Formula	Compound	Volume	Reference
$O_{59}P_2W_{18}$	$P_2O_5 \cdot 18\ WO_3 \cdot 42\ H_2O$		
	$= H_{10}[P_2(OH)_2(W_2O_7)_9] \cdot 36\ H_2O$	54 (W):	Hb-365
$O_{59}V_2W_{18}$	$V_2O_5 \cdot 18\ WO_3 \cdot 36\ H_2O$	54 (W):	Hb-385
$O_{60}P_2Rb_{10}V_{20}$	$5\ Rb_2O \cdot 10\ V_2O_5 \cdot P_2O_5 \cdot 46\ H_2O$	48 (V):	Hb/B2-491/2
$O_{62}PtW_{20}$	$PtO_2 \cdot 20\ WO_3 \cdot xH_2O$	68 (Pt):	Hb/C3-321
$O_{65}P_2W_{20}$	$P_2O_5 \cdot 20\ WO_3 \cdot xH_2O$	54 (W):	Hb-364
$O_{65}P_{14}V_{12}$	$6\ V_2O_5 \cdot 7\ P_2O_5 \cdot 37\ H_2O$	48 (V):	Hb/B1-363
$O_{67}P_2Pb_2W_{20}$	$2\ PbO \cdot P_2O_5 \cdot 20\ WO_3 \cdot 6\ H_2O$	54 (W):	Hb-370
$O_{68}P_2W_{21}$	$P_2O_5 \cdot 21\ WO_3 \cdot xH_2O$	54 (W):	Hb-364
$O_{68}V_8W_{16}$	$4\ V_2O_5 \cdot 16\ WO_3 \cdot 41\ H_2O$		
	$= V_2O_5 \cdot 4\ WO_3 \cdot 10.25\ H_2O$	54 (W):	Hb-385/6
$O_{71}P_2W_{22}$	$P_2O_5 \cdot 22\ WO_3 \cdot xH_2O$	54 (W):	Hb-363
$O_{72}P_2Rb_{14}V_{24}$	$7\ Rb_2O \cdot 12\ V_2O_5 \cdot P_2O_5 \cdot 32\ H_2O$	48 (V):	Hb/B2-491/2
$O_{72}P_{20}Zr_{11}$	$11\ ZrO_2 \cdot 10\ P_2O_5 \cdot 13\ H_2O$	42 (Zr):	Hb-393
$O_{75}W_{22}Zn_9$	$9\ ZnO \cdot 22\ WO_3 \cdot 66\ H_2O$	54 (W):	Hb-306
$O_{77}P_2W_{24}$	$P_2O_5 \cdot 24\ WO_3 \cdot 45\ H_2O$		
	$= H_7[P(W_2O_7)_6] \cdot 19\ H_2O$	54 (W):	Hb-363
–	$P_2O_5 \cdot 24\ WO_3 \cdot 51\ H_2O$	54 (W):	Hb-363
–	$P_2O_5 \cdot 24\ WO_3 \cdot 63\ H_2O$	54 (W):	Hb-359/63
$O_{78}V_{30}Yb_2$	$Yb_2O_3 \cdot 15\ V_2O_5$	48 (V):	Hb/B2-556
$O_{79}Rb_6Si_2W_{24}$	$3\ Rb_2O \cdot 2\ (SiO_2 \cdot 12\ WO_3) \cdot 23\ H_2O$		
	$= Rb_3H_5[Si(W_2O_7)_6] \cdot 9\ H_2O$	54 (W):	Hb-349
$O_{79}Si_2UW_{24}$	$3\ UO_3 \cdot 6(SiO_2 \cdot 12\ WO_3) \cdot 177\ H_2O$	55 (U):	Hb-244
$O_{100}Ta_{34}U_6$	$3\ U_2O_5 \cdot 17\ Ta_2O_5$	50 (Ta):	Hb/B1-51
$O_{105}P_2V_{40}$	$20\ V_2O_5 \cdot P_2O_5 \cdot 59\ H_2O$	48 (V):	Hb/B1-363
$O_{116}Si_3ThW_{36}$	$ThO_2 \cdot 3\ (SiO_2 \cdot 12\ WO_3) \cdot 49\ H_2O$		
	$= ThH_{20}[Si(W_2O_7)_6]_3 \cdot 39\ H_2O$	54 (W):	Hb-353
$O_{116}Si_3UW_{36}$	$UO_2 \cdot 3(SiO_2 \cdot 12\ WO_3) \cdot 90\ H_2O$	54 (W):	Hb-354
$O_{118}Si_3U_2W_{36}$	$2\ UO_2 \cdot 3(SiO_2 \cdot 12\ WO_3) \cdot 87\ H_2O$	54 (W):	Hb-354
$O_{120}Si_3U_2W_{36}$	$2\ UO_3 \cdot 3(SiO_2 \cdot 12\ WO_3) \cdot 87\ H_2O$	55 (U):	Hb-244
$O_{123}Si_3U_3W_{36}$	$3\ UO_3 \cdot 3(SiO_2 \cdot 12\ WO_3) \cdot 88\ H_2O$	55 (U):	Hb-244
$O_{129}Si_3U_5W_{36}$	$5\ UO_3 \cdot 3(SiO_2 \cdot 12\ WO_3) \cdot 90\ H_2O$	55 (U):	Hb-244
$O_{132}Si_3U_6W_{36}$	$6\ UO_3 \cdot 3(SiO_2 \cdot 12\ WO_3) \cdot 91\ H_2O$	55 (U):	Hb-244
$O_{177}Sr_{19}V_{20}W_{36}$	$19\ SrO \cdot 10\ V_2O_5 \cdot 36\ WO_3 \cdot 122\ H_2O$	54 (W):	Hb-389
$O_{237}Si_6U_3W_{72}$	$3\ UO_3 \cdot 6(SiO_2 \cdot 12\ WO_3) \cdot 177\ H_2O$	55 (U):	Hb-244
$O_{354}Si_9U_4W_{108}$	$4\ UO_3 \cdot 9\ (SiO_2 \cdot 12\ WO_3) \cdot 267\ H_2O$	55 (U):	Hb-244
$O_{357}Si_9U_5W_{108}$	$5\ UO_3 \cdot 9\ (SiO_2 \cdot 12\ WO_3) \cdot 264\ H_2O$	55 (U):	Hb-244
O_xPb	PbO_x	47 (Pb):	Hb/C1-139/47
O_xPr	PrO_x (x = 1.5 to 2)	39 (SE):	Hb/C1-279/90
–	PrO_x solid solutions:		
	PrO_x-TbO_x	39 (SE):	Hb/C1-402/4
–	PrO_x systems:		
	$PrO_x-La_2O_3$	39 (SE):	Hb/C1-294/5
	$PrO_x-Nd_2O_3$	39 (SE):	Hb/C1-294/5, 310/1
	PrO_x-O	39 (SE):	Hb/C1-275/8
	$PrO_x-Sm_2O_3$	39 (SE):	Hb/C1-294/5
	$PrO_x-Y_2O_3$	39 (SE):	Hb/C1-294/5
O_xRuZr_2	Zr_2RuO_x	63 (Ru):	Eb-259

Hb = Main Series Volume; GD = Gmelin-Durrer; Ab = Appended Volume to Main Series;
Eb = Supplement Volume to Main Series; Erg. W. = New Supplement Series Volume

O_xSm . . . . SmO_x (x = 0.4 to 0.6) . . . . 39 (SE): Hb/C1–314
O_xTa . . . . TaO_x . . . . 50 (Ta): Hb/B1–17/26, 30
O_xTb . . . . TbO_x (x = 1.5 to 2) . . . . 39 (SE): Hb/C1–107, 387, 391/8
–. . . . . TbO_x solid solutions:
TbO_x–CeO_2 . . . . 39 (SE): Hb/C1–401/2
TbO_x–Gd_2O_3 . . . . 39 (SE): Hb/C1–404
TbO_x–PrO_x . . . . 39 (SE): Hb/C1–402/4
TbO_x–Y_2O_3 . . . . 39 (SE): Hb/C1–400/1

–. . . . . TbO_x systems:
TbO_x–Er_2O_3 . . . . 39 (SE): Hb/C1–421/2
TbO_x–O . . . . 39 (SE): Hb/C1–391/4

O_xV . . . . VO_x . . . . 48 (V): Hb/B1–11/20
$O_{x+6}Ta_2Tl_2$ . . . . . $Tl_2Ta_2O_{6+x}$ . . . . 50 (Ta): Hb/B2–201
$O_{x+7}Sn_{x+2}Ta_2$ . . . . $Sn_{2+x}Ta_2O_{7+x}$ . . . . 50 (Ta): Hb/B2–231
$O_{2x+3}S_xSb_2$ . . . . . $Sb_2O_3 \cdot xSO_2$ . . . . 18 (Sb): Hb/B3–541

Os. . . . . Osmium:
Abundance . . . . 68 (Pt): Hb/A1–24/7
Atom. . . . . 66 (Os): Hb–3/7
Atomic nucleus . . . . 66 (Os): Hb–1/3
Catalytic properties . . . . 66 (Os): Hb–23/4
68 (Pt): Hb/A3–429/30
Chemical reactions . . . . 66 (Os): Hb–21/4
Colloidal forms . . . . 68 (Pt): Hb/A3–427/30
Cosmochemistry . . . . 68 (Pt): Hb/A1–15/22
Crystallographic properties . . . . . . 66 (Os): Hb–8/9
Detection . . . . 68 (Pt): Hb/A4–432, 447/51, 490/1
– in iron . . . . 59 (Fe): Hb/FII1–41
Determination. . . . . 68 (Pt): Hb/A4–472/6, 490/1
– in special substances . . . . 68 (Pt): Hb/A4–525/32
Economic deposits
see Pt ... Platinum

Economics. . . . . 68 (Pt): Hb/A2–302/10
Electrical properties . . . . 66 (Os): Hb–20
Electrochemical behavior . . . . 66 (Os): Hb–20/1
Geochemical character . . . . 68 (Pt): Hb/A1–23/8
Geochemistry . . . . 68 (Pt): Hb/A1–22/49
Hb/A2–275/6
History. . . . . 68 (Pt): Hb/A1–12/3
Magnetic properties . . . . 66 (Os): Hb–19/20
Mechanical properties . . . . 66 (Os): Hb–9/11
Mineralogy . . . . 68 (Pt): Hb/A2–276/301
Occurrence, extraterrestrial . . . . . . 68 (Pt): Hb/A1–15/22
–, terrestrial . . . . 68 (Pt): Hb/A1–22/49
Hb/A2–275/6
– in meteorites . . . . 68 (Pt): Hb/A1–16/22

See the "Instructions for the user" at the beginning of the volume

Os	Osmium:		
	Occurrence in the biosphere	68 (Pt):	Hb/A2–275
	– in the hydrosphere	68 (Pt):	Hb/A2–275
	– in the lithosphere	68 (Pt):	Hb/A1–28/49
	Optical properties	66 (Os):	Hb–13/9
	Physiological action	68 (Pt):	Hb/A2–275/6
	Position in the periodic system	66 (Os):	Hb–8
	Preparation	68 (Pt):	Hb/A3–311/430
	– from minerals	68 (Pt):	Hb/A3–312/50
	– from wastes and residues	68 (Pt):	Hb/A3–350/62
	– in pure state	68 (Pt):	Hb/A3–375/6
	– of dense osmium	68 (Pt):	Hb/A3–383
	Purity testing	68 (Pt):	Hb/A4–520
	Separation, analytical	68 (Pt):	Hb/A4–493/5, 517/8
	Special forms	68 (Pt):	Hb/A3–401/4, 427/30
	Spectra	66 (Os):	Hb–13/9
	Thermal properties	66 (Os):	Hb–11/3
	Toxicity	68 (Pt):	Hb/A2–275/6
	Uses	68 (Pt):	Hb/A3–385/91
	Valence	66 (Os):	Hb–7/8
–	Osmium ions:		
	Properties of the ion	66 (Os):	Hb–5/7
	Spectra	66 (Os):	Hb–14/5
–	Osmium isotopes	66 (Os):	Hb–1/2
–	Osmium systems:		
	Os–Fe	68 (Pt):	Hb/A6–725/6
	Os–Fe–S	68 (Pt):	Hb/A6–726
	Os–H	66 (Os):	Hb–25
	Os–P	66 (Os):	Hb–62/3
	Os–S	66 (Os):	Hb–58
	Os–Ti	68 (Pt):	Hb/A6–721
Os^+	Os^+	66 (Os):	Hb–5
Os^{2+}	Os^{2+}	66 (Os):	Hb–5
Os^{3+}	Os^{3+}	66 (Os):	Hb–5/6
Os^{4+}	Os^{4+}	66 (Os):	Hb–5/7
Os^{6+}	Os^{6+}	66 (Os):	Hb–6
Os^{8+}	Os^{8+}	66 (Os):	Hb–5
OsP_2	OsP_2	66 (Os):	Hb–63
OsS	OsS	66 (Os):	Hb–58
OsS_2	OsS_2	66 (Os):	Hb–59
OsS_3	OsS_3	66 (Os):	Hb–58
OsS_4	OsS_4	66 (Os):	Hb–59/60
$OsSe_2$	$OsSe_2$	66 (Os):	Hb–61
$OsTe_2$	$OsTe_2$	66 (Os):	Hb–61
OsTi	OsTi	68 (Pt):	Hb/A6–721
Os_2S_3	Os_2S_3	66 (Os):	Hb–58
Os_2Zr	$ZrOs_2$	68 (Pt):	Hb/A6–722

Hb = Main Series Volume; GD = Gmelin-Durrer; Ab = Appended Volume to Main Series; Eb = Supplement Volume to Main Series; Erg. W. = New Supplement Series Volume

P . . . . Phosphorus:

Absorption of X- and gamma rays . .	16 (P):	Hb/B–191/8
Abundance	16 (P):	Hb/A–104
Atom.	16 (P):	Hb/B–161/208
Atomic nucleus	16 (P):	Hb/B–159/61
Atomic radius	16 (P):	Hb/B–203/4
Atomic refraction	16 (P):	Hb/B–207/8
Atomic scattering factors.	16 (P):	Hb/B–162/3
Atomic volume	16 (P):	Hb/B–204
Atomic weight	16 (P):	Hb/B–99/104
Behavior as solvent.	16 (P):	Hb/B–339
Bond properties.	16 (P):	Hb/B–199/203
Chemical reactions	16 (P):	Hb/B–258/339
– in electrical discharges	16 (P):	Hb/B–259
– on heating.	16 (P):	Hb/B–260/3
– on irradiation	16 (P):	Hb/B–258/9
– with acids	16 (P):	Hb/B–279/81
– with air	16 (P):	Hb/B–263/76
– with metal compounds.	16 (P):	Hb/B–310/6
– with metals	16 (P):	Hb/B–288/304
– with nonmetal compounds	16 (P):	Hb/B–276/81, 304/10
– with nonmetals	16 (P):	Hb/B–281/8
– with organic compounds.	16 (P):	Hb/B–329/31
– with oxygen	16 (P):	Hb/B–263/76
– with solutions of metal compounds	16 (P):	Hb/B–316/29
– with water.	16 (P):	Hb/B–276/8
Chemical shift.	16 (P):	Hb/B–200/1
Colloidal forms	16 (P):	Hb/B–107
Constitution	16 (P):	Hb/B–215/7
– of phosphorus vapor	16 (P):	Hb/B–233/4
Cosmochemistry	16 (P):	Hb/A–53/68
Crystal chemistry	16 (P):	Hb/A–69/103
Crystallographic properties	16 (P):	Hb/B–215/8, 227/32, 238/40, 247/8, 251/2
Detection	16 (P):	Hb/B–339/52
– in iron	59 (Fe):	Hb/FI–369/72
– in special substances	16 (P):	Hb/B–352/9
Determination.	16 (P):	Hb/B–359/81
– in inorganic substances	16 (P):	Hb/B–396/420, 449/54
– in iron	16 (P):	Hb/B–399/400
	59 (Fe):	Hb/FI–372/411
		Hb/FII2–314, 349
– in organic substances	16 (P):	Hb/B–396/7, 420/39
– in soils	16 (P):	Hb/B–396/7, 439/49
Diadochy	16 (P):	Hb/A–85/101
Diffusion in iron.	59 (Fe):	Hb/A8–1797

See the "Instructions for the user" at the beginning of the volume

P . . . . Phosphorus:

Subject	Vol.	Reference
Diffusion in silicon	16 (P):	Hb/C–635/6
Discovery	16 (P):	Hb/A–6/14
Economics	16 (P):	Hb/A–240/2
	28 (Ca):	Hb/B1–202/4
Electrical properties	16 (P):	Hb/B–238, 245, 250/1, 254/5
Electrochemical behavior	16 (P):	Hb/B–256/8
Electron affinity	16 (P):	Hb/B–205
Electron configuration	16 (P):	Hb/B–161
Electronegativity	16 (P):	Hb/B–201/3
Formation	16 (P):	Hb/B–104/5
g factor	16 (P):	Hb/B–161/2
Geochemical character	16 (P):	Hb/A–104
Geochemical cycle	16 (P):	Hb/A–237/40
Geochemistry	16 (P):	Hb/A–69/240
Handling	16 (P):	Hb/B–84/5
History	16 (P):	Hb/A–1/52
Hydrosols	16 (P):	Hb/B–107
Increments	16 (P):	Hb/B–207/8
Industrial hazards	16 (P):	Hb/B–84/5
Interaction with electrons	16 (P):	Hb/B–205/7
– with ions	16 (P):	Hb/B–207
Ionization potentials	16 (P):	Hb/B–205/7
Magnetic moment	16 (P):	Hb/B–161/2
Magnetic properties	16 (P):	Hb/B–244/5, 250/1, 254
Mass absorption coefficient	16 (P):	Hb/B–192/7
Mass photoabsorption coefficient	16 (P):	Hb/B–197
Mass scattering coefficient	16 (P):	Hb/B–197/8
Mechanical properties	16 (P):	Hb/B–235, 239/43, 248/9, 252/3
Mesonic atom	16 (P):	Hb/B–188/90
Mineralogy:		
Anhydrous phosphates with foreign anions	16 (P):	Hb/A–350/431
– without foreign anions	16 (P):	Hb/A–251/98
Apatite group	16 (P):	Hb/A–350/82
Hydrogen phosphates	16 (P):	Hb/A–243/51
Hydrous phosphates with foreign anions	16 (P):	Hb/A–431/509
– without foreign anions	16 (P):	Hb/A–298/350
Structures of phosphorus minerals	16 (P):	Hb/A–101/3
Unclassified phosphates	16 (P):	Hb/A–509/10
Modifications:		
Black phosphorus:		
Formation	16 (P):	Hb/B–105, 227/9
Historical data	16 (P):	Hb/A–52

Hb = Main Series Volume; GD = Gmelin-Durrer; Ab = Appended Volume to Main Series; Eb = Supplement Volume to Main Series; Erg. W. = New Supplement Series Volume

P . . . . Phosphorus:
Modifications:
Black phosphorus:
Oxidation. . . . . 16 (P): Hb/B–276
Phase transitions . . . . 16 (P): Hb/B–224
Physical properties . . . . 16 (P): Hb/B–98, 247/51
Brown phosphorus. . . . . 16 (P): Hb/B–254/5
Gray phosphorus. . . . . 16 (P): Hb/B–98, 247/51
Orange-red phosphorus . . . . 16 (P): Hb/B–97, 216, 255
Purple phosphorus. . . . . 16 (P): Hb/B–232, 255
Red phosphorus:
Constitution . . . . 16 (P): Hb/B–216/7
Deposition from waste gases. . . 16 (P): Hb/B–42
Determination . . . . 16 (P): Hb/B–361/2
Formation . . . . 16 (P): Hb/B–104/5
Historical data . . . . 16 (P): Hb/A–46/52
Oxidation. . . . . 16 (P): Hb/B–275/6
Phase transitions . . . . 16 (P): Hb/B–224/7
Physical properties . . . . 16 (P): Hb/B–251/6
Preparation . . . . 16 (P): Hb/B–81/3, 86
Purification. . . . . 16 (P): Hb/B–83
Purity testing. . . . . 16 (P): Hb/B–83/4
Stabilization . . . . 16 (P): Hb/B–83
Ruby-red phosphorus . . . . 16 (P): Hb/B–232
Violet phosphorus:
Constitution . . . . 16 (P): Hb/B–216/7
Oxidation. . . . . 16 (P): Hb/B–275/6
Phase transitions . . . . 16 (P): Hb/B–224/7
Physical properties . . . . 16 (P): Hb/B–98, 251/6
White phosphorus:
Aftertreatment. . . . . 16 (P): Hb/B–80/1
Constitution . . . . 16 (P): Hb/B–215
Deposition from waste gases. . . 16 (P): Hb/B–40/2
Determination . . . . 16 (P): Hb/B–81, 382, 398
Formation . . . . 16 (P): Hb/B–104
Luminescence . . . . 16 (P): Hb/B–265/74
Oxidation. . . . . 16 (P): Hb/B–263/75
Phase transitions . . . . 16 (P): Hb/B–217/23
Physical properties . . . . 16 (P): Hb/B–238/47
Preparation . . . . 16 (P): Hb/B–40/3, 80/1, 85/6, 105/7
Purification. . . . . 16 (P): Hb/B–80, 106/7
Purity testing. . . . . 16 (P): Hb/B–81
Yellow phosphorus . . . . 16 (P): Hb/B–97, 240, 245/6, 255
Molecule. . . . . 16 (P): Hb/B–208/15

See the "Instructions for the user"
at the beginning of the volume

P . . . . Phosphorus:

Occurrence, extraterrestrial . . . . . . 16 (P): Hb/A–53/68
–, terrestrial . . . . 16 (P): Hb/A–69/240
– in meteorites . . . . 16 (P): Hb/A–55/68
– in minerals . . . . 16 (P): Hb/A–69/85
– in soils . . . . 16 (P): Hb/A–156/76
– in the atmosphere . . . . 16 (P): Hb/A–214/5
– in the biosphere. . . . . 16 (P): Hb/A–215/37
– in the hydrosphere . . . . 16 (P): Hb/A–182/214
– in the lithosphere . . . . 16 (P): Hb/A–104/82
– in the magmatic cycle . . . . 16 (P): Hb/A–104/23
– in the metamorphic cycle . . . . . . 16 (P): Hb/A–176/82
– in the sedimentary cycle . . . . 16 (P): Hb/A–123/56
Optical properties. . . . . 16 (P): Hb/B–238, 245/7, 251, 255/6
Organosols . . . . 16 (P): Hb/B–107
Oxidation . . . . 16 (P): Hb/B–263/76
Parachor . . . . 16 (P): Hb/B–207
Phase diagram . . . . 16 (P): Hb/B–215
Phase transitions . . . . 16 (P): Hb/B–215/33
Polymorphism . . . . 16 (P): Hb/B–215/33
Position in the electrochemical series . . . . 16 (P): Hb/B–256
– in the periodic system . . . . 16 (P): Hb/B–98
Preparation . . . . 16 (P): Hb/B–105/7
–, historical data . . . . 16 (P): Hb/A–23/37
–, industrial . . . . 16 (P): Hb/B–1/86
–, –, oxidation of PH_3 . . . . 16 (P): Hb/B–58/9
–, –, thermal methods . . . . 16 (P): Hb/B–21/50
–, –, wet dressing . . . . 16 (P): Hb/B–3/21
– of special forms. . . . . 16 (P): Hb/B–107
Purification . . . . 16 (P): Hb/B–106/7
Scattering of X- and gamma rays . . . 16 (P): Hb/B–191/8
Separation, analytical . . . . 16 (P): Hb/B–382/96
Solubility . . . . 16 (P): Hb/B–331/9
Spectra:
Arc spectrum . . . . 16 (P): Hb/B–163/7
Band spectrum . . . . 16 (P): Hb/B–210/1
Radiation of the mesonic atom . . . 16 (P): Hb/B–188/90
Spark spectra . . . . 16 (P): Hb/B–167/74
X-ray spectrum . . . . 16 (P): Hb/B–174/88

Technology
see Preparation, industrial

Thermal properties . . . . 16 (P): Hb/B–218/27, 232/8, 240, 243/4, 249, 253/4
Toxicity . . . . 16 (P): Hb/B–87/91
Uses, historical data . . . . 16 (P): Hb/A–37/46

Hb = Main Series Volume; GD = Gmelin-Durrer; Ab = Appended Volume to Main Series;
Eb = Supplement Volume to Main Series; Erg. W. = New Supplement Series Volume

P . . . . Phosphorus:
Valence . . . . 16 (P): Hb/B–98
Vapor pressure . . . . 16 (P): Hb/B–221, 226

–. . . . . Phosphorus ions
see P^+ ... P^+, etc.

–. . . . . Phosphorus isotopes:
Detection and determination. . . . . . 16 (P): Hb/B–455/8
Enrichment, preparation,
and separation . . . . 16 (P): Hb/B–108/59, 455/6
Nuclear properties . . . . 16 (P): Hb/B–159/61
Toxicity . . . . 16 (P): Hb/B–97
Uses . . . . 16 (P): Hb/B–108, 149

–. . . . . Phosphorus solutions:
P–inorganic solvents. . . . . 16 (P): Hb/B–332/3
P–organic solvents . . . . 16 (P): Hb/B–333/9

–. . . . . Phosphorus systems:
P–Ag . . . . 61 (Ag): Hb/B4–1/3
P–Ag–Cu . . . . 61 (Ag): Hb/B4–484/5
P–As. 17 (As): Hb–473
P–Au . . . . 62 (Au): Hb/3–728
P–Bi . . . . 19 (Bi): Eb–847
P–Br . . . . 16 (P): Hb/C–490/1
P–C–Fe . . . . 59 (Fe): Hb/A5–1095/6
P–C–Fe–Si . . . . 59 (Fe): Hb/A8–1798
P–Co. 58 (Co): Hb/A2–388
Eb/A–738
P–Cr . . . . 52 (Cr): Hb/B–437
P–Cr–Cu . . . . 60 (Cu): Hb/B3–1228
P–Cr–Ni. . . . . 57 (Ni): Hb/B1–222
Hb/B3–1215
P–Cu. 60 (Cu): Hb/B2–904/6
P–Cu–Fe . . . . 60 (Cu): Hb/B3–1315
P–Cu–Mn . . . . 60 (Cu): Hb/B3–1248
P–Cu–Ni . . . . 60 (Cu): Hb/B3–1258/9
P–Cu–O. . . . . 60 (Cu): Hb/B2–918
P–Cu–Si . . . . 60 (Cu): Hb/B2–933
P–Cu–Sn . . . . 60 (Cu): Hb/B3–1196
P–Cu–Zn . . . . 60 (Cu): Hb/B3–1151
P–Fe. 59 (Fe): Hb/A8–1781/4
P–Fe–O . . . . 59 (Fe): Hb/A5–923/35
P–Fe–Si. . . . . 59 (Fe): Hb/A8–1797/8
P–Ge . . . . 45 (Ge): Eb–558
P–Hg . . . . 34 (Hg): Hb/B4–1357
P–I. . . . . 16 (P): Hb/C–518
P–Ir . . . . 68 (Pt): Hb/A6–729/30
P–Mn–Ni . . . . 57 (Ni): Hb/B1–311
P–Ni . . . . 57 (Ni): Hb/B3–946/8

*See the "Instructions for the user"
at the beginning of the volume*

P	Phosphorus systems:		
	P–Os	66 (Os):	Hb–62/3
	P–Pd	65 (Pd):	Hb/2–302/5
	P–Pt	68 (Pt):	Hb/C1–133/5
	P–Re	70 (Re):	Hb–123
	P–Rh	68 (Pt):	Hb/A5–547/9
	P–Ru	68 (Pt):	Hb/A5–534/5
	P–S	16 (P):	Hb/C–567
	P–Se	16 (P):	Hb/C–605/6
	P–Si	16 (P):	Hb/C–635/6
	P–Ti	41 (Ti):	Hb–378
	P–Tl	38 (Tl):	Hb/3–425/6
	P–V	48 (V):	Hb/B1–358
	P–Zr	42 (Zr):	Hb–389/90
P^{+}	P^{+}:		
	Formation	16 (P):	Hb/B–208 Hb/C–24
	Ionization potentials	16 (P):	Hb/B–205/6
	Properties of the ion	16 (P):	Hb/B–161/2, 204/5
	Spectrum	16 (P):	Hb/B–167/9
P^{2+}	P^{2+}	16 (P):	Hb/B–161/2, 169/70, 205/6
P^{3+}	P^{3+}:		
	Ionic susceptibility	16 (P):	Hb/B–208
	Ionization potentials	16 (P):	Hb/B–205/6
	Properties of the ion	16 (P):	Hb/B–161/2
	Spectrum	16 (P):	Hb/B–170/1
P^{4+}	P^{4+}	16 (P):	Hb/B–161/2, 171/2, 205/6
P^{5+}	P^{5+}:		
	Increments	16 (P):	Hb/B–208
	Ionic mobility	16 (P):	Hb/B–256
	Ionization potentials	16 (P):	Hb/B–205/6
	Properties of the ion	16 (P):	Hb/B–162, 204/5
	Scattering factors	16 (P):	Hb/B–163
	Spectrum	16 (P):	Hb/B–172
P^{n+}	P^{n+} (n = 6 to 12)	16 (P):	Hb/B–162, 173/4, 205/6
P^{-}	P^{-}	16 (P):	Hb/B–161, 208 Hb/C–24
P^{2-}	P^{2-}	16 (P):	Hb/B–161
P^{3-}	P^{3-}	16 (P):	Hb/B–161, 163, 205, 208
PPd_3	Pd_3P	65 (Pd):	Hb/2–303/4, 306
PPd_5	Pd_5P	65 (Pd):	Hb/2–303, 305
PPt	PtP	68 (Pt):	Hb/C1–135
PPt_2	Pt_2P	68 (Pt):	Hb/C1–135
PPu	PuP	Erg. W. 4–249/51	

Hb = Main Series Volume; GD = Gmelin-Durrer; Ab = Appended Volume to Main Series;
Eb = Supplement Volume to Main Series; Erg. W. = New Supplement Series Volume

PRe	ReP	70 (Re):	Hb–123/4
PRe_2	Re_2P	70 (Re):	Hb–123/4
PRh_2	Rh_2P	68 (Pt):	Hb/A5–547/9
PRu	RuP	63 (Ru):	Eb–335
		68 (Pt):	Hb/A5–534/5
PRuS	RuPS	63 (Ru):	Eb–335
PRuSe	RuPSe	63 (Ru):	Eb–335
PRu_2	Ru_2P	63 (Ru):	Eb–334
		68 (Pt):	Hb/A5–534/5
$PS_{0.25}$	$PS_{0.25} = P_4S$	16 (P):	Hb/C–581
$PS_{0.5}$	$PS_{0.5} = P_4S_2$	16 (P):	Hb/C–581
PS	PS.	16 (P):	Hb/C–572/3
–	$(PS)_n$	16 (P):	Hb/C–573
PS^+	PS^+	16 (P):	Hb/C–572/3
PSTa	TaPS	50 (Ta):	Hb/B1–271
–	TaPS solid solutions:		
	TaPS–NbPS	50 (Ta):	Hb/B1–271
	TaPS–NbPSe	49 (Nb):	Hb/B1–414
		50 (Ta):	Hb/B1–271
$PS_{1.5}$	$PS_{1.5} = P_2S_3$	16 (P):	Hb/B–96/7
			Hb/C–581
$PS_{1.66}$	$PS_{1.66} = P_3S_5$	16 (P):	Hb/C–581
PS_2	PS_2	16 (P):	Hb/C–582
PS_3	PS_3	16 (P):	Hb/C–582
PS_4	PS_4	16 (P):	Hb/C–582
PS_4Sb	$SbPS_4$	18 (Sb):	Hb/B3–561/2
PS_4Tl_3	Tl_3PS_4	38 (Tl):	Hb/3–432
PS_5	PS_5	16 (P):	Hb/C–582
PSe_3	PSe_3	16 (P):	Hb/C–606
PSe_5	PSe_5	16 (P):	Hb/C–607
PSi	PSi	16 (P):	Hb/C–636/7
PSi_2	PSi_2	16 (P):	Hb/C–637
PTa	TaP	50 (Ta):	Hb/B1–267
PTa_2	Ta_2P	50 (Ta):	Hb/B1–266
PTa_3	Ta_3P	50 (Ta):	Hb/B1–266
PTi_2	Ti_2P	41 (Ti):	Hb–378
PTl_3	Tl_3P	38 (Tl):	Hb/3–426
PU	UP	55 (U):	Hb–179
PV	VP	48 (V):	Hb/B1–358/9
PV_2	V_2P	48 (V):	Hb/B1–358
PV_3	V_3P	48 (V):	Hb/B1–358
PW	WP	54 (W):	Hb–206
PW_2	W_2P	54 (W):	Hb–206
PZn	ZnP	32 (Zn):	Hb–275
PZr	ZrP	42 (Zr):	Hb–391
PZr_3	Zr_3P	42 (Zr):	Hb–391
$P_{1.33}Zr$	$ZrP_{1.33}$	42 (Zr):	Hb–390
$P_{1.46}Zr$	$ZrP_{1.46}$	42 (Zr):	Hb–389, 390

See the "Instructions for the user"
at the beginning of the volume

P_2	P_2	16 (P):	Hb/B–209/12
P_2^+	P_2^+	16 (P):	Hb/B–212
P_2^-	P_2^-	16 (P):	Hb/B–208
$P_2Pb_2S_6$	$Pb_2P_2S_6$	47 (Pb):	Hb/C3–884
$P_2Pb_2S_7$	$Pb_2P_2S_7$	47 (Pb):	Hb/C3–885
P_2Pb_3	Pb_3P_2	47 (Pb):	Hb/C3–861
$P_2Pb_3S_8$	$Pb_3(PS_4)_2$	47 (Pb):	Hb/C3–884
P_2Pd	PdP_2	65 (Pd):	Hb/2–303/4, 306
P_2Pd_5	Pd_5P_2	65 (Pd):	Hb/2–303/4, 306
P_2Pt	PtP_2	68 (Pt):	Hb/C1–133, 135/6
P_2Re	ReP_2	70 (Re):	Hb–123/4
P_2Rh	RhP_2	68 (Pt):	Hb/A5–547/9
P_2Ru	RuP_2	63 (Ru):	Eb–335
		68 (Pt):	Hb/A5–534/5
P_2S_3	$P_2S_3 = P_4S_6$	16 (P):	Hb/B–96/7
			Hb/C–581
P_2S_4	P_2S_4	16 (P):	Hb/C–582
P_2S_5	P_2S_5 see P_4S_{10} ... P_4S_{10}		
$P_2S_6Zn_2$	$Zn_2P_2S_6$	32 (Zn):	Hb–281
$P_2S_6Zn_3$	$Zn_3(PS_3)_2$	32 (Zn):	Hb–281
$P_2S_7Zn_2$	$Zn_2P_2S_7$	32 (Zn):	Hb–281
$P_2S_8Zn_3$	$Zn_3(PS_4)_2$	32 (Zn):	Hb–281
P_2S_{11}	P_2S_{11}	16 (P):	Hb/C–582
P_2Se	P_2Se	16 (P):	Hb/C–606
P_2Se_3	P_2Se_3	16 (P):	Hb/C–606
P_2Se_5	P_2Se_5	16 (P):	Hb/C–607/8
P_2Sr_3	Sr_3P_2	29 (Sr):	Hb–211
			Eb–283
P_2Ta	TaP_2	50 (Ta):	Hb/B1–268
P_2Te_3	P_2Te_3	16 (P):	Hb/C–608/9
P_2V	VP_2	48 (V):	Hb/B1–359
P_2W	WP_2	54 (W):	Hb–206
P_2Zn	ZnP_2	32 (Zn):	Hb–275/6
			Eb–995
P_2Zn_3	Zn_3P_2	16 (P):	Hb/B–92
		32 (Zn):	Hb–275
			Eb–994/5
P_2Zr	ZrP_2	42 (Zr):	Hb–391/2
P_3	P_3	16 (P):	Hb/B–213
P_3^+	P_3^+	16 (P):	Hb/B–208
P_3^-	P_3^-	16 (P):	Hb/B–208
P_3Pb	PbP_3	47 (Pb):	Hb/C3–861
P_3Re	ReP_3	70 (Re):	Hb–123/4
P_3Rh	RhP_3	68 (Pt):	Hb/A5–547/9
P_3S_2Zn	ZnP_3S_2	32 (Zn):	Hb–281
P_3S_5	$P_3S_5 = PS_{1.66}$	16 (P):	Hb/C–581
P_3S_6	P_3S_6	16 (P):	Hb/C–582

Hb = Main Series Volume; GD = Gmelin-Durrer; Ab = Appended Volume to Main Series;
Eb = Supplement Volume to Main Series; Erg. W. = New Supplement Series Volume

P_4	P_4	16 (P):	Hb/B–213/5
P_4^+	P_4^+	16 (P):	Hb/B–208
P_4^-	P_4^-	16 (P):	Hb/B–208
P_4Rh_5	Rh_5P_4	68 (Pt):	Hb/A5–547/9
P_4S	$P_4S = PS_{0.25}$	16 (P):	Hb/C–581
P_4SSe_2	P_4SSe_2	16 (P):	Hb/C–608
P_4S_2	$P_4S_2 = PS_{0.5}$	16 (P):	Hb/C–581
P_4S_2Se	P_4S_2Se	16 (P):	Hb/C–608
P_4S_3	P_4S_3	16 (P):	Hb/B–96 Hb/C–568/72
P_4S_5	P_4S_5	16 (P):	Hb/C–573/4
P_4S_6	$P_4S_6 = P_2S_3$	16 (P):	Hb/B–96/7 Hb/C–581
P_4S_7	P_4S_7	16 (P):	Hb/C–574/6
P_4S_{10}	$P_4S_{10} = P_2S_5$	16 (P):	Hb/B–97 Hb/C–576/81
–	P_4S_{10} systems:		
	$P_2S_5–POCl_3–PSCl_3$	16 (P):	Hb/C–597/8
	$P_2S_5–POCl_3–PSCl_3–P_2O_5$	16 (P):	Hb/C–597/8
P_4S_{11}	P_4S_{11}	16 (P):	Hb/C–582
P_4Se	P_4Se	16 (P):	Hb/C–606
P_4Se_3	P_4Se_3	16 (P):	Hb/C–606/7
P_4Ta_5	Ta_5P_4	50 (Ta):	Hb/B1–266
P_4Th_3	Th_3P_4	44 (Th):	Hb–315
P_4U_3	U_3P_4	55 (U):	Hb–179
P_4W_3	W_3P_4	54 (W):	Hb–206
P_4Zn	ZnP_4	32 (Zn):	Hb–276
P_4Zn_3	Zn_3P_4	32 (Zn):	Hb–276
P_5Pb	PbP_5	47 (Pb):	Hb/C3–861
P_5Pt_3	Pt_3P_5	68 (Pt):	Hb/C1–135
P_5Rb_2	Rb_2P_5	24 (Rb):	Hb–233
P_5Sr	SrP_5	29 (Sr):	Hb–211
P_6	P_6	16 (P):	Hb/B–208
P_6Zn	ZnP_6	32 (Zn):	Hb–276
P_7Pt_{20}	$Pt_{20}P_7$	68 (Pt):	Hb/C1–133, 135
P_8	P_8	16 (P):	Hb/B–208
P_8^+	P_8^+	16 (P):	Hb/B–208
P_8S_7	P_8S_7	16 (P):	Hb/C–575
P_8S_{11}	$P_8S_{11} = P_4S_3 \cdot 2\ P_2S_4$	16 (P):	Hb/C–581
$P_{14}PbZn$	$PbZnP_{14}$	47 (Pb):	Hb/C4–1229
P_xTh	ThP_x	44 (Th):	Hb–314
P_xTi	TiP_x	41 (Ti):	Hb–378/9
Pa	Protactinium:		
	Atom	51 (Pa): Erg. W. 8–1/34	Hb–43/7, 77, 94
	Atomic nucleus	51 (Pa):	Hb–37/43, 47/51, 74/87, 93/7

See the "Instructions for the user"
at the beginning of the volume

Pa . . . . Protactinium:

Atomic radius . . . . Erg. W. 8–4/5

Chemical reactions . . . . 51 (Pa): Hb–53/61, 87/8, 91/3

Detection . . . . 51 (Pa): Hb–62/4, 88, 97/8

Determination. . . . . 51 (Pa): Hb–64/6, 88, 98/9

Electrochemical behavior . . . . 51 (Pa): Hb–52/3, 87, 97

Electron configuration . . . . Erg. W. 8–3/23

Enrichment . . . . 51 (Pa): Hb–13/28

Formation . . . . 51 (Pa): Hb–11/2, 37, 71/2, 90, 93

History. . . . . 51 (Pa): Hb–1/10, 70/1, 89/90

Ionization potentials . . . . Erg. W. 8–23

Nuclear decay,

alpha radioactivity . . . . 51 (Pa): Hb–49
Erg. W. 7a–85/96

–, beta radioactivity. . . . . 51 (Pa): Hb–49/50, 75/6, 79/84, 94/6

–, decay scheme . . . . Erg. W. 7a–99

–, gamma radioactivity . . . . 51 (Pa): Hb–50/1, 75/7, 84/7, 96/7
Erg. W. 7a–80

Nuclear fission by neutrons . . . . . . Erg. W. 7a–140

Occurrence . . . . 51 (Pa): Hb–10/1, 71, 90

Oxidation states. . . . . Erg. W. 8–8/11

Position in the periodic system . . . . 51 (Pa): Hb–46/7, 77, 94

Preparation . . . . 51 (Pa): Hb–12/37, 72/4, 91/3

– in charged particle reactions. . . . . Erg. W. 7b–145

Purification . . . . 51 (Pa): Hb–29/32

Spectra:

Emission spectra . . . . 51 (Pa): Hb–51

Mössbauer spectra. . . . . Erg. W. 8–89/90

X-ray spectra . . . . 51 (Pa): Hb–51/2

Uses . . . . 51 (Pa): Hb–36/7, 74, 93

Valence . . . . 51 (Pa): Hb–46, 77, 94

–. . . . . Protactinium ions:

Chemical reactions . . . . 51 (Pa): Hb–56/61, 87/8, 91/3

Electrochemical behavior . . . . 51 (Pa): Hb–52/3, 87

Ionization potentials . . . . Erg. W. 8–23

Properties of the ion . . . . 51 (Pa): Hb–44
Erg. W. 8–5/28

–. . . . . Protactinium isotopes
see Pa ... Protactinium

Pa^+ . . . . Pa^+ . . . . Erg. W. 8–23

Hb = Main Series Volume; GD = Gmelin-Durrer; Ab = Appended Volume to Main Series;
Eb = Supplement Volume to Main Series; Erg. W. = New Supplement Series Volume

Pa^{2+} . . . . Pa^{2+} . . . . Erg. W. 8–11/23
Pa^{3+} . . . . Pa^{3+} . . . . Erg. W. 8–5, 11/25
Pa^{4+} . . . . Pa^{4+} . . . . Erg. W. 8–5, 11/23
Pa^{5+} . . . . Pa^{5+} . . . . 51 (Pa): Hb–44, 52/3
Erg. W. 8–5, 23
Pa^{6+} . . . . Pa^{6+} . . . . Erg. W. 8–23
Pa^{n+} . . . . Pa^{n+} (n = 7 to 10) . . . . Erg. W. 8–23

Pb . . . . Lead:
Absorption of light . . . . 47 (Pb): Hb/B1–300/7
– of X-rays . . . . 47 (Pb): Hb/B1–94/9
Abundance . . . . 47 (Pb): Hb/A2a–1/2, 278/80
Anodic dissolution . . . . 47 (Pb): Hb/B2–219/26, 227
Anodic protective layers . . . . 47 (Pb): Hb/B2–197/219
Atom . . . . 47 (Pb): Hb/B1–55/102
Atomic field . . . . 47 (Pb): Hb/B1–55/8
Atomic nucleus . . . . 47 (Pb): Hb/B1–47/55
Atomic radius . . . . 47 (Pb): Hb/B1–101/2
Atomic refraction . . . . 47 (Pb): Hb/B1–138
Atomic scattering factors . . . . 47 (Pb): Hb/B1–160, 207/8
Atomic susceptibility . . . . 47 (Pb): Hb/B1–102
Atomic volume . . . . 47 (Pb): Hb/B1–101/2
Atomic weight . . . . 47 (Pb): Hb/A1–49/50
Hb/B1–3/9
Behavior as anode . . . . 47 (Pb): Hb/B2–150/227
– as cathode . . . . 47 (Pb): Hb/B2–133/50
– as electrode . . . . 47 (Pb): Hb/B2–108/33
Boiling point . . . . 47 (Pb): Hb/B1–239
Chemical constant . . . . 47 (Pb): Hb/B1–251
Chemical reactions . . . . 47 (Pb): Hb/B1–307/54
– with acids . . . . 47 (Pb): Hb/B1–335/53
– with air . . . . 47 (Pb): Hb/B1–308/14
– with metal compounds . . . . 47 (Pb): Hb/B1–325/35
– with metals . . . . 47 (Pb): Hb/B1–321
– with nonmetal compounds . . . . . 47 (Pb): Hb/B1–314/8, 321/5
– with nonmetals . . . . 47 (Pb): Hb/B1–308/14, 318/21
– with organic substances . . . . 47 (Pb): Hb/B1–350/4
– with water . . . . 47 (Pb): Hb/B1–314/8
Compressibility . . . . 47 (Pb): Hb/B1–222/3
Condensation . . . . 47 (Pb): Hb/B1–237
Conduction mechanism . . . . 47 (Pb): Hb/B1–260/2
Constitution of vapor . . . . 47 (Pb): Hb/B1–237
Corrosion, electrochemical . . . . 47 (Pb): Hb/B2–415
Cosmochemistry . . . . 47 (Pb): Hb/A2a–1/56
Creep properties . . . . 47 (Pb): Hb/B1–230/2
Critical constants . . . . 47 (Pb): Hb/B1–239
Crystal chemistry . . . . 47 (Pb): Hb/A2a–63/244
Crystallization . . . . 47 (Pb): Hb/B1–142/58

See the "Instructions for the user"
at the beginning of the volume

Pb . . . . Lead:

Entry	Volume	Reference
Crystallographic properties	47 (Pb):	Hb/B1–140/212
Debye temperature	47 (Pb):	Hb/B1–246/8
Deformation	47 (Pb):	Hb/B1–174/85
Density	47 (Pb):	Hb/B1–214/6
Deposition equivalent	47 (Pb):	Hb/B2–229
Detection	47 (Pb):	Hb/B1–377/89
– in iron	59 (Fe):	Hb/FII1–54/6
– in special substances	47 (Pb):	Hb/B1–390/9
Determination	47 (Pb):	Hb/B1–399/400
– in iron	59 (Fe):	Hb/FII1–56/7 Hb/FII2–300, 350
– in special substances	47 (Pb):	Hb/B1–400/77
Diffusion in Bi	47 (Pb):	Hb/C3–942/3
– in Bi–Pb alloys	47 (Pb):	Hb/C3–942/3
– in Cd–Pb alloys	47 (Pb):	Hb/C4–1232
– in Hg	47 (Pb):	Hb/C4–1261
– in Hg–Pb alloys	47 (Pb):	Hb/C4–1261
– in In	47 (Pb):	Hb/C4–1307
– in In–Pb alloys	47 (Pb):	Hb/C4–1307
– in Pb–Sb alloys	47 (Pb):	Hb/C3–905
– in Pb–Sn alloys	47 (Pb):	Hb/C4–1482/6
– in Pb–Tl alloys	47 (Pb):	Hb/C4–1327/9
– in Pb–Zn alloys	47 (Pb):	Hb/C4–1213
– in Sn	47 (Pb):	Hb/C4–1482/6
Disorder	47 (Pb):	Hb/B1–164/74
Economic deposits:		
Africa	47 (Pb):	Hb/A4–1/29
Arctic	47 (Pb):	Hb/A4–101/2
Asia	47 (Pb):	Hb/A4–62/89
Australia	47 (Pb):	Hb/A4–89/100
Central America and Mexico	47 (Pb):	Hb/A4–159/65
Europe	47 (Pb):	Hb/A3–132/89
North America	47 (Pb):	Hb/A4–102/58
Oceania	47 (Pb):	Hb/A4–101
South America	47 (Pb):	Hb/A4–165/80
Soviet Union	47 (Pb):	Hb/A4–29/62
Economics	47 (Pb):	Hb/A3–115/32
Elasticity	47 (Pb):	Hb/B1–219/22
Electrical conductivity	47 (Pb):	Hb/B1–262/72
Electrical double layer	47 (Pb):	Hb/B2–119/33
Electrical properties	47 (Pb):	Hb/B1–251/99
Electrochemical behavior	47 (Pb):	Hb/B2–1/416
Electrokinetics	47 (Pb):	Hb/B2–67/8
Electrolytic deposition	47 (Pb):	Hb/B2–227/414 Hb/C1–31/3
– from aqueous solutions	47 (Pb):	Hb/B2–229/397 Hb/C1–31/3
– from aqueous suspensions	47 (Pb):	Hb/B2–397/400

Hb = Main Series Volume; GD = Gmelin-Durrer; Ab = Appended Volume to Main Series;
Eb = Supplement Volume to Main Series; Erg. W. = New Supplement Series Volume

Pb. . . . . Lead:
Electrolytic deposition from melts. . . 47 (Pb): Hb/B2–402/14
Hb/C1–33
– from nonaqueous solutions . . . . . 47 (Pb): Hb/B2–400/1
– from solid lead compounds . . . . . 47 (Pb): Hb/B2–414
Electrolytic solution pressure . . . . . 47 (Pb): Hb/B2–11
Electron configuration . . . . 47 (Pb): Hb/B1–58/9
Electronegativity . . . . 47 (Pb): Hb/B1–100/1
Electron emission. . . . . 47 (Pb): Hb/B1–296/9
Electropolishing. . . . . 47 (Pb): Hb/B2–225/6
Energy band structure . . . . 47 (Pb): Hb/B1–251/6
Equation of state . . . . 47 (Pb): Hb/B1–212/4
Expansion coefficient. . . . . 47 (Pb): Hb/B1–217/8
Galvanomagnetic effects. . . . . 47 (Pb): Hb/B1–286/90
Geochemical character. . . . . 47 (Pb): Hb/A2a–277/8
Geochemical cycle . . . . 47 (Pb): Hb/A2a–57/62
Geochemistry . . . . 47 (Pb): Hb/A2a–57/280
Hb/A2b–1/276
Hb/A2c–1/185
Hall effect . . . . 47 (Pb): Hb/B1–286/9
Hardness . . . . 47 (Pb): Hb/B1–228/30
Heat of fusion . . . . 47 (Pb): Hb/B1–241
– of sublimation. . . . . 47 (Pb): Hb/B1–241
– of vaporization . . . . 47 (Pb): Hb/B1–241
History:
Elemental lead . . . . 47 (Pb): Hb/A1–6/50
Lead compounds. . . . . 47 (Pb): Hb/A1–99/218
Lead ores, dressing and recovery. . 47 (Pb): Hb/A1–50/75
Lead recovery, ore smelting. . . . . 47 (Pb): Hb/A1–75/87
Names and symbols . . . . 47 (Pb): Hb/A1–1/6
Preparation in pure state . . . . . . 47 (Pb): Hb/A1–87/98
Interaction with electrons . . . . 47 (Pb): Hb/B1–99/100
Interfacial tension. . . . . 47 (Pb): Hb/B1–234/7
Internal friction . . . . 47 (Pb): Hb/B1–224/8
Ionic mobility . . . . 47 (Pb): Hb/B2–11/3
Ionization . . . . 47 (Pb): Hb/B1–99
Lattice defects. . . . . 47 (Pb): Hb/B1–164/74
Lattice energy. . . . . 47 (Pb): Hb/B1–160/2
Lattice structure. . . . . 47 (Pb): Hb/B1–158/63
Lattice vibrations . . . . 47 (Pb): Hb/B1–196/200
Lead storage battery . . . . 47 (Pb): Hb/B2–415/6
Magnetic properties . . . . 47 (Pb): Hb/B1–251/99
Mass absorption coefficient . . . . . . 47 (Pb): Hb/B1–94/7
Mass energy transfer coefficient. . . . 47 (Pb): Hb/B1–98
Mechanical properties . . . . 47 (Pb): Hb/B1–212/37
Melting point . . . . 47 (Pb): Hb/B1–239/40
Mineral deposits
see Economic deposits
Mineralogy . . . . 47 (Pb): Hb/A3–1/114

See the "Instructions for the user"
at the beginning of the volume

Pb . . . . Lead:

Modifications . . . . 47 (Pb): Hb/B1–140/2
Molecule. . . . . 47 (Pb): Hb/B1–138/9
Names and symbols . . . . 47 (Pb): Hb/A1–1/6
Nuclear magnetic resonance. . . . . . 47 (Pb): Hb/B1–257/8
Nuclear moments. . . . . 47 (Pb): Hb/B1–48/9
Occurrence, extraterrestrial . . . . . . 47 (Pb): Hb/A2a–1/56
–, terrestrial . . . . 47 (Pb): Hb/A2a–57/280
Hb/A2b–1/276
Hb/A2c–1/185
– in meteorites . . . . 47 (Pb): Hb/A2a–2/25
– in minerals . . . . 47 (Pb): Hb/A2a–63/244
– in pegmatites and pneumatolytes . . 47 (Pb): Hb/A2b–169/91
– in recent volcanism . . . . 47 (Pb): Hb/A2b–270/6
– in the atmosphere . . . . 47 (Pb): Hb/A2c–171/85
– in the hydrosphere . . . . 47 (Pb): Hb/A2c–143/70
– in the hydrothermal phase . . . . . . 47 (Pb): Hb/A2b–192/269
– in the lithosphere . . . . 47 (Pb): Hb/A2b–1/276
– in the magmatic cycle . . . . 47 (Pb): Hb/A2b–1/276
– in the metamorphic cycle . . . . . . 47 (Pb): Hb/A2c–75/142
– in the orthomagmatic phase . . . . . 47 (Pb): Hb/A2b–1/168
– in the sedimentary cycle . . . . 47 (Pb): Hb/A2c–1/74
– on the moon . . . . 47 (Pb): Hb/A2a–25/56
Optical properties. . . . . 47 (Pb): Hb/B1–300/7
Overvoltage of hydrogen. . . . . 47 (Pb): Hb/B2–69/102
– of oxygen . . . . 47 (Pb): Hb/B2–102/8
Passivation by anodic
protective layers . . . . 47 (Pb): Hb/B2–197/219
Photoabsorption coefficient . . . . . . 47 (Pb): Hb/B1–98/9
Photoelectric effect . . . . 47 (Pb): Hb/B1–297/9
Photoelectron emission . . . . 47 (Pb): Hb/B1–297/8
Plasma vibrations. . . . . 47 (Pb): Hb/B1–256/7
Plasticity. . . . . 47 (Pb): Hb/B1–230/2
Polymorphism . . . . 47 (Pb): Hb/B1–140/2
Position in the periodic system . . . . 47 (Pb): Hb/B1–1
Potentials vs. acid solutions . . . . . . 47 (Pb): Hb/B2–28/35
– vs. bases. . . . . 47 (Pb): Hb/B2–35/6
– vs. foreign ionic salt solutions . . . . 47 (Pb): Hb/B2–37/45
– vs. lead-free gels . . . . 47 (Pb): Hb/B2–47
– vs. lead-free melts . . . . 47 (Pb): Hb/B2–62/3
– vs. lead salt solutions. . . . . 47 (Pb): Hb/B2–13/28
– vs. melts containing lead salts . . . . 47 (Pb): Hb/B2–48/62
– vs. nonaqueous solutions
and fluids . . . . 47 (Pb): Hb/B2–47
– vs. solid lead salts . . . . 47 (Pb): Hb/B2–63/4
vs. water. . . . . 47 (Pb): Hb/B2–45/7
Preparation, industrial . . . . 47 (Pb): Hb/C1–1/46
–, –, dry processes . . . . 47 (Pb): Hb/C1–1/30
–, –, electrolytic processes . . . . 47 (Pb): Hb/C1–31/3

Pb. . . . . Lead:

Preparation, industrial,
historical data . . . . 47 (Pb): Hb/A1–50/98
–, –, recovery from residues . . . . . . 47 (Pb): Hb/C1–33/5
–, –, refining . . . . 47 (Pb): Hb/C1–35/46
–, –, wet processes . . . . 47 (Pb): Hb/C1–30/1
– in pure state . . . . 47 (Pb): Hb/A1–87/98
Raffination . . . . 47 (Pb): Hb/C1–35/46
Recrystallization . . . . 47 (Pb): Hb/B1–186/96
Redox potentials
in aqueous solutions . . . . 47 (Pb): Hb/B2–47/8
– in melts . . . . 47 (Pb): Hb/B2–63
Redox reactions . . . . 47 (Pb): Hb/B1–375/7
Reflectivity . . . . 47 (Pb): Hb/B1–300/7
Refractive index . . . . 47 (Pb): Hb/B1–300/7
Scattering of electrons . . . . 47 (Pb): Hb/B1–100
Self-diffusion . . . . 47 (Pb): Hb/B1–208/12
Separation, analytical . . . . 47 (Pb): Hb/B1–400
Solidification point . . . . 47 (Pb): Hb/B1–240
Sound velocity . . . . 47 (Pb): Hb/B1–223/4
Spectra:
Arc spectrum . . . . 47 (Pb): Hb/B1–65/80
Band spectrum . . . . 47 (Pb): Hb/B1–139
Spark spectra . . . . 47 (Pb): Hb/B1–109/15
Terms . . . . 47 (Pb): Hb/B1–58/65
X-ray spectrum . . . . 47 (Pb): Hb/B1–81/99
Standard potential . . . . 47 (Pb): Hb/B2–1/11
Strain release . . . . 47 (Pb): Hb/B1–185/96
Structure of the melt . . . . 47 (Pb): Hb/B1–200/8
Superconductivity . . . . 47 (Pb): Hb/B1–272/86
Surface structure . . . . 47 (Pb): Hb/B1–163
Surface tension . . . . 47 (Pb): Hb/B1–234/7
Susceptibility . . . . 47 (Pb): Hb/B1–258/60
Thermal conductivity . . . . 47 (Pb): Hb/B1–248/51
Thermal diffusivity . . . . 47 (Pb): Hb/B1–251
Thermal expansion . . . . 47 (Pb): Hb/B1–217/8
Thermal properties . . . . 47 (Pb): Hb/B1–212/51
Thermal radiation . . . . 47 (Pb): Hb/B1–307
Thermodynamic functions . . . . 47 (Pb): Hb/B1–242/6
Thermoelectric effects . . . . 47 (Pb): Hb/B1–290/6
Thermomagnetic effects . . . . 47 (Pb): Hb/B1–296
Toxicity . . . . 47 (Pb): Hb/B1–482/97
Ultrasound attenuation . . . . 47 (Pb): Hb/B1–224/8
Vaporization . . . . 47 (Pb): Hb/B1–237
Vapor pressure . . . . 47 (Pb): Hb/B1–237/8
Viscosity of the melt . . . . 47 (Pb): Hb/B1–232/4
Wave functions . . . . 47 (Pb): Hb/B1–55/8
Wetting . . . . 47 (Pb): Hb/B1–236/7
Zero charge potential . . . . 47 (Pb): Hb/B2–64/7

See the "Instructions for the user"
at the beginning of the volume

Pb	Lead ions *see* Pb^{+} ... Pb^{+}, *etc.*		
–	Lead isotopes:		
	Atomic nucleus	47 (Pb):	Hb/B1–47/55
	Detection and determination	47 (Pb):	Hb/B1–477/82
	Enrichment and separation of stable Pb isotopes	47 (Pb):	Hb/B1–9/12
	–, preparation, and separation of unstable Pb isotopes	47 (Pb):	Hb/B1–12/47
	Formation of natural lead isotopes	47 (Pb):	Hb/A2a–246/8
	Isotope geochemistry	47 (Pb):	Hb/A2a–245/77
	Isotopic abundance	47 (Pb):	Hb/B1–1/2
	Isotopic ratios in lunar rocks	47 (Pb):	Hb/A2a–42/51
	– in meteorites	47 (Pb):	Hb/A2a–8/25
	– in minerals	47 (Pb):	Hb/A2a–136/52, 241/4
	– in rocks	47 (Pb):	Hb/A2b–33/4, 116/26
			Hb/A2c–66/9
	Occurrence in fumarolic products and vapor condensates	47 (Pb):	Hb/A2b–272/6
	– in mineral waters	47 (Pb):	Hb/A2c–169/70
	– in the atmosphere	47 (Pb):	Hb/A2c–171/85
–	Lead systems:		
	Pb–Ag	61 (Ag):	Hb/C–292/6
	Pb–Ag–Ag_2S–PbS	61 (Ag):	Hb/B4–263/5
	Pb–Ag–Al	35 (Al):	Hb/A6–1087/9
		61 (Ag):	Hb/C–310/1
	Pb–Ag–As	61 (Ag):	Hb/B4–273
	Pb–Ag–Cu–S	61 (Ag):	Hb/B4–470
	Pb–Ag–Se	61 (Ag):	Hb/B4–267/8
	Pb–Ag–Te	61 (Ag):	Hb/B4–269
	Pb–Ag–Zn	61 (Ag):	Hb/C–308/10
	Pb–Al	35 (Al):	Hb/A5–858/9
		47 (Pb):	Hb/C4–1278/82
	Pb–Al–Bi	47 (Pb):	Hb/C4–1287
	Pb–Al–Bi–Sb	47 (Pb):	Hb/C4–1287/8
	Pb–Al–Ce	47 (Pb):	Hb/C4–1365
	Pb–Al–Er	47 (Pb):	Hb/C4–1371
	Pb–Al–Ga	47 (Pb):	Hb/C4–1305
	Pb–Al–Mg	35 (Al):	Hb/A5–863
		47 (Pb):	Hb/C4–1288/90
	Pb–Al–Sb	35 (Al):	Hb/A5–861/3
		47 (Pb):	Hb/C4–1286/7
	Pb–Al–Sn	35 (Al):	Hb/A5–863/4
		47 (Pb):	Hb/C4 1529
	Pb–Al–Zn	47 (Pb):	Hb/C4–1290/1
	Pb–As	47 (Pb):	Hb/C3–887/9
	Pb–As–S	47 (Pb):	Hb/C3–901/3

Hb = Main Series Volume; GD = Gmelin-Durrer; Ab = Appended Volume to Main Series;
Eb = Supplement Volume to Main Series; Erg. W. = New Supplement Series Volume

Pb . . . . Lead systems:

Pb–As–Sb	47 (Pb):	Hb/C3–941/2
Pb–As–Se	47 (Pb):	Hb/C3–904/5
Pb–As–Zn	47 (Pb):	Hb/C4–1223
Pb–Au	62 (Au):	Hb/3–848/9
Pb–Au–Zn	62 (Au):	Hb/3–798
Pb–B	47 (Pb):	Hb/C2–685
Pb–Ba	47 (Pb):	Hb/C3–1200/2
Pb–Bi	47 (Pb):	Hb/C3–943/4
Pb–Bi–Cd	47 (Pb):	Hb/C4–1246/8
Pb–Bi–Cd–Sn	47 (Pb):	Hb/C4–1525/6
Pb–Bi–Cd–Sn–Zn	47 (Pb):	Hb/C4–1528/9
Pb–Bi–Cd–Zn	47 (Pb):	Hb/C4–1251
Pb–Bi–Cl	47 (Pb):	Hb/C3–968
Pb–Bi–Cu	60 (Cu):	Hb/B3–1206
Pb–Bi–F	47 (Pb):	Hb/C3–968
Pb–Bi–Ga	47 (Pb):	Hb/C4–1304
Pb–Bi–Ga–Sn	47 (Pb):	Hb/C4–1529
Pb–Bi–O	47 (Pb):	Hb/C3–996/7
Pb–Bi–S	47 (Pb):	Hb/C3–969/70
Pb–Bi–Sb	47 (Pb):	Hb/C3–977/8
Pb–Bi–Se	47 (Pb):	Hb/C3–971/2
Pb–Bi–Sn	47 (Pb):	Hb/C4–1518/9
Pb–Bi–Te	47 (Pb):	Hb/C3–973/5
Pb–Bi–Th	47 (Pb):	Hb/C4–1470
Pb–Bi–Tl	47 (Pb):	Hb/C4–1347
Pb–Bi–Zn	47 (Pb):	Hb/C4–1224/5
Pb–Br–Sn	47 (Pb):	Hb/C4–1584
Pb–C–Ni	57 (Ni):	Hb/B1–152
Pb–Ca	47 (Pb):	Hb/C3–1179/80
Pb–Cd	47 (Pb):	Hb/C4–1232/3
Pb–Cd–Ga	47 (Pb):	Hb/C4–1304
Pb–Cd–Hg	47 (Pb):	Hb/C4–1275
Pb–Cd–Mg	27 (Mg):	Hb/A4–655/6
	47 (Pb):	Hb/C4–1248
Pb–Cd–Ni	57 (Ni):	Hb/B1–153
Pb–Cd–Sb	47 (Pb):	Hb/C4–1245/6
Pb–Cd–Sn	47 (Pb):	Hb/C4–1524/5
Pb–Cd–Sn–Zn	47 (Pb):	Hb/C4–1528/9
Pb–Cd–Tl	47 (Pb):	Hb/C4–1347
Pb–Cd–Zn	47 (Pb):	Hb/C4–1248/9
Pb–Ce	47 (Pb):	Hb/C4–1364/5
Pb–Ce–Zn	47 (Pb):	Hb/C4–1363, 1365
Pb–Cl–S–Sn	47 (Pb):	Hb/C4–1586
Pb–Cl–Sn	47 (Pb):	Hb/C4–1583/4
Pb–Co	58 (Co):	Hb/A1–196
		Eb/A–415
Pb–Cr	52 (Cr):	Hb/A2–727
Pb–Cs–Zn	47 (Pb):	Hb/C4–1225

See the "Instructions for the user"
at the beginning of the volume

Pb . . . . Lead systems:			
	Pb–Cu	60 (Cu):	Hb/B3–1197
	Pb–Cu–Cu_2S–PbS	60 (Cu):	Hb/B3–1201
	Pb–Cu–O	60 (Cu):	Hb/B3–1197/8
	Pb–Cu–S	60 (Cu):	Hb/B3–1200/2
	Pb–Cu–Sb	60 (Cu):	Hb/B3–1204
	Pb–Cu–Te	60 (Cu):	Hb/B3–1202
	Pb–Dy	47 (Pb):	Hb/C4–1369/70
	Pb–Dy–Zn	47 (Pb):	Hb/C4–1363, 1370
	Pb–Er	47 (Pb):	Hb/C4–1370/1
	Pb–Er–Zn	47 (Pb):	Hb/C4–1371
	Pb–Eu	47 (Pb):	Hb/C4–1367/8
	Pb–Eu–Zn	47 (Pb):	Hb/C4–1363, 1368
	Pb–Ga	36 (Ga):	Hb–63
		47 (Pb):	Hb/C4–1303/4
	Pb–Ga–Ge	47 (Pb):	Hb/C4–1473
	Pb–Ga–Hg	47 (Pb):	Hb/C4–1305
	Pb–Ga–Sb	47 (Pb):	Hb/C4–1305
	Pb–Ga–Se	47 (Pb):	Hb/C4–1306
	Pb–Ga–Sn–Zn	47 (Pb):	Hb/C4–1529
	Pb–Ga–Zn	47 (Pb):	Hb/C4–1305
	Pb–Gd	47 (Pb):	Hb/C4–1368/9
	Pb–Gd–Zn	47 (Pb):	Hb/C4–1363, 1369
	Pb–Ge	47 (Pb):	Hb/C4–1472/3
	Pb–Ge–Mg	47 (Pb):	Hb/C4–1473
	Pb–Ge–Ni	57 (Ni):	Hb/B1–153
	Pb–Ge–Ni–Sn	57 (Ni):	Hb/B1–153
	Pb–H	47 (Pb):	Hb/C1–47/8
	Pb–Hg	47 (Pb):	Hb/C4–1261/2
	Pb–Hg–Na	47 (Pb):	Hb/C4–1274/5
	Pb–Ho	47 (Pb):	Hb/C4–1370
	Pb–Ho–Zn	47 (Pb):	Hb/C4–1363, 1370
	Pb–In	37 (In):	Hb–60/1
		47 (Pb):	Hb/C4–1307/9
	Pb–In–Sb	47 (Pb):	Hb/C4–1324
	Pb–In–Sn	47 (Pb):	Hb/C4–1529/30
	Pb–In–Tl	47 (Pb):	Hb/C4–1347
	Pb–K	47 (Pb):	Hb/C3–1052/5
	Pb–K–Na	47 (Pb):	Hb/C3–1055/6
	Pb–La	47 (Pb):	Hb/C4–1361/3
	Pb–La–Zn	47 (Pb):	Hb/C4–1363
	Pb–Li	47 (Pb):	Hb/C3–979/80
	Pb–Lu	47 (Pb):	Hb/C4–1372/3
	Pb–Lu–Zn	47 (Pb):	Hb/C4–1363, 1373
	Pb–Mg	27 (Mg):	Hb/A4–637/40
		47 (Pb):	Hb/C3–1161/2
	Pb–Mg–Na	27 (Mg):	Hb/A4–654
	Pb–Mg–Sb	27 (Mg):	Hb/A4–653

Hb = Main Series Volume; GD = Gmelin-Durrer; Ab = Appended Volume to Main Series;
Eb = Supplement Volume to Main Series; Erg. W. = New Supplement Series Volume

Pb . . . . Lead systems:

Pb–Mg–Sn	27 (Mg):	Hb/A4–656/7
	47 (Pb):	Hb/C4–1522
Pb–Mg–Zn	47 (Pb):	Hb/C4–1225
Pb–Na	47 (Pb):	Hb/C3–993/7
Pb–Na–Zn	47 (Pb):	Hb/C4–1225
Pb–Nb	49 (Nb):	Hb/B2–241/2
Pb–Nd	47 (Pb):	Hb/C4–1366/7
Pb–Nd–Zn	47 (Pb):	Hb/C4–1363, 1367
Pb–Ni	57 (Ni):	Hb/B1–148/9
Pb–Ni–S	57 (Ni):	Hb/B3–1195
Pb–Ni–Sb	57 (Ni):	Hb/B1–152
Pb–O	47 (Pb):	Hb/C1–51/2
Pb–O–S	9 (S):	Hb/B1–183
	47 (Pb):	Hb/C2–533/4
Pb–O–Sb	47 (Pb):	Hb/C3–926/9
Pb–O–Ta	50 (Ta):	Hb/B2–232
Pb–O–Te	47 (Pb):	Hb/C2–678/9
Pb–$PbBr_2$	47 (Pb):	Hb/C2–348
Pb–$PbCl_2$	47 (Pb):	Hb/C1–282/3
Pb–PbI_2–Tl–TlI	47 (Pb):	Hb/C4–1358
Pb–PbO	47 (Pb):	Hb/C1–52/4
Pb–PbS–Sn–SnS	47 (Pb):	Hb/C4–1584
Pb–PbSe–PbTe	47 (Pb):	Hb/C2–681
Pb–Pd	68 (Pt):	Hb/A5–588/93
Pb–Po	47 (Pb):	Hb/C2–685
Pb–Pr	47 (Pb):	Hb/C4–1365/6
Pb–Pr–Zn	47 (Pb):	Hb/C4–1363, 1366
Pb–Pt	68 (Pt):	Hb/A6–763/7
Pb–Pt–Tl	68 (Pt):	Hb/A6–767
Pb–S	47 (Pb):	Hb/C2–413/4
Pb–S–Sb	47 (Pb):	Hb/C3–935/6
Pb–S–Se	47 (Pb):	Hb/C2–639/40
Pb–S–Si	47 (Pb):	Hb/C3–855
Pb–S–Sn	47 (Pb):	Hb/C4–1584
Pb–Sb	47 (Pb):	Hb/C3–905/7
Pb–Sb–Se	47 (Pb):	Hb/C3–940
Pb–Sb–Si	47 (Pb):	Hb/C3–941
Pb–Sb–Sn	47 (Pb):	Hb/C4–1515
Pb–Sb–Te	47 (Pb):	Hb/C3–940/1
Pb–Sb–Zn	47 (Pb):	Hb/C4–1223/4
Pb–Sc	47 (Pb):	Hb/C4–1360
Pb–Se	47 (Pb):	Hb/C2–608/9
Pb–Se–Sn–Te	47 (Pb):	Hb/C4–1594/5
Pb–Si	47 (Pb):	Hb/C3–807/8
Pb–Sm	47 (Pb):	Hb/C4–1367
Pb–Sm–Zn	47 (Pb):	Hb/C4–1363, 1367
Pb–Sn	47 (Pb):	Hb/C4–1487/9

Pb . . . . Lead systems:

	Pb–Sn–Te	47 (Pb):	Hb/C4–1588
	Pb–Sn–Tl	47 (Pb):	Hb/C4–1530
	Pb–Sn–Zn	47 (Pb):	Hb/C4–1522/4
	Pb–Sr	47 (Pb):	Hb/C3–1193/4
	Pb–Sr–Zn	47 (Pb):	Hb/C4–1225
	Pb–Tb	47 (Pb):	Hb/C4–1369
	Pb–Te	47 (Pb):	Hb/C2–640/3
	Pb–Th	47 (Pb):	Hb/C4–1468/70
	Pb–Ti	47 (Pb):	Hb/C4–1377/8
	Pb–Ti–Zr	47 (Pb):	Hb/C4–1423
	Pb–Tl	47 (Pb):	Hb/C4–1329/32
	Pb–Tm	47 (Pb):	Hb/C4–1371
	Pb–V	48 (V):	Hb/B2–687
	Pb–W	54 (W):	Hb–104
	Pb–Y	47 (Pb):	Hb/C4–1360/1
	Pb–Y–Zn	47 (Pb):	Hb/C4–1361, 1363
	Pb–Yb	47 (Pb):	Hb/C4–1371/2
	Pb–Zn	47 (Pb):	Hb/C4–1213/4
	Pb–Zr	47 (Pb):	Hb/C4–1420/2
Pb^+	Pb^+:		
	Ionic radius	47 (Pb):	Hb/B1–136
	Ionization	47 (Pb):	Hb/B1–115
	Scattering of electrons	47 (Pb):	Hb/B1–116
	Spectra:		
	Crystal spectra	21 (Na):	Eb/6–337
	Spark spectrum	47 (Pb):	Hb/B1–109/12
	Standard potential	47 (Pb):	Hb/B2–4
	Terms	47 (Pb):	Hb/B1–103/5
Pb^{2+}	Pb^{2+}:		
	Chemical reactions	47 (Pb):	Hb/B1–354/75
	– with inorganic compounds	47 (Pb):	Hb/B1–361/70
	– with organic compounds	47 (Pb):	Hb/B1–370
	Diffusion in sodium halides	21 (Na):	Eb/6–293
	Electrochemical deposition equivalent	47 (Pb):	Hb/B2–229
	Electrolysis	47 (Pb):	Hb/B2–227/414
	Exchange reactions	47 (Pb):	Hb/B1–370/5
	Faraday effect	47 (Pb):	Hb/B1–138
	Hydration	47 (Pb):	Hb/B1–354/8
	Hydrolysis	47 (Pb):	Hb/B1–359/61
	Ionic activity	47 (Pb):	Hb/B1–358/9
	Ionic mobility	47 (Pb):	Hb/B2–11/3
	Ionic radius	47 (Pb):	Hb/B1–136/7
	Ionic refraction	47 (Pb):	Hb/B1–138
	Ionic susceptibility	47 (Pb):	Hb/B1–137/8
	Ionic volume	47 (Pb):	Hb/B1–136/7
	Ionization	47 (Pb):	Hb/B1–115

Pb^{2+} . . . . Pb^{2+}:
Redox reactions . . . . 47 (Pb): Hb/B1–375/7
Solvation . . . . 47 (Pb): Hb/B1–354/8
Sorption . . . . 47 (Pb): Hb/B1–371/5
Spectra:
Absorption spectra . . . . 47 (Pb): Hb/B1–121/6, 132/6
Crystal spectra . . . . 21 (Na): Eb/6–333/4, 337
Luminescence spectra . . . . 21 (Na): Eb/6–347
47 (Pb): Hb/B1–116/21, 126/30
Spark spectrum . . . . 47 (Pb): Hb/B1–112/4
Standard potential . . . . 47 (Pb): Hb/B2–4/5, 10
Terms . . . . 47 (Pb): Hb/B1–105/7
–. . . . . $[Pb(H_2O)_n]^{2+}$ (n = 1 to 5) . . . . 47 (Pb): Hb/C4–1609/10
Pb^{3+} . . . . Pb^{3+}:
Ionization . . . . 47 (Pb): Hb/B1–115
Paramagnetic resonance absorption . 47 (Pb): Hb/B1–130/2
Spectra . . . . 47 (Pb): Hb/B1–114
Terms . . . . 47 (Pb): Hb/B1–107/8
Pb^{4+} . . . . Pb^{4+}:
Chemical reactions . . . . 47 (Pb): Hb/B1–354
Ionic radius . . . . 47 (Pb): Hb/B1–136/7
Ionic refraction . . . . 47 (Pb): Hb/B1–138
Ionic susceptibility . . . . 47 (Pb): Hb/B1–138
Ionization . . . . 47 (Pb): Hb/B1–115
Spectra . . . . 47 (Pb): Hb/B1–115, 136
Standard potential . . . . 47 (Pb): Hb/B2–10
Terms . . . . 47 (Pb): Hb/B1–108/9
Pb^{5+} . . . . Pb^{5+} . . . . 47 (Pb): Hb/B1–115
Pb^{n+} . . . . Pb^{n+} (n = 6 to 81) . . . . 47 (Pb): Hb/B1–115/6
Pb^{-} . . . . Pb^{-} . . . . 47 (Pb): Hb/B1–136
Pb^{4-} . . . . Pb^{4-} . . . . 47 (Pb): Hb/B1–137
PbPd . . . . PdPb . . . . 68 (Pt): Hb/A5–589/93
$PbPd_2$ . . . . Pd_2Pb . . . . 68 (Pt): Hb/A5–589/93
$PbPd_3$ . . . . Pd_3Pb . . . . 68 (Pt): Hb/A5–589/93
PbPo . . . . PbPo . . . . 47 (Pb): Hb/C2–685
PbPr . . . . PrPb . . . . 47 (Pb): Hb/C4–1366
$PbPr_2$ . . . . Pr_2Pb . . . . 47 (Pb): Hb/C4–1366
$PbPr_2S_4$ . . . . $PbPr_2S_4$ . . . . 47 (Pb): Hb/C4–1374
$PbPr_2Se_4$ . . . . $PbPr_2Se_4$ . . . . 47 (Pb): Hb/C4–1375
PbPt . . . . PtPb . . . . 68 (Pt): Hb/A6–764, 766/7
$PbPt_3$ . . . . Pt_3Pb . . . . 68 (Pt): Hb/A6–764, 766
PbRb . . . . RbPb . . . . 47 (Pb): Hb/C3–1137
PbS . . . . PbS:
Chemical reactions . . . . 47 (Pb): Hb/C2–486/529
– on heating . . . . 47 (Pb): Hb/C2–486
– with acids . . . . 47 (Pb): Hb/C2–516/9
– with alkaline solutions . . . . 47 (Pb): Hb/C2–519

See the "Instructions for the user"
at the beginning of the volume

PbS . . . . PbS:

Chemical reactions

with metal compounds . . . . 47 (Pb): Hb/C2–511/6, 520/7

– with metal compounds . . . . 47 (Pb): Hb/C2–511/6, 520/7

– with metals . . . . 47 (Pb): Hb/C2–509/11, 520

– with nonmetal compounds . . . . . 47 (Pb): Hb/C2–506/9, 520

– with nonmetals . . . . 47 (Pb): Hb/C2–503/6, 519/20

– with organic compounds . . . . 47 (Pb): Hb/C2–527/9

– with oxygen . . . . 47 (Pb): Hb/C2–486/503

Colloidal forms . . . . 47 (Pb): Hb/C2–440/3

Crystallographic properties . . . . . . 47 (Pb): Hb/C2–444/51

Electrical properties . . . . 47 (Pb): Hb/C2–456/76

Electrochemical behavior . . . . 47 (Pb): Hb/B2–403, 413
Hb/C2–481/5

Formation . . . . 47 (Pb): Hb/C2–414/23

History . . . . 47 (Pb): Hb/A1–131/9

Hydrosols . . . . 47 (Pb): Hb/C2–440/1

Magnetic properties . . . . 47 (Pb): Hb/C2–458, 472, 475

Mechanical properties . . . . 47 (Pb): Hb/C2–451/3

Molecule . . . . 47 (Pb): Hb/C2–443/4

Optical properties . . . . 47 (Pb): Hb/C2–476/80

Organosols . . . . 47 (Pb): Hb/C2–441

Photoconductivity . . . . 47 (Pb): Hb/C2–464/71

Photosensitive PbS films . . . . 47 (Pb): Hb/C2–438/40

Photovoltaic effect . . . . 47 (Pb): Hb/C2–471/2

Preparation . . . . 47 (Pb): Hb/C2–414/23

– in pure state . . . . 47 (Pb): Hb/C2–423/6

– of special forms . . . . 47 (Pb): Hb/C2–426/32

Rectifier effect . . . . 47 (Pb): Hb/C2–463/4

Semiconductor properties . . . . 47 (Pb): Hb/C2–426/8, 434, 438/40, 456/80

Single crystals . . . . 47 (Pb): Hb/C2–432/4

Sorption of oxygen . . . . 47 (Pb): Hb/C2–498/9

Spectra . . . . 47 (Pb): Hb/C2–443/4

Suspensions . . . . 47 (Pb): Hb/C2–442/3

Thermal properties . . . . 47 (Pb): Hb/C2–451/6

Thin layers . . . . 47 (Pb): Hb/C2–435/8, 449/50

Toxicity . . . . 47 (Pb): Hb/B1–494

–. . . . . PbS melts:

$PbS-Ag_2S-Cu_2S$ . . . . 61 (Ag): Hb/B4–471

–. . . . . PbS solid solutions:

PbS–AgBr . . . . 61 (Ag): Hb/B4–265

$PbS-Ce_2S_3$ . . . . 47 (Pb): Hb/C4–1374

$PbS-Dy_2S_3$ . . . . 47 (Pb): Hb/C4–1374

$PbS-Er_2S_3$ . . . . 47 (Pb): Hb/C4–1374

$PbS-Gd_2S_3$ . . . . 47 (Pb): Hb/C4–1374

$PbS-Ho_2S_3$ . . . . 47 (Pb): Hb/C4–1374

$PbS-La_2S_3$ . . . . 47 (Pb): Hb/C4–1374

$PbS-Lu_2S_3$ . . . . 47 (Pb): Hb/C4–1374

PbS . . . . PbS solid solutions:

$PbS-Nd_2S_3$ . . . . 47 (Pb): Hb/C4-1374
$[(Ni,Pb)S]_n$ . . . . 57 (Ni): Hb/B3-1196
PbS-PbSe. . . . . 47 (Pb): Hb/C2-639/40
PbS-PbTe-SnTe . . . . 47 (Pb): Hb/C4-1594
$PbS-Pr_2S_3$. . . . . 47 (Pb): Hb/C4-1374
$PbS-Sm_2S_3$ . . . . 47 (Pb): Hb/C4-1374
PbS-SnS . . . . 47 (Pb): Hb/C4-1585
$PbS-Tb_2S_3$ . . . . 47 (Pb): Hb/C4-1374
$PbS-Tm_2S_3$ . . . . 47 (Pb): Hb/C4-1374
$PbS-Y_2S_3$ . . . . 47 (Pb): Hb/C4-1374
$PbS-Yb_2S_3$ . . . . 47 (Pb): Hb/C4-1374

–. . . . . PbS solutions:

$PbS-H_2O$ . . . . 47 (Pb): Hb/C2-529/30
$PbS-H_2S-H_2O$ . . . . 47 (Pb): Hb/C2-530/1

–. . . . . PbS systems:

$PbS-Ag-Ag_2S-Pb$. . . . . 61 (Ag): Hb/B4-263/5
$PbS-AgBiS_2$. . . . . 61 (Ag): Hb/B4-266
$PbS-AgSbS_2$ . . . . 61 (Ag): Hb/B4-265/6
$PbS-Ag_2S$. . . . . 61 (Ag): Hb/B4-264
$PbS-Ag_2S-Bi_2S_3$ . . . . 61 (Ag): Hb/B4-266
$PbS-Ag_2S-Cu_2S$ . . . . 61 (Ag): Hb/B4-470/1
$PbS-Al_2S_3$. . . . . 47 (Pb): Hb/C4-1294
$PbS-As_2S_3$ . . . . 47 (Pb): Hb/C3-901/3
$PbS-B_2S_3$ . . . . 47 (Pb): Hb/C2-696
$PbS-Bi_2S_3$. . . . . 47 (Pb): Hb/C3-969/70
$PbS-Cu-Cu_2S-Pb$ . . . . 60 (Cu): Hb/B3-1201
$PbS-Cu_2S-Cu_3As-Cu_3Sb$ . . . . 60 (Cu): Hb/B3-1206
$PbS-Cu_2S-FeS$. . . . . 60 (Cu): Hb/B3-1329
$PbS-Cu_2S-FeS-ZnS$ . . . . 60 (Cu): Hb/B3-1329
$PbS-Cu_2S-ZnS$. . . . . 60 (Cu): Hb/B3-1208
PbS-FeS . . . . 59 (Fe): Hb/B5-1132
$PbS-FeS-FeSiO_3-PbSiO_3$. . . . . . . 59 (Fe): Hb/B5-1137
$PbS-GeS_2$. . . . . 47 (Pb): Hb/C4-1477
PbS-GeSe . . . . 47 (Pb): Hb/C4-1478
$PbS-KCl-PbCl_2$. . . . . 47 (Pb): Hb/C3-1096
$PbS-NaCl-PbCl_2$ . . . . 47 (Pb): Hb/C3-1032
PbS-Pb-Sn-SnS . . . . 47 (Pb): Hb/C4-1584
$PbS-PbCl_2$ . . . . 47 (Pb): Hb/C2-605
$PbS-PbCl_2-SnCl_2-SnS$ . . . . 47 (Pb): Hb/C4-1586
PbS-PbO . . . . 47 (Pb): Hb/C2-534
PbS-PbSe-PbTe . . . . 47 (Pb): Hb/C2-684
PbS-PbTe. . . . . 47 (Pb): Hb/C2-679/80
$PbS-Sb_2S_3$ . . . . 47 (Pb): Hb/C3-935/6
PbS-SnS . . . . 47 (Pb): Hb/C4-1585
PbS-SnSe. . . . . 47 (Pb): Hb/C4-1588
$PbS-Sn_2S_3$ . . . . 47 (Pb): Hb/C4-1585
$PbS-Tl_2S$ . . . . 47 (Pb): Hb/C4-1352/3
PbS-ZnS . . . . 47 (Pb): Hb/C4-1227

See the "Instructions for the user"
at the beginning of the volume

PbS_2	PbS_2	47 (Pb):	Hb/C2–532
PbS_2^{2-}	$[PbS_2]^{2-}$	47 (Pb):	Hb/A2b–226
PbS_2Sn	$PbSnS_2$	47 (Pb):	Hb/C4–1585/6
PbS_3	PbS_3	47 (Pb):	Hb/C2–532
PbS_3Ta	$PbTaS_3$	50 (Ta):	Hb/B2–318
PbS_3Ti	$PbTiS_3$	47 (Pb):	Hb/C4–1391
PbS_3Zr	$PbZrS_3$	47 (Pb):	Hb/C4–1431
PbS_4Sb_2	$Pb(SbS_2)_2 = PbS \cdot Sb_2S_3$	47 (Pb):	Hb/C3–939
PbS_4Sm_2	$PbSm_2S_4$	47 (Pb):	Hb/C4–1374
PbS_4Tm_2	$PbTm_2S_4$	47 (Pb):	Hb/C4–1374
PbS_4Y_2	PbY_2S_4	47 (Pb):	Hb/C2–515
PbS_4Yb_2	$PbYb_2S_4$	47 (Pb):	Hb/C4–1374
PbS_5	PbS_5	47 (Pb):	Hb/C2–533
PbS_5Sn_4	$PbSn_4S_5$	47 (Pb):	Hb/C4–1586
$PbSb_2Se_4$	$Pb(SbSe_2)_2 = PbSe \cdot Sb_2Se_3$	47 (Pb):	Hb/C3–940
PbSe	PbSe:		
	Chemical reactions	47 (Pb):	Hb/C2–630/2
	Crystallographic properties	47 (Pb):	Hb/C2–614/7
	Electrical properties	47 (Pb):	Hb/C2–619/29
	Formation and preparation	47 (Pb):	Hb/C2–610/3
	Magnetic properties	47 (Pb):	Hb/C2–622, 627, 628/9
	Mechanical properties	47 (Pb):	Hb/C2–617/9
	Molecule	47 (Pb):	Hb/C2–613/4
	Optical properties	47 (Pb):	Hb/C2–629/30
	Special forms	47 (Pb):	Hb/C2–612/3
	Spectra	47 (Pb):	Hb/C2–614
	Thermal properties	47 (Pb):	Hb/C2–617/9
–	PbSe solid solutions:		
	PbSe–BiSe	47 (Pb):	Hb/C3–972
	$PbSe–Bi_2Se_3$	47 (Pb):	Hb/C3–972
	PbSe–CdSe	47 (Pb):	Hb/C4–1256
	$PbSe–Ce_2Se_3$	47 (Pb):	Hb/C4–1374
	$PbSe–Dy_2Se_3$	47 (Pb):	Hb/C4–1374/5
	$PbSe–Er_2Se_3$	47 (Pb):	Hb/C4–1374/5
	$PbSe–Gd_2Se_3$	47 (Pb):	Hb/C4–1374/5
	$PbSe–Ho_2Se_3$	47 (Pb):	Hb/C4–1374/5
	$PbSe–La_2Se_3$	47 (Pb):	Hb/C4–1374
	$PbSe–Lu_2Se_3$	47 (Pb):	Hb/C4–1374/5
	$PbSe–Nd_2Se_3$	47 (Pb):	Hb/C4–1374
	PbSe–PbS	47 (Pb):	Hb/C2–639/40
	$PbSe–Pr_2Se_3$	47 (Pb):	Hb/C4–1374
	$PbSe–Sm_2Se_3$	47 (Pb):	Hb/C4–1374
	PbSe–SnTe	47 (Pb):	Hb/C4–1595
	$PbSe–Tb_2Se_3$	47 (Pb):	Hb/C4–1374/5
	$PbSe–Tm_2Se_3$	47 (Pb):	Hb/C4–1374/5
	$PbSe–Yb_2Se_3$	47 (Pb):	Hb/C4–1374/5

PbSe . . . . PbSe systems:
$PbSe-AgBiSe_2$ . . . . 61 (Ag): Hb/B4-268/9
$PbSe-AgSbSe_2$ . . . . 61 (Ag): Hb/B4-268
$PbSe-AgSbTe_2$ . . . . 61 (Ag): Hb/B4-273
$PbSe-Ag_2Se$ . . . . 61 (Ag): Hb/B4-268
$PbSe-Bi_2Se_3$ . . . . 47 (Pb): Hb/C3-971/2
PbSe-CdSe . . . . 47 (Pb): Hb/C4-1255/6
PbSe-GeS . . . . 47 (Pb): Hb/C4-1479
PbSe-GeSe . . . . 47 (Pb): Hb/C4-1477/8
PbSe-GeTe . . . . 47 (Pb): Hb/C4-1480
PbSe-InSe . . . . 47 (Pb): Hb/C4-1326/7
PbSe-Pb-PbTe . . . . 47 (Pb): Hb/C2-681
PbSe-PbO . . . . 47 (Pb): Hb/C2-632
PbSe-PbS-PbTe . . . . 47 (Pb): Hb/C2-684
$PbSe-Sb_2Se_3$ . . . . 47 (Pb): Hb/C3-940
PbSe-SnS . . . . 47 (Pb): Hb/C4-1587/8
PbSe-SnSe . . . . 47 (Pb): Hb/C4-1586/7

$PbSe_2$ . . . . $PbSe_2$ . . . . 47 (Pb): Hb/C2-632
$PbSe_4Sm_2$ . . . . . . $PbSm_2Se_4$ . . . . 47 (Pb): Hb/C4-1375
$PbSe_4Tm_2$ . . . . . . $PbTm_2Se_4$ . . . . 47 (Pb): Hb/C4-1375
$PbSe_4Y_2$ . . . . PbY_2Se_4 . . . . 47 (Pb): Hb/C4-1375
$PbSe_4Yb_2$ . . . . $PbYb_2Se_4$ . . . . 47 (Pb): Hb/C4-1375
PbSr . . . . SrPb . . . . 47 (Pb): Hb/C3-1194
$PbSr_2$ . . . . Sr_2Pb . . . . 47 (Pb): Hb/C3-1193/4
PbTe . . . . PbTe:
Chemical reactions . . . . 47 (Pb): Hb/C2-677/8
Crystallographic properties . . . . . . 47 (Pb): Hb/C2-651/4
Electrical properties . . . . 47 (Pb): Hb/C2-658/74
Electrochemical behavior . . . . 47 (Pb): Hb/C2-677
Formation and preparation . . . . . . 47 (Pb): Hb/C2-648/51
Magnetic properties . . . . 47 (Pb): Hb/C2-663, 669/70, 674
Mechanical properties . . . . 47 (Pb): Hb/C2-654/5
Molecule . . . . 47 (Pb): Hb/C2-651
Optical properties . . . . 47 (Pb): Hb/C2-675/7
Spectra . . . . 47 (Pb): Hb/C2-651
Thermal properties . . . . 47 (Pb): Hb/C2-654/8

– . . . . PbTe solid solutions:
$PbTe-AgSbTe_2-GeTe$ . . . . 61 (Ag): Hb/B4-273
$PbTe-AgSbTe_2-SnTe$ . . . . 61 (Ag): Hb/B4-272
PbTe-BiTe . . . . 47 (Pb): Hb/C3-475
$PbTe-Bi_2Te_3$ . . . . 47 (Pb): Hb/C3-975
PbTe-CdTe . . . . 47 (Pb): Hb/C4-1256
PbTe-EuTe . . . . 47 (Pb): Hb/C4-1375
PbTe-PbS-SnTe . . . . 47 (Pb): Hb/C4-1594
PbTe-SnSe . . . . 47 (Pb): Hb/C4-1595
PbTe-SnSe-SnTe . . . . 47 (Pb): Hb/C4-1595
PbTe-SnTe . . . . 47 (Pb): Hb/C4-1589/94

See the "Instructions for the user" at the beginning of the volume

Formula	Compound	Volume	Reference
PbTe	PbTe systems:		
	$PbTe-AgBiSe_2$	61 (Ag):	Hb/B4–273
	$PbTe-AgBiTe_2$	61 (Ag):	Hb/B4–270/2
	$PbTe-AgSbSe_2$	61 (Ag):	Hb/B4–273
	$PbTe-AgSbTe_2$	61 (Ag):	Hb/B4–269/70
	$PbTe-Ag_2Te-Sb_2Te_3$	61 (Ag):	Hb/B4–269/70
	$PbTe-Bi_2Te_3$	47 (Pb):	Hb/C3–973/5
	PbTe–CdTe	47 (Pb):	Hb/C4–1256
	PbTe–GeS	47 (Pb):	Hb/C4–1480
	PbTe–GeSe	47 (Pb):	Hb/C4–1480
	PbTe–GeTe	47 (Pb):	Hb/C4–1479/80
	PbTe–InTe	47 (Pb):	Hb/C4–1327
	PbTe–Pb–PbSe	47 (Pb):	Hb/C2–681
	PbTe–PbS	47 (Pb):	Hb/C2–679/80
	PbTe–PbS–PbSe	47 (Pb):	Hb/C2–684
	PbTe–SnSe–SnTe	47 (Pb):	Hb/C4–1594/5
	$PbTe-Tl_2Te$	47 (Pb):	Hb/C4–1353/4
$PbTe_3Tl_4$	Tl_4PbTe_3	47 (Pb):	Hb/C4–1354
PbTh	ThPb	47 (Pb):	Hb/C4–1468, 1470
$PbTi_2$	Ti_2Pb	47 (Pb):	Hb/C4–1377
$PbTi_4$	Ti_4Pb	47 (Pb):	Hb/C4–1378
–	Ti_4Pb solid solutions:		
	$(Zr,Ti)_4Pb$	47 (Pb):	Hb/C4–1423
$PbTl_2$	$PbTl_2$	47 (Pb):	Hb/C4–1330, 1347
PbV_3	V_3Pb	48 (V):	Hb/B2–687
PbYb	YbPb	47 (Pb):	Hb/C4–1372
$PbYb_2$	Yb_2Pb	47 (Pb):	Hb/C4–1372
$PbZr_4$	Zr_4Pb solid solutions:		
	$(Zr,Ti)_4Pb$	47 (Pb):	Hb/C4–1423
$PbZr_5$	Zr_5Pb	47 (Pb):	Hb/C4–1420/1
Pb_2	Pb_2	47 (Pb):	Hb/B1–138/9
Pb_2^+	Pb_2^+	47 (Pb):	Hb/B1–237
Pb_2Pd	$PdPb_2$	68 (Pt):	Hb/A5–589/93
Pb_2Pd_3	Pd_3Pb_2	68 (Pt):	Hb/A5–589/93
Pb_2Pt	$PtPb_2$	68 (Pt):	Hb/A6–764, 767
$Pb_2Pt_3S_6$	$2\,PbS \cdot 2\,PtS \cdot PtS_2 = Pb_2Pt_2[PtS_6]$	68 (Pt):	Hb/C3–315
Pb_2Rh	$RhPb_2$	68 (Pt):	Hb/A5–557
Pb_2S	Pb_2S	47 (Pb):	Hb/C2–414
Pb_2S_3	Pb_2S_3	47 (Pb):	Hb/C2–414
Pb_2S_4Si	Pb_2SiS_4	47 (Pb):	Hb/C3–855/6
$Pb_2S_5Sb_2$	$Pb_2Sb_2S_5 = 2\,PbS \cdot Sb_2S_3$	47 (Pb):	Hb/C3–938
$Pb_2S_{11}Sb_6$	$Pb_2Sb_6S_{11} = 2\,PbS \cdot 3\,Sb_2S_3$	47 (Pb):	Hb/C3–939
Pb_2Te_3	$Pb_2Te_3 \cdot 4\,H_2O$	47 (Pb):	Hb/C2–678
Pb_2Th	$ThPb_2$	47 (Pb):	Hb/C4–1468
Pb_2Y	YPb_2	47 (Pb):	Hb/C4–1360
Pb_3Pr	$PrPb_3$	47 (Pb):	Hb/C4–1360, 1366
Pb_3Pr_5	Pr_5Pb_3	47 (Pb):	Hb/C4–1359, 1365/6

Hb = Main Series Volume; GD = Gmelin-Durrer; Ab = Appended Volume to Main Series; Eb = Supplement Volume to Main Series; Erg. W. = New Supplement Series Volume

$Pb_3S_6Sb_2$	$Pb_3(SbS_3)_2 = 3\,PbS \cdot Sb_2S_3$	47 (Pb):	Hb/C3–937
$Pb_3S_7Si_2$	$Pb_3Si_2S_7$	47 (Pb):	Hb/C3–855/6
$Pb_3S_8Sb_2$	$Pb_3(SbS_4)_2 = 3\,PbS \cdot Sb_2S_5$	47 (Pb):	Hb/C3–939
$Pb_3S_9Sb_4$	$Pb_3Sb_4S_9 = 3\,PbS \cdot 2\,Sb_2S_3$	47 (Pb):	Hb/C3–938
$Pb_3S_{14}Sb_2Sn_4$	$Pb_3Sn_4Sb_2S_{14}$	47 (Pb):	Hb/C4–1595
Pb_3Sc	$ScPb_3$	47 (Pb):	Hb/C4–1360
Pb_3Sc_5	Sc_5Pb_3	47 (Pb):	Hb/C4–1359/60
Pb_3Se	Pb_3Se	47 (Pb):	Hb/C2–632
Pb_3Sm	$SmPb_3$	47 (Pb):	Hb/C4–1360, 1367
Pb_3Sm_5	Sm_5Pb_3	47 (Pb):	Hb/C4–1359, 1367
Pb_3Sr	$SrPb_3$	47 (Pb):	Hb/C3–1194
Pb_3Tb	$TbPb_3$	47 (Pb):	Hb/C4–1360, 1369
Pb_3Tb_5	Tb_5Pb_3	47 (Pb):	Hb/C4–1359, 1369
Pb_3Te	Pb_3Te	47 (Pb):	Hb/C2–678
Pb_3Th	$ThPb_3$	47 (Pb):	Hb/C4–1468, 1470
Pb_3Ti_5	Ti_5Pb_3	47 (Pb):	Hb/C4–1377/8
–	Ti_5Pb_3 solid solutions:		
	$(Zr,Ti)_5Pb_3$	47 (Pb):	Hb/C4–1423
Pb_3Tm	$TmPb_3$	47 (Pb):	Hb/C4–1360, 1371
Pb_3Tm_5	Tm_5Pb_3	47 (Pb):	Hb/C4–1359, 1371
Pb_3Y	YPb_3	47 (Pb):	Hb/C4–1360
Pb_3Y_5	Y_5Pb_3	47 (Pb):	Hb/C4–1359, 1360/1
Pb_3Yb	$YbPb_3$	47 (Pb):	Hb/C4–1360, 1372
Pb_3Yb_5	Yb_5Pb_3	47 (Pb):	Hb/C4–1359, 1372
Pb_3Zr_5	Zr_5Pb_3	47 (Pb):	Hb/C4–1420/1
–	Zr_5Pb_3 solid solutions:		
	$(Zr,Ti)_5Pb_3$	47 (Pb):	Hb/C4–1423
Pb_4Pt	$PtPb_4$	68 (Pt):	Hb/A6–764, 767
Pb_4S	Pb_4S	47 (Pb):	Hb/C2–414
$Pb_4S_7Sb_2$	$Pb_4Sb_2S_7 = 4\,PbS \cdot Sb_2S_3$	47 (Pb):	Hb/C3–937
Pb_4Th	$ThPb_4$	47 (Pb):	Hb/C4–1468
Pb_4Y_5	Y_5Pb_4	47 (Pb):	Hb/C4–1360
$Pb_5S_8Sb_2$	$Pb_5(SbS_4)_2 = 5\,PbS \cdot Sb_2S_3$	47 (Pb):	Hb/C3–937
$Pb_5S_{11}Sb_4$	$Pb_5Sb_4S_{11} = 5\,PbS \cdot 2\,Sb_2S_3$	47 (Pb):	Hb/C3–937/8
$Pb_5S_{14}Sb_3Sn_3$	$Pb_5Sn_3Sb_3S_{14}$	47 (Pb):	Hb/C4–1595
$Pb_5S_{14}Sb_6$	$Pb_5Sb_6S_{14} = 5\,PbS \cdot 3\,Sb_2S_3$	47 (Pb):	Hb/C3–938
$Pb_5S_{17}Sb_8$	$Pb_5Sb_8S_{17} = 5\,PbS \cdot 4\,Sb_2S_3$	47 (Pb):	Hb/C3–938
$Pb_6S_9Sb_2$	$Pb_6Sb_2S_9 = 6\,PbS \cdot Sb_2S_3$	47 (Pb):	Hb/C3–937
$Pb_7S_{22}Sb_{10}$	$Pb_7Sb_{10}S_{22} = 7\,PbS \cdot 5\,Sb_2S_3$	47 (Pb):	Hb/C3–938
$Pb_7S_{25}Sb_{12}$	$Pb_7Sb_{12}S_{25} = 7\,PbS \cdot 6\,Sb_2S_3$	47 (Pb):	Hb/C3–939
$Pb_9S_{21}Sb_8$	$Pb_9Sb_8S_{21} = 9\,PbS \cdot 4\,Sb_2S_3$	47 (Pb):	Hb/C3–938
$Pb_9S_{30}Sb_{14}$	$Pb_9Sb_{14}S_{30} = 9\,PbS \cdot 7\,Sb_2S_3$	47 (Pb):	Hb/C3–938
Pd	Palladium:		
	Absorption of light	65 (Pd):	Hb/1–50
	– of X-rays	65 (Pd):	Hb/1–62/3
	Abundance	68 (Pt):	Hb/A1–24/7
	Atom	65 (Pd):	Hb/1–7/13

See the "Instructions for the user" at the beginning of the volume

Pd . . . . Palladium:

Topic	Volume	Reference
Atomic nucleus	65 (Pd):	Hb/1–1/7
Atomic radius	65 (Pd):	Hb/1–8/9
Atomic volume	65 (Pd):	Hb/1–11/2
Atomic weight	65 (Pd):	Hb/1–12/3
Behavior as anode	65 (Pd):	Hb/1–95/9, 105/8
– as cathode	65 (Pd):	Hb/1–94/5, 99/105
– at alternating current electrolysis	65 (Pd):	Hb/1–99
Catalytic properties	68 (Pt):	Hb/A3–429/30
Cells	65 (Pd):	Hb/1–89/94
Chemical constant	65 (Pd):	Hb/1–48
Chemical reactions	65 (Pd):	Hb/1–109/14
Colloidal forms	68 (Pt):	Hb/A3–425/7, 429/30
Color	65 (Pd):	Hb/1–49
Compressibility	65 (Pd):	Hb/1–35/6
Contact electricity	65 (Pd):	Hb/1–77
Cosmochemistry	68 (Pt):	Hb/A1–15/22
Critical constants	65 (Pd):	Hb/1–39
Crystallographic properties	65 (Pd):	Hb/1–14/25
Debye temperature	65 (Pd):	Hb/1–47
Deformation	65 (Pd):	Hb/1–22/3
Density	65 (Pd):	Hb/1–25/6
Detection	68 (Pt):	Hb/A4–432, 439/46, 489/90
– in iron	59 (Fe):	Hb/FII1–41
Determination	68 (Pt):	Hb/A4–467/71, 489/90
– in special substances	68 (Pt):	Hb/A4–525/32
Diffusion	65 (Pd):	Hb/1–37/9
– of hydrogen	65 (Pd):	Hb/1–116/33
Discharge phenomena	65 (Pd):	Hb/1–81/2
Economic deposits *see Pt ... Platinum*		
Economics	68 (Pt):	Hb/A2–302/10
Elasticity	65 (Pd):	Hb/1–31/3
Electrical conductivity	65 (Pd):	Hb/1–69/77
Electrical properties	65 (Pd):	Hb/1–69/86
Electrochemical behavior	65 (Pd):	Hb/1–87/109
Electrode sputtering	65 (Pd):	Hb/1–82/3
Electrolytic deposition	68 (Pt):	Hb/A3–390
Electron configuration	65 (Pd):	Hb/1–7/8
Electron emission	65 (Pd):	Hb/1–83/5
Equation of state	65 (Pd):	Hb/1–39
Ettingshausen effect	65 (Pd):	Hb/1–67/8
Faraday effect	65 (Pd):	Hb/1–51
Geochemical character	68 (Pt):	Hb/A1–23/8
Geochemistry	68 (Pt):	Hb/A1–22/49 Hb/A2–275/6

Hb = Main Series Volume; GD = Gmelin-Durrer; Ab = Appended Volume to Main Series; Eb = Supplement Volume to Main Series; Erg. W. = New Supplement Series Volume

Pd . . . . Palladium:

Hall effect	65 (Pd):	Hb/1–67
Hardness	65 (Pd):	Hb/1–29/31
Heat of transformation	65 (Pd):	Hb/1–42/4
History	68 (Pt):	Hb/A1–9/11
Interfacial tension	65 (Pd):	Hb/1–37
Internal friction	65 (Pd):	Hb/1–37
Internal pressure	65 (Pd):	Hb/1–36
Ion emission	65 (Pd):	Hb/1–85/6
Ionization potentials	65 (Pd):	Hb/1–86
Lattice structure	65 (Pd):	Hb/1–15/20
Leduc effect	65 (Pd):	Hb/1–68/9
Magnetic moment	65 (Pd):	Hb/1–9/11
Magnetic properties	65 (Pd):	Hb/1–63/9
Mechanical properties	65 (Pd):	Hb/1–25/39
Mineralogy	68 (Pt):	Hb/A2–276/301
Nernst effect	65 (Pd):	Hb/1–68
Occurrence, extraterrestrial	68 (Pt):	Hb/A1–15/22
–, terrestrial	68 (Pt):	Hb/A1–22/49 Hb/A2–275/6
– in the biosphere	68 (Pt):	Hb/A2–275
– in the hydrosphere	68 (Pt):	Hb/A2–275
– in the lithosphere	68 (Pt):	Hb/A1–28/49
Optical properties	65 (Pd):	Hb/1–49/63
Overvoltage phenomena	65 (Pd):	Hb/1–99/109
Passage of elementary particles	65 (Pd):	Hb/1–85/6
Physiological action	68 (Pt):	Hb/A2–275/6
Polymorphism	65 (Pd):	Hb/1–14/5
Position in the electrochemical series	65 (Pd):	Hb/1–87
– in the periodic system	65 (Pd):	Hb/1–14
Potentials	65 (Pd):	Hb/1–87/9
Preparation	68 (Pt):	Hb/A3–311/430
– from minerals	68 (Pt):	Hb/A3–312/50
– from wastes and residues	68 (Pt):	Hb/A3–350/62
– in pure state	68 (Pt):	Hb/A3–374/5
– of dense palladium	68 (Pt):	Hb/A3–384/5
Purity testing	68 (Pt):	Hb/A4–519/20
Recrystallization	65 (Pd):	Hb/1–24/5
Reflection	65 (Pd):	Hb/1–49/50
Refraction	65 (Pd):	Hb/1–50
Scattering of X-rays	65 (Pd):	Hb/1–62/3
Separation, analytical	68 (Pt):	Hb/A4–497/500, 517/8
Sorption	65 (Pd):	Hb/1–39
– of hydrogen	65 (Pd):	Hb/2–133/256
– of nitrogen	65 (Pd):	Hb/2–268
– of oxygen	65 (Pd):	Hb/2–257/9
– of rare gases	65 (Pd):	Hb/2–115/6

See the "Instructions for the user"
at the beginning of the volume

Pd . . . . Palladium:

Special forms . . . . 68 (Pt): Hb/A3–399/404, 425/7, 429/30

Specific heat . . . . 65 (Pd): Hb/1–44/6

Spectra:

Arc spectrum . . . . 65 (Pd): Hb/1–51/8

Spark spectra . . . . 65 (Pd): Hb/1–51/8

Terms . . . . 65 (Pd): Hb/1–54/6

X-ray spectrum . . . . 65 (Pd): Hb/1–58/62

Standard potential . . . . 65 (Pd): Hb/1–87

Strength . . . . 65 (Pd): Hb/1–33/5

Structure . . . . 65 (Pd): Hb/1–23/4

– of thin layers . . . . 65 (Pd): Hb/1–20/2

Surface tension . . . . 65 (Pd): Hb/1–37

Susceptibility . . . . 65 (Pd): Hb/1–64/7

Thermal conductivity . . . . 65 (Pd): Hb/1–48/9

Thermal expansion . . . . 65 (Pd): Hb/1–26/9

Thermal properties . . . . 65 (Pd): Hb/1–39/49

Thermal radiation . . . . 65 (Pd): Hb/1–51

Thermodynamic functions . . . . 65 (Pd): Hb/1–47/8

Thermoelectric effects . . . . 65 (Pd): Hb/1–78/81

Toxicity . . . . 68 (Pt): Hb/A2–275/6

Transformation points . . . . 65 (Pd): Hb/1–40/2

Triboelectricity . . . . 65 (Pd): Hb/1–77

Uses . . . . 68 (Pt): Hb/A3–385/91

Valence . . . . 65 (Pd): Hb/1–13/4

Vapor pressure . . . . 65 (Pd): Hb/1–39/40

Vibration frequency . . . . 65 (Pd): Hb/1–46/7

–. . . . . Palladium ions:

Diffusion . . . . 21 (Na): Eb/6–293

Ionic refraction . . . . 65 (Pd): Hb/1–50

Ionization potentials . . . . 65 (Pd): Hb/1–86

Properties of the ion . . . . 65 (Pd): Hb/1–9/10, 12

Spectra . . . . 65 (Pd): Hb/1–56

–. . . . . Palladium isotopes . . . . 65 (Pd): Hb/1–1/2

–. . . . . Palladium systems:

Pd–Ag . . . . 68 (Pt): Hb/A5–653

Pd–Ag–Au . . . . 68 (Pt): Hb/A5–710

Pd–Ag–Cu . . . . 68 (Pt): Hb/A5–686/8

Pd–Ag–H . . . . 68 (Pt): Hb/A5–666/78

Pd–Au . . . . 68 (Pt): Hb/A5–691

Pd–Au–Cu . . . . 68 (Pt): Hb/A5–709/10

Pd–Au–H . . . . 68 (Pt): Hb/A5–701/9

Pd–Au–Ni . . . . 68 (Pt): Hb/A5–709

Pd–Au–Pt . . . . 68 (Pt): Hb/A6–840/1

Pd–Be . . . . 68 (Pt): Hb/A5–579/80

Pd–Bi . . . . 68 (Pt): Hb/A5–575/8

Pd–Cd . . . . 68 (Pt): Hb/A5–583

Pd	Palladium systems:		
	Pd–Co	68 (Pt):	Hb/A5–613/5
	Pd–Cr	68 (Pt):	Hb/A5–594/6
	Pd–Cu	68 (Pt):	Hb/A5–624/31
	Pd–Fe	68 (Pt):	Hb/A5–617/9
	Pd–Fe–S	68 (Pt):	Hb/A5–624
	Pd–H	65 (Pd):	Hb/2–133/216
	Pd–D	65 (Pd):	Hb/2–209/16
	Pd–H–Pt	68 (Pt):	Hb/A6–839/40
	Pd–In	68 (Pt):	Hb/A5–584/5
	Pd–Mn	68 (Pt):	Hb/A5–598/607
	Pd–Ni	68 (Pt):	Hb/A5–607/8
	Pd–O–S	9 (S):	Hb/B1–185
	Pd–P	65 (Pd):	Hb/2–302/5
	Pd–Pb	68 (Pt):	Hb/A5–588/93
	Pd–PdS	65 (Pd):	Hb/2–289/91
	Pd–Pt	68 (Pt):	Hb/A6–836
	Pd–Ru	68 (Pt):	Hb/A5–715/6
	Pd–S	65 (Pd):	Hb/2–289
	Pd–Sb	68 (Pt):	Hb/A5–570/5
	Pd–Si	65 (Pd):	Hb/2–302
	Pd–Sn	68 (Pt):	Hb/A5–586/8
	Pd–Ti	68 (Pt):	Hb/A5–585
Pd^+	Pd^+	65 (Pd):	Hb/1–9, 56, 86
Pd^{2+}	Pd^{2+}	21 (Na):	Eb/6–293
		65 (Pd):	Hb/1–9/10, 12, 50
Pd^{3+}	Pd^{3+}	65 (Pd):	Hb/1–9
Pd^{4+}	Pd^{4+}	65 (Pd):	Hb/1–9/10
PdS	PdS	65 (Pd):	Hb/2–292/4
–	PdS systems:		
	PdS–Pd	65 (Pd):	Hb/2–289/91
PdS_2	PdS_2	65 (Pd):	Hb/2–294/5
PdSb	PdSb	68 (Pt):	Hb/A5–574
$PdSb_2$	$PdSb_2$	68 (Pt):	Hb/A5–574/5
PdSe	PdSe	65 (Pd):	Hb/2–296
$PdSe_2$	$PdSe_2$	65 (Pd):	Hb/2–296
PdSi	PdSi	65 (Pd):	Hb/2–302
PdSn	PdSn	68 (Pt):	Hb/A5–587/8
$PdSn_2$	$PdSn_2$	68 (Pt):	Hb/A5–587/8
$PdSn_4$	$PdSn_4$	68 (Pt):	Hb/A5–587/8
PdTe	PdTe	65 (Pd):	Hb/2–297
$PdTe_2$	$PdTe_2$	65 (Pd):	Hb/2–297/8
PdTi	PdTi	68 (Pt):	Hb/A5–585
$PdTi_2$	$PdTi_2$	68 (Pt):	Hb/A5–585
Pd_2S	Pd_2S	65 (Pd):	Hb/2–292
Pd_2Si	Pd_2Si	65 (Pd):	Hb/2–302
Pd_2Sn	Pd_2Sn	68 (Pt):	Hb/A5–587/8
Pd_2Tl	Pd_2Tl	68 (Pt):	Hb/A5–585

See the "Instructions for the user" at the beginning of the volume

Pd_3Pu . . . . $PuPd_3$ . . . . Erg. W. 4–209
Pd_3S . . . . Pd_3S . . . . 65 (Pd): Hb/2–292
Pd_3S_4 . . . . $Pd_3S_4 = 2\ PdS \cdot PdS_2$ . . . . 65 (Pd): Hb/2–294
Pd_3Sb . . . . Pd_3Sb . . . . 68 (Pt): Hb/A5–573/4
Pd_3Sn . . . . Pd_3Sn . . . . 68 (Pt): Hb/A5–587/8
Pd_3Sn_2 . . . . Pd_3Sn_2 . . . . 68 (Pt): Hb/A5–587/8
Pd_3Ti . . . . Pd_3Ti . . . . 68 (Pt): Hb/A5–585
Pd_3Ti_2 . . . . Pd_3Ti_2 . . . . 68 (Pt): Hb/A5–585
Pd_3Zr . . . . Pd_3Zr . . . . 68 (Pt): Hb/A5–585
Pd_4S . . . . Pd_4S . . . . 65 (Pd): Hb/2–292
Pd_4Se . . . . Pd_4Se . . . . 65 (Pd): Hb/2–296
Pd_5Sb_3 . . . . Pd_5Sb_3 . . . . 68 (Pt): Hb/A5–574
Pd_5Zn_{21} . . . . Pd_5Zn_{21} . . . . 68 (Pt): Hb/A5–582

Pm . . . . Promethium:
Abundance . . . . 39 (SE): Hb/A1–34
Crystallographic properties . . . . . . 39 (SE): Hb/B3–69, 77/8
Diffusion . . . . 47 (Pb): Hb/C4–1364
Electrical properties . . . . 39 (SE): Hb/B3–236, 282/3
History . . . . 39 (SE): Hb/A1–18
Magnetic properties . . . . 39 (SE): Hb/B3–294
Mechanical properties . . . . 39 (SE): Hb/B3–77/8, 208/9
Occurrence . . . . 39 (SE): Hb/A1–34/42
Position in the periodic system . . . . 39 (SE): Hb/A1–2/4
Thermal properties . . . . 39 (SE): Hb/B3– 84/5, 103/12, 118/20, 139/40, 187, 198/9

–. . . . . Promethium ions . . . . 39 (SE): Hb/B3–284, 294
–. . . . . Promethium isotopes . . . . Erg. W. 8–207
–. . . . . Promethium systems:
Pm–N–O . . . . 39 (SE): Hb/C2–180
Pm^{3+} . . . . Pm^{3+} . . . . 39 (SE): Hb/B3–284, 294

Po . . . . Polonium:
Atom . . . . 12 (Po): Hb–32/7, 60/1, 145, 174
Atomic nucleus . . . . 12 (Po): Hb–29/32, 141/5, 153/4, 158/64, 176/7, 185
Chemical reactions . . . . 12 (Po): Hb–93/9, 147/8
Crystallographic properties . . . . . . 12 (Po): Hb–61/2
Detection . . . . 12 (Po): Hb–124/6
Determination . . . . 12 (Po): Hb–126/30
Electrical properties . . . . 12 (Po): Hb–69, 146/7, 152, 156, 183
Electrochemical behavior . . . . 12 (Po): Hb–69/93, 147
Electrochemical deposition . . . . . . 12 (Po): Hb–72/93, 147
Enrichment . . . . 12 (Po): Hb–7/18, 140/1

Po . . . . Polonium:
Formation . . . . 12 (Po): Hb–6, 29, 139, 149, 153, 157/8, 175, 184
History . . . . 12 (Po): Hb–1/5, 138, 148, 152/3, 157, 175, 184
Mechanical properties . . . . 12 (Po): Hb–62/4
Molecule. . . . . 12 (Po): Hb–60
Nuclear decay,
alpha radiation . . . . 12 (Po): Hb–38/56, 143/5, 150/2, 154/6, 164/73, 177/87
Erg. W. 7a–85/95
–, beta radiation. . . . . 12 (Po): Hb–56, 183
–, decay scheme . . . . 12 (Po): Hb–139
–, gamma radiation. . . . . 12 (Po): Hb–56, 183
Occurrence . . . . 12 (Po): Hb–5/6, 138/40, 149, 153, 157, 175, 184
Optical properties. . . . . 12 (Po): Hb–66/8
Position in the periodic system . . . . 12 (Po): Hb–37/8
Potentials . . . . 12 (Po): Hb–69/80
Preparation . . . . 12 (Po): Hb–6/29, 140/1, 149, 153
– in pure state. . . . . 12 (Po): Hb–19/26
– of special forms. . . . . 12 (Po): Hb–27/8
Separation, analytical . . . . 12 (Po): Hb–130
Spectra . . . . 12 (Po): Hb–67/8
Thermal properties . . . . 12 (Po): Hb–64/6, 146
Toxicity . . . . 12 (Po): Hb–123/4
Uses . . . . 12 (Po): Hb–28/9
Erg. W. 8–207, 224
Valence . . . . 12 (Po): Hb–34/7

–. . . . . Polonium ions:
Chemical reactions . . . . 12 (Po): Hb–99/123
Colloidchemical behavior . . . . 12 (Po): Hb–106/9
Electrochemical behavior . . . . 12 (Po): Hb–72/93
Properties of the ion . . . . 12 (Po): Hb–33/4
Sorption . . . . 12 (Po): Hb–108/9

–. . . . . Polonium isotopes
see Po ... Polonium

–. . . . . Polonium systems:
Po–Bi . . . . 19 (Bi): Eb–828/9
Po–Hg . . . . 34 (Hg): Hb/B3–1180
Po–Ni . . . . 57 (Ni): Hb/B2–764
Po–Pb . . . . 47 (Pb): Hb/C2–685

*See the "Instructions for the user"
at the beginning of the volume*

Po^{n+} . . . . Po^{n+} (n = 1 to 6)
see Po ... Polonium ions

PoS . . . . PoS . . . . 12 (Po): Hb–135
PoSe . . . . PoSe . . . . 12 (Po): Hb–135
PoTe . . . . PoTe . . . . 12 (Po): Hb–135
Po_2 . . . . Po_2 . . . . 12 (Po): Hb–60
Po_2S_3 . . . . Po_2S_3 . . . . 12 (Po): Hb–135

Pr . . . . Praseodymium:
Abundance . . . . 39 (SE): Hb/A1–34/5
Band model . . . . 39 (SE): Hb/B3–211/5
Boiling point . . . . 39 (SE): Hb/B3–105
Characteristic electron energy
losses . . . . 39 (SE): Hb/B3–225/6
Charge carrier concentration. . . . . . 39 (SE): Hb/B3–228
Cosmochemistry . . . . 39 (SE): Hb/A1–23/7
Crystal chemistry . . . . 39 (SE): Hb/A1–28/32
Crystal field parameters . . . . 39 (SE): Hb/B3–219
Crystallographic properties . . . . . . 39 (SE): Hb/B3–60/1, 66/9, 76/8
Crystal structure . . . . 39 (SE): Hb/B3–66/9
Debye temperature . . . . 39 (SE): Hb/B3–185/6
Density . . . . 39 (SE): Hb/B3–85/7
Diffusion. . . . . 39 (SE): Hb/B3–77/8
Disorder . . . . 39 (SE): Hb/B3–76
Economic deposits . . . . 39 (SE): Hb/A1–110/22
Elasticity. . . . . 39 (SE): Hb/B3–92/8
Electrical properties . . . . 39 (SE): Hb/B3–209/83
Electrical resistance. . . . . 39 (SE): Hb/B3–232/4, 258
Electron density. . . . . 39 (SE): Hb/B3–216
Electron emission . . . . 39 (SE): Hb/B3–278/83
Enthalpy of fusion . . . . 39 (SE): Hb/B3–120
– of sublimation. . . . . 39 (SE): Hb/B3–110/1
– of transformation . . . . 39 (SE): Hb/B3–118
– of vaporization . . . . 39 (SE): Hb/B3–110/1
Equation of state . . . . 39 (SE): Hb/B3–83/4
Geochemistry . . . . 39 (SE): Hb/A1–28/54
Grüneisen constant. . . . . 39 (SE): Hb/B3–84/5
Hall effect . . . . 39 (SE): Hb/B3–265
Hardness . . . . 39 (SE): Hb/B3–103
History. . . . . 39 (SE): Hb/A1–4/22
Hyperfine interactions . . . . 39 (SE): Hb/B3–219/20
Isomer shift . . . . 39 (SE): Hb/B3–216
Magnetic properties . . . . 39 (SE): Hb/B3–289/92
Mechanical properties . . . . 39 (SE): Hb/B3–85/103, 193/5, 207/9
Melting point . . . . 39 (SE): Hb/B3–118/9
Mineralogy . . . . 39 (SE): Hb/A1–54/110

Pr	Praseodymium:		
	Modifications	39 (SE):	Hb/B3-60/1
	Molar volume	39 (SE):	Hb/B3-85/7
	Occurrence, extraterrestrial	39 (SE):	Hb/A1-22/7
	-, terrestrial	39 (SE):	Hb/A1-27/54
	- in the biosphere	39 (SE):	Hb/A1-53/4
	- in the lithosphere	39 (SE):	Hb/A1-27/54
	Optical properties	39 (SE):	Hb/B3-338/9, 343
	Polymorphism	39 (SE):	Hb/B3-60/1
	Position in the periodic system	39 (SE):	Hb/A1-2/4
	Spectral emissivity	39 (SE):	Hb/B3-343
	Superconductivity	39 (SE):	Hb/B3-254
	Surface tension	39 (SE):	Hb/B3-208/9
	Susceptibility	39 (SE):	Hb/B3-289/92
	Thermal conductivity	39 (SE):	Hb/B3-198/9
	Thermal diffusivity	39 (SE):	Hb/B3-206
	Thermal expansion	39 (SE):	Hb/B3-88/92
	Thermal properties	39 (SE):	Hb/B3-83/92, 103/20, 129/34, 185, 198/9, 206
	Thermodynamic functions	39 (SE):	Hb/B3-129/34
	Thermoelectric power	39 (SE):	Hb/B3-274
	Ultrasound absorption	39 (SE):	Hb/B3-193
	Valence	39 (SE):	Hb/A1-4
	Vapor pressure	39 (SE):	Hb/B3-103/5
	Viscosity	39 (SE):	Hb/B3-207/8
-	Praseodymium ions *see Pr^{3+} ... Pr^{3+}*		
-	Praseodymium systems:		
	Pr-Ag	61 (Ag):	Hb/C-217/8
	Pr-Al	35 (Al):	Hb/A5-825/6
	Pr-Gd-N	39 (SE):	Hb/C2-175
	Pr-H	39 (SE):	Hb/C1-2/3, 50/1
	Pr-Mg	27 (Mg):	Hb/A4-599/601
	Pr-Ni	57 (Ni):	Hb/B1-73/4
	Pr-Pb	47 (Pb):	Hb/C4-1365/6
	Pr-Pb-Zn	47 (Pb):	Hb/C4-1363, 1366
	Pr-Sn	46 (Sn):	Hb/D-307
	Pr-V	48 (V):	Hb/B2-600/1
Pr^{3+}	Pr^{3+}:		
	Magnetic moment	39 (SE):	Hb/B3-284
	Spectra	49 (Nb):	Hb/B3-133, 135
		50 (Ta):	Hb/B2-203
$PrRu_2$	Ru_2Pr	63 (Ru):	Eb-183
$PrSn_3$	$PrSn_3$	46 (Sn):	Hb/D-307/10
-	$PrSn_3$ solid solutions:		
	$La_{1-x}Pr_xSn_3$	46 (Sn):	Hb/D-310

See the "Instructions for the user" at the beginning of the volume

Pr_2S_3	Pr_2S_3 solid solutions:		
	Pr_2S_3–Ag_2S	61 (Ag):	Hb/B4–232
	Pr_2S_3–PbS	47 (Pb):	Hb/C4–1374
Pr_2Se_3	Pr_2Se_3 solid solutions:		
	Pr_2Se_3–Ag_2Se	61 (Ag):	Hb/B4–234/5
	Pr_2Se_3–PbSe	47 (Pb):	Hb/C4–1374
Pr_2Sn	Pr_2Sn	46 (Sn):	Hb/D–307
Pr_2Sn_3	Pr_2Sn_3	46 (Sn):	Hb/D–307
Pr_3Sn	Pr_3Sn	46 (Sn):	Hb/D–307/10
Pr_5Sn_3	Pr_5Sn_3	46 (Sn):	Hb/D–307/10
Pr_5Sn_4	Pr_5Sn_4	46 (Sn):	Hb/D–307/10
Pt	Platinum:		
	Abrasion	68 (Pt):	Hb/B1–45
	Absorption of light	68 (Pt):	Hb/B2–77/8
	– of X- and gamma rays	68 (Pt):	Hb/B2–101/3
	Abundance	68 (Pt):	Hb/A1–24/7
	Aerosols	68 (Pt):	Hb/A3–405
	Anodic polarization	68 (Pt):	Hb/B3–205/8
	Atom	68 (Pt):	Hb/B1–7/12
	Atomic nucleus	68 (Pt):	Hb/B1–1/7 Hb/B2–180
	Atomic number	68 (Pt):	Hb/B1–2
	Atomic radius	68 (Pt):	Hb/B1–8
	Atomic volume	68 (Pt):	Hb/B1–10/1
	Atomic weight	68 (Pt):	Hb/B1–11/2
	Becquerel effect	68 (Pt):	Hb/B4–272/9
	Behavior as anode	68 (Pt):	Hb/B4–298/308 Hb/C1–34/5
	– as cathode	68 (Pt):	Hb/B4–295/8 Hb/C1–9
	– at alternating current electrolysis	68 (Pt):	Hb/B4–308/14
	Benedicks effect	68 (Pt):	Hb/B2–153
	Boiling point	68 (Pt):	Hb/B1–57
	Catalytic properties	68 (Pt):	Hb/A3–429/30 Hb/B4–338
	Cathodic polarization	68 (Pt):	Hb/B3–228/9
	Cells	68 (Pt):	Hb/B4–267/70
	Chemical constant	68 (Pt):	Hb/B1–70
	Chemical reactions	68 (Pt):	Hb/B4–319/38
	– on heating	68 (Pt):	Hb/B4–319/20
	– with air	68 (Pt):	Hb/B4–319
	– with elements	68 (Pt):	Hb/B4–320/3
	– with metal compounds	68 (Pt):	Hb/B4–331/8
	– with nonmetal compounds	68 (Pt):	Hb/B4–323/31
	– with organic compounds	68 (Pt):	Hb/B4–329, 330/1
	Colloidal forms	68 (Pt):	Hb/A3–405/23, 429/30

Pt . . . . Platinum:

Color	68 (Pt):	Hb/B2–73
Compressibility	68 (Pt):	Hb/B1–45
Contact electricity	68 (Pt):	Hb/B2–138/40
Corbino effect	68 (Pt):	Hb/B2–108
Cosmochemistry	68 (Pt):	Hb/A1–15/22
Critical constants	68 (Pt):	Hb/B1–55
Crystallographic properties	68 (Pt):	Hb/B1–12/26
Debye temperature	68 (Pt):	Hb/B1–68/9
Deformation	68 (Pt):	Hb/B1–21/2
Density	68 (Pt):	Hb/B1–26/9
Deposition potential	68 (Pt):	Hb/B4–285/6
Detection	68 (Pt):	Hb/A4–432, 455/62, 491/2
– in iron	59 (Fe):	Hb/FII1–41
Determination	68 (Pt):	Hb/A4–480/8, 491/2
– in special substances	68 (Pt):	Hb/A4–525/32
Diffusion	68 (Pt):	Hb/B1–47/55
Discharge phenomena	68 (Pt):	Hb/B2–154/7
Economic deposits of platinum and platinum metals:		
Africa	68 (Pt):	Hb/A2–190/263
Asia	68 (Pt):	Hb/A1–143/4 Hb/A2–145/6
Australia and Oceania	68 (Pt):	Hb/A2–263/74
Central America	68 (Pt):	Hb/A2–179
Europe	68 (Pt):	Hb/A1–50/3
North America	68 (Pt):	Hb/A2–146/79
South America	68 (Pt):	Hb/A2–179/90
Soviet Union	68 (Pt):	Hb/A1–53/143
Economics	68 (Pt):	Hb/A2–302/10
Elasticity	68 (Pt):	Hb/B1–38/41
Electrical conductivity	68 (Pt):	Hb/B2–109/38
– of thin layers	68 (Pt):	Hb/B2–132/7
Electrical properties	68 (Pt):	Hb/B2–109/79
Electrochemical behavior	68 (Pt):	Hb/B3–181/262 Hb/B4–263/318
Electrode sputtering	68 (Pt):	Hb/B1–20/1 Hb/B2–157/9
Electrolytic deposition	68 (Pt):	Hb/A3–387/9 Hb/B4–279/95
Electrolytic solution pressure	68 (Pt):	Hb/B4–266
Electron configuration	68 (Pt):	Hb/B1–7/8
Electron emission	68 (Pt):	Hb/B2–159/72
Enthalpy	68 (Pt):	Hb/B1–69
Entropy	68 (Pt):	Hb/B1–69/70
Equation of state	68 (Pt):	Hb/B1–55
Faraday effect	68 (Pt):	Hb/B2–78
Free energy	68 (Pt):	Hb/B1–69

See the "Instructions for the user" at the beginning of the volume

Pt . . . . Platinum:

Geochemical character	68 (Pt):	Hb/A1–23/8
Geochemistry	68 (Pt):	Hb/A1–22/49 Hb/A2–275/6
Hall effect	68 (Pt):	Hb/B2–107/8
Hardness	68 (Pt):	Hb/B1–34/8
Heat content.	68 (Pt):	Hb/B1–69
Heat of fusion	68 (Pt):	Hb/B1–60
– of sublimation.	68 (Pt):	Hb/B1–59/60
– of vaporization	68 (Pt):	Hb/B1–59
History.	68 (Pt):	Hb/A1–1/9
Hydrosols	68 (Pt):	Hb/A3–406/21
Interaction with electrons	68 (Pt):	Hb/B2–84, 172/6
– with positive ions	68 (Pt):	Hb/B2–176/9
– with positrons.	68 (Pt):	Hb/B2–176
Interfacial tension.	68 (Pt):	Hb/B1–46/7
Internal energy	68 (Pt):	Hb/B1–69
Internal pressure	68 (Pt):	Hb/B1–46
Ionic refraction	68 (Pt):	Hb/B2–77
Ionization potentials	68 (Pt):	Hb/B2–179
Lattice structure.	68 (Pt):	Hb/B1–14/7
Lattice vibrations	68 (Pt):	Hb/B1–67/8
Lorenz number	68 (Pt):	Hb/B2–137/8
Magnetic moment	68 (Pt):	Hb/B1–9/10
Magnetic properties	68 (Pt):	Hb/B2–103/7
Mechanical properties	68 (Pt):	Hb/B1–26/55
Melting point	68 (Pt):	Hb/B1–57/9
Mineralogy	68 (Pt):	Hb/A2–276/301
Nuclear properties	68 (Pt):	Hb/B1–1/7 Hb/B2–180
Occurrence, extraterrestrial	68 (Pt):	Hb/A1–15/22
–, terrestrial	68 (Pt):	Hb/A1–22/49 Hb/A2–275/6
– in meteorites	68 (Pt):	Hb/A1–16/22
– in the biosphere.	68 (Pt):	Hb/A2–275
– in the hydrosphere	68 (Pt):	Hb/A2–275
– in the lithosphere	68 (Pt):	Hb/A1–28/49
Optical properties.	68 (Pt):	Hb/B2–73/103
Organosols	68 (Pt):	Hb/A3–421/3
Orientation of thin layers.	68 (Pt):	Hb/B1–17/20
Overvoltage phenomena, theoretical .	68 (Pt):	Hb/B3–232/51, 262
– during anodic deposition of halogen	68 (Pt):	Hb/B3–252/62
– – of oxygen	68 (Pt):	Hb/B3–209/28
– during anodic solution of hydrogen	68 (Pt):	Hb/B3–205/8
– during cathodic deposition of hydrogen	68 (Pt):	Hb/B3–181/204
– during cathodic solution of oxygen .	68 (Pt):	Hb/B3–228/9

Hb = Main Series Volume; GD = Gmelin-Durrer; Ab = Appended Volume to Main Series; Eb = Supplement Volume to Main Series; Erg. W. = New Supplement Series Volume

Pt . . . . Platinum:

Overvoltage phenomena
during oxygen electrode-
hydrogen electrode transition. . . . 68 (Pt): Hb/B3–229/31
Peltier effect. . . . . 68 (Pt): Hb/B2–153/4
Photoelectric electron emission . . . . 68 (Pt): Hb/B2–159/67
Physiological action . . . . 68 (Pt): Hb/A2–275/6
Plasticity. . . . . 68 (Pt): Hb/B1–41/2
Platinum oxygen electrode. . . . . . . 68 (Pt): Hb/B3–229/31
Platinum resistance thermometer . . . 68 (Pt): Hb/B2–115/24
Polarization . . . . 68 (Pt): Hb/B4–314/8
Polymorphism . . . . 68 (Pt): Hb/B1–12/3
Position in the electrochemical series. 68 (Pt): Hb/B2–140
Hb/B4–265
– in the periodic system . . . . 68 (Pt): Hb/B1–12
Potentials . . . . 68 (Pt): Hb/B4–266/7
Preparation . . . . 68 (Pt): Hb/A3–311/430
– from minerals. . . . . 68 (Pt): Hb/A3–312/50
– from wastes and residues . . . . . . 68 (Pt): Hb/A3–350/62
– in pure state. . . . . 68 (Pt): Hb/A3–367/71
– of dense platinum . . . . 68 (Pt): Hb/A3–378/83
Purity testing . . . . 68 (Pt): Hb/A3–365/7
Hb/A4–521/4
Recrystallization . . . . 68 (Pt): Hb/B1–23/6
Reflection of light. . . . . 68 (Pt): Hb/B2–73/5
– of X-rays. . . . . 68 (Pt): Hb/B2–100/1
Refraction of light. . . . . 68 (Pt): Hb/B2–75/7
– of X-rays. . . . . 68 (Pt): Hb/B2–100/1
Scattering of X- and gamma rays . . . 68 (Pt): Hb/B2–101/3
Separation, analytical . . . . 68 (Pt): Hb/A4–504/18
Sorption of hydrogen. . . . . 68 (Pt): Hb/C1–5/25
– of nitrogen . . . . 68 (Pt): Hb/C1–54/6
– of oxygen . . . . 68 (Pt): Hb/C1–26/41
– of rare gases . . . . 68 (Pt): Hb/C1–1/4
Special forms . . . . 68 (Pt): Hb/A3–391/8, 401/23,
429/30
Specific heat . . . . 68 (Pt): Hb/B1–60/7
Spectra:
Arc spectrum . . . . 68 (Pt): Hb/B2–84/92
Spark spectra. . . . . 68 (Pt): Hb/B2–92/3
X-ray spectrum. . . . . 68 (Pt): Hb/B2–93/100
Standard potential . . . . 68 (Pt): Hb/B4–263/5
Static electricity. . . . . 68 (Pt): Hb/B2–138
Strength . . . . 68 (Pt): Hb/B1–42/4
Structure . . . . 68 (Pt): Hb/B1–22/3
– of thin layers . . . . 68 (Pt): Hb/B1–17/20
Superconductivity . . . . 68 (Pt): Hb/B2–114/5
Surface tension . . . . 68 (Pt): Hb/B1–46/7
Susceptibility . . . . 68 (Pt): Hb/B2–103/7

*See the "Instructions for the user"
at the beginning of the volume*

Pt . . . . Platinum:

	Thermal conductivity	68 (Pt):	Hb/B1–70/2
	Thermal electron emission	68 (Pt):	Hb/B2–167/9
	Thermal expansion	68 (Pt):	Hb/B1–29/34
	Thermal properties	68 (Pt):	Hb/B1–55/72
	Thermal radiation	68 (Pt):	Hb/B2–78/84
	Thermocouples	68 (Pt):	Hb/B2–141/51
	Thermoelectric power	68 (Pt):	Hb/B2–140/53
	Thomson effect	68 (Pt):	Hb/B2–154
	Toxicity	68 (Pt):	Hb/A2–275/6
	Uses	68 (Pt):	Hb/A3–385/91
			Hb/B2–115/24
			Hb/B4–270/1
	Valence	68 (Pt):	Hb/B1–12
	Vapor pressure	68 (Pt):	Hb/B1–55/6
	Vibration frequency	68 (Pt):	Hb/B1–67/8
	Viscosity	68 (Pt):	Hb/B1–47
	Wiedemann-Franz number	68 (Pt):	Hb/B2–137/8
	Work function of electrons	68 (Pt):	Hb/B2–170/2
	X-ray photoelectric effect	68 (Pt):	Hb/B2–166/7
–	Platinum ions *see* Pt^+ ... Pt^+, *etc.*		
–	Platinum isotopes	68 (Pt):	Hb/B1–1/2
			Hb/B2–179, 180
–	Platinum systems:		
	Pt–Ag	68 (Pt):	Hb/A6–810/2
	Pt–Al	35 (Al):	Hb/A6–1107/8
	Pt–As	68 (Pt):	Hb/C1–137/8
	Pt–Au	68 (Pt):	Hb/A6–818/9
	Pt–Au–Ge	68 (Pt):	Hb/A6–825/6
	Pt–Au–Pd	68 (Pt):	Hb/A6–840/1
	Pt–Au–Rh	68 (Pt):	Hb/A6–836
	Pt–C	68 (Pt):	Hb/C1–115
	Pt–Cd	68 (Pt):	Hb/A6–751/2
	Pt–Co	68 (Pt):	Hb/A6–785/6
	Pt–Cr	68 (Pt):	Hb/A6–769
	Pt–Cu	68 (Pt):	Hb/A6–799
	Pt–Cu–Fe	68 (Pt):	Hb/A6–809
	Pt–Cu–Ni	68 (Pt):	Hb/A6–809
	Pt–Fe	68 (Pt):	Hb/A6–791/3
	Pt–Fe–S	68 (Pt):	Hb/A6–798
	Pt–H	68 (Pt):	Hb/C1–5/25
	Pt–H–Pd	68 (Pt):	Hb/A6–839/40
	Pt–Hg	68 (Pt):	Hb/A6–752/3
	Pt–Ir	68 (Pt):	Hb/A6–842/3
	Pt–Mn	68 (Pt):	Hb/A6–778
	Pt–Mo	68 (Pt):	Hb/A6–773/4
	Pt–N	68 (Pt):	Hb/C1–54/6

Hb = Main Series Volume; GD = Gmelin-Durrer; Ab = Appended Volume to Main Series;
Eb = Supplement Volume to Main Series; Erg. W. = New Supplement Series Volume

Pt	Platinum systems:		
	Pt–Nb	68 (Pt):	Hb/A6–768
	Pt–Ni	68 (Pt):	Hb/A6–780/1
	Pt–O	68 (Pt):	Hb/C1–26/41
	Pt–P	68 (Pt):	Hb/C1–133/5
	Pt–Pb	68 (Pt):	Hb/A6–763/7
	Pt–Pb–Tl	68 (Pt):	Hb/A6–767
	Pt–Pd	68 (Pt):	Hb/A6–836
	Pt–Rh	68 (Pt):	Hb/A6–829/30
	Pt–Ru	68 (Pt):	Hb/A6–826
	Pt–S	68 (Pt):	Hb/C1–103/4
	Pt–Sb	68 (Pt):	Hb/A6–743/4
	Pt–Si	68 (Pt):	Hb/C1–128/31
	Pt–Sn	68 (Pt):	Hb/A6–758/63
	Pt–Tl	68 (Pt):	Hb/A6–755/7
	Pt–W	68 (Pt):	Hb/A6–774/5
Pt^{+}	Pt^{+}	68 (Pt):	Hb/B1–8 Hb/B2–179
Pt^{2+}	Pt^{2+}	68 (Pt):	Hb/B1–7/9 Hb/B2–92/3, 107, 179
Pt^{3+}	Pt^{3+}	68 (Pt):	Hb/B1–7/8
Pt^{4+}	Pt^{4+}	68 (Pt):	Hb/B1–7/8, 9/10 Hb/B2–77
Pt^{5+}	Pt^{5+}	68 (Pt):	Hb/B1–7/8
Pt^{6+}	Pt^{6+}	68 (Pt):	Hb/B1–7/8
PtS	PtS	68 (Pt):	Hb/C1–103/5
PtS_2	PtS_2	68 (Pt):	Hb/C1–106/8
PtSb	PtSb	68 (Pt):	Hb/A6–743/4, 746
$PtSb_2$	$PtSb_2$	68 (Pt):	Hb/A6–743/4, 746
PtSe	PtSe	68 (Pt):	Hb/C1–112
$PtSe_2$	$PtSe_2$	68 (Pt):	Hb/C1–112/3
$PtSe_3$	$PtSe_3$	68 (Pt):	Hb/C1–113
PtSi	PtSi	68 (Pt):	Hb/C1–129, 132
$PtSi_{12}$	$PtSi_{12}$	68 (Pt):	Hb/C1–131
PtSn	PtSn	68 (Pt):	Hb/A6–758/60, 762
$PtSn_2$	$PtSn_2$	68 (Pt):	Hb/A6–758/60, 763
$PtSn_3$	$PtSn_3$	68 (Pt):	Hb/A6–759
$PtSn_4$	$PtSn_4$	68 (Pt):	Hb/A6–758/60, 763
PtTe	PtTe	68 (Pt):	Hb/C1–113
$PtTe_2$	$PtTe_2$	68 (Pt):	Hb/C1–114
PtTi	PtTi	68 (Pt):	Hb/A6–757
$PtTi_2$	$PtTi_2$	68 (Pt):	Hb/A6–757
PtTl	PtTl	68 (Pt):	Hb/A6–755/7
PtZn	PtZn	68 (Pt):	Hb/A6–750/1
$PtZn_2$	$PtZn_2$	68 (Pt):	Hb/A6–750/1
Pt_2S_3	$Pt_2S_3 = Pt_3[PtS_6]$	68 (Pt):	Hb/C1–106
Pt_2Si	Pt_2Si	68 (Pt):	Hb/C1–129, 132
Pt_2Sn_3	Pt_2Sn_3	68 (Pt):	Hb/A6–758/60, 762/3 Hb/C3–310

See the "Instructions for the user"
at the beginning of the volume

Pt_2Te . . . . Pt_2Te . . . . 68 (Pt): Hb/C1–113
Pt_2Zn_3 . . . . Pt_2Zn_3 . . . . 68 (Pt): Hb/A6–750/1
Pt_3S_6Sn . . . . 2 PtS · PtS_2 · SnS_2 = $Pt_3[SnS_6]$ . . . . . . 68 (Pt): Hb/C3–313
$Pt_3S_6Tl_4$ . . . . 2 Tl_2S · 2 PtS · PtS_2 = $Tl_4Pt_2[PtS_6]$ . . . . 68 (Pt): Hb/C3–307
$Pt_3S_6Zn_2$ . . . . 2 ZnS · 2 PtS · PtS_2 = $Zn_2Pt_2[PtS_6]$ . . . . 68 (Pt): Hb/C3–296
Pt_3Si_2 . . . . Pt_3Si_2 . . . . 68 (Pt): Hb/C1–131
Pt_3Sn . . . . Pt_3Sn . . . . 68 (Pt): Hb/A6–758/62
Pt_3Sn_8 . . . . Pt_3Sn_8 . . . . 68 (Pt): Hb/A6–759, 763
Pt_3Ti . . . . Pt_3Ti . . . . 68 (Pt): Hb/A6–757
Pt_3Ti_2 . . . . Pt_3Ti_2 . . . . 68 (Pt): Hb/A6–757
Pt_3Zr . . . . Pt_3Zr . . . . 68 (Pt): Hb/A6–758
Pt_4S_6 . . . . $Pt_3[PtS_6]$ = Pt_2S_3 . . . . 68 (Pt): Hb/C1–106
Pt_4S_6Sn . . . . 4 PtS · SnS_2 = $Pt_4[SnS_6]$ . . . . 68 (Pt): Hb/C3–312
Pt_4Sb . . . . Pt_4Sb . . . . 68 (Pt): Hb/A6–743/4
Pt_4Se_6Sn . . . . 4 PtSe · $SnSe_2$ = $Pt_4[SnSe_6]$ . . . . 68 (Pt): Hb/C3–313
Pt_4Si_3 . . . . Pt_4Si_3 . . . . 68 (Pt): Hb/C1–131
Pt_4Sn_3 . . . . Pt_4Sn_3 . . . . 68 (Pt): Hb/C3–310
Pt_5S_6 . . . . Pt_5S_6 . . . . 68 (Pt): Hb/C1–105/6
Pt_5Sb_2 . . . . Pt_5Sb_2 . . . . 68 (Pt): Hb/A6–743/4
Pt_5Si_2 . . . . Pt_5Si_2 . . . . 68 (Pt): Hb/C1–129, 131
Pt_5Zn_2 . . . . Pt_5Zn_2 . . . . 68 (Pt): Hb/A6–750
Pt_5Zn_{21} . . . . Pt_5Zn_{21} . . . . 68 (Pt): Hb/A6–751

Pu . . . . Plutonium:
Atom . . . . Erg. W. 8–1/34
Atomic mass . . . . Erg. W. 7a–123/7, 128, 133
Atomic nucleus . . . . Erg. W. 7a–24/30, 61/170
Atomic radius . . . . Erg. W. 8–4/5
Biological behavior . . . . Erg. W. 8–403/15
Criticality . . . . Erg. W. 8–361/84
Decomposition and
solution methods . . . . Erg. W. 8–124/6
Detection in soils and fallout . . . . . . Erg. W. 7b–353/8
Determination:
Complexometric methods . . . . . . Erg. W. 8–139/41
Electrochemical methods . . . . . . Erg. W. 8–132/9
General . . . . Erg. W. 8–119/30
Gravimetric methods . . . . Erg. W. 8–130/1
Isotope analysis . . . . Erg. W. 8–158/61
Isotope dilution analysis . . . . Erg. W. 8–162/3
Optical methods . . . . Erg. W. 8–141/7
Radiometric methods . . . . Erg. W. 8–151/8
Redox methods . . . . Erg. W. 8–132/5
Separation methods . . . . Erg. W. 8–126/30
X-ray optical methods . . . . Erg. W. 8–147/51
Other methods . . . . Erg. W. 8–163/4
Determination in biological
materials . . . . Erg. W. 8–179/85
Diffusion . . . . Erg. W. 8–266/70

Hb = Main Series Volume; GD = Gmelin-Durrer; Ab = Appended Volume to Main Series;
Eb = Supplement Volume to Main Series; Erg. W. = New Supplement Series Volume

Pu . . . . Plutonium:
Electron configuration . . . . Erg. W. 8–3/23
Fabrication processes . . . . Erg. W. 8–234/50
Fire prevention . . . . Erg. W. 8–331/2
Formation in nature . . . . Erg. W. 7b–1/2
Handling techniques . . . . Erg. W. 8–123, 314, 349/61
Health physics . . . . Erg. W. 8–320/49
History . . . . Erg. W. 7a–1/11
– (Ekaosmium) . . . . 66 (Os): Hb–98/100
In-pile behavior of nuclear fuels . . . . Erg. W. 8–259/87
Ionization potentials . . . . Erg. W. 8–23
Isotope separation . . . . Erg. W. 7b–359/70
Nuclear binding energies . . . . Erg. W. 7a–128
Nuclear decay,
alpha radioactivity . . . . Erg. W. 7a–82/96
–, decay energies . . . . Erg. W. 7a–95, 123/6
–, decay properties . . . . Erg. W. 7a–24/30
–, decay scheme . . . . Erg. W. 7a–98/103, 107/12
–, gamma radioactivity . . . . Erg. W. 7a–78/81
Nuclear fission,
fission by charged particles . . . . . Erg. W. 7a–151/63
–, – by neutrons . . . . Erg. W. 7a–138/50
–, spontaneous fission . . . . Erg. W. 7a–134/8, 140
–, theoretical . . . . Erg. W. 7a–163/70
Nuclear models . . . . Erg. W. 7a–61/82
Nuclear moments . . . . Erg. W. 8–37
Nuclear properties . . . . Erg. W. 7a–24/30, 61/170
Nuclear radiation,
effect on gases . . . . Erg. W. 8–309/12
–, – on organic liquids . . . . Erg. W. 8–308/9
–, – on solids . . . . Erg. W. 8–292/300
–, – on solutions . . . . Erg. W. 8–302/7
–, – on the chemical properties . . . . Erg. W. 8–312/4
Occurrence, extraterrestrial . . . . . . Erg. W. 7a–14/7
–, terrestrial . . . . Erg. W. 7a–12/4
Oxidation states . . . . Erg. W. 8–8/11
Performance of heat sources . . . . . Erg. W. 8–288/90
Production, aqueous processing
of irradiated fuels . . . . Erg. W. 7b–177/240
–, – –, chemistry of
reprocessing solutions . . . . Erg. W. 7b–193/6
–, – –, general . . . . Erg. W. 7b–177/83
–, – –, precipitation . . . . Erg. W. 7b–196/7
–, – –, preparation of feed solutions . . Erg. W. 7b–184/93
–, – –, solvent extraction . . . . Erg. W. 7b–197/235
–, – –, sorption and ion exchange . . . Erg. W. 7b–235/40
–, nonaqueous processing
of irradiated fuels . . . . Erg. W. 7b–240/65
–, recovery from scrap and wastes . . Erg. W. 7b–279/88

See the "Instructions for the user"
at the beginning of the volume

Pu . . . . Plutonium:
Production in charged particle reactions, cross sections. . . . . . . Erg. W. 7b–104/16, 121/76
– –, experiments . . . . Erg. W. 7b–37/104, 116/21
– –, general features . . . . Erg. W. 7b–28/37
– in nuclear explosions. . . . . Erg. W. 7b–21/8
– in nuclear reactors . . . . Erg. W. 7b–3/18
– of metallic plutonium. . . . . Erg. W. 8–234/42
Radiotoxicity . . . . Erg. W. 8–320/5
Recoil reactions. . . . . Erg. W. 8–315/9
Recovery of plutonium-238 . . . . . . Erg. W. 7b–305/13
– of plutonium-242 . . . . Erg. W. 7b–313/5
Safety measures . . . . Erg. W. 8–385/93
Sampling . . . . Erg. W. 8–123
Separation from nuclear reaction products . . . . Erg. W. 7b–336/42, 347
–, analytical . . . . Erg. W. 8–126/30
Specifications . . . . Erg. W. 8–122
Spectra:
Absorption spectra. . . . . Erg. W. 8–51/7, 63/4, 67/76
Emission spectra . . . . Erg. W. 8–36/41
Luminescence spectra . . . . Erg. W. 8–57, 68
Mössbauer spectra . . . . Erg. W. 8–89/90, 97/8
X-ray spectra . . . . Erg. W. 8–80/8
Standards . . . . Erg. W. 8–120/1
Storage and control of wastes . . . . . Erg. W. 8–393/402
– of samples. . . . . Erg. W. 8–123
Therapy after contamination. . . . . . Erg. W. 8–415/24
Transition reactions . . . . Erg. W. 8–305/6
Treatment and storage of wastes . . . Erg. W. 7b–265/79
Uses of fissile plutonium isotopes. . . Erg. W. 8–189/201
– of other plutonium isotopes . . . . . Erg. W. 8–201/13, 250

–. . . . . Plutonium fuel elements:
In-pile behavior . . . . Erg. W. 8–259/83
Production. . . . . Erg. W. 8–239/42

–. . . . . Plutonium ions:
Ionization potentials . . . . Erg. W. 8–23
Properties of the ion . . . . Erg. W. 8–5/28
Spectra:
Absorption spectra. . . . . Erg. W. 8–51/3, 55/7, 63/4, 67/72, 74/6
Emission spectra . . . . Erg. W. 8–40/1
Luminescence spectra . . . . Erg. W. 8–57, 68

–. . . . . Plutonium isotopes
see Pu ... Plutonium

–. . . . . Plutonium systems:
Pu–C. . . . . Erg. W. 4–201

Hb = Main Series Volume; GD = Gmelin-Durrer; Ab = Appended Volume to Main Series;
Eb = Supplement Volume to Main Series; Erg. W. = New Supplement Series Volume

Pu . . . . Plutonium systems:
$Pu-C-Fe$ . . . . Erg. W. 4–209
$Pu-C-N$ . . . . Erg. W. 4–225/8
$Pu-C-O$ . . . . Erg. W. 4–224/5
$Pu-C-Si$ . . . . Erg. W. 4–209
$Pu-C-Th$ . . . . Erg. W. 4–210/1
$Pu-C-U$ . . . . Erg. W. 4–211/3
$Pu-H$ . . . . Erg. W. 4–3/5
$Pu-Nb-Si$ . . . . Erg. W. 4–247/8
$Pu-O$ . . . . Erg. W. 4–10/2
$Pu-O-H_2O$ . . . . Erg. W. 4–10/2
$Pu-O-U$ . . . . Erg. W. 4–59/60
$Pu-S$ . . . . Erg. W. 4–159/61
$Pu-Si$ . . . . Erg. W. 4–245/6

Pu^{+} . . . . Pu^{+} . . . . Erg. W. 8–23, 40/1
Pu^{2+} . . . . Pu^{2+} . . . . Erg. W. 8–11/23
Pu^{3+} . . . . Pu^{3+} . . . . Erg. W. 8–5, 11/27, 51/3, 55/7
Pu^{4+} . . . . Pu^{4+} . . . . Erg. W. 8–5, 11/23, 63/4, 67/8
Pu^{5+} . . . . Pu^{5+} . . . . Erg. W. 8–5, 11/23
Pu^{6+} . . . . Pu^{6+} . . . . Erg. W. 8–5, 11/23
Pu^{7+} . . . . Pu^{7+} . . . . Erg. W. 8–11/25
Pu^{n+} . . . . Pu^{n+} (n = 8 to 10) . . . . Erg. W. 8–23

$PuRh_3$ . . . . $PuRh_3$ . . . . Erg. W. 4–209
$PuRu_2$ . . . . $PuRu_2$ . . . . Erg. W. 4–209
PuS . . . . PuS . . . . Erg. W. 4–159/66
–. . . . . PuS solid solutions:
(Pu,U)S . . . . Erg. W. 4–168/71

PuS_2 . . . . PuS_2 . . . . Erg. W. 4–159/66
–. . . . . PuS_2 solid solutions:
$(Pu,U)S_2$ . . . . Erg. W. 4–170

PuS_{2-x} . . . . PuS_{2-x} . . . . Erg. W. 4–159/66
PuSe . . . . PuSe . . . . Erg. W. 4–189/90
$PuSe_{2-x}$ . . . . $PuSe_{2-x}$ . . . . Erg. W. 4–189/90
PuSi. . . . . PuSi . . . . Erg. W. 4–245/8
–. . . . . PuSi systems:
PuSi–PuC . . . . Erg. W. 4–248

$PuSi_2$ . . . . $PuSi_2$ . . . . Erg. W. 4–245/8
PuTe . . . . PuTe . . . . Erg. W. 4–192
$PuTe_2$ . . . . $PuTe_2$ . . . . Erg. W. 4–192/3
$PuTe_{2-x}$ . . . . $PuTe_{2-x}$ . . . . Erg. W. 4–192/3
Pu_2S_3 . . . . Pu_2S_3 . . . . Erg. W. 4–159/66
–. . . . . Pu_2S_3 solid solutions:
$(Pu,U)_2S_3$ . . . . Erg. W. 4–169/70

Pu_2S_5Th . . . . $ThPu_2S_5$ . . . . Erg. W. 4–170
Pu_2S_5U . . . . UPu_2S_5 . . . . Erg. W. 4–170

See the "Instructions for the user"
at the beginning of the volume

Pu_2Se_3 . . . . Pu_2Se_3 . . . . Erg. W. 4–189/90
Pu_2Te_3 . . . . Pu_2Te_3 . . . . Erg. W. 4–192/3
Pu_3S_4 . . . . Pu_3S_4 . . . . Erg. W. 4–159/66
Pu_3S_5 . . . . Pu_3S_5 . . . . Erg. W. 4–170
Pu_3Se_4 . . . . Pu_3Se_4 . . . . Erg. W. 4–189/90
Pu_3Si_2 . . . . Pu_3Si_2 . . . . Erg. W. 4–245/8
Pu_3Si_5 . . . . Pu_3Si_5 . . . . Erg. W. 4–245/8
Pu_5S_7 . . . . Pu_5S_7 . . . . Erg. W. 4–159/66
Pu_5Si_3 . . . . Pu_5Si_3 . . . . Erg. W. 4–245/8

Ra . . . . Radium:
Abundance . . . . 31 (Ra): Hb–4
Atom . . . . 31 (Ra): Hb–3, 50/1
Atomic nucleus . . . . 31 (Ra): Hb–2/3, 51/4, 68/9, 74, 78/9
Chemical reactions . . . . 31 (Ra): Hb–56, 69, 75/6, 79/80
Detection . . . . 31 (Ra): Hb–56, 69, 76, 80
Determination . . . . 31 (Ra): Hb–56/7, 70, 76, 80
Economic deposits . . . . 31 (Ra): Hb–4/7, 65
Economics . . . . 31 (Ra): Hb–49/50, 67/8
Electrochemical behavior . . . . 31 (Ra): Hb–56, 75, 79
Formation . . . . 31 (Ra): Hb–50, 68, 72, 77
History . . . . 31 (Ra): Hb–3, 64, 72, 77
Mineralogy . . . . 31 (Ra): Hb–7/26
Nuclear decay,
alpha radiation . . . . 31 (Ra): Hb–51/4, 74, 78
Erg. W. 7a–85/95
–, beta radiation . . . . 31 (Ra): Hb–53, 68/9, 79
–, decay scheme . . . . 31 (Ra): Hb–2
–, gamma radiation . . . . 31 (Ra): Hb–53/4, 68/9, 79
Erg. W. 7a–80
Occurrence . . . . 31 (Ra): Hb–3/27, 64/5
Ore dressing . . . . 31 (Ra): Hb–28/33, 66/7
Physical properties . . . . 31 (Ra): Hb–50/6, 68/9, 74, 78
Preparation . . . . 31 (Ra): Hb–50, 72/3, 77/8
–, industrial . . . . 31 (Ra): Hb–27/48, 65/7
– in pure state . . . . 31 (Ra): Hb–40/7
Standard preparations . . . . 31 (Ra): Hb–47/8
Storage . . . . 31 (Ra): Hb–47
Uses . . . . 31 (Ra): Hb–63, 71, 76
Erg. W. 8–224, 230/1
Valence . . . . 31 (Ra): Hb–51
– . . . . Radium ions . . . . 31 (Ra): Hb–56, 75, 79
– . . . . Radium isotopes
see Ra ... Radium
Ra^{2+} . . . . Ra^{2+} . . . . 31 (Ra): Hb–56, 75, 79

Hb = Main Series Volume; GD = Gmelin-Durrer; Ab = Appended Volume to Main Series;
Eb = Supplement Volume to Main Series; Erg. W. = New Supplement Series Volume

RaA . . . . Radium A = ^{218}Po
see Po ... Polonium

RaB . . . . Radium B = ^{214}Pb
see Pb ... Lead isotopes

RaC . . . . Radium C = ^{214}Bi
see Bi ... Bismuth isotopes

RaC′. . . . . Radium C′ = ^{214}Po
see Po ... Polonium

RaC″ . . . . Radium C″ = ^{210}Tl
see Tl ... Thallium isotopes

RaD . . . . Radium D = ^{210}Pb
see Pb ... Lead isotopes

RaE . . . . Radium E = ^{210}Bi
see Bi ... Bismuth isotopes

RaE″ . . . . Radium E″ = ^{206}Tl
see Tl ... Thallium isotopes

RaEm . . . . Radium emanation = ^{222}Rn
see Rn ... Radon

RaF . . . . Radium F = ^{210}Po
see Po ... Polonium

RaG . . . . Radium G = ^{206}Pb
see Pb ... Lead isotopes

Rb. . . . . Rubidium:

Absorption of light	24 (Rb):	Hb–54
– of X-rays	24 (Rb):	Hb–71
Abundance	24 (Rb):	Hb–2/3
Affinity	24 (Rb):	Hb–42
Atom	24 (Rb):	Hb–34/42
Atomic nucleus	24 (Rb):	Hb–27/34
Chemical reactions	24 (Rb):	Hb–88/9
Cosmochemistry	24 (Rb):	Hb–1
Crystallographic properties	24 (Rb):	Hb–44
Detection	24 (Rb):	Hb–90/6
– in iron	59 (Fe):	Hb/FII1–4
– in special substances	24 (Rb):	Hb–102/3
Determination	24 (Rb):	Hb–96/100
– in special substances	24 (Rb):	Hb–102/3
Economics	24 (Rb):	Hb–23
Electrical properties	24 (Rb):	Hb–73/85
Electrochemical behavior	24 (Rb):	Hb–85/8
Formation	24 (Rb):	Hb–23/6
Geochemistry	24 (Rb):	Hb–2/13
History	24 (Rb):	Hb–1
Ionization	24 (Rb):	Hb–83/5

Rb . . . . Rubidium:

Magnetic properties . . . . 24 (Rb): Hb–71/3
Mechanical properties . . . . 24 (Rb): Hb–45/7
Mineralogy . . . . 24 (Rb): Hb–13
Molecule . . . . 24 (Rb): Hb–42/3
Occurrence, extraterrestrial . . . . . . 24 (Rb): Hb–1
–, terrestrial . . . . 24 (Rb): Hb–2/13
– in the biosphere . . . . 24 (Rb): Hb–10/3
– in the hydrosphere . . . . 24 (Rb): Hb–8/10
– in the lithosphere . . . . 24 (Rb): Hb–2/8
Optical properties . . . . 24 (Rb): Hb–53/71
Position in the periodic system . . . . 24 (Rb): Hb–42
Preparation . . . . 24 (Rb): Hb–23/6
–, industrial . . . . 24 (Rb): Hb–13/20
– of special forms . . . . 24 (Rb): Hb–26
Purification . . . . 24 (Rb): Hb–20/3, 26
Radioactivity . . . . 24 (Rb): Hb–3, 30/4
Scattering of X-rays . . . . 24 (Rb): Hb–71
Separation, analytical . . . . 24 (Rb): Hb–100/2
Spectra:
Arc spectrum . . . . 24 (Rb): Hb–55/65
Band spectrum . . . . 24 (Rb): Hb–68/9
Spark spectra . . . . 24 (Rb): Hb–65/7
X-ray spectrum . . . . 24 (Rb): Hb–69/71
Storage . . . . 24 (Rb): Hb–26
Thermal properties . . . . 24 (Rb): Hb–47/53
Valence . . . . 24 (Rb): Hb–42

–. . . . . Rubidium ions
see Rb^+ .... Rb^+, etc.

–. . . . . Rubidium isotopes . . . . 24 (Rb): Hb–27/8

–. . . . . Rubidium systems:

Rb–Al . . . . 35 (Al): Hb/A4–637
Rb–B–Ta . . . . 50 (Ta): Hb/B2–320
Rb–Cs . . . . 25 (Cs): Hb/2–258
Rb–Cu . . . . 60 (Cu): Hb/B3–1107
Rb–Hg . . . . 34 (Hg): Hb/A2–1010/2
Rb–K . . . . 24 (Rb): Hb–245/6
Rb–Na . . . . 24 (Rb): Hb–242/3
Rb–S . . . . 24 (Rb): Hb–205
Rb–V . . . . 48 (V): Hb/B2–578

Rb^+ . . . . Rb^+:

Chemical reactions . . . . 24 (Rb): Hb–89/102
Diffusion in sodium halides . . . . . . 21 (Na): Eb/6–291
Electrochemical behavior . . . . 24 (Rb): Hb–86/8
Interaction with gases . . . . 24 (Rb): Hb–83/4
Ionic susceptibility . . . . 24 (Rb): Hb–72/3
Ionization potentials . . . . 24 (Rb): Hb–84/5

Hb = Main Series Volume; GD = Gmelin-Durrer; Ab = Appended Volume to Main Series;
Eb = Supplement Volume to Main Series; Erg. W. = New Supplement Series Volume

Rb^+ . . . . Rb^+:
Properties of the ion . . . . 24 (Rb): Hb–37/9, 41
Spectra . . . . 24 (Rb): Hb–65/7
Rb^{n+} . . . . Rb^{n+} (n = 2 to 8) . . . . 24 (Rb): Hb–39, 67, 84/5
RbSn . . . . RbSn . . . . 46 (Sn): Hb/D–78
Rb_2 . . . . Rb_2 . . . . 24 (Rb): Hb–42/3
Rb_2S . . . . Rb_2S . . . . 24 (Rb): Hb–204
–. . . . . $Rb_2S \cdot 2\,H_2O$ . . . . 24 (Rb): Hb–205
–. . . . . $Rb_2S \cdot 4\,H_2O$ . . . . 24 (Rb): Hb–205
Rb_2S_2 . . . . Rb_2S_2 . . . . 24 (Rb): Hb–206
–. . . . . $Rb_2S_2 \cdot H_2O$. . . . . 24 (Rb): Hb–206
–. . . . . $Rb_2S_2 \cdot 2\,H_2O$. . . . . 24 (Rb): Hb–206
–. . . . . $Rb_2S_2 \cdot 3\,H_2O$. . . . . 24 (Rb): Hb–206
Rb_2S_3 . . . . Rb_2S_3 . . . . 24 (Rb): Hb–206
–. . . . . $Rb_2S_3 \cdot H_2O$. . . . . 24 (Rb): Hb–206
Rb_2S_4 . . . . Rb_2S_4 . . . . 24 (Rb): Hb–206
–. . . . . $Rb_2S_4 \cdot 2\,H_2O$. . . . . 24 (Rb): Hb–206
Rb_2S_5 . . . . Rb_2S_5 . . . . 24 (Rb): Hb–206
Rb_2S_6 . . . . Rb_2S_6 . . . . 24 (Rb): Hb–205
Rb_2Se . . . . Rb_2Se . . . . 24 (Rb): Hb–216
Rb_2Te . . . . Rb_2Te . . . . 24 (Rb): Hb–219

RdAc . . . . Radioactinium = ^{227}Th
see Th ... Thorium
RdTh . . . . Radiothorium = ^{228}Th
see Th ... Thorium

Re . . . . Rhenium:
Atom. . . . . 70 (Re): Hb–44/6
Atomic nucleus . . . . 70 (Re): Hb–40/4
Chemical reactions . . . . 70 (Re): Hb–59/61
Colloidal forms . . . . 70 (Re): Hb–36
Cosmochemistry . . . . 70 (Re): Hb–11/2
Crystallographic properties . . . . . . 70 (Re): Hb–47/8
Detection . . . . 70 (Re): Hb–62/9
Determination. . . . . 70 (Re): Hb–69/75, 79/81
Economics. . . . . 70 (Re): Hb–30
Electrical properties . . . . 70 (Re): Hb–57/8
Electrochemical behavior . . . . 70 (Re): Hb–59
Formation . . . . 70 (Re): Hb–31/2
Geochemical character . . . . 70 (Re): Hb–13/4
Geochemistry . . . . 70 (Re): Hb–12/25
History. . . . . 70 (Re): Hb–1/10
Magnetic properties . . . . 70 (Re): Hb–56/7
Mechanical properties . . . . 70 (Re): Hb–48/9
Occurrence, extraterrestrial . . . . . . 70 (Re): Hb–11/2
–, terrestrial . . . . 70 (Re): Hb–12/25

See the "Instructions for the user"
at the beginning of the volume

Re	Rhenium:		
	Occurrence in industrial products	70 (Re):	Hb–24
	– in the hydrosphere	70 (Re):	Hb–25
	– in the lithosphere	70 (Re):	Hb–14/24
	Optical properties	70 (Re):	Hb–50/6
	Physiological action	70 (Re):	Hb–61
	Position in the periodic system	70 (Re):	Hb–47
	Preparation	70 (Re):	Hb–32/3
	–, industrial	70 (Re):	Hb–25/30
	– of special forms	70 (Re):	Hb–34/6
	Purity testing	70 (Re):	Hb–33/4
	Separation, analytical	70 (Re):	Hb–75/9
	Spectra	70 (Re):	Hb–51/6
	Thermal properties	70 (Re):	Hb–49/50
	Uses	70 (Re):	Hb–36/40, 83
	Valence	70 (Re):	Hb–46/7
–	Rhenium ions	70 (Re):	Hb–45/6, 57
–	Rhenium isotopes	70 (Re):	Hb–40/1
–	Rhenium systems:		
	Re–As	70 (Re):	Hb–124
	Re–Fe	70 (Re):	Hb–85/6
	Re–Nb	49 (Nb):	Hb/A–161
	Re–P	70 (Re):	Hb–123
	Re–Rh	70 (Re):	Hb–86/7
	Re–S	70 (Re):	Hb–117
	Re–W	70 (Re):	Hb–84/5
Re^{n+}	Re^{n+} (n = 1 to 7)	70 (Re):	Hb–45/6, 57
ReS_2	ReS_2	70 (Re):	Hb–117/8
ReS_4Tl	$TlReS_4$	70 (Re):	Hb–149
$ReSe_2$	$ReSe_2$	70 (Re):	Hb–120
Re_2S_7	Re_2S_7	70 (Re):	Hb–118/9
Re_2Se_7	Re_2Se_7	70 (Re):	Hb–120
Re_3W_2	W_2Re_3	70 (Re):	Hb–84/5
Rf	Rutherfordium (Element 104)	Erg. W. 7a–4	
	see also this Volume, Appendix		
Rh	Rhodium:		
	Abundance	68 (Pt):	Hb/A1–24/7
	Atom	64 (Rh):	Hb–9/14
	Atomic nucleus	64 (Rh):	Hb–1/9
	Catalytic properties	68 (Pt):	Hb/A3–429/30
	Chemical reactions	64 (Rh):	Hb–42/4
	Colloidal forms	68 (Pt):	Hb/A3–424, 429/30
	Cosmochemistry	68 (Pt):	Hb/A1–15/22
	Crystallographic properties	64 (Rh):	Hb–14/6
	Detection	68 (Pt):	Hb/A4–436/8, 488/9
	– in iron	59 (Fe):	Hb/FII1–41

Hb = Main Series Volume; GD = Gmelin-Durrer; Ab = Appended Volume to Main Series; Eb = Supplement Volume to Main Series; Erg. W. = New Supplement Series Volume

Rh. . . . . Rhodium:

Determination.	68 (Pt):	Hb/A4–464/7, 488/9
– in special substances	68 (Pt):	Hb/A4–525/32
Economic deposits *see Pt ... Platinum*		
Economics.	68 (Pt):	Hb/A2–302/10
Electrical properties	64 (Rh):	Hb–35/40
Electrochemical behavior	64 (Rh):	Hb–40/2
Electrolytic deposition	68 (Pt):	Hb/A3–389/90
Geochemical character.	68 (Pt):	Hb/A1–23/8
Geochemistry	68 (Pt):	Hb/A1–22/49 Hb/A2–275/6
History.	68 (Pt):	Hb/A1–11/2
Magnetic properties	64 (Rh):	Hb–33/4
Mechanical properties	64 (Rh):	Hb–16/9
Mineralogy	68 (Pt):	Hb/A2–276/301
Occurrence, extraterrestrial	68 (Pt):	Hb/A1–15/22
–, terrestrial	68 (Pt):	Hb/A1–22/49 Hb/A2–275/6
– in meteorites	68 (Pt):	Hb/A1–16/22
– in the biosphere.	68 (Pt):	Hb/A2–275
– in the hydrosphere	68 (Pt):	Hb/A2–275
– in the lithosphere	68 (Pt):	Hb/A1–28/49
Optical properties.	64 (Rh):	Hb–22/33
Physiological action	68 (Pt):	Hb/A2–275/6
Position in the periodic system	64 (Rh):	Hb–14
Preparation	68 (Pt):	Hb/A3–311/430
– from minerals.	68 (Pt):	Hb/A3–312/50
– from wastes and residues	68 (Pt):	Hb/A3–350/62
– in pure state.	68 (Pt):	Hb/A3–372/4
– of dense rhodium.	68 (Pt):	Hb/A3–383/4
Purity testing	68 (Pt):	Hb/A3–365/7 Hb/A4–519
Separation, analytical	68 (Pt):	Hb/A4–495/6, 517/8
Special forms	68 (Pt):	Hb/A3–399, 401/4, 424, 429/30
Spectra	64 (Rh):	Hb–23/33
Thermal properties	64 (Rh):	Hb–19/22
Toxicity	68 (Pt):	Hb/A2–275/6
Uses	68 (Pt):	Hb/A3–385/91
Valence	64 (Rh):	Hb–13/4

–. . . . . Rhodium ions:

Electrochemical behavior	64 (Rh):	Hb–41/2
Ionic susceptibility	64 (Rh):	Hb–34
Properties of the ion	64 (Rh):	Hb–9/12
Spectra	64 (Rh):	Hb–26/7

–. . . . . Rhodium isotopes . . . . 64 (Rh): Hb–1/2
Erg. W. 7b–355/6

See the "Instructions for the user"
at the beginning of the volume

Formula	Entry	Gmelin Syst. No.	Reference
Rh	Rhodium systems:		
	Rh–Ag	68 (Pt):	Hb/A5–566
	Rh–Au	68 (Pt):	Hb/A5–566/8
	Rh–Au–Pt	68 (Pt):	Hb/A6–836
	Rh–Be	68 (Pt):	Hb/A5–552
	Rh–Bi	68 (Pt):	Hb/A5–550/2
	Rh–Cd	68 (Pt):	Hb/A5–554
	Rh–Cu	68 (Pt):	Hb/A5–563/5
	Rh–Fe	68 (Pt):	Hb/A5–560/2
	Rh–Fe–S	68 (Pt):	Hb/A5–563
	Rh–P	68 (Pt):	Hb/A5–547/9
	Rh–Pt	68 (Pt):	Hb/A6–829/30
	Rh–Re	70 (Re):	Hb–86/7
	Rh–S	64 (Rh):	Hb–62/4
	Rh–Sn	68 (Pt):	Hb/A5–555/6
Rh^{+}	Rh^{+}	64 (Rh):	Hb–10/1, 26/7
Rh^{2+}	Rh^{2+}	64 (Rh):	Hb–9, 11, 26
Rh^{3+}	Rh^{3+}	64 (Rh):	Hb–9/12, 34, 41/2
–	$[Rh(H_2O)_6]^{3+}$	64 (Rh):	Hb–58
Rh^{4+}	Rh^{4+}	64 (Rh):	Hb–9, 11
Rh^{6+}	Rh^{6+}	64 (Rh):	Hb–9
RhS	RhS	64 (Rh):	Hb–64
RhS_2	RhS_2	64 (Rh):	Hb–66
RhSn	RhSn	68 (Pt):	Hb/A5–555/6
$RhSn_2$	$RhSn_2$	68 (Pt):	Hb/A5–555/6
$RhSn_3$	$RhSn_3$	68 (Pt):	Hb/A5–555
$RhSn_4$	$RhSn_4$	68 (Pt):	Hb/A5–555/6
$RhZn_2$	$RhZn_2$	68 (Pt):	Hb/A5–553/4
Rh_2S_3	Rh_2S_3	64 (Rh):	Hb–65
Rh_2S_5	Rh_2S_5	64 (Rh):	Hb–66
Rh_2Se_5	Rh_2Se_5	64 (Rh):	Hb–69
Rh_2Sn	Rh_2Sn	68 (Pt):	Hb/A5–555/6
Rh_2Te_5	Rh_2Te_5	64 (Rh):	Hb–69
Rh_2Ti	Rh_2Ti	68 (Pt):	Hb/A5–555
Rh_3S_4	Rh_3S_4	64 (Rh):	Hb–64
Rh_3Sn_2	Rh_3Sn_2	68 (Pt):	Hb/A5–555/6
Rh_3Ti	Rh_3Ti	68 (Pt):	Hb/A5–555
Rh_5Zn_{21}	Rh_5Zn_{21}	68 (Pt):	Hb/A5–554
Rh_9S_8	Rh_9S_8	64 (Rh):	Hb–64
Rn	Radon:		
	Affinity	1 (EG):	Hb–223/4
	Atom	1 (EG):	Hb–207/9
	Atomic nucleus	1 (EG):	Hb–199/207
	Chemical reactions	1 (EG):	Hb–223/4, 229/31
	Cosmochemistry	1 (EG):	Hb–6/7
	Detection	1 (EG):	Hb–241/8
	Determination	1 (EG):	Hb–241/8

Hb = Main Series Volume; GD = Gmelin-Durrer; Ab = Appended Volume to Main Series;
Eb = Supplement Volume to Main Series; Erg. W. = New Supplement Series Volume

Rn . . . . Radon:
Diffusion . . . . 1 (EG): Hb–211/2
Electrical properties . . . . 1 (EG): Hb–222
Formation . . . . 1 (EG): Hb–193
Geochemistry . . . . 1 (EG): Hb–7/35
History . . . . 1 (EG): Hb–3/6
Ionization . . . . 1 (EG): Hb–222
Mechanical properties . . . . 1 (EG): Hb–209/21
Molecule . . . . 1 (EG): Hb–209
Nuclear decay, alpha radiation . . . . Erg. W. 7a–87/95
Occurrence, extraterrestrial . . . . . . 1 (EG): Hb–6/7
–, terrestrial . . . . 1 (EG): Hb–7/35
– in the atmosphere . . . . 1 (EG): Hb–7/17
– in the hydrosphere . . . . 1 (EG): Hb–17/28
– in the lithosphere . . . . 1 (EG): Hb–28/35
Optical properties . . . . 1 (EG): Hb–221/2
Position in the periodic system . . . . 1 (EG): Hb–209, 223
Preparation . . . . 1 (EG): Hb–193/8
Solubility . . . . 1 (EG): Hb–212/7
Sorption . . . . 1 (EG): Hb–217/21
– on palladium . . . . 65 (Pd): Hb/2–115
– on platinum . . . . 68 (Pt): Hb/C1–1/3
Spectra . . . . 1 (EG): Hb–221/2
Thermal properties . . . . 1 (EG): Hb–210/1
Uses . . . . 1 (EG): Hb–251
Valence . . . . 1 (EG): Hb–223/4

–. . . . . $Rn \cdot xH_2O$ . . . . Erg. W. 1–160
–. . . . . Radon compounds . . . . Erg. W. 1–129/30, 160
–. . . . . Radon isotopes:
Analysis . . . . 1 (EG): Hb–245/7
Preparation . . . . 1 (EG): Hb–198

–. . . . . Radon solid solutions:
$Rn-CO_2$ . . . . 14 (C): Hb/C1–507
$Rn-H_2S$ . . . . 9 (S): Hb/B1–19, 107
$Rn-SO_2$ . . . . 9 (S): Hb/B1–210
$Rn \cdot 6\,H_2O-SO_2 \cdot 6\,H_2O$ . . . . 9 (S): Hb/B2–1120/1

–. . . . . Radon systems:
Rn–HCl . . . . 6 (Cl): Eb/B1–1

Rn^+ . . . . Rn^+ . . . . 1 (EG): Hb–222
Rn_2 . . . . Rn_2 . . . . 1 (EG): Hb–209

Ru . . . . Ruthenium:
Absorption of X-rays . . . . 63 (Ru): Hb–20
Eb–31/2
Abundance . . . . 68 (Pt): Hb/A1–24/7
Atom . . . . 63 (Ru): Hb–3/7
Eb–10/4

See the "Instructions for the user"
at the beginning of the volume

Ru . . . . Ruthenium:
Atomic nucleus . . . . 63 (Ru): Hb–1/3
Eb–3/10
Atomic refraction . . . . 63 (Ru): Eb–32
Atomic scattering factors. . . . . 63 (Ru): Eb–11
Atomic susceptibility . . . . 63 (Ru): Eb–32
Atomic weight . . . . 63 (Ru): Hb–6/7
Eb–2
Chemical reactions . . . . 63 (Ru): Hb–22/4
Eb–69/73
Colloidal forms . . . . 68 (Pt): Hb/A3–423/4
Cosmochemistry . . . . 68 (Pt): Hb/A1–15/22
Crystallographic properties . . . . . . 63 (Ru): Hb–7/9
Eb–32/5
Detection . . . . 63 (Ru): Eb–91/6
68 (Pt): Hb/A4–432/6, 488
– in iron . . . . 59 (Fe): Hb/FII1–41
– in special substances . . . . 63 (Ru): Eb–96
Determination. . . . . 63 (Ru): Eb–96/102
68 (Pt): Hb/A4–463/4, 488
– in special substances . . . . 63 (Ru): Eb–104/7
68 (Pt): Hb/A4–525/32
Economic deposits
see Pt ... Platinum
Economics. . . . . 68 (Pt): Hb/A2–306/10
Electrical properties . . . . 63 (Ru): Hb–21
Eb–45/51
Electrochemical behavior . . . . 63 (Ru): Hb–22
Eb–51/69
Electrolytic deposition . . . . 63 (Ru): Eb–65/9
68 (Pt): Hb/A3–389
Electron configuration . . . . 63 (Ru): Hb–3/4
Eb–1, 10
Geochemical character . . . . 68 (Pt): Hb/A1–23/8
Geochemistry . . . . 68 (Pt): Hb/A1–22/49
Hb/A2–275/6
History. . . . . 68 (Pt): Hb/A1–13/4
Ionization . . . . 63 (Ru): Hb–21
Eb–11/2
Magnetic properties . . . . 63 (Ru): Hb–20/1
Eb–45/51
Mass absorption coefficient . . . . . . 63 (Ru): Eb–31/2
Mechanical properties . . . . 63 (Ru): Hb–9/10
Eb–35/45
Mineralogy . . . . 68 (Pt): Hb/A2–276/301
Nuclear properties . . . . 63 (Ru): Hb–1/3
Eb–3/10
Occurrence, extraterrestrial . . . . . . 68 (Pt): Hb/A1–15/22

Hb = Main Series Volume; GD = Gmelin-Durrer; Ab = Appended Volume to Main Series;
Eb = Supplement Volume to Main Series; Erg. W. = New Supplement Series Volume

Ru . . . . Ruthenium:

Occurrence, terrestrial . . . . 68 (Pt): Hb/A1–22/49
Hb/A2–275/6
– in meteorites . . . . 68 (Pt): Hb/A1–16/22
– in the biosphere. . . . . 68 (Pt): Hb/A2–275
– in the hydrosphere . . . . 68 (Pt): Hb/A2–275
– in the lithosphere . . . . 68 (Pt): Hb/A1–28/49
Optical properties. . . . . 63 (Ru): Hb–12/20
Eb–14/32, 51
Physiological action . . . . 68 (Pt): Hb/A2–275/6
Position in the periodic system . . . . 63 (Ru): Hb–7
Eb–1
Preparation . . . . 68 (Pt): Hb/A3–311/430
– from minerals. . . . . 68 (Pt): Hb/A3–312/50
– from wastes and residues . . . . . . 68 (Pt): Hb/A3–350/62
– in pure state. . . . . 68 (Pt): Hb/A3–372
– of dense ruthenium. . . . . 68 (Pt): Hb/A3–383
Redox reactions. . . . . 63 (Ru): Eb–74/91
Scattering of X-rays . . . . 63 (Ru): Hb–20
Eb–31
Separation, analytical . . . . 63 (Ru): Eb–102/4
68 (Pt): Hb/A4–493/5, 517/8
Sorption of hydrogen. . . . . 63 (Ru): Eb–231/2
Special forms . . . . 68 (Pt): Hb/A3–398/9, 401/4, 423/4
Spectra:
Arc spectrum . . . . 63 (Ru): Hb–12/5
Eb–14/9
Band spectrum . . . . 63 (Ru): Hb–15/6
Crystal spectra . . . . 63 (Ru): Eb–21/6
Spark spectra . . . . 63 (Ru): Hb–12/5
Eb–19/21
X-ray spectrum . . . . 63 (Ru): Hb–16/20
Eb–26/32
Thermal properties . . . . 63 (Ru): Hb–10/2
Eb–35/45
Toxicity . . . . 68 (Pt): Hb/A2–275/6
Uses . . . . 63 (Ru): Eb–172/6
68 (Pt): Hb/A3–385/91
Valence . . . . 63 (Ru): Hb–7

–. . . . . Ruthenium ions
see Ru^+ *...* Ru^+*, etc.*

–. . . . . Ruthenium isotopes:

Abundance . . . . 63 (Ru): Hb–1/2
Eb–2/3
Atomic nucleus . . . . 63 (Ru): Hb–1/3
Eb–3/10
Deposition from radioactive wastes. . 63 (Ru): Eb–69

See the "Instructions for the user"
at the beginning of the volume

Ru . . . . Ruthenium isotopes:

Detection . . . . 63 (Ru): Eb–107/16
Erg. W. 7b–355/6

Determination. . . . . 63 (Ru): Eb–107/16

Enrichment and separation of stable Ru isotopes. . . . . 63 (Ru): Eb–117

Handling. . . . . 63 (Ru): Eb–118

Preparation, separation, and enrichment of unstable Ru isotopes from fallout . . . . . 63 (Ru): Eb–166/72

–, –, – – from nuclear fuels. . . . . . . 63 (Ru): Eb–136/42

–, –, – – from other elements . . . . . 63 (Ru): Eb–118/36

–, –, – – from radioactive wastes . . . 63 (Ru): Eb–69, 142/64

–, –, – – from reactor material . . . . . 63 (Ru): Eb–164/6

– . . . . Ruthenium systems:

Ru–Ag. . . . . 63 (Ru): Eb–228/30

Ru–Al . . . . 63 (Ru): Eb–178/80

Ru–Au. . . . . 63 (Ru): Eb–230

Ru–B . . . . 63 (Ru): Eb–318/9

Ru–Be. . . . . 63 (Ru): Eb–177/8
68 (Pt): Hb/A5–536

Ru–Bi . . . . 63 (Ru): Eb–337

Ru–C . . . . 63 (Ru): Eb–321/2

Ru–Ce. . . . . 63 (Ru): Eb–181/3

Ru–Co. . . . . 63 (Ru): Eb–223/4

Ru–Co–U . . . . 63 (Ru): Eb–224

Ru–Cr . . . . 63 (Ru): Eb–202/4

Ru–Cu. . . . . 63 (Ru): Eb–228

Ru–Fe . . . . 63 (Ru): Eb–224/7
68 (Pt): Hb/A5–543

Ru–Fe–S . . . . 68 (Pt): Hb/A5–544/5

Ru–Ga. . . . . 63 (Ru): Eb–180

Ru–Ge. . . . . 63 (Ru): Eb–193

Ru–Hf . . . . 63 (Ru): Eb–191

Ru–In . . . . 63 (Ru): Eb–180

Ru–La–Th. . . . . 63 (Ru): Eb–192/3

Ru–Mn . . . . 63 (Ru): Eb–219/20

Ru–Mo . . . . 63 (Ru): Eb–206/8

Ru–N . . . . 63 (Ru): Eb–260

Ru–Nb. . . . . 63 (Ru): Eb–195/6

Ru–Ni . . . . 63 (Ru): Eb–220/2
68 (Pt): Hb/A5–541

Ru–O . . . . 63 (Ru): Eb–232/3

Ru–O–H_2O . . . . 63 (Ru): Eb–232/3

Ru–P. . . . . 68 (Pt): Hb/A5–534/5

Ru–Pd . . . . 68 (Pt): Hb/A5–715/6

Ru–Pt . . . . 68 (Pt): Hb/A6–826

Ru–$RuCl_3$ . . . . 63 (Ru): Eb–275

Ru	Ruthenium systems:		
	Ru–S	63 (Ru):	Hb–71 Eb–313
	Ru–Si	63 (Ru):	Eb–331/2
	Ru–Sn	63 (Ru):	Eb–193/4
		68 (Pt):	Hb/A5–538/9
	Ru–Ta	63 (Ru):	Eb–198/9
	Ru–Th	63 (Ru):	Eb–191/2
	Ru–Ti	63 (Ru):	Eb–187/9
		68 (Pt):	Hb/A5–538
	Ru–U	63 (Ru):	Eb–213/6
	Ru–V	63 (Ru):	Eb–194/5
	Ru–W	63 (Ru):	Eb–210/3
	Ru–Zr	63 (Ru):	Eb–189/91
Ru^+	Ru^+:		
	Ionic susceptibility	63 (Ru):	Eb–32
	Ionization energy	63 (Ru):	Eb–12
	Properties of the ion	63 (Ru):	Hb–3/4
	Spectra	63 (Ru):	Hb–13/5 Eb–19/21
	Thermodynamic functions	63 (Ru):	Eb–44
Ru^{2+}	Ru^{2+}:		
	Chemical reactions	63 (Ru):	Eb–74/7
	Electrochemical behavior	63 (Ru):	Eb–65/9
	Ionic refraction	63 (Ru):	Eb–32
	Ionic susceptibility	63 (Ru):	Eb–32
	Ionization energy	63 (Ru):	Eb–12
	Magnetic moment	63 (Ru):	Eb–22
	Properties of the ion	63 (Ru):	Hb–3/4
	Spectra	63 (Ru):	Hb–14 Eb–21/2
	Thermodynamic functions	63 (Ru):	Eb–44
–	$[Ru(H_2O)_6]^{2+}$	63 (Ru):	Eb–247/8
Ru^{3+}	Ru^{3+}:		
	Chemical reactions	63 (Ru):	Eb–77/81
	Electrochemical behavior	63 (Ru):	Eb–65/9
	Ionic refraction	63 (Ru):	Eb–32
	Ionic susceptibility	63 (Ru):	Hb–21 Eb–32
	Ionization energy	63 (Ru):	Eb–12
	Paramagnetic resonance	63 (Ru):	Eb–23/4
	Properties of the ion	63 (Ru):	Hb–3/5 Eb–13/4, 21/3
	Spectra	63 (Ru):	Eb–21, 22/4
–	$[Ru(H_2O)_6]^{3+}$	63 (Ru):	Eb–285
Ru^{4+}	Ru^{4+}:		
	Chemical reactions	63 (Ru):	Eb–81/6
	Electrochemical behavior	63 (Ru):	Eb–65/9

See the "Instructions for the user"
at the beginning of the volume

Ru^{4+} . . . . Ru^{4+}:
Ionic refraction . . . . 63 (Ru): Eb–32
Ionic susceptibility . . . . 63 (Ru): Eb–32
Ionization energy . . . . 63 (Ru): Eb–12
Paramagnetic resonance absorption . 63 (Ru): Eb–25
Properties of the ion . . . . 63 (Ru): Hb–3/4, Eb–13/4, 21, 24/5
Spectra . . . . 63 (Ru): Eb–21, 25

–. . . . . $[Ru(H_2O)_6]^{4+}$ . . . . 63 (Ru): Eb–248

Ru^{5+} . . . . Ru^{5+}:
Chemical reactions . . . . 63 (Ru): Eb–86
Ionic susceptibility . . . . 63 (Ru): Eb–32
Ionization energy . . . . 63 (Ru): Eb–12
Magnetic moment . . . . 63 (Ru): Eb–25
Properties of the ion . . . . 63 (Ru): Hb–3
Spectra . . . . 63 (Ru): Eb–21, 25

Ru^{6+} . . . . Ru^{6+}:
Chemical reactions . . . . 63 (Ru): Eb–86/7
Ionic susceptibility . . . . 63 (Ru): Eb–32
Ionization energy . . . . 63 (Ru): Eb–12
Properties of the ion . . . . 63 (Ru): Hb–3
Spectra . . . . 63 (Ru): Eb–21, 25/6

Ru^{7+} . . . . Ru^{7+}:
Chemical reactions . . . . 63 (Ru): Eb–87/8
Ionic susceptibility . . . . 63 (Ru): Eb–32
Ionization energy . . . . 63 (Ru): Eb–12
Properties of the ion . . . . 63 (Ru): Hb–3
Spectra . . . . 63 (Ru): Eb–26

Ru^{8+} . . . . Ru^{8+}:
Chemical reactions . . . . 63 (Ru): Eb–88/91
Ionic refraction . . . . 63 (Ru): Eb–32
Ionization energy . . . . 63 (Ru): Eb–12
Properties of the ion . . . . 63 (Ru): Hb–3
Spectra . . . . 63 (Ru): Eb–26

Ru^- . . . . Ru^- . . . . 63 (Ru): Eb–22, 32

RuS . . . . RuS . . . . 63 (Ru): Eb–313
RuSSb . . . . RuSbS . . . . 63 (Ru): Eb–337
RuS_2 . . . . RuS_2 . . . . 63 (Ru): Hb–71/3, Eb–313/4
RuS_3 . . . . RuS_3 . . . . 63 (Ru): Hb–73
RuS_6 . . . . RuS_6 . . . . 63 (Ru): Hb–73
RuSb . . . . RuSb . . . . 63 (Ru): Eb–336
RuSbSe. . . . . RuSbSe. . . . . 63 (Ru): Eb–337
RuSbTe. . . . . RuSbTe. . . . . 63 (Ru): Eb–337
$RuSb_2$. . . . . $RuSb_2$. . . . . 63 (Ru): Eb–336/7
RuSc . . . . RuSc . . . . 63 (Ru): Eb–181
RuSeTe . . . . RuSeTe. . . . . 63 (Ru): Eb–318

Hb = Main Series Volume; GD = Gmelin-Durrer; Ab = Appended Volume to Main Series; Eb = Supplement Volume to Main Series; Erg. W. = New Supplement Series Volume

$RuSe_2$	$RuSe_2$	63 (Ru):	Hb–79
			Eb–317
RuSi	RuSi	63 (Ru):	Hb–85
			Eb–333
$RuSi_{1.8}$	$RuSi_{1.8}$	63 (Ru):	Eb–332
$RuSi_2$	$RuSi_2$	63 (Ru):	Eb–332
$RuSn_2$	$RuSn_2$	63 (Ru):	Eb–193/4
		68 (Pt):	Hb/A5–539
$RuSn_2^+$	Sn_2Ru^+	63 (Ru):	Eb–562
RuTa	RuTa	63 (Ru):	Eb–199
RuTe	RuTe	63 (Ru):	Hb–79
$RuTe_2$	$RuTe_2$	63 (Ru):	Hb–79
			Eb–317/8
RuTh	RuTh	63 (Ru):	Eb–191/2
RuTi	RuTi	63 (Ru):	Eb–187/9
		68 (Pt):	Hb/A5–538
RuTm	RuTm	63 (Ru):	Eb–186/7
RuU	RuU	63 (Ru):	Eb–216
RuU_2	RuU_2	63 (Ru):	Eb–216
RuV	RuV	63 (Ru):	Eb–194
RuY	RuY	63 (Ru):	Eb–181
RuZr	RuZr	63 (Ru):	Eb–189/91
Ru_2S_3	Ru_2S_3	63 (Ru):	Eb–313
–	$Ru_2S_3 \cdot 2\,H_2O$	63 (Ru):	Eb–313
Ru_2S_6Tl	$TlRu_2S_6$	63 (Ru):	Eb–337/8
Ru_2Sc	Ru_2Sc	63 (Ru):	Eb–181
Ru_2Se_3	$Ru_2Se_3 \cdot 2\,H_2O$	63 (Ru):	Eb–317
Ru_2Si	Ru_2Si	63 (Ru):	Eb–332
Ru_2Si_3	Ru_2Si_3	63 (Ru):	Eb–334
Ru_2Sm	Ru_2Sm	63 (Ru):	Eb–184
Ru_2Sn_3	Ru_2Sn_3	63 (Ru):	Eb–193/4
Ru_2Tb	Ru_2Tb	63 (Ru):	Eb–186
Ru_2Th	Ru_2Th	63 (Ru):	Eb–191/2
Ru_2Th_3	Ru_2Th_3	63 (Ru):	Eb–191/2
Ru_2Ti	Ru_2Ti	68 (Pt):	Hb/A5–538
Ru_2Tm	Ru_2Tm	63 (Ru):	Eb–186/7
Ru_2V	Ru_2V	63 (Ru):	Eb–194
Ru_2Y	Ru_2Y	63 (Ru):	Eb–181
Ru_2Yb	Ru_2Yb	63 (Ru):	Eb–187
Ru_2Zr	Ru_2Zr	63 (Ru):	Eb–189/91
		68 (Pt):	Hb/A5–538
Ru_3Si_2	Ru_3Si_2	63 (Ru):	Eb–332
Ru_3Sn_7	Ru_3Sn_7	63 (Ru):	Eb–193/4
		68 (Pt):	Hb/A5–538/9
Ru_3Th_7	Ru_3Th_7	63 (Ru):	Eb–191/2
Ru_3Ti	Ru_3Ti	68 (Pt):	Hb/A5–538
Ru_3U	Ru_3U	63 (Ru):	Eb–216
Ru_3U_2	Ru_3U_2	63 (Ru):	Eb–216
Ru_4Si_3	Ru_4Si_3	63 (Ru):	Eb–332/3

See the "Instructions for the user"
at the beginning of the volume

Ru_4U_3	Ru_4U_3	63 (Ru):	Eb–216
Ru_5U_3	Ru_5U_3	63 (Ru):	Eb–216
$S_{0.95}V$	$VS_{0.95}$	48 (V):	Hb/B1–268, 270/1
S	Sulfur:		
	Absorption of light	9 (S):	Hb/A3–643/4
	– of X-rays	9 (S):	Hb/A3–674/6
	Abundance	9 (S):	Hb/A2–81
	Adhesion	9 (S):	Hb/A3–612
	Aerosols	9 (S):	Hb/A2–486
	Atom	9 (S):	Hb/A3–546/60
	Atomic nucleus	9 (S):	Hb/A3–537/46
	Atomic radius	9 (S):	Hb/A3–548/53
	Atomic refraction	9 (S):	Hb/A3–642/3
	Atomic scattering factors	9 (S):	Hb/A3–600/1
	Atomic volume	9 (S):	Hb/A3–553/6
	Atomic weight	9 (S):	Hb/A3–556/9
	Becquerel effect	9 (S):	Hb/A3–687
	Behavior as anode	9 (S):	Hb/A3–689
	– as cathode	9 (S):	Hb/A3–688/9
	– as solvent	9 (S):	Hb/A3–762
	– at the dropping mercury electrode	9 (S):	Hb/A3–688
	Boiling point	9 (S):	Hb/A3–623/5
	Bond energy of the molecule	9 (S):	Hb/A3–569
	Cells	9 (S):	Hb/A3–686
	Chemical constant	9 (S):	Hb/A3–637
	Chemical reactions	9 (S):	Hb/A3–689/727
	– on irradiation	9 (S):	Hb/A3–689
	– with acids	9 (S):	Hb/A3–693/4
	– with air	9 (S):	Hb/A3–689/90
	– with alkaline solutions	9 (S):	Hb/A3–692
	– with metal compounds	9 (S):	Hb/A3–705/26
	– with metals	9 (S):	Hb/A3–696/9
	– with nonmetal compounds	9 (S):	Hb/A3–699/705
	– with nonmetals	9 (S):	Hb/A3–694/5
	– with organic compounds	9 (S):	Hb/A3–723, 726/7
	– with oxygen	9 (S):	Hb/B1–155/66
	– with water	9 (S):	Hb/A3–690/2
	Cleavage	9 (S):	Hb/A3–591
	Coagels	9 (S):	Hb/A2–501/2
	Colloidal forms	9 (S):	Hb/A2–485/503
	Color	9 (S):	Hb/A3–638/40
	Commercial products	9 (S):	Hb/A2–283
	Compressibility	9 (S):	Hb/A3–609
	Cosmochemistry	9 (S):	Hb/A2–61/8
	Critical constants	9 (S):	Hb/A3–622
	Cryoscopic constant	9 (S):	Hb/A3–626
	Crystal chemistry	9 (S):	Hb/A2–68/81

S . . . . Sulfur:

Crystal form	9 (S):	Hb/A3–586/9
Crystallization	9 (S):	Hb/A3–589/91
Crystallographic properties	9 (S):	Hb/A3–571/601
Debye temperature	9 (S):	Hb/A3–635
Deformation	9 (S):	Hb/A3–601
Density	9 (S):	Hb/A3–602/5
Detection in iron	59 (Fe):	Hb/FI–341/2
Determination in iron	59 (Fe):	Hb/FI–343/68
		Hb/FII2–315/6, 349/50
Dielectric constant	9 (S):	Hb/A3–679/80
Dielectric relaxation	9 (S):	Hb/A3–683
Diffraction	9 (S):	Hb/A3–644
Diffusion	9 (S):	Hb/A3–616/7
Dissociation of the molecule	9 (S):	Hb/A3–560/7, 569/71
Dressing of raw materials	9 (S):	Hb/A2–204/11
Ebullioscopic constant	9 (S):	Hb/A3–625
Economic deposits:		
Africa	9 (S):	Hb/A2–183/4
America	9 (S):	Hb/A2–184/90
Asia	9 (S):	Hb/A2–178/82
Australia	9 (S):	Hb/A2–183
Europe	9 (S):	Hb/A2–156/74
Soviet Union	9 (S):	Hb/A2–174/8
Economics	9 (S):	Hb/A2–199/203
Elasticity	9 (S):	Hb/A3–605/7
Electrical conductivity	9 (S):	Hb/A3–680/3
Electrical moment of the molecule	9 (S):	Hb/A3–569
Electrical properties	9 (S):	Hb/A3–679/85
Electrochemical behavior	9 (S):	Hb/A3–685/9
Electrokinetic phenomena	9 (S):	Hb/A3–687/8
Electrolytic deposition	9 (S):	Hb/A3–688
Electron configuration	9 (S):	Hb/A3–546/7
Electronegativity	9 (S):	Hb/A3–559/60
Electron emission	9 (S):	Hb/A3–683/4
Enthalpy	9 (S):	Hb/A3–635/7
Entropy	9 (S):	Hb/A3–635/7
Equation of state	9 (S):	Hb/A3–622
Expansion coefficient	9 (S):	Hb/A3–620/2
Formation	9 (S):	Hb/A3–511, 517/23
		Hb/B1–278/83
Free energy	9 (S):	Hb/A3–635/7
Geochemical character	9 (S):	Hb/A2–81/2
Geochemical cycle	9 (S):	Hb/A2–82/3
Geochemistry	9 (S):	Hb/A2–68/155
Hardness	9 (S):	Hb/A3–605
Heat content	9 (S):	Hb/A3–635/6
Heat of fusion	9 (S):	Hb/A3–628/9
– of sublimation	9 (S):	Hb/A3–627/8

*See the "Instructions for the user"
at the beginning of the volume*

S . . . . Sulfur:

Heat of transformation	9 (S):	Hb/A3–629/31
– of vaporization	9 (S):	Hb/A3–627
History:		
Antiquity	9 (S):	Hb/A1–1/7
Middle ages	9 (S):	Hb/A1–8/21
Modern times	9 (S):	Hb/A1–22/60
Hydrosols	9 (S):	Hb/A2–486/501
Internal pressure	9 (S):	Hb/A3–610
Ionization	9 (S):	Hb/A3–684/5
Isomorphism	9 (S):	Hb/A3–591
Lattice structure	9 (S):	Hb/A3–592/5
Luminescence	9 (S):	Hb/A3–645
Magnetic moment of the atom	9 (S):	Hb/A3–547/8
– of the molecule	9 (S):	Hb/A3–569
Magnetic properties	9 (S):	Hb/A3–676/9
Magnetooptical effects	9 (S):	Hb/A3–644
Mechanical properties	9 (S):	Hb/A3–602/20
Melting point	9 (S):	Hb/A3–625
Mineralogy	9 (S):	Hb/A2–190/9
Modifications	9 (S):	Hb/A3–517/23, 530/7, 571/86
Molar refraction	9 (S):	Hb/A3–642/3
Molecule	9 (S):	Hb/A3–560/71
Nuclear properties	9 (S):	Hb/A3–537/46
Occurrence, extraterrestrial	9 (S):	Hb/A2–61/8
–, terrestrial	9 (S):	Hb/A2–68/155
– in meteorites	9 (S):	Hb/A2–62/8
– in soils	9 (S):	Hb/A2–149/55
– in the atmosphere	9 (S):	Hb/A2–127/34
– in the biosphere	9 (S):	Hb/A2–135/49
– in the hydrosphere	9 (S):	Hb/A2–116/27
– in the lithosphere	9 (S):	Hb/A2–83/115
Optical properties	9 (S):	Hb/A3–638/76
Organosols	9 (S):	Hb/A2–502/3
Phase transitions	9 (S):	Hb/A3–525/37
Phosphorescence	9 (S):	Hb/A3–645
Photoelectric effect	9 (S):	Hb/A3–683/4
Photophoresis	9 (S):	Hb/A3–645
Piezoelectricity	9 (S):	Hb/A3–683
Plasticity	9 (S):	Hb/A3–605/7
Polarizability	9 (S):	Hb/A3–547
Polarization	9 (S):	Hb/A3–644
Polymorphism	9 (S):	Hb/A3–530/7, 571/86
Position in the electrochemical series	9 (S):	Hb/A3–686
– in the periodic system	9 (S):	Hb/A3–559
Potentials	9 (S):	Hb/A3–685/6
Preparation	9 (S):	Hb/A3–511/2

Hb = Main Series Volume; GD = Gmelin-Durrer; Ab = Appended Volume to Main Series;
Eb = Supplement Volume to Main Series; Erg. W. = New Supplement Series Volume

S . . . . Sulfur:

Entry	Volume	Reference
Preparation, industrial	9 (S):	Hb/A2–199/284
–, –, from coal	9 (S):	Hb/A2–258/78
–, –, from gases	9 (S):	Hb/A2–258
–, –, from hydrogen sulfide	9 (S):	Hb/A2–241/58
–, –, from ores	9 (S):	Hb/A2–211/23
–, –, from sulfur dioxide	9 (S):	Hb/A2–224/41
–, –, from other raw materials	9 (S):	Hb/A2–278/80
– of colloidal forms	9 (S):	Hb/A2–487/91
– of modifications	9 (S):	Hb/A3–517/23
– of special forms	9 (S):	Hb/A3–516/7
Purification	9 (S):	Hb/A2–280/3 Hb/A3–512/5
Purity testing	9 (S):	Hb/A2–283/4 Hb/A3–515/6
Reflection	9 (S):	Hb/A3–640
Refraction	9 (S):	Hb/A3–640/3
Scattering of X-rays	9 (S):	Hb/A3–674/6
Solidification point	9 (S):	Hb/A3–625
Solubility in inorganic solvents	9 (S):	Hb/A3–731/3 Hb/B3–1748/52
– in organic solvents	9 (S):	Hb/A3–733/62
Sorption	9 (S):	Hb/A3–617/20
Special forms	9 (S):	Hb/A3–516/7
Specific heat	9 (S):	Hb/A3–631/5
Spectra:		
Arc spectrum	9 (S):	Hb/A3–645/9
Band spectrum	9 (S):	Hb/A3–654/63
Continuous spectrum	9 (S):	Hb/A3–663/4
Raman spectrum	9 (S):	Hb/A3–664/5
Spark spectra	9 (S):	Hb/A3–649/54
X-ray spectrum	9 (S):	Hb/A3–665/76
Strength	9 (S):	Hb/A3–607/8
Structure	9 (S):	Hb/A3–595/600
Surface tension	9 (S):	Hb/A3–610/2
Susceptibility	9 (S):	Hb/A3–676/9
Technology	9 (S):	Hb/A2–199/284
Texture	9 (S):	Hb/A3–601
Thermal conductivity	9 (S):	Hb/A3–637/8
Thermal expansion	9 (S):	Hb/A3–620/2
Thermal properties	9 (S):	Hb/A3–525/37, 620/38
Toxicity	9 (S):	Hb/A2–503
Transition point	9 (S):	Hb/A3–626/7
Twinning	9 (S):	Hb/A3–591
Valence	9 (S):	Hb/A3–559
Vapor pressure	9 (S):	Hb/A3–525/9, 622/3
Vibrational frequency	9 (S):	Hb/A3–635
Viscosity	9 (S):	Hb/A3–612/6

See the "Instructions for the user"
at the beginning of the volume

S	Sulfur:		
	Wettability	9 (S):	Hb/A3–611/2
	X-ray photoelectric effect	9 (S):	Hb/A3–684
–	Sulfur ions		
	see S^+ ... S^+, etc.		
–	Sulfur isotopes:		
	Abundance	9 (S):	Hb/A3–538
	Enrichment and separation	9 (S):	Hb/A3–523/5
	Formation	9 (S):	Hb/A3–541/2
	Nuclear properties	9 (S):	Hb/A3–537/46
–	Sulfur solutions:		
	S–$AlBr_3$	9 (S):	Hb/A3–728
	S–$AsCl_3$	9 (S):	Hb/A3–729, 732
	S–Br	9 (S):	Hb/A3–727, 731
	S–CS_2	9 (S):	Hb/A3–729/31, 743/8
	S–Cl	9 (S):	Hb/A3–729, 731
	S–$H_2S_2O_7$	9 (S):	Hb/A3–728
	S–$HgBr_2$	9 (S):	Hb/A3–728
	S–$HgCl_2$	9 (S):	Hb/A3–728, 732
	S–I	9 (S):	Hb/A3–727, 729, 731
	S–NH_3	9 (S):	Hb/A3–732
	S–organic solvents	9 (S):	Hb/A3–728/30, 733/62
	S–P	9 (S):	Hb/A3–727, 731
	S–SO_3	9 (S):	Hb/A3–731
	S–S_2Cl_2	9 (S):	Hb/A3–727, 729, 731/2
			Hb/B3–1751/2
	S–$SbCl_3$	9 (S):	Hb/A3–714, 727, 732
	S–$SeOCl_2$	9 (S):	Hb/A3–732
	S–$SnCl_4$	9 (S):	Hb/A3–729, 732
–	Sulfur systems:		
	S–Ag	61 (Ag):	Hb/B3–2/8
	S–Ag–As	61 (Ag):	Hb/B4–53/6
	S–Ag–As–Cu	61 (Ag):	Hb/B4–487
	S–Ag–Au	62 (Au):	Hb/3–766
	S–Ag–Bi	61 (Ag):	Hb/B4–127/8
	S–Ag–Cu	61 (Ag):	Hb/B4–458/63
	S–Ag–Cu–Fe	61 (Ag):	Hb/B4–472
	S–Ag–Cu–Ni	61 (Ag):	Hb/B4–471/2
	S–Ag–Cu–Pb	61 (Ag):	Hb/B4–470
	S–Ag–Cu–Sb	61 (Ag):	Hb/B4–491
	S–Ag–Fe	61 (Ag):	Hb/B4–411/3
	S–Ag–O	9 (S):	Hb/B1–185
		61 (Ag):	Hb/B3–61/2
	S–Ag–Sb	61 (Ag):	Hb/B4–90/2
	S–Ag–Tl	61 (Ag):	Hb/B4–225
	S–Al–C–Cu–Fe	60 (Cu):	Hb/B3–1327
	S–As	17 (As):	Hb–415

S . . . . Sulfur systems:

System	Volume	Reference
S–As–Bi	19 (Bi):	Eb–858
S–As–Co–Cu	60 (Cu):	Hb/B3–1269
S–As–Cu	60 (Cu):	Hb/B2–959/60
S–As–Cu–Fe	60 (Cu):	Hb/B3–1315
S–As–Cu–Ni	60 (Cu):	Hb/B3–1259
S–As–Cu–Sb	60 (Cu):	Hb/B2–973
S–As–Ni	57 (Ni):	Hb/B3–989/91
S–As–Pb	47 (Pb):	Hb/C3–901/3
S–Au–Cu	62 (Au):	Hb/3–764
S–Au–Fe	62 (Au):	Hb/3–764
S–Ba–O	9 (S):	Hb/B1–183
S–BaO–BaS–$BaSO_4$–SO_2	30 (Ba):	Eb–448
S–BaS	30 (Ba):	Hb–257
S–BaS–$BaSO_4$–SO_2	30 (Ba):	Eb–448
S–Bi	19 (Bi):	Hb–161
		Eb–725/9
S–Bi–Co	58 (Co):	Eb/A–756
S–Bi–Co–H	58 (Co):	Eb/A–756
S–Bi–Cu	60 (Cu):	Hb/B2–974/5
S–Bi–H	19 (Bi):	Hb–161
S–Bi–Ni	57 (Ni):	Hb/B1–14
		Hb/B3–996
S–Bi–O	9 (S):	Hb/B1–182
	19 (Bi):	Eb–745
S–Bi–Pb	47 (Pb):	Hb/C3–969/70
S–Br	9 (S):	Hb/B3–1864
S–Br–I	9 (S):	Hb/B3–1873
S–Ca–O	9 (S):	Hb/B1–182/3
	28 (Ca):	Hb/B3–657/60
S–CaO	28 (Ca):	Hb/B3–657
S–$CaSO_4$	28 (Ca):	Hb/B3–658
S–Cd	33 (Cd):	Eb–584
S–Cl	9 (S):	Hb/B3–1748/59
S–Cl–O–Ta	50 (Ta):	Hb/B1–178
S–Cl–Pb–Sn	47 (Pb):	Hb/C4–1586
S–Cl–Ta	50 (Ta):	Hb/B1–178
S–Co	58 (Co):	Hb/A2–318
		Eb/A–605/7
S–Co–Cu	60 (Cu):	Hb/B3–1266
S–Co–H–Sb	58 (Co):	Eb/A–755/6
S–Co–Ni	58 (Co):	Hb/A1–210
		Eb/A–881
S–Co–O	9 (S):	Hb/B1–183
	58 (Co):	Eb/A–626
S–Co–Sb	58 (Co):	Eb/A–754/5
S–CoS	58 (Co):	Eb/A–607
S–Cr	52 (Cr):	Hb/B–295/7
S–Cs	25 (Cs):	Hb/2–217

See the "Instructions for the user"
at the beginning of the volume

S . . . . Sulfur systems:

S–Cu	60 (Cu):	Hb/B1–432/4
S–Cu–Fe	60 (Cu):	Hb/B3–1282/4
S–Cu–Fe–Mn	60 (Cu):	Hb/B3–1330
S–Cu–Fe–O	60 (Cu):	Hb/B3–1295
S–Cu–Fe–Sb	60 (Cu):	Hb/B3–1315
S–Cu–Fe–Sn	60 (Cu):	Hb/B3–1327
S–Cu–Fe–Zn	60 (Cu):	Hb/B3–1325/6
S–Cu–Mn	60 (Cu):	Hb/B3–1246
S–Cu–Mn–O	60 (Cu):	Hb/B3–1246
S–Cu–Ni	60 (Cu):	Hb/B3–1255/6
S–Cu–O	9 (S):	Hb/B1–184/5
S–Cu–Pb	60 (Cu):	Hb/B3–1200/2
S–Cu–Sb	60 (Cu):	Hb/B2–971
S–Cu–Sn	60 (Cu):	Hb/B3–1195/6
S–Cu–Zn	60 (Cu):	Hb/B3–1148
S–Fe	59 (Fe):	Hb/A6–1168/70
S–Fe–FeO–Ni–NiO–SiO_2	59 (Fe):	Hb/B5–1160
S–Fe–H	59 (Fe):	Hb/A5–939/40
S–Fe–Ir	68 (Pt):	Hb/A6–738
S–Fe–Mn	59 (Fe):	Hb/A5–936/9
S–Fe–O	9 (S):	Hb/B1–183/4
S–Fe–Os	68 (Pt):	Hb/A6–726
S–Fe–Pd	68 (Pt):	Hb/A5–624
S–Fe–Pt	68 (Pt):	Hb/A6–798
S–Fe–Rh	68 (Pt):	Hb/A5–563
S–Fe–Ru	68 (Pt):	Hb/A5–544/5
S–FeS	59 (Fe):	Hb/A6–1170
S–H_2S	9 (S):	Hb/B1–107/8
S–I	9 (S):	Hb/B3–1870/1
S–I–Se	10 (Se):	Hb/B–193
S–In	37 (In):	Hb–96
S–Ir	67 (Ir):	Hb–74
S–Ir_2S_3	67 (Ir):	Hb–75
S–K	22 (K):	Hb/3–691/2
S–Li_2S	20 (Li):	Eb–473/4
S–Na	21 (Na):	Hb–484/5
S–Na–Ni	57 (Ni):	Hb/B3–1016
S–Na_2S	21 (Na):	Eb/3–1059/60
S–Nb	49 (Nb):	Hb/B1–257/8
S–Ni	57 (Ni):	Hb/B2–627/31
S–Ni–O	9 (S):	Hb/B1–183
	57 (Ni):	Hb/B2–676/7
S–Ni–Pb	57 (Ni):	Hb/B3–1195
S–Ni–Sb	57 (Ni):	Hb/B1–10
		Hb/B3–994/5
S–O–Pb	9 (S):	Hb/B1–183
	47 (Pb):	Hb/C2–533/4

Hb = Main Series Volume; GD = Gmelin-Durrer; Ab = Appended Volume to Main Series;
Eb = Supplement Volume to Main Series; Erg. W. = New Supplement Series Volume

S . . . . Sulfur systems:

S–O–Pd	9 (S):	Hb/B1–185
S–O–Sb	9 (S):	Hb/B1–182
S–O–Sr	9 (S):	Hb/B1–183
S–O–Ti	41 (Ti):	Hb–343
S–Os	66 (Os):	Hb–58
S–P	16 (P):	Hb/C–567
S–Pb	47 (Pb):	Hb/C2–413/4
S–Pb–Sb	47 (Pb):	Hb/C3–935/6
S–Pb–Se	47 (Pb):	Hb/C2–639/40
S–Pb–Si	47 (Pb):	Hb/C3–855
S–Pb–Sn	47 (Pb):	Hb/C4–1584
S–Pd	65 (Pd):	Hb/2–289
S–Pt	68 (Pt):	Hb/C1–103/4
S–Pu	Erg. W. 4–159/61	
S–Rb	24 (Rb):	Hb–205
S–Re	70 (Re):	Hb–117
S–Rh	64 (Rh):	Hb–62/4
S–Ru	63 (Ru):	Hb–71
		Eb–313
S–SO_2–SrO–SrS–$SrSO_4$	29 (Sr):	Eb–257
S–SO_2–SrS–$SrSO_4$	29 (Sr):	Eb–257
S–S_2Br_2	9 (S):	Hb/B3–1864
S–S_2Cl_2	9 (S):	Hb/B3–1748/52
S–Sb	18 (Sb):	Hb/B3–499/500
S–Se	10 (Se):	Hb/B–160/70
S–Se–Te	11 (Te):	Hb–363
S–Se_2Cl_2	10 (Se):	Hb/B–193
S–Ta	50 (Ta):	Hb/B1–169/70
S–Te	11 (Te):	Hb–350/1
S–Th	44 (Th):	Hb–277/8
S–Ti	41 (Ti):	Hb–337/9
S–Tl	38 (Tl):	Hb/3–339
S–V	48 (V):	Hb/B1–267/8
S–Zr	42 (Zr):	Hb–326

S^+ . . . . S^+:

Ionization potential	9 (S):	Hb/A3–684/5
Properties of the ion	9 (S):	Hb/A3–547/50, 554
Spectra	9 (S):	Hb/A3–649/50

S^{2+} . . . . S^{2+}:

Ionization potential	9 (S):	Hb/A3–684/5
Properties of the ion	9 (S):	Hb/A3–547/50, 554
Spectra	9 (S):	Hb/A3–650/1

S^{3+} . . . . S^{3+}:

Ionization potential	9 (S):	Hb/A3–684/5
Properties of the ion	9 (S):	Hb/A3–547/50, 554
Spectra	9 (S):	Hb/A3–651/2

See the "Instructions for the user"
at the beginning of the volume

Formula	Entry	Volume	Reference
S^{4+}	S^{4+}:		
	Electrochemical behavior	9 (S):	Hb/A3–688
	Formation	9 (S):	Hb/B1–310
	Ionic refraction	9 (S):	Hb/A3–642/3
	Ionization potentials	9 (S):	Hb/A3–684/5
	Properties of the ion	9 (S):	Hb/A3–547/50, 554/6
	Scattering factors	9 (S):	Hb/A3–601
	Spectra	9 (S):	Hb/A3–652
S^{5+}	S^{5+}:		
	Ionization potential	9 (S):	Hb/A3–684/5
	Properties of the ion	9 (S):	Hb/A3–547/50, 554
	Spectra	9 (S):	Hb/A3–652/3
S^{6+}	S^{6+}:		
	Ionic refraction	9 (S):	Hb/A3–642/3
	Ionization potential	9 (S):	Hb/A3–684/5
	Properties of the ion	9 (S):	Hb/A3–547/56
	Scattering factors	9 (S):	Hb/A3–600
	Spectra	9 (S):	Hb/A3–653
S^{n+}	S^{n+} (n = 7 to 11)	9 (S):	Hb/A3–547, 653/4, 684/5
S^-	S^-	9 (S):	Hb/A3–547/50, 554, 684/5
S^{2-}	S^{2-}:		
	Chemical reactions	9 (S):	Hb/B1–86/106
	Electrochemical behavior	9 (S):	Hb/A3–688 Hb/B1–83/6
	Formation	9 (S):	Hb/B1–81
	Hydration	9 (S):	Hb/B1–89
	Ionic mobility	9 (S):	Hb/B1–85
	Ionic refraction	9 (S):	Hb/A3–642/3 Hb/B1–37/8
	Ionization potential	9 (S):	Hb/A3–684/5
	Properties of the ion	9 (S):	Hb/A3–547/56
	Scattering factors	9 (S):	Hb/A3–600
SSb	SbS	18 (Sb):	Hb/B3–500
SSe	SeS	10 (Se):	Hb/B–172/3
SSe^{2-}	$[SeS]^{2-}$	10 (Se):	Hb/B–175/6
SSeTa	TaSeS	50 (Ta):	Hb/B1–189
SSi	SiS	15 (Si):	Hb/B–743/6
SSiZn	ZnSiS	32 (Zn):	Hb–274
SSn	SnS solid solutions:		
	SnS–PbS	47 (Pb):	Hb/C4–1585
–	SnS systems:		
	SnS–FeS	59 (Fe):	Hb/B5–1130
	SnS–Pb–PbS–Sn	47 (Pb):	Hb/C4–1584
	$SnS–PbCl_2–PbS–SnCl_2$	47 (Pb):	Hb/C4–1586
	SnS–PbS	47 (Pb):	Hb/C4–1585
	SnS–PbSe	47 (Pb):	Hb/C4–1587/8

SSr . . . . . SrS . . . . . 29 (Sr): Hb–17/9, 21/2, 165/9
Eb–146, 244/8
–. . . . . SrS solutions:
$SrS-H_2O$ . . . . . 29 (Sr): Hb–169
–. . . . . SrS systems:
SrS–CaS . . . . . 29 (Sr): Eb–304
SrS–MgS . . . . . 29 (Sr): Eb–303
$SrS-S-SO_2-SrO-SrSO_4$ . . . . . 29 (Sr): Eb–257
$SrS-S-SO_2-SrSO_4$ . . . . . 29 (Sr): Eb–257
$SrS-SrCl_2-SrO$ . . . . . 29 (Sr): Eb–249
SSr_2 . . . . . Sr_2S. . . . . 29 (Sr): Hb–165
STaTe. . . . . TaTeS . . . . . 50 (Ta): Hb/B1–192
STa_2 . . . . . Ta_2S . . . . . 50 (Ta): Hb/B1–170, 171/2
STa_6. . . . . Ta_6S . . . . . 50 (Ta): Hb/B1–170, 171/2
STe . . . . . TeS . . . . . 11 (Te): Hb–350, 353
STh . . . . . ThS . . . . . 44 (Th): Hb–277, 279
STi . . . . . TiS . . . . . 41 (Ti): Hb–337/9, 340
–. . . . . TiS systems:
$TiS-CoS-TiS_2$ . . . . . 58 (Co): Eb/A–870
$TiS-Fe-FeS-Fe_2Ti$. . . . . 59 (Fe): Hb/A9–1874
$TiS-NiS-TiS_2$. . . . . 57 (Ni): Hb/B3–1181/2
STi_2 . . . . . Ti_2S. . . . . 41 (Ti): Hb–337/40
STl_2 . . . . . Tl_2S . . . . . 38 (Tl): Hb/1–23
Hb/3–339/45, 346/9
–. . . . . Tl_2S systems:
$Tl_2S-As_2S_3$ . . . . . 38 (Tl): Hb/3–435
Tl_2S-PbS . . . . . 47 (Pb): Hb/C4–1352/3
SU . . . . . US . . . . . 55 (U): Hb–142
–. . . . . US solid solutions:
(U,Pu)S . . . . . Erg. W. 4–168/71
SV. . . . . VS . . . . . 48 (V): Hb/B1–271/3
–. . . . . VS systems:
VS–Fe–FeS–V . . . . . 59 (Fe): Hb/A9–1924/5
SV_3 . . . . . V_3S . . . . . 48 (V): Hb/B1–268, 269/70
SZn . . . . . ZnS:
Chemical reactions . . . . . 32 (Zn): Hb–201/6
Eb–924/34
Colloidal forms . . . . . 32 (Zn): Hb–206/7
Eb–905
Electrochemical behavior . . . . . 32 (Zn): Eb–924
Formation . . . . . 32 (Zn): Hb–193/7, 198
Eb–903/4, 914
Lithopone . . . . . 32 (Zn): Hb–202/3, 217/8
Eb–908/14,
926/7, 929

See the "Instructions for the user"
at the beginning of the volume

SZn	ZnS:		
	Phosphors	32 (Zn):	Hb-202/3, 207/17 Eb-904/5, 923, 927, 929
	Physical properties	32 (Zn):	Hb-198/201, 209/17 Eb-914/24
	Pigments	32 (Zn):	Hb-217 Eb-906/8
	Preparation	32 (Zn):	Hb-193/7 Eb-903/5
	–, industrial	32 (Zn):	Hb-197 Eb-905/14
	– of special forms	32 (Zn):	Hb-206/9, 217 Eb-904/5, 906/14
	Sidot's blende	32 (Zn):	Hb-207/17
	Solubility	32 (Zn):	Hb-201
	Special forms	32 (Zn):	Hb-202/3, 206/18 Eb-904/14, 923, 926/7, 929
–	$7 ZnS \cdot H_2O$	32 (Zn):	Hb-219
–	ZnS solid solutions:		
	ZnS-HgS	34 (Hg):	Hb/B4-1501/2
	$Zn_{1-x}Hg_xS_ySe_{1-y}$	34 (Hg):	Hb/B4-1502
	(Ni,Zn)S	57 (Ni):	Hb/B3-1136
–	ZnS systems:		
	$ZnS-AgAlS_2$	61 (Ag):	Hb/B4-185/6
	$ZnS-AgGaS_2$	61 (Ag):	Hb/B4-194
	$ZnS-AgInS_2$	61 (Ag):	Hb/B4-207
	$ZnS-Ag_2S$	61 (Ag):	Hb/B4-141
	ZnS-CdS	33 (Cd):	Eb-711/2
	$ZnS-CuGaS_2$	60 (Cu):	Hb/B3-1181
	$ZnS-Cu_2S-FeS-PbS$	60 (Cu):	Hb/B3-1329
	$ZnS-Cu_2S-PbS$	60 (Cu):	Hb/B3-1208
	ZnS-FeS	59 (Fe):	Hb/B5-1101/2
	ZnS-PbS	47 (Pb):	Hb/C4-1227
SZn_2	Zn_2S	32 (Zn):	Hb-193
SZr_4	Zr_4S	42 (Zr):	Hb-329
$S_{1.02}V$	$VS_{1.02}$	48 (V):	Hb/B1-268, 270/3
$S_{1.075}V$	$VS_{1.075}$	48 (V):	Hb/B1-273
$S_{1.1}V$	$VS_{1.1}$	48 (V):	Hb/B1-270/3
$S_{1.4}V$	$VS_{1.4}$	48 (V):	Hb/B1-273/4
$S_{1.67}V$	$VS_{1.67}$	48 (V):	Hb/B1-268, 273, 276
$S_{1.8}V$	$VS_{1.8}$	48 (V):	Hb/B1-268, 273, 276/7
S_2	S_2	9 (S):	Hb/A3-560/71, 654/63
	see also S ... Sulfur		
–	$^{32}S^{34}S$	9 (S):	Hb/A3-567/8

Hb = Main Series Volume; GD = Gmelin-Durrer; Ab = Appended Volume to Main Series; Eb = Supplement Volume to Main Series; Erg. W. = New Supplement Series Volume

S_2^{2-}	S_2^{2-}	9 (S):	Hb/B1–83
S_2Sb_2	Sb_2S_2	18 (Sb):	Hb/B3–500
S_2Sb_2Se	Sb_2S_2Se	18 (Sb):	Hb/B3–553
S_2Sb_3	Sb_3S_2	18 (Sb):	Hb/B3–517
S_2Se	SeS_2	10 (Se):	Hb/B–173/4
S_2Si	SiS_2	15 (Si):	Hb/B–746/52
–	SiS_2 systems:		
	SiS_2–Ag_2S	61 (Ag):	Hb/B3–387/8
S_2Sn	SnS_2	46 (Sn):	Hb/A–170/1
–	SnS_2 systems:		
	SnS_2–Ag_2S	61 (Ag):	Hb/B4–247
S_2Ta	TaS_2	50 (Ta):	Hb/A2–517
			Hb/B1–169/70, 172/5
S_2Ta_{x+1}	$Ta_{1+x}S_2$	50 (Ta):	Hb/B1–170/1
S_2Te	TeS_2	11 (Te):	Hb–350, 353/4
S_2Th	ThS_2	44 (Th):	Hb–277, 280/2
S_2Ti	TiS_2	41 (Ti):	Hb–337/9, 341/3
–	TiS_2 systems:		
	TiS_2–CoS–TiS	58 (Co):	Eb/A–870
	TiS_2–NiS–TiS	57 (Ni):	Hb/B3–1181/2
	TiS_2–TiO_2	41 (Ti):	Hb–343
S_2Tl_2	$Tl_2S_2 = Tl_2S \cdot Tl_2S_3$	38 (Tl):	Hb/3–345/6
S_2U	US_2	55 (U):	Hb–143
–	US_2 solid solutions:		
	$(U,Pu)S_2$	Erg. W. 4–170	
S_2V	VS_2	48 (V):	Hb/B1–268, 273, 277
S_2W	WS_2	54 (W):	Hb–184/5
S_2Zr	ZrS_2	42 (Zr):	Hb–326/7, 330/2
S_3^{2-}	S_3^{2-}	9 (S):	Hb/B1–83
S_3Sb_2	Sb_2S_3:		
	Chemical reactions	18 (Sb):	Hb/B3–517/24, 532/3
	Colloidal forms	18 (Sb):	Hb/B3–525/33
	Electrochemical behavior	18 (Sb):	Hb/B3–516/7, 532
	Formation	18 (Sb):	Hb/B3–503/8, 525/6
	History	18 (Sb):	Hb/A1–7/14, 32/3, 40/2
	Physical properties	18 (Sb):	Hb/B3–508/16, 531
	Preparation	18 (Sb):	Hb/B3–503/8, 525/7
	–, industrial	18 (Sb):	Hb/A3–311/2
			Hb/B3–500/2
	Purity testing	18 (Sb):	Hb/B3–503
	Sols	18 (Sb):	Hb/B3–525/33
	Toxicity	18 (Sb):	Hb/B2–276
	Uses	18 (Sb):	Hb/A3–305
			Hb/B3–502/3
–	Sb_2S_3 systems:		
	Sb_2S_3–Ag_2S	61 (Ag):	Hb/B4–90/2
	Sb_2S_3–Ag_2S–Cu_2S	61 (Ag):	Hb/B4–491

Formula	Compound	Gmelin No.	Reference
S_3Sb_2	Sb_2S_3 systems:		
	$Sb_2S_3-As_2S_3$	18 (Sb):	Hb/B3–564
	Sb_2S_3-BiS	19 (Bi):	Hb–202
	$Sb_2S_3-Bi_2S_3$	19 (Bi):	Eb–859/61
	Sb_2S_3-CaS	28 (Ca):	Hb/B3–1240/1
	Sb_2S_3-HgS	34 (Hg):	Hb/B4–1383
	$Sb_2S_3-Na_2S-H_2O$	21 (Na):	Eb/4–1720/1
	$Sb_2S_3-Na_2SO_4$	21 (Na):	Eb/5–2025
	$Sb_2S_3-PbO-Sb_2O_3$	47 (Pb):	Hb/C3–939
	Sb_2S_3-PbS	47 (Pb):	Hb/C3–935/6
	$Sb_2S_3-Sb_2O_3$	18 (Sb):	Hb/B3–539/40
$S_3Sb_2Se_2$	$Sb_2S_3Se_2$	18 (Sb):	Hb/B3–553
S_3Sc_2	Sc_2S_3 solid solutions:		
	$Sc_2S_3-Ag_2S$	61 (Ag):	Hb/B4–232
S_3Se	SeS_3	10 (Se):	Hb/B–174
S_3Se^{2-}	$[SeS_3]^{2-}$	10 (Se):	Hb/B–159
S_3Sm_2	Sm_2S_3 solid solutions:		
	$Sm_2S_3-Ag_2S$	61 (Ag):	Hb/B4–232
	Sm_2S_3-PbS	47 (Pb):	Hb/C4–1374
S_3Sn^{2-}	$[SnS_3]^{2-}$	46 (Sn):	Hb/B–393
S_3Sn_2	Sn_2S_3 systems:		
	Sn_2S_3-PbS	47 (Pb):	Hb/C4–1585
S_3SrTa	$SrTaS_3$	50 (Ta):	Hb/B2–317
S_3Ta	TaS_3	50 (Ta):	Hb/B1–175/6
–	TaS_3 systems:		
	$TaS_3-Ta-TaCl_5$	50 (Ta):	Hb/B1–178
S_3Tb_2	Tb_2S_3 solid solutions:		
	$Tb_2S_3-Ag_2S$	61 (Ag):	Hb/B4–232
	Tb_2S_3-PbS	47 (Pb):	Hb/C4–1374
S_3Te	TeS_3	11 (Te):	Hb–350, 354/5
S_3Th_2	Th_2S_3	44 (Th):	Hb–277, 279
S_3Ti	TiS_3	41 (Ti):	Hb–337/9, 343
S_3Ti_2	Ti_2S_3	41 (Ti):	Hb–337/9, 340/1
S_3Tl_2	Tl_2S_3	38 (Tl):	Hb/3–345
S_3Tm_2	Tm_2S_3 solid solutions:		
	$Tm_2S_3-Ag_2S$	61 (Ag):	Hb/B4–232
	Tm_2S_3-PbS	47 (Pb):	Hb/C4–1374
S_3U_2	U_2S_3	55 (U):	Hb–143
–	U_2S_3 solid solutions:		
	$(U,Pu)_2S_3$	Erg. W. 4–169/70	
S_3V^-	$[VS_3]^-$	48 (V):	Hb/B1–280
–	$[VS_3]_3^{3-}$	48 (V):	Hb/B1–280
–	$[VS_3(H_2O)]^-$	48 (V):	Hb/B1–281
S_3V_2	V_2S_3	48 (V):	Hb/A1–278, 294/7 Hb/A2–380 Hb/B1–267/8, 274/6

Hb = Main Series Volume; GD = Gmelin-Durrer; Ab = Appended Volume to Main Series;
Eb = Supplement Volume to Main Series; Erg. W. = New Supplement Series Volume

S_3W	WS_3	54 (W):	Hb–185/6
S_3Y_2	Y_2S_3 solid solutions:		
	Y_2S_3–Ag_2S	61 (Ag):	Hb/B4–232
	Y_2S_3–PbS	47 (Pb):	Hb/C4–1374
S_3Yb_2	Yb_2S_3 solid solutions:		
	Yb_2S_3–Ag_2S	61 (Ag):	Hb/B4–232
	Yb_2S_3–PbS	47 (Pb):	Hb/C4–1374
S_3Zr	ZrS_3	42 (Zr):	Hb–326/7, 332
S_3Zr_2	Zr_2S_3	42 (Zr):	Hb–329/30
S_3Zr_4	Zr_4S_3	42 (Zr):	Hb–329
S_4	S_4	9 (S):	Hb/A3–560/71
	see also S ... Sulfur		
S_4^{2-}	S_4^{2-}	9 (S):	Hb/B1–83
S_4Sb^{3-}	$[SbS_4]^{3-}$	18 (Sb):	Hb/B3–547
S_4Sb_2	Sb_2S_4	18 (Sb):	Hb/B3–500/3, 533
S_4Sb_2Sr	$Sr[Sb_2S_4]$	29 (Sr):	Eb–295
S_4Sn^{4-}	$[SnS_4]^{4-}$	46 (Sn):	Hb/B–342, 393
S_4Sr	SrS_4	29 (Sr):	Hb–169/70
–	$SrS_4 \cdot 2\ H_2O$	29 (Sr):	Hb–170
–	$SrS_4 \cdot 6\ H_2O$	29 (Sr):	Hb–170
S_4Ta^{3-}	$[TaS_4]^{3-}$	50 (Ta):	Hb/B2–318
S_4TaTi_2	Ti_2TaS_4	50 (Ta):	Hb/B2–318
S_4TaTl_3	Tl_3TaS_4	50 (Ta):	Hb/B2–318
S_4Ti_3	Ti_3S_4	41 (Ti):	Hb–337, 341
S_4Tl_3V	Tl_3VS_4	48 (V):	Hb/B2–549
S_4Tl_4	$Tl_2S \cdot Tl_2S_3 = Tl_2S_2$	38 (Tl):	Hb/3–345/6
S_4V	VS_4	48 (V):	Hb/B1–267/8, 278/80
S_4V^{3-}	$[VS_4]^{3-}$	48 (V):	Hb/B1–280
S_4V_3	V_3S_4	48 (V):	Hb/B1–273/4
S_4V_5	V_5S_4	48 (V):	Hb/B1–268, 270
S_5Sb_2	Sb_2S_5:		
	Chemical reactions	18 (Sb):	Hb/B3–536/9
	Colloidal forms	18 (Sb):	Hb/B3–539
	Formation	18 (Sb):	Hb/B3–534/6
	Physical properties	18 (Sb):	Hb/B3–536
	Preparation	18 (Sb):	Hb/B3–534/6
	–, industrial	18 (Sb):	Hb/B3–500/2
	Purity testing	18 (Sb):	Hb/B3–503
	Toxicity	18 (Sb):	Hb/B2–276
	Uses	18 (Sb):	Hb/A3–305
			Hb/B3–502/3
$S_5Sb_2Sr_2$	$Sr_2[Sb_2S_5]$	29 (Sr):	Eb–295
–	$Sr_2[Sb_2S_5] \cdot 15\ H_2O$	29 (Sr):	Hb–221
S_5Sr	SrS_5	29 (Sr):	Hb–169
			Eb–248
S_5Sr_3Te	Sr_3TeS_5	29 (Sr):	Hb–187

See the "Instructions for the user"
at the beginning of the volume

Formula	Compound / Subject	System No.	Reference
S_5Ti_3	Ti_3S_5	41 (Ti):	Hb–337
S_5Ti_4	Ti_4S_5	41 (Ti):	Hb–337
S_5Tl_2	Tl_2S_5	38 (Tl):	Hb/3–346
S_5Tl_4	Tl_4S_5	38 (Tl):	Hb/3–346
S_5V	VS_5	48 (V):	Hb/B1–268, 280
S_5V^-	$[VS_5]_3^{3-}$	48 (V):	Hb/B1–280
S_5V_2	V_2S_5	48 (V):	Hb/A1–278, 282 Hb/A2–380 Hb/B1–268, 277/8
S_5V_4	V_4S_5	48 (V):	Hb/B1–273/4
S_5Zn	ZnS_5	32 (Zn):	Hb–219
S_5Zr_3	Zr_3S_5	42 (Zr):	Hb–330
S_6	S_6 *see also S ... Sulfur*	9 (S):	Hb/A3–560/71
$S_6Sb_2Sr_3$	$Sr_3Sb_2S_6 \cdot 10\ H_2O$	29 (Sr):	Hb–221
$S_6Sb_2Zn_3$	$Zn_3(SbS_3)_2$	32 (Zn):	Hb–291
$S_7Sb_2Sr_2$	$Sr_2Sb_2S_7 \cdot 11\ H_2O$	29 (Sr):	Hb–221
S_7Th_3	Th_3S_7	44 (Th):	Hb–277, 282
S_7Tl_6	$Tl_6S_7 = Tl_2S \cdot 2\ Tl_2S_3$	38 (Tl):	Hb/3–346
S_7Tl_8	$Tl_8S_7 = 5\ Tl_2S \cdot 3\ Tl_2S_3$	38 (Tl):	Hb/3–345/6
S_8	S_8 *see also S ... Sulfur*	9 (S):	Hb/A3–560/71
$S_8Sb_2Zn_3$	$Zn_3(SbS_4)_2$	32 (Zn):	Hb–291
S_8V_5	V_5S_8	48 (V):	Hb/B1–268, 276
$S_9V_2^{4-}$	$[V_2S_9]^{4-}$	48 (V):	Hb/B1–280
$S_9V_3^{3-}$	$[VS_3]_3^{3-}$	48 (V):	Hb/B1–280
$S_{12}Th_7$	Th_7S_{12}	44 (Th):	Hb–277, 279/80
$S_{14}Tl_{16}$	$5\ Tl_2S \cdot 3\ Tl_2S_3 = Tl_8S_7$	38 (Tl):	Hb/3–345/6
$S_{15}V_3^{3-}$	$[VS_5]_3^{3-}$	48 (V):	Hb/B1–280
S_xSr	SrS_x	29 (Sr):	Eb–248/9
S_xV	VS_x	48 (V):	Hb/B1–267/80
Sb	Antimony:		
	Absorption of light	18 (Sb):	Hb/B1–87/8 Hb/B2–183/4
	– of X-rays	18 (Sb):	Hb/B1–111/2 Hb/B2–185
	Abundance	18 (Sb):	Hb/A1–55/60
	Aerosols	18 (Sb):	Hb/A3–349
	Affinity	18 (Sb):	Hb/B1–26 Hb/B2–168
	Antimony electrode:		
	Effects on applicability	18 (Sb):	Hb/B2–226/47
	Field of employment	18 (Sb):	Hb/B2–250/9
	General	18 (Sb):	Hb/B2–216/7
	Handling	18 (Sb):	Hb/B2–225/6

Sb. . . . . Antimony:

Antimony electrode:		
Manufacture and types	18 (Sb):	Hb/B2–217/24
Potential	18 (Sb):	Hb/B2–241/7
Theory	18 (Sb):	Hb/B2–247/50
Atom	18 (Sb):	Hb/B1–17/26 Hb/B2–166/8
Atomic moments	18 (Sb):	Hb/B1–21
Atomic nucleus	18 (Sb):	Hb/B1–1/17 Hb/B2–156/66
Atomic number	18 (Sb):	Hb/B1–3
Atomic radius	18 (Sb):	Hb/B1–18/9 Hb/B2–166
Atomic refraction	18 (Sb):	Hb/B1–87
Atomic volume	18 (Sb):	Hb/B1–21 Hb/B2–167
Atomic weight	18 (Sb):	Hb/B1–22/5 Hb/B2–167
Behavior as anode	18 (Sb):	Hb/B2–209/11
– as cathode	18 (Sb):	Hb/B2–207/9
– as electrode	18 (Sb):	Hb/B2–211/3, 215/59
– at alternating current electrolysis	18 (Sb):	Hb/B2–213/5
Birefringence	18 (Sb):	Hb/B1–86/7
Black antimony	18 (Sb):	Hb/A3–343/4
Boiling point	18 (Sb):	Hb/B1–69 Hb/B2–180
Bond energy	18 (Sb):	Hb/B1–27 Hb/B2–168
Cells	18 (Sb):	Hb/B2–196/201
Chemical constant	18 (Sb):	Hb/B1–83
Chemical reactions	18 (Sb):	Hb/B2–260/74
– on heating	18 (Sb):	Hb/B2–260
– with acids	18 (Sb):	Hb/B2–263/7
– with air	18 (Sb):	Hb/B2–260
– with elements	18 (Sb):	Hb/B2–261/3
– with metal compounds	18 (Sb):	Hb/B2–269/72
– with nonmetal compounds	18 (Sb):	Hb/B2–267/9
– with organic compounds	18 (Sb):	Hb/B2–272/4
– with water	18 (Sb):	Hb/B2–260/1
Cleavage	18 (Sb):	Hb/B1–37/8 Hb/B2–171
Colloidal forms	18 (Sb):	Hb/A3–349/51
Color	18 (Sb):	Hb/B1–85 Hb/B2–183
Compressibility	18 (Sb):	Hb/B1–62/4 Hb/B2–178
Contact electricity	18 (Sb):	Hb/B2–141

See the "Instructions for the user"
at the beginning of the volume

Sb	Antimony:		
	Corbino effect	18 (Sb):	Hb/B1–118
	Cosmochemistry	18 (Sb):	Hb/A1–49/50
	Critical constants	18 (Sb):	Hb/B1–67
	Cryoscopic constant	18 (Sb):	Hb/B1–71
	Crystal chemistry	18 (Sb):	Hb/A1–51/3
	Crystallization	18 (Sb):	Hb/B1–32/7
			Hb/B2–170/1
	Crystallographic properties	18 (Sb):	Hb/B1–30/51
			Hb/B2–169/76
	Debye temperature	18 (Sb):	Hb/B1–81/2
			Hb/B2–181
	Deformation	18 (Sb):	Hb/B1–46/7
			Hb/B2–174/5
	Density	18 (Sb):	Hb/B1–52/5
			Hb/B2–176
	Detection	18 (Sb):	Hb/B2–277/307
	– in special substances	18 (Sb):	Hb/B2–305/7
		59 (Fe):	Hb/FI–424
	Determination	18 (Sb):	Hb/B2–307/26
	– in special substances	18 (Sb):	Hb/B2–329/33
		59 (Fe):	Hb/FI–423/6
			Hb/FII–314/5, 349
	Dielectric properties	18 (Sb):	Hb/B2–186
	Diffusion	18 (Sb):	Hb/B1–66/7
			Hb/B2–179
		46 (Sn):	Hb/D–1
		47 (Pb):	Hb/C3–905
	Discharge phenomena	18 (Sb):	Hb/B2–149/50
	Dispersion	18 (Sb):	Hb/B1–87
	Dissociation of the molecule	18 (Sb):	Hb/B1–28/9
			Hb/B2–169
	Dressing of ores	18 (Sb):	Hb/A3–308/11
	Economic deposits:		
	Africa	18 (Sb):	Hb/A1–144/8
	Asia	18 (Sb):	Hb/A1–123/39
	Australia and Oceania	18 (Sb):	Hb/A1–139/44
	Central America	18 (Sb):	Hb/A1–161/3
	Europe	18 (Sb):	Hb/A1–88/123
	North America	18 (Sb):	Hb/A1–148/61
	South America	18 (Sb):	Hb/A1–163/72
	Soviet Union	18 (Sb):	Hb/A1–172/94
	Economics	18 (Sb):	Hb/A1–194/7
			Hb/A3–305/7
	Elasticity	18 (Sb):	Hb/B1–60/1
			Hb/B2–177
	Electrical conductivity	18 (Sb):	Hb/B1–120/8
			Hb/B2–129/40, 187/9

Hb = Main Series Volume; GD = Gmelin-Durrer; Ab = Appended Volume to Main Series;
Eb = Supplement Volume to Main Series; Erg. W. = New Supplement Series Volume

Sb. . . . . Antimony:

Electrical properties . . . . 18 (Sb): Hb/B1–116/8, 120/8
Hb/B2–129/56, 186/9
Electrical relaxation. . . . . 18 (Sb): Hb/B2–141/2
Electrocapillarity . . . . 18 (Sb): Hb/B2–201
Electrochemical behavior . . . . 18 (Sb): Hb/B2–189/259
Electrode sputtering . . . . 18 (Sb): Hb/B2–150/1
Electrolytic deposition . . . . 18 (Sb): Hb/B2–201/7
Electrolytic solution pressure . . . . . 18 (Sb): Hb/B2–190
Electron configuration . . . . 18 (Sb): Hb/B1–17/8
Hb/B2–166
Electron emission. . . . . 18 (Sb): Hb/B2–153/4
Enthalpy . . . . 18 (Sb): Hb/B1–82/3
Entropy . . . . 18 (Sb): Hb/B1–82/3
Equation of state . . . . 18 (Sb): Hb/B1–67
Hb/B2–179
Ettingshausen effect . . . . 18 (Sb): Hb/B1–118
Explosive antimony. . . . . 18 (Sb): Hb/A3–344/8
Formation . . . . 18 (Sb): Hb/A3–339/42
Free energy . . . . 18 (Sb): Hb/B1–82/3
Frictional electricity. . . . . 18 (Sb): Hb/B2–141
Galvanomagnetic effects. . . . . 18 (Sb): Hb/B1–118/9
Geochemical character. . . . . 18 (Sb): Hb/A1–54
Geochemistry . . . . 18 (Sb): Hb/A1–51/192
Hall effect . . . . 18 (Sb): Hb/B1–116/8
Hb/B2–186
Hardness . . . . 18 (Sb): Hb/B1–58/60
Hb/B2–176/7
Heat content. . . . . 18 (Sb): Hb/B1–82/3
Heat of fusion . . . . 18 (Sb): Hb/B1–72/3
Hb/B2–180
– of sublimation. . . . . 18 (Sb): Hb/B1–72
Hb/B2–180
– of transformation . . . . 18 (Sb): Hb/B1–73/5
– of vaporization . . . . 18 (Sb): Hb/B1–71/2
History. . . . . 18 (Sb): Hb/A1–1/48
Hydrosols . . . . 18 (Sb): Hb/A3–349/50
Interaction with ion beams. . . . . . . 18 (Sb): Hb/B2–155
Interfacial tension. . . . . 18 (Sb): Hb/B1–64/5
Hb/B2–178/9
Internal pressure . . . . 18 (Sb): Hb/B1–64
Ionization potentials . . . . 18 (Sb): Hb/B2–155/6
Isomorphism . . . . 18 (Sb): Hb/B1–38
Hb/B2–171
Lattice structure. . . . . 18 (Sb): Hb/B1–38/42
Hb/B2–171/2
Leduc effect . . . . 18 (Sb): Hb/B1–119
Hb/B2–186
Luminescence. . . . . 18 (Sb): Hb/B1–89

See the "Instructions for the user" at the beginning of the volume

Sb. . . . . Antimony:

Machinability . . . . 18 (Sb): Hb/B1–66
Magnetic moment . . . . 18 (Sb): Hb/B1–21
Magnetic properties . . . . 18 (Sb): Hb/B1–112/9
Hb/B2–186
Magnetostriction . . . . 18 (Sb): Hb/B1–119
Mechanical properties . . . . 18 (Sb): Hb/B1–52/67
Hb/B2–176/9
Melting point . . . . 18 (Sb): Hb/B1–69/71
Hb/B2–180
Mineralogy . . . . 18 (Sb): Hb/A1–197/226
Hb/A2–227/302
Molar polarization . . . . 18 (Sb): Hb/B2–186
Molecule. . . . . 18 (Sb): Hb/B1–26/9
Hb/B2–168/9
Nernst effect . . . . 18 (Sb): Hb/B1–118/9
Nuclear properties . . . . 18 (Sb): Hb/B1–1/17
Hb/B2–156/66
Occurrence, extraterrestrial . . . . . . 18 (Sb): Hb/A1–49/50
–, terrestrial . . . . 18 (Sb): Hb/A1–51/192
– in meteorites . . . . 18 (Sb): Hb/A1–50
– in the biosphere. . . . . 18 (Sb): Hb/A1–59/60
– in the hydrosphere . . . . 18 (Sb): Hb/A1–58/9
– in the lithosphere . . . . 18 (Sb): Hb/A1–56/8, 81/8
Optical properties. . . . . 18 (Sb): Hb/B1–85/112
Hb/B2–183/5
Organosols . . . . 18 (Sb): Hb/A3–350/1
Orientation of thin layers. . . . . 18 (Sb): Hb/B1–42/6
Hb/B2–172/3
Overvoltage phenomena. . . . . 18 (Sb): Hb/B2–211/3
Passage of alpha particles . . . . 18 (Sb): Hb/B2–154/5
– of electrons . . . . 18 (Sb): Hb/B2–153/4
Photoelectric effect . . . . 18 (Sb): Hb/B2–152/3
Plasticity. . . . . 18 (Sb): Hb/B1–61
Polymorphism . . . . 18 (Sb): Hb/B1–30/2
Hb/B2–169/70
Position in the electrochemical
series . . . . 18 (Sb): Hb/B2–190
– in the periodic system . . . . 18 (Sb): Hb/B1–26
Hb/B2–168
Potentials . . . . 18 (Sb): Hb/B2–189/95
Preparation . . . . 18 (Sb): Hb/A3–342/3
–, industrial . . . . 18 (Sb): Hb/A3–303/39
–, –, dry methods . . . . 18 (Sb): Hb/A3–312/30
–, –, wet methods. . . . . 18 (Sb): Hb/A3–330/4
– of special forms. . . . . 18 (Sb): Hb/A3–343/51
Purification . . . . 18 (Sb): Hb/A3–334/8
Purity testing . . . . 18 (Sb): Hb/A3–339
Recrystallization . . . . 18 (Sb): Hb/B1–51

Hb = Main Series Volume; GD = Gmelin-Durrer; Ab = Appended Volume to Main Series;
Eb = Supplement Volume to Main Series; Erg. W. = New Supplement Series Volume

Sb. . . . . Antimony:
Rectifying effect. . . . . 18 (Sb): Hb/B2–142
Reflection . . . . 18 (Sb): Hb/B1–85/6
Hb/B2–183
Refraction . . . . 18 (Sb): Hb/B1–86/7
Hb/B2–183
Rotational constant. . . . . 18 (Sb): Hb/B1–28
Scattering of light. . . . . 18 (Sb): Hb/B1–87
– of X-rays. . . . . 18 (Sb): Hb/B1–111/2
Hb/B2–185
Secondary electron emission . . . . . 18 (Sb): Hb/B2–154, 155
Semiconductor properties . . . . 18 (Sb): Hb/B2–137/9
Separation, analytical . . . . 18 (Sb): Hb/B2–326/9
Shrinkage . . . . 18 (Sb): Hb/B1–58
Hb/B2–176
Sorption . . . . 18 (Sb): Hb/B1–67
Hb/B2–179
Special forms . . . . 18 (Sb): Hb/B2–45/6, 173/4
Specific heat . . . . 18 (Sb): Hb/B1–75/81
Hb/B2–180/1
Spectra:
Arc spectrum . . . . 18 (Sb): Hb/B1–89/96
Hb/B2–184
Band spectrum . . . . 18 (Sb): Hb/B1–104/7
Hb/B2–184/5
Continuous spectrum . . . . 18 (Sb): Hb/B1–107
Spark spectra. . . . . 18 (Sb): Hb/B1–96/102
Hb/B2–184
X-ray spectrum. . . . . 18 (Sb): Hb/B1–107/11
Hb/B2–185
Standard potential . . . . 18 (Sb): Hb/B2–189/90
Strength . . . . 18 (Sb): Hb/B1–61/2
Structure . . . . 18 (Sb): Hb/B1–47/51
Hb/B2–175/6
– of thin layers . . . . 18 (Sb): Hb/B1–42/6
Hb/B2–172/4
Superconductivity . . . . 18 (Sb): Hb/B1–125
Hb/B2–188
Surface tension . . . . 18 (Sb): Hb/B1–64/5
Hb/B2–178/9
Susceptibility . . . . 18 (Sb): Hb/B1–112/6
Hb/B2–186
Thermal conductivity. . . . . 18 (Sb): Hb/B1–83/5
Hb/B2–181/3
Thermal expansion . . . . 18 (Sb): Hb/B1–55/8
Hb/B2–176
Thermal properties . . . . 18 (Sb): Hb/B1–55/8, 67/85
Hb/B2–176, 179/83
Thermal radiation. . . . . 18 (Sb): Hb/B1–88

See the "Instructions for the user"
at the beginning of the volume

Sb	Antimony:		
	Thermoelectric effects	18 (Sb):	Hb/B2–142/9
	Thermomagnetic effects	18 (Sb):	Hb/B1–119
			Hb/B2–186
	Toxicity	18 (Sb):	Hb/B2–274/7
	Transformation point	18 (Sb):	Hb/B1–71
	Uses	18 (Sb):	Hb/A1–43/6
			Hb/A3–304/5
	Valence	18 (Sb):	Hb/B1–25/6
			Hb/B2–167/8
	Vapor pressure	18 (Sb):	Hb/B1–68/9
			Hb/B2–179
	Vibrational frequency	18 (Sb):	Hb/B1–81/2
			Hb/B2–181
	Viscosity	18 (Sb):	Hb/B1–65/6
	Wiedemann-Franz number	18 (Sb):	Hb/B2–139/40
	Yellow antimony	18 (Sb):	Hb/A3–344
–	Antimony ions *see* Sb^{+} ... Sb^{+}, *etc.*		
–	Antimony isotopes	18 (Sb):	Hb/B1–1/3, 27
			Hb/B2–156/8
		Erg. W. 8–223/4	
–	Antimony systems:		
	Sb–Ag	61 (Ag):	Hb/B4–69/73
	Sb–Ag–Bi	61 (Ag):	Hb/B4–135
	Sb–Ag–Bi–Zn	61 (Ag):	Hb/B4–144
	Sb–Ag–Cd	61 (Ag):	Hb/B4–156/7
	Sb–Ag–Cu–S	61 (Ag):	Hb/B4–491
	Sb–Ag–S	61 (Ag):	Hb/B4–90/2
	Sb–Ag–Se	61 (Ag):	Hb/B4–101/2
	Sb–Ag–Sn	61 (Ag):	Hb/B4–252
	Sb–Ag–Zn	61 (Ag):	Hb/B4–142/3
	Sb–Al	35 (Al):	Hb/A4–629/32
	Sb–Al–Bi–Pb	47 (Pb):	Hb/C4–1287/8
	Sb–Al–Cu	60 (Cu):	Hb/B3–1177
	Sb–Al–Cu–Sn	60 (Cu):	Hb/B3–1196
	Sb–Al–Mg	35 (Al):	Hb/A4–665/6
	Sb–Al–Nb	49 (Nb):	Hb/B2–45
	Sb–Al–Pb	35 (Al):	Hb/A5–861/3
		47 (Pb):	Hb/C4–1286/7
	Sb–Al–Sn	35 (Al):	Hb/A5–854
		46 (Sn):	Hb/D–218
	Sb–As	18 (Sb):	Hb/B3–562/3
	Sb–As–Cu–S	60 (Cu):	Hb/B2–973
	Sb–As–Ni	57 (Ni):	Hb/B1–10
	Sb–As–Pb	47 (Pb):	Hb/C3–941/2
	Sb–Au	62 (Au):	Hb/3–770/1
	Sb–B–Ni	57 (Ni):	Hb/B3–996

Hb = Main Series Volume; GD = Gmelin-Durrer; Ab = Appended Volume to Main Series;
Eb = Supplement Volume to Main Series; Erg. W. = New Supplement Series Volume

Sb. . . . . Antimony systems:

Sb–Ba . . . . 30 (Ba): Eb–272
Sb–Bi . . . . 19 (Bi): Hb–198/202
Eb–601/2
Sb–Bi–Cd. . . . . 33 (Cd): Hb–167
Sb–Bi–Pb. . . . . 47 (Pb): Hb/C3–977/8
Sb–Bi–Se. . . . . 19 (Bi): Eb–861
Sb–Bi–Se–Te . . . . 19 (Bi): Eb–865/6
Sb–Bi–Sn. . . . . 46 (Sn): Hb/D–58/9
Sb–Bi–Te . . . . 19 (Bi): Eb–861/5
Sb–Bi–Zn. . . . . 32 (Zn): Hb–293
Eb–617
Sb–Ca . . . . 28 (Ca): Hb/A2–479
Sb–Cd . . . . 33 (Cd): Hb–161/4
Eb–353/5
Sb–Cd–Pb . . . . 47 (Pb): Hb/C4–1245/6
Sb–Cd–Sn . . . . 46 (Sn): Hb/D–177/8
Sb–Cd–Zn . . . . 33 (Cd): Hb–213
Sb–Co. . . . . 58 (Co): Hb/A1–185/7
Eb/A–376/87
Sb–Co–H–S . . . . 58 (Co): Eb/A–755/6
Sb–Co–S . . . . 58 (Co): Eb/A–754/5
Sb–Cr . . . . 52 (Cr): Hb/A2–710/2
Sb–Cu–Fe . . . . 60 (Cu): Hb/B3–1315
Sb–Cu–Fe–S. . . . . 60 (Cu): Hb/B3–1315
Sb–Cu–Ge . . . . 60 (Cu): Hb/B3–1193
Sb–Cu–Mg . . . . 27 (Mg): Hb/A4–716/8
60 (Cu): Hb/B3–1128
Sb–Cu–Mn–Ni. . . . . 60 (Cu): Hb/B3–1249
Sb–Cu–Ni. . . . . 60 (Cu): Hb/B3–1259/60
Sb–Cu–Pb . . . . 60 (Cu): Hb/B3–1204
Sb–Cu–S . . . . 60 (Cu): Hb/B2–971
Sb–Cu–Sn . . . . 60 (Cu): Hb/B3–1196
Sb–Fe . . . . 59 (Fe): Hb/A8–1805/6
Sb–Ga–Ge–Sn. . . . . 46 (Sn): Hb/D–397
Sb–Ga–In–Sn . . . . 46 (Sn): Hb/D–277
Sb–Ga–Pb . . . . 47 (Pb): Hb/C4–1305
Sb–Ge. . . . . 45 (Ge): Eb–475/7
Sb–I . . . . 18 (Sb): Hb/B2–486
Sb–In–Pb. . . . . 47 (Pb): Hb/C4–1324
Sb–In–Sn. . . . . 46 (Sn): Hb/D–264/5
Sb–K . . . . 22 (K): Hb/5–1035/6
Sb–Mg . . . . 27 (Mg): Hb/A3–406/7
Sb–Mg–Pb . . . . 27 (Mg): Hb/A4–653
Sb–Mn–Ni . . . . 57 (Ni): Hb/B1–311
Sb–NH_3–Na . . . . 21 (Na): Eb/4–1709/10
Sb–Na. . . . . 21 (Na): Hb–965/6
Eb/4–1707/8
Sb–Nb. . . . . 49 (Nb): Hb/B1–417/8

Sb	Antimony systems:		
	Sb–Ni	57 (Ni):	Hb/B1–2/10
	Sb–Ni–Pb	57 (Ni):	Hb/B1–152
	Sb–Ni–S	57 (Ni):	Hb/B1–10 Hb/B3–994/5
	Sb–O–Pb	47 (Pb):	Hb/C3–926/9
	Sb–O–S	9 (S):	Hb/B1–182
	Sb–Pb	47 (Pb):	Hb/C3–905/7
	Sb–Pb–S	47 (Pb):	Hb/C3–935/6
	Sb–Pb–Se	47 (Pb):	Hb/C3–940
	Sb–Pb–Si	47 (Pb):	Hb/C3–941
	Sb–Pb–Sn	47 (Pb):	Hb/C4–1515
	Sb–Pb–Te	47 (Pb):	Hb/C3–940/1
	Sb–Pb–Zn	47 (Pb):	Hb/C4–1223/4
	Sb–Pd	68 (Pt):	Hb/A5–570/5
	Sb–Pt	68 (Pt):	Hb/A6–743/4
	Sb–S	18 (Sb):	Hb/B3–499/500
	Sb–Se	18 (Sb):	Hb/B3–550/2
	Sb–Si	18 (Sb):	Hb/B3–560
	Sb–Sn	46 (Sn):	Hb/D–1/4
	Sb–Sn–Zn	46 (Sn):	Hb/D–153
	Sb–Te	18 (Sb):	Hb/B3–554/5
	Sb–Tl	38 (Tl):	Hb/2–187/91
	Sb–V	48 (V):	Hb/B2–575/7
	Sb–Zn	32 (Zn):	Hb–286/90 Eb–606/8
Sb^{+}	Sb^{+}	18 (Sb):	Hb/B1–20/2, 96/8, 103/4 Hb/B2–155, 167
Sb^{2+}	Sb^{2+}	18 (Sb):	Hb/B1–20/2, 98/100, 103/4 Hb/B2–155
Sb^{3+}	Sb^{3+}:		
	Chemical reactions	18 (Sb):	Hb/B2–279/303
	Electrochemical behavior	18 (Sb):	Hb/B2–201/7
	Ionization potential	18 (Sb):	Hb/B2–155/6
	Properties of the ion	18 (Sb):	Hb/B1–20, 21/2 Hb/B2–167
	Spectra	18 (Sb):	Hb/B1–100/1
Sb^{4+}	Sb^{4+}	18 (Sb):	Hb/B1–20/2, 101/2 Hb/B2–155/6, 167
Sb^{5+}	Sb^{5+}:		
	Chemical reactions	18 (Sb):	Hb/B2–279/303
	Ionization potential	18 (Sb):	Hb/B2–155/6
	Molar polarization	18 (Sb):	Hb/B2–186
	Properties of the ion	18 (Sb):	Hb/B1–20, 21/2 Hb/B2–167
	Spectra	18 (Sb):	Hb/B1–102
	Susceptibility	18 (Sb):	Hb/B1–116

Hb = Main Series Volume; GD = Gmelin-Durrer; Ab = Appended Volume to Main Series; Eb = Supplement Volume to Main Series; Erg. W. = New Supplement Series Volume

Sb^{6+}	Sb^{6+}	18 (Sb):	Hb/B2–155
Sb^{7+}	Sb^{7+}	18 (Sb):	Hb/B2–155
Sb^{8+}	Sb^{8+}	18 (Sb):	Hb/B2–155
Sb^{22+}	Sb^{22+}	18 (Sb):	Hb/B2–184
Sb^{-}	Sb^{-}	18 (Sb):	Hb/B1–20, 22
			Hb/B2–167
Sb^{2-}	Sb^{2-}	18 (Sb):	Hb/B1–22
Sb^{3-}	Sb^{3-}	18 (Sb):	Hb/B1–116
			Hb/B2–167, 186
SbSe	SbSe	18 (Sb):	Hb/B3–550
SbSn	SnSb	46 (Sn):	Hb/D–4/5
$SbTa_3$	Ta_3Sb	50 (Ta):	Hb/B1–273
$SbTl_3$	Tl_3Sb	38 (Tl):	Hb/2–189
SbV	VSb	48 (V):	Hb/B2–575/7
SbV_3	V_3Sb	48 (V):	Hb/B2–575/7
SbZn	ZnSb	32 (Zn):	Hb–286/8, 290
			Eb–615
Sb_2	Sb_2	18 (Sb):	Hb/B1–26/9
			Hb/B2–168/9
		46 (Sn):	Hb/D–22
Sb_2Se_3	Sb_2Se_3	18 (Sb):	Hb/B3–552/3
–	Sb_2Se_3 solid solutions:		
	$(Bi_{1-x}Sb_x)_2(Te_{1-y}Se_y)_3$	19 (Bi):	Eb–865
–	Sb_2Se_3 systems:		
	Sb_2Se_3–Ag_2Se	61 (Ag):	Hb/B4–101/2
	Sb_2Se_3–PbSe	47 (Pb):	Hb/C3–940
Sb_2Se_5	Sb_2Se_5	18 (Sb):	Hb/B3–553
$Sb_2Se_8Zn_3$	$Zn_3(SbSe_4)_2$	32 (Zn):	Hb–291
Sb_2Sn_3	Sb_2Sn_3	46 (Sn):	Hb/D–2/3
Sb_2Ta	$TaSb_2$	50 (Ta):	Hb/B1–272, 273
Sb_2Te_3	Sb_2Te_3	18 (Sb):	Hb/B3–555
–	Sb_2Te_3 solid solutions:		
	$(Bi_{1-x}Sb_x)_2(Te_{1-y}Se_y)_3$	19 (Bi):	Eb–865
	$(Bi_{1-x}Sb_x)_2Te_3$	19 (Bi):	Eb–862
–	Sb_2Te_3 systems:		
	Sb_2Te_3–Ag_2Te	61 (Ag):	Hb/B4–105/8
	Sb_2Te_3–Ag_2Te–PbTe	61 (Ag):	Hb/B4–269/70
Sb_2Tl_7	Tl_7Sb_2	38 (Tl):	Hb/2–187/91
Sb_2V	VSb_2	48 (V):	Hb/B2–575/7
Sb_2Zn	$ZnSb_2$	32 (Zn):	Hb–287
Sb_2Zn_3	Zn_3Sb_2	32 (Zn):	Hb–286/8, 290
			Eb–615
Sb_3	Sb_3	18 (Sb):	Hb/B1–26
Sb_3Se_4	Sb_3Se_4	18 (Sb):	Hb/B3–550, 553
Sb_3Zn_4	Zn_4Sb_3	32 (Zn):	Eb–615
Sb_4	Sb_4	18 (Sb):	Hb/B1–26
			Hb/B2–168/9
		46 (Sn):	Hb/D–22

See the "Instructions for the user" at the beginning of the volume

Sb_4Se_5	Sb_4Se_5	18 (Sb):	Hb/B3–550, 553
Sb_4Ta_5	Ta_5Sb_4	50 (Ta):	Hb/B1–273
Sb_4U_3	U_3Sb_4	55 (U):	Hb–185
Sb_8	Sb_8	18 (Sb):	Hb/B1–27
Sc	Scandium:		
	Abundance	39 (SE):	Hb/A1–34/5
			Hb/A2–17/8, 88
	Cosmochemistry	39 (SE):	Hb/A1–23/7
			Hb/A2–3/46
	Crystal chemistry	39 (SE):	Hb/A1–28/32
			Hb/A2–47/88
	Crystallographic properties	39 (SE):	Hb/B3–1/4
	Diadochy	39 (SE):	Hb/A2–48/55
	Economic deposits	39 (SE):	Hb/A1–110/22
			Hb/A2–164/78
	Economics	39 (SE):	Hb/A2–180/1
	Electrical properties	39 (SE):	Hb/B3–18/27
	Geochemical character	39 (SE):	Hb/A2–88
	Geochemical cycle	39 (SE):	Hb/A2–149/51
	Geochemistry	39 (SE):	Hb/A1–28/54
			Hb/A2–47/151
	History	39 (SE):	Hb/A1–4/22
			Hb/A2–1/2
	Magnetic properties	39 (SE):	Hb/B3–18/27
	Mechanical properties	39 (SE):	Hb/B3–5/10
	Mineralogy	39 (SE):	Hb/A1–54/110
			Hb/A2–152/9
	Occurrence, extraterrestrial	39 (SE):	Hb/A1–22/7
			Hb/A2–3/46
	–, terrestrial	39 (SE):	Hb/A1–27/54
			Hb/A2–47/151
	– in meteorites and tektites	39 (SE):	Hb/A1–25/7
			Hb/A2–18/34
	– in minerals	39 (SE):	Hb/A1–50/1
			Hb/A2–55/88
	– in the atmosphere	39 (SE):	Hb/A2–149
	– in the biosphere	39 (SE):	Hb/A1–53/4
	– in the hydrosphere	39 (SE):	Hb/A2–147/9
	– in the lithosphere	39 (SE):	Hb/A1–27/54
			Hb/A2–88/147
	– in the magmatic cycle	39 (SE):	Hb/A1–28/32
			Hb/A2–88/120
	– in the metamorphic cycle	39 (SE):	Hb/A2–133/47
	– in the sedimentary cycle	39 (SE):	Hb/A1–32/3
			Hb/A2–121/33
	– on the moon	39 (SE):	Hb/A2–34/46
	Position in the periodic system	39 (SE):	Hb/A1–2/4
	Preparation	39 (SE):	Hb/A2–160/4

Hb = Main Series Volume; GD = Gmelin-Durrer; Ab = Appended Volume to Main Series;
Eb = Supplement Volume to Main Series; Erg. W. = New Supplement Series Volume

Sc	Scandium:		
	Thermal properties	39 (SE):	Hb/B3–4/18
	Uses	39 (SE):	Hb/A2–178/80
	Valence	39 (SE):	Hb/A1–4
–	Scandium ions	39 (SE):	Hb/A1–3
			Hb/B3–15/6
–	Scandium systems:		
	Sc–Ag	61 (Ag):	Hb/C–210/1
	Sc–H	39 (SE):	Hb/C1–2/3, 18/9
	Sc–H–Y	39 (SE):	Hb/C1–31/3
	Sc–Np–O	Erg. W. 4–45/6	
	Sc–O	39 (SE):	Hb/C1–180
	Sc–Pb	47 (Pb):	Hb/C4–1360
	Sc–Ta	50 (Ta):	Hb/B2–18/9
	Sc–V	48 (V):	Hb/B2–592/3
Sc^{+}	Sc^{+}	39 (SE):	Hb/B3–16
Sc^{2+}	Sc^{2+}	39 (SE):	Hb/B3–16
Sc^{3+}	Sc^{3+}	39 (SE):	Hb/A1–3
			Hb/B3–16
–	$[Sc(H_2O)_6]^{3+}$	39 (SE):	Hb/C2–87
$ScSn_3$	$ScSn_3$	46 (Sn):	Hb/D–293
Sc_2Se_3	Sc_2Se_3 systems:		
	Sc_2Se_3–Ag_2Se	61 (Ag):	Hb/B4–234/5
Sc_5Sn_3	Sc_5Sn_3	46 (Sn):	Hb/D–293
Se	Selenium:		
	Absorption of light	10 (Se):	Hb/A1–206/10
	– of X-rays	10 (Se):	Hb/A1–210, 231/2
	Abundance	10 (Se):	Hb/A1–8/9
	Aerosols	10 (Se):	Hb/A1–111/6
	Atom	10 (Se):	Hb/A1–140/50
	Atomic moments	10 (Se):	Hb/A1–145
	Atomic nucleus	10 (Se):	Hb/A1–131/40
	Atomic number	10 (Se):	Hb/A1–132
	Atomic radius	10 (Se):	Hb/A1–142/5
	Atomic refraction	10 (Se):	Hb/A1–206
	Atomic volume	10 (Se):	Hb/A1–145/6
	Atomic weight	10 (Se):	Hb/A1–146/9
	Becquerel effect	10 (Se):	Hb/A1–238/42
	Behavior as electrode	10 (Se):	Hb/A1–237/8
	Boiling point	10 (Se):	Hb/A1–188/9
	Cells	10 (Se):	Hb/A1–235/6
	Chemical constant	10 (Se):	Hb/A1–197
	Chemical reactions	10 (Se):	Hb/A1–242/59
	– with elements	10 (Se):	Hb/A1–242/6
	– with inorganic compounds	10 (Se):	Hb/A1–246/55
	– with organic compounds	10 (Se):	Hb/A1–255/9

Se . . . . Selenium:

Coagula	10 (Se):	Hb/A1–129/30
Colloidal forms	10 (Se):	Hb/A1–118/30
Color	10 (Se):	Hb/A1–199/200
Compressibility	10 (Se):	Hb/A1–185
Contact electricity	10 (Se):	Hb/A2–295/6
Cosmochemistry	10 (Se):	Hb/A1–4/5
Critical constants	10 (Se):	Hb/A1–186
Crystal form	10 (Se):	Hb/A1–160/3
Crystallographic properties	10 (Se):	Hb/A1–155/75
Debye temperature	10 (Se):	Hb/A1–196
Deformation	10 (Se):	Hb/A1–174
Density	10 (Se):	Hb/A1–175/80
Detection	10 (Se):	Hb/A1–261/9
– in iron	59 (Fe):	Hb/FII1–27/8
– in special substances	10 (Se):	Hb/A1–269/71
Determination	10 (Se):	Hb/A1–271/83
– in iron	59 (Fe):	Hb/FII1–28/30
– in special substances	10 (Se):	Hb/A1–284/92
Dielectric constant	10 (Se):	Hb/A2–293/4
Diffusion	10 (Se):	Hb/A1–186
Discharge phenomena	10 (Se):	Hb/A2–298/9
Dissociation of the molecule	10 (Se):	Hb/A1–150/2, 154/5
Economic deposits:		
America	10 (Se):	Hb/A1–43/8
Asia	10 (Se):	Hb/A1–35/42
Australia	10 (Se):	Hb/A1–42
Europe	10 (Se):	Hb/A1–28/37
Elasticity	10 (Se):	Hb/A1–183
Electrical conductivity	10 (Se):	Hb/A2–301/414
Electrical properties	10 (Se):	Hb/A2–293/414
Electrochemical behavior	10 (Se):	Hb/A1–235/42
Electron configuration	10 (Se):	Hb/A1–140/1
Electron emission	10 (Se):	Hb/A2–299/300
Entropy	10 (Se):	Hb/A1–197
Equation of state	10 (Se):	Hb/A1–186
Faraday effect	10 (Se):	Hb/A1–210
Formation	10 (Se):	Hb/A1–76/7
Free energy	10 (Se):	Hb/A1–197
Frictional electricity	10 (Se):	Hb/A2–295
Geochemical character	10 (Se):	Hb/A1–5/8
Geochemistry	10 (Se):	Hb/A1–5/28
Grain structure	10 (Se):	Hb/A1–174/5
Hall effect	10 (Se):	Hb/A1–234
Hardness	10 (Se):	Hb/A1–182
Heat content	10 (Se):	Hb/A1–197
Heat of fusion	10 (Se):	Hb/A1–192
– of sublimation	10 (Se):	Hb/A1–191

Se . . . . Selenium:

Heat of transition	10 (Se):	Hb/A1–192/3
– of vaporization	10 (Se):	Hb/A1–190/1
History	10 (Se):	Hb/A1–1/3
Hydrosols	10 (Se):	Hb/A1–116/27
Ionization	10 (Se):	Hb/A2–300/1
Isomorphism	10 (Se):	Hb/A1–166/7
Lattice structure	10 (Se):	Hb/A1–167/72
Magnetic properties	10 (Se):	Hb/A1–232/4
Mechanical properties	10 (Se):	Hb/A1–175/86
Melting point	10 (Se):	Hb/A1–189/90
Mineralogy	10 (Se):	Hb/A1–48/75
Molecule	10 (Se):	Hb/A1–150/5
Nuclear properties	10 (Se):	Hb/A1–131/40
Occurrence, extraterrestrial	10 (Se):	Hb/A1–4/5
–, terrestrial	10 (Se):	Hb/A1–5/28
– in industrial products	10 (Se):	Hb/A1–25/8
– in soils	10 (Se):	Hb/A1–19/22
– in the biosphere	10 (Se):	Hb/A1–22/5
– in the hydrosphere	10 (Se):	Hb/A1–25
– in the lithosphere	10 (Se):	Hb/A1–10/9
Optical properties	10 (Se):	Hb/A1–199/232
Passage of radiation	10 (Se):	Hb/A2–299/300
Photoelectric effect	10 (Se):	Hb/A2–301/414
Piezoelectricity	10 (Se):	Hb/A2–294
Plasticity	10 (Se):	Hb/A1–183/4
Polymorphism	10 (Se):	Hb/A1–155/60
Position in the electrochemical series	10 (Se):	Hb/A1–235
– in the periodic system	10 (Se):	Hb/A1–150
Preparation	10 (Se):	Hb/A1–95/9
–, industrial	10 (Se):	Hb/A1–78/95
– in pure state	10 (Se):	Hb/A1–95/9
– of modifications	10 (Se):	Hb/A1–100/7
– of special forms	10 (Se):	Hb/A1–99/100
Radioactivity	10 (Se):	Hb/A1–140
Reflection	10 (Se):	Hb/A1–200/2
Refraction	10 (Se):	Hb/A1–202/6
Scattering of X-rays	10 (Se):	Hb/A1–231/2
Selenium photocell:		
Action of gamma rays and corpuscular radiation	10 (Se):	Hb/A3–562/3
– of infrared light	10 (Se):	Hb/A3–557/8
– of spectrally decomposed visible light	10 (Se):	Hb/A3–543/57
– of ultraviolet light	10 (Se):	Hb/A3–558/9
– of visible light	10 (Se):	Hb/A3–529/43
– of X-rays	10 (Se):	Hb/A3–559/62
Aging	10 (Se):	Hb/A3–578/80

See the "Instructions for the user"
at the beginning of the volume

Se . . . . Selenium:

Selenium photocell:

Application	10 (Se):	Hb/A3–590
Creep effect	10 (Se):	Hb/A3–571
Electrical properties	10 (Se):	Hb/A3–525/63
Fatigue	10 (Se):	Hb/A3–571/7
General remarks	10 (Se):	Hb/A3–524/5
Inertia	10 (Se):	Hb/A3–564/70
Instability	10 (Se):	Hb/A3–564/80
Optical properties	10 (Se):	Hb/A3–525/63
Recovery	10 (Se):	Hb/A3–577/8
Service	10 (Se):	Hb/A3–590/1
Technology	10 (Se):	Hb/A3–580/90
Theory	10 (Se):	Hb/A3–591/8

Selenium photoresistive cell:

Action of light	10 (Se):	Hb/A2–315/30
– of X- and gamma rays	10 (Se):	Hb/A2–330/4
– of other factors	10 (Se):	Hb/A2–334/59
Aging	10 (Se):	Hb/A2–393/4
Fabrication	10 (Se):	Hb/A2–395/414
Fatigue	10 (Se):	Hb/A2–387/92
General remarks	10 (Se):	Hb/A2–304/15
Industrial types	10 (Se):	Hb/A2–396/400
Inertia	10 (Se):	Hb/A2–362/77
Lifetime	10 (Se):	Hb/A2–393/4
Mechanism of dark conductivity	10 (Se):	Hb/A2–394/5
Negative photoeffect	10 (Se):	Hb/A2–359/61
Polarization	10 (Se):	Hb/A2–361/2
Recovery	10 (Se):	Hb/A2–377/87

Selenium rectifier:

Aging	10 (Se):	Hb/A3–437/8
Capacity	10 (Se):	Hb/A3–426/9
Current-voltage characteristics	10 (Se):	Hb/A3–417/9
Dependence on various factors	10 (Se):	Hb/A3–439
Electrical properties	10 (Se):	Hb/A3–416/39
Fabrication	10 (Se):	Hb/A3–480
General remarks	10 (Se):	Hb/A3–415/6
Instability	10 (Se):	Hb/A3–435/8
Lifetime	10 (Se):	Hb/A3–438/9
Load characteristics	10 (Se):	Hb/A3–430/3
Magnitude of rectifier action	10 (Se):	Hb/A3–433/5
Rectifier circuits	10 (Se):	Hb/A3–419/20
Resistance	10 (Se):	Hb/A3–420/6
Service	10 (Se):	Hb/A3–481/2
Special types	10 (Se):	Hb/A3–478/80
Technology	10 (Se):	Hb/A3–440/80
Theory	10 (Se):	Hb/A3–482/523
Utilization	10 (Se):	Hb/A3–480/1

Se . . . . Selenium:

Separation, analytical . . . . 10 (Se): Hb/A1–283/4
Sols . . . . 10 (Se): Hb/A1–111/29
Solubility . . . . 10 (Se): Hb/A1–257/9
Specific heat . . . . 10 (Se): Hb/A1–193/6
Spectra:
Arc spectrum . . . . 10 (Se): Hb/A1–210/5
Band spectrum . . . . 10 (Se): Hb/A1–219/27
Continuous spectrum . . . . 10 (Se): Hb/A1–227
Spark spectra . . . . 10 (Se): Hb/A1–215/9
X-ray spectrum . . . . 10 (Se): Hb/A1–228/32
Standard potential . . . . 10 (Se): Hb/A1–235
Strength . . . . 10 (Se): Hb/A1–184/5
Structure . . . . 10 (Se): Hb/A1–172/4
Surface tension . . . . 10 (Se): Hb/A1–185
Susceptibility . . . . 10 (Se): Hb/A1–232/4
Thermal conductivity . . . . 10 (Se): Hb/A1–197/9
Thermal expansion . . . . 10 (Se): Hb/A1–180/2
Thermal properties . . . . 10 (Se): Hb/A1–186/99
Thermal radiation . . . . 10 (Se): Hb/A1–210
Thermoelectric effects . . . . 10 (Se): Hb/A2–296/8
Toxicity . . . . 10 (Se): Hb/A1–259/60
Transition point . . . . 10 (Se): Hb/A1–190
Triboluminescence . . . . 10 (Se): Hb/A1–210
Valence . . . . 10 (Se): Hb/A1–149/50
Vapor pressure . . . . 10 (Se): Hb/A1–187/8
Vibration frequencies . . . . 10 (Se): Hb/A1–152/3, 196
Viscosity . . . . 10 (Se): Hb/A1–186
Uses . . . . 10 (Se): Hb/A1–108/10

–. . . . . Selenium ions
see Se^{+} *...* Se^{+}*, etc.*

–. . . . . Selenium isotopes . . . . 10 (Se): Hb/A1–131/2

–. . . . . Selenium systems:

Se–Ag . . . . 61 (Ag): Hb/B3–152/4
Se–Ag–As . . . . 61 (Ag): Hb/B4–66/7
Se–Ag–Cu . . . . 61 (Ag): Hb/B4–474/5
Se–Ag–Pb . . . . 61 (Ag): Hb/B4–267/8
Se–Ag–Sb . . . . 61 (Ag): Hb/B4–101/2
Se–Al . . . . 35 (Al): Hb/B1–279
Se–As–Bi . . . . 19 (Bi): Eb–858
Se–As–Hg . . . . 34 (Hg): Hb/B4–1378
Se–As–Pb . . . . 47 (Pb): Hb/C3–904/5
Se–Bi . . . . 19 (Bi): Hb–173/4
Eb–755/60
Se–Bi–H . . . . 19 (Bi): Hb–174
Se–Bi–Pb . . . . 47 (Pb): Hb/C3–971/2
Se–Bi–Sb . . . . 19 (Bi): Eb–861
Se–Bi–Sb–Te . . . . 19 (Bi): Eb–865/6

See the "Instructions for the user"
at the beginning of the volume

Se . . . . Selenium systems:

	Se–Co	58 (Co):	Eb/A–658/9
	Se–Cr	52 (Cr):	Hb/B–337/8
	Se–Cr–Te	52 (Cr):	Hb/B–343
	Se–Cu	60 (Cu):	Hb/B1–600
	Se–Cu–Fe	60 (Cu):	Hb/B3–1297
	Se–Fe	59 (Fe):	Hb/A6–1171/2
	Se–Ga–Pb	47 (Pb):	Hb/C4–1306
	Se–Hg	34 (Hg):	Hb/B3–1046/7
	Se–$HgBr_2$	34 (Hg):	Hb/B2–757
	Se–I	10 (Se):	Hb/B–157/8
	Se–I–S	10 (Se):	Hb/B–193
	Se–Na	21 (Na):	Hb–634
	Se–Nb	49 (Nb):	Hb/B1–279/81
	Se–Ni	57 (Ni):	Hb/B2–740/3
	Se–P	16 (P):	Hb/C–605/6
	Se–Pb	47 (Pb):	Hb/C2–608/9
	Se–Pb–S	47 (Pb):	Hb/C2–639/40
	Se–Pb–Sb	47 (Pb):	Hb/C3–940
	Se–Pb–Sn–Te	47 (Pb):	Hb/C4–1594/5
	Se–S	10 (Se):	Hb/B–160/70
	Se–S–Te	11 (Te):	Hb–363
	Se–S_2Cl_2	10 (Se):	Hb/B–193
	Se–Sb	18 (Sb):	Hb/B3–550/2
	Se–Ta	50 (Ta):	Hb/B1–179/80
	Se–Te	11 (Te):	Hb–361/2
	Se–Th	44 (Th):	Hb–294
	Se–Ti	41 (Ti):	Hb–352
	Se–Tl	38 (Tl):	Hb/3–370/1
	Se–V	48 (V):	Hb/B1–308/9
	Se–Zn	32 (Zn):	Eb–967
Se^{+}	Se^{+}	10 (Se):	Hb/A1–144, 215/6 Hb/A2–300/1
Se^{2+}	Se^{2+}	10 (Se):	Hb/A1–144, 216/7 Hb/A2–300/1
Se^{3+}	Se^{3+}	10 (Se):	Hb/A1–144, 217/8 Hb/A2–300/1
Se^{4+}	Se^{4+}	10 (Se):	Hb/A1–145/6, 218, 234 Hb/A2–300/1
Se^{5+}	Se^{5+}	10 (Se):	Hb/A1–218/9 Hb/A2–300/1
Se^{6+}	Se^{6+}:		
	Increments	10 (Se):	Hb/A1–206, 234
	Ionization potential	10 (Se):	Hb/A2–300/1
	Properties of the ion	10 (Se):	Hb/A1–145/6
	Spectra	10 (Se):	Hb/A1–219
Se^{7+}	Se^{7+}	10 (Se):	Hb/A2–300/1

Se^- . . . . Se^- . . . . 10 (Se): Hb/A1–144
Se^{2-} . . . . Se^{2-}:
Chemical reactions . . . . 10 (Se): Hb/B–17/9
Electrochemical behavior . . . . 10 (Se): Hb/A1–236
Ionic refraction . . . . 10 (Se): Hb/A1–206
Properties of the ion . . . . 10 (Se): Hb/A1–143/4, 146
Hb/B–7

SeSi. . . . . SiSe . . . . 15 (Si): Hb/B–754/5
SeSn . . . . SnSe solid solutions:
SnSe–PbTe . . . . 47 (Pb): Hb/C4–1595
SnSe–PbTe–SnTe . . . . 47 (Pb): Hb/C4–1595

–. . . . . SnSe systems:
$SnSe–Ag_2Se–SnSe_2$ . . . . 61 (Ag): Hb/B4–249/50
SnSe–PbS. . . . . 47 (Pb): Hb/C4–1588
SnSe–PbSe . . . . 47 (Pb): Hb/C4–1586/7
SnSe–PbTe–SnTe . . . . 47 (Pb): Hb/C4–1594/5

SeSr . . . . SrSe . . . . 29 (Sr): Hb–185/6
Eb–260/1

–. . . . . SrSe systems:
$SrSe–SrCl_2–SrO$ . . . . 29 (Sr): Eb–261

SeTa . . . . TaSe . . . . 50 (Ta): Hb/B1–181/7
SeTaTe . . . . TaTeSe . . . . 50 (Ta): Hb/B1–192
SeTe . . . . TeSe . . . . 11 (Te): Hb–361
SeTh . . . . ThSe . . . . 44 (Th): Hb–294
SeTi. . . . . TiSe. . . . . 41 (Ti): Hb–352/3
SeTl. . . . . $TlSe = Tl_2Se \cdot Tl_2Se_3$ . . . . 38 (Tl): Hb/3–372
$SeTl_2$ . . . . Tl_2Se . . . . 38 (Tl): Hb/3–371/2
–. . . . . Tl_2Se solid solutions:
$(Tl,Ag,Cu)_2Se$ . . . . 61 (Ag): Hb/B4–477

SeV . . . . VSe . . . . 48 (V): Hb/B1–308, 309/11
–. . . . . VSe solid solutions:
(V,Cr)Se . . . . 48 (V): Hb/B1–310

–. . . . . VSe systems:
VSe–V . . . . 48 (V): Hb/B1–308

SeZn . . . . ZnSe . . . . 32 (Zn): Hb–246
Eb–967/8

–. . . . . ZnSe solid solutions:
$Zn_{1-x}Hg_xS_ySe_{1-y}$ . . . . 34 (Hg): Hb/B4–1502
ZnSe–HgSe . . . . 34 (Hg): Hb/B4–1502

Se_2 . . . . Se_2 . . . . 10 (Se): Hb/A1–150/5
Se_2Si . . . . $SiSe_2$ . . . . 15 (Si): Hb/B–755/6
–. . . . . $SiSe_2$ systems:
$SiSe_2–Ag_2Se$ . . . . 61 (Ag): Hb/B3–388

Se_2Sn. . . . . $SnSe_2$ systems:
$SnSe_2–Ag_2Se$ . . . . 61 (Ag): Hb/B4–249
$SnSe_2–Ag_2Se–SnSe$ . . . . 61 (Ag): Hb/B4–249/50

See the "Instructions for the user"
at the beginning of the volume

Se_2Ta . . . . $TaSe_2$ . . . . 50 (Ta): Hb/B1–181/7
Se_2Ta_{x+1} . . . . . . $Ta_{1+x}Se_2$ . . . . 50 (Ta): Hb/B1–181/7
Se_2Th . . . . $ThSe_2$ . . . . 44 (Th): Hb–294
Se_2Ti . . . . $TiSe_2$ . . . . 41 (Ti): Hb–352
Se_2Ti_x . . . . Ti_xSe_2 . . . . 41 (Ti): Hb–353
Se_2U . . . . USe_2 . . . . 55 (U): Hb–157
Se_2V . . . . VSe_2 . . . . 48 (V): Hb/B1–309, 312/3
Se_2W . . . . WSe_2 . . . . 54 (W): Hb–186
Se_2Zr . . . . $ZrSe_2$ . . . . 42 (Zr): Hb–354
Se_3 . . . . Se_3 . . . . 10 (Se): Hb/A1–150
Se_3Sm_2 . . . . Sm_2Se_3 solid solutions:
Sm_2Se_3–Ag_2Se . . . . 61 (Ag): Hb/B4–234/5
Sm_2Se_3–PbSe. . . . . 47 (Pb): Hb/C4–1374

Se_3Ta . . . . $TaSe_3$ . . . . 50 (Ta): Hb/B1–179/80, 187/9
Se_3Ta_2 . . . . Ta_2Se_3 . . . . 50 (Ta): Hb/B1–182, 187
Se_3Tb_2 . . . . Tb_2Se_3 solid solutions:
Tb_2Se_3–Ag_2Se . . . . 61 (Ag): Hb/B4–234/5
Tb_2Se_3–PbSe . . . . 47 (Pb): Hb/C4–1374/5

Se_3Th_2 . . . . Th_2Se_3 . . . . 44 (Th): Hb–294
Se_3Tl_2 . . . . Tl_2Se_3. 38 (Tl): Hb/3–372/3
Se_3Tm_2 . . . . Tm_2Se_3 solid solutions:
Tm_2Se_3–Ag_2Se . . . . 61 (Ag): Hb/B4–234/5
Tm_2Se_3–PbSe. . . . . 47 (Pb): Hb/C4–1374/5

Se_3U_2 . . . . U_2Se_3 . . . . 55 (U): Hb–157
Se_3V_2 . . . . V_2Se_3 . . . . 48 (V): Hb/B1–308, 311/2
Se_3W . . . . WSe_3 . . . . 54 (W): Hb–186
Se_3Y_2 . . . . Y_2Se_3 solid solutions:
Y_2Se_3–Ag_2Se . . . . 61 (Ag): Hb/B4–234/5

Se_3Yb_2 . . . . Yb_2Se_3 solid solutions:
Yb_2Se_3–Ag_2Se . . . . 61 (Ag): Hb/B4–234/5
Yb_2Se_3–PbSe . . . . 47 (Pb): Hb/C4–1374/5

Se_4 . . . . Se_4 . . . . 10 (Se): Hb/A1–151
Se_4TaTl_3 . . . . Tl_3TaSe_4 . . . . 50 (Ta): Hb/B2–319
Se_4Ti_2V . . . . Ti_2VSe_4 . . . . 48 (V): Hb/B2–561
Se_4Tl_3V . . . . Tl_3VSe_4 . . . . 48 (V): Hb/B2–551
Se_4Tl_4 . . . . $Tl_2Se \cdot Tl_2Se_3$ = TlSe . . . . 38 (Tl): Hb/3–372
Se_4V_3 . . . . V_3Se_4 . . . . 48 (V): Hb/B1–308, 311
Se_4V_5 . . . . V_5Se_4 . . . . 48 (V): Hb/B1–308, 309
Se_5Tl_2 . . . . Tl_2Se_5. 38 (Tl): Hb/3–373
Se_6 . . . . Se_6 . . . . 10 (Se): Hb/A1–150/1, 155
Se_8 . . . . Se_8 . . . . 10 (Se): Hb/A1–150/1
Se_8V_5 . . . . V_5Se_8 . . . . 48 (V): Hb/B1–308, 312
Se_{10} . . . . Se_{10} . . . . 10 (Se): Hb/A1–150
$Se_{12}Th_7$ . . . . Th_7Se_{12} . . . . 44 (Th): Hb–294
$Se_{32}V_{29}$ . . . . $V_{29}Se_{32}$ . . . . 48 (V): Hb/B1–308, 311
Se_xTi . . . . $TiSe_x$ . . . . 41 (Ti): Hb–352/3
Se_xV . . . . VSe_x . . . . 48 (V): Hb/B1–308/13

Hb = Main Series Volume; GD = Gmelin-Durrer; Ab = Appended Volume to Main Series;
Eb = Supplement Volume to Main Series; Erg. W. = New Supplement Series Volume

Si . . . . Silicon:

Absorption of light	15 (Si):	Hb/B–78
– of X-rays.	15 (Si):	Hb/B–91
Atom.	15 (Si):	Hb/B–17/33
Atomic nucleus	15 (Si):	Hb/B–16/7
Atomic radius	15 (Si):	Hb/B–19/23
Atomic scattering factors.	15 (Si):	Hb/B–46/8
Atomic volume	15 (Si):	Hb/B–26/7
Atomic weight	15 (Si):	Hb/B–27/31
Behavior as electrode	15 (Si):	Hb/B–115/6
Boiling point	15 (Si):	Hb/B–67
Brittleness.	15 (Si):	Hb/B–52
Catalytic properties	15 (Si):	Hb/B–130
Cells	15 (Si):	Hb/B–112/4
Chemical constant	15 (Si):	Hb/B–76
Chemical reactions	15 (Si):	Hb/B–116/30, 922
Cleavage.	15 (Si):	Hb/B–37
Colloidal forms	15 (Si):	Hb/B–9/10
Color.	15 (Si):	Hb/B–76
Compressibility	15 (Si):	Hb/B–57/8
Critical constants	15 (Si):	Hb/B–62/3
Crystal form	15 (Si):	Hb/B–35/7
Crystallographic properties	15 (Si):	Hb/B–34/52
Debye temperature	15 (Si):	Hb/B–74/5
Deformation.	15 (Si):	Hb/B–51/2
Density	15 (Si):	Hb/B–53/4
Detection	15 (Si):	Hb/B–135/6
– in iron	59 (Fe):	Hb/FI–314
– in special substances.	15 (Si):	Hb/B–136/49
Determination in iron.	59 (Fe):	Hb/FI–314/40 Hb/FII2–308/10, 345/7, 369
– in special substances.	15 (Si):	Hb/B–149/226
Dielectric constant	15 (Si):	Hb/B–110
Diffusion.	15 (Si):	Hb/B–59/62
Economics.	15 (Si):	Hb/B–11
Elasticity.	15 (Si):	Hb/B–56
Electrical conductivity	15 (Si):	Hb/B–93/104, 922/3
Electrical properties	15 (Si):	Hb/B–93/111, 921/3
Electrochemical behavior	15 (Si):	Hb/B–111/6
Electrolytic deposition	15 (Si):	Hb/B–116
Electron affinity	15 (Si):	Hb/B–110
Electron configuration	15 (Si):	Hb/B–17
Electronegativity	15 (Si):	Hb/B–32/3
Electron emission.	15 (Si):	Hb/B–108/10
Enthalpy	15 (Si):	Hb/B–75
Entropy	15 (Si):	Hb/B–75/6
Equation of state	15 (Si):	Hb/B–62
Expansion coefficient.	15 (Si):	Hb/B–63/5

See the "Instructions for the user" at the beginning of the volume

Si . . . . Silicon:

Formability	15 (Si):	Hb/B–56
Formation	15 (Si):	Hb/B–1/2
Free energy	15 (Si):	Hb/B–75
Galvanomagnetic effects	15 (Si):	Hb/B–104/5
Hall effect	15 (Si):	Hb/B–104/5
Hardness	15 (Si):	Hb/B–55/6
Heat of fusion	15 (Si):	Hb/B–69/70
– of sublimation	15 (Si):	Hb/B–68/9
– of vaporization	15 (Si):	Hb/B–68/9
Interfacial tension	15 (Si):	Hb/B–58/9
Internal pressure	15 (Si):	Hb/B–58
Ionization potentials	15 (Si):	Hb/B–110/1
Isomorphism	15 (Si):	Hb/B–38
Lattice structure	15 (Si):	Hb/B–38/42
Magnetic moment	15 (Si):	Hb/B–18/9
Magnetic properties	15 (Si):	Hb/B–92/3
Mechanical properties	15 (Si):	Hb/B–53/62
Melting point	15 (Si):	Hb/B–67/8
Modifications	15 (Si):	Hb/B–34/5
Molecule	15 (Si):	Hb/B–33/4
Nuclear properties	15 (Si):	Hb/B–16/7
Optical constants	15 (Si):	Hb/B–76/8
Optical properties	15 (Si):	Hb/B–76/91
Parachor	15 (Si):	Hb/B–59
Photoelectric conductivity	15 (Si):	Hb/B–103/4
Plasticity	15 (Si):	Hb/B–56
Polarizability	15 (Si):	Hb/B–17/8
Polymorphism	15 (Si):	Hb/B–34/5
Position in the electrochemical series	15 (Si):	Hb/B–111
– in the periodic system	15 (Si):	Hb/B–31
Potentials	15 (Si):	Hb/B–112
Preparation	15 (Si):	Hb/B–2/14
–, industrial	15 (Si):	Hb/B–10/4
– in the purest state	15 (Si):	Hb/B–7
– of special forms	15 (Si):	Hb/B–7/10
Purification	15 (Si):	Hb/B–6/7, 13/4
Purity testing	15 (Si):	Hb/B–15
Scattering of X-rays	15 (Si):	Hb/B–91
Semiconductor properties	15 (Si):	Hb/B–93/102, 921/3
Silicon transistors	15 (Si):	Hb/B–921/3
Sorption	15 (Si):	Hb/B–62
Specific heat	15 (Si):	Hb/B–70/3
Spectra:		
Arc spectrum	15 (Si):	Hb/B–79/82
Band spectrum	15 (Si):	Hb/B–91
Microwave spectrum	15 (Si):	Hb/B–78

Si . . . . Silicon:

Spectra:

Spark spectra . . . . 15 (Si): Hb/B-83/8
X-ray spectrum . . . . 15 (Si): Hb/B-88/91

Standard potential . . . . 15 (Si): Hb/B-111
Strength . . . . 15 (Si): Hb/B-57
Structure . . . . 15 (Si): Hb/B-48/51
– of thin layers . . . . 15 (Si): Hb/B-42/6
Susceptibility . . . . 15 (Si): Hb/B-92/3
Thermal conductivity . . . . 15 (Si): Hb/B-76
Thermal expansion . . . . 15 (Si): Hb/B-63/5
Thermal properties . . . . 15 (Si): Hb/B-62/76
Thermoelectric effects . . . . 15 (Si): Hb/B-106/8
Thermomagnetic effects . . . . 15 (Si): Hb/B-105/6
Toxicity . . . . 15 (Si): Hb/B-130/5
Twinning . . . . 15 (Si): Hb/B-37
Uses . . . . 15 (Si): Hb/B-15
Valence . . . . 15 (Si): Hb/B-31
Vapor pressure . . . . 15 (Si): Hb/B-65/7
Vibrational frequency . . . . 15 (Si): Hb/B-73/4
Wettability . . . . 15 (Si): Hb/B-58

–. . . . . Silicon ions
see Si^{+} ... Si^{+}, etc.

–. . . . . Silicon isotopes . . . . 15 (Si): Hb/B-16/7
–. . . . . Silicon systems:

Si-Ag . . . . 61 (Ag): Hb/B3-372/4
Si-Ag-Fe . . . . 61 (Ag): Hb/B4-436
Si-Ag-Ni . . . . 61 (Ag): Hb/B4-386
Si-Al . . . . 35 (Al): Hb/A4-536/40
Si-Al-Be . . . . 35 (Al): Hb/A4-644/5
Si-Al-Ca . . . . 35 (Al): Hb/A4-681
Si-Al-Cr . . . . 35 (Al): Hb/A5-880
52 (Cr): Hb/B-895
Si-Al-Cu . . . . 60 (Cu): Hb/B3-1176
Si-Al-Cu-Fe . . . . 60 (Cu): Hb/B3-1327
Si-Al-Cu-Mg . . . . 60 (Cu): Hb/B3-1178/9
Si-Al-Cu-Mn . . . . 60 (Cu): Hb/B3-1249
Si-Al-Fe . . . . 35 (Al): Hb/A8-1334/62
Si-Al-Mg . . . . 35 (Al): Hb/A4-666/78
Si-Al-Mg-Zn . . . . 35 (Al): Hb/A5-783
Si-Al-Mn . . . . 35 (Al): Hb/A6-928/9
Si-Al-Na . . . . 35 (Al): Hb/A4-636
Si-Al-Nb . . . . 49 (Nb): Hb/B2-43/4
Si-Al-Ni . . . . 35 (Al): Hb/A6-1000
Si-Al-Sn . . . . 35 (Al): Hb/A5-853
46 (Sn): Hb/D-217/8
Si-Al-Ta . . . . 50 (Ta): Hb/B2-16/7

*See the "Instructions for the user"
at the beginning of the volume*

Si . . . . Silicon systems:

Si–Al–Zn	35 (Al):	Hb/A5–760/1
Si–As	17 (As):	Hb–471/2
Si–Au	62 (Au):	Hb/3–727
Si–B–Cr	52 (Cr):	Hb/B–436
Si–B–Nb	49 (Nb):	Hb/B1–402/3
Si–B–Ta	50 (Ta):	Hb/B1–264
Si–B–V	48 (V):	Hb/B1–357
Si–Be	26 (Be):	Hb–161
Si–Be–Cu	60 (Cu):	Hb/B3–1121/2
Si–Bi	19 (Bi):	Eb–843/4
Si–C–Cr	52 (Cr):	Hb/B–436
Si–C–Fe	59 (Fe):	Hb/A8–1761/79
Si–C–Fe–P	59 (Fe):	Hb/A8–1798
Si–C–Fe–V	59 (Fe):	Hb/A9–1936
Si–C–Ge–V	48 (V):	Hb/B2–684
Si–C–Nb	49 (Nb):	Hb/B1–403/4
Si–C–Nb–V	49 (Nb):	Hb/B2–271
Si–C–Pu	Erg. W. 4–209	
Si–C–Ta	50 (Ta):	Hb/B1–264/5
Si–Ca	28 (Ca):	Hb/B3–1029/30
Si–Ca–N	28 (Ca):	Hb/B3–1099
Si–Cd	33 (Cd):	Eb–569
Si–Co	58 (Co):	Hb/A2–384/5
		Eb/A–725/8
Si–Co–Cu	60 (Cu):	Hb/B3–1269
Si–Co–Mn	58 (Co):	Eb/A–880
Si–Co–Ni	58 (Co):	Eb/A–470
Si–Cr	52 (Cr):	Hb/B–426/8
Si–Cr–N	52 (Cr):	Hb/B–436
Si–Cr–Ni	57 (Ni):	Hb/B1–221/2
		Hb/B3–1214/5
Si–Cu	60 (Cu):	Hb/B2–883/6
Si–Cu–Fe	60 (Cu):	Hb/B3–1314
Si–Cu–H–Mg	60 (Cu):	Hb/B3–1127
Si–Cu–Mg	27 (Mg):	Hb/A4–714/5
	60 (Cu):	Hb/B3–1126
Si–Cu–Mn	60 (Cu):	Hb/B3–1248
Si–Cu–Ni	60 (Cu):	Hb/B3–1258
Si–Cu–P	60 (Cu):	Hb/B2–933
Si–Cu–Th	60 (Cu):	Hb/B3–1192
Si–Cu–Ti	60 (Cu):	Hb/B3–1190
Si–Cu–Zn	60 (Cu):	Hb/B3–1150
Si–Cu–Zr	60 (Cu):	Hb/B3–1191
Si–Fe	59 (Fe):	Hb/A8–1740/50
Si–Fe–Mg	59 (Fe):	Hb/A9–1821/2
Si–Fe–Ni–O	59 (Fe):	Hb/A5–943
Si–Fe–O	59 (Fe):	Hb/A5–876/81

Si . . . . Silicon systems:

Si–Fe–P . . . . 59 (Fe): Hb/A8–1797/8
Si–Fe–Ti . . . . 59 (Fe): Hb/A9–1888/93
Si–Ga–O–V . . . . 48 (V): Hb/B2–592
Si–Ga–V . . . . 48 (V): Hb/B2–592
Si–Ge . . . . 45 (Ge): Eb–557
Si–Ge–V . . . . 48 (V): Hb/B2–683/4
Si–Hg . . . . 34 (Hg): Hb/B4–1355/6
Si–Li. . . . . 20 (Li): Eb–513
Si–Mg . . . . 27 (Mg): Hb/A3–374/8
Si–Mg–Zn . . . . 27 (Mg): Hb/A4–520/1
Si–Mn–Ni. . . . . 57 (Ni): Hb/B1–310/1
Hb/B3–1240/1
Si–Mn–O . . . . 59 (Fe): Hb/A5–941/2
Si–Mo–Ni. . . . . 57 (Ni): Hb/B1–250
Hb/B3–1222/3
Si–N–Nb . . . . 49 (Nb): Hb/B1–402
Si–N–Ta . . . . 50 (Ta): Hb/B1–263/4
Si–Nb . . . . 49 (Nb): Hb/B1–389/95
Si–Nb–Pu. . . . . Erg. W. 4–247/8
Si–Nb–Ta. . . . . 49 (Nb): Hb/B2–104
Si–Nb–V . . . . 49 (Nb): Hb/B2–268
Si–NbB_2–Nb_5Si_3 . . . . 49 (Nb): Hb/B1–402/3
Si–Ni . . . . 57 (Ni): Hb/B3–914/9
Si–Ni–U. . . . . 57 (Ni): Hb/B1–277
Si–Ni–V. . . . . 57 (Ni): Hb/B1–159
Si–Ni–W . . . . 57 (Ni): Hb/B1–269
Si–O–Ta . . . . 50 (Ta): Hb/B1–263
Si–P . . . . 16 (P): Hb/C–635/6
Si–Pb . . . . 47 (Pb): Hb/C3–807/8
Si–Pb–S . . . . 47 (Pb): Hb/C3–855
Si–Pb–Sb. . . . . 47 (Pb): Hb/C3–941
Si–Pd . . . . 65 (Pd): Hb/2–302
Si–Pt. . . . . 68 (Pt): Hb/C1–128/31
Si–Pu . . . . Erg. W. 4–245/6
Si–Ru . . . . 63 (Ru): Eb–331/2
Si–Sb . . . . 18 (Sb): Hb/B3–560
Si–Ta . . . . 50 (Ta): Hb/B1–250/5
Si–Ta–Ti . . . . 50 (Ta): Hb/B2–39
Si–Ta–V. . . . . 50 (Ta): Hb/B2–77/8
Si–Ti. . . . . 41 (Ti): Hb–375
Si–Ti–V . . . . 48 (V): Hb/B2–659/61
Si–Tl. . . . . 38 (Tl): Hb/3–424
Si–V . . . . 48 (V): Hb/B1–352/4
Si–Zn . . . . 32 (Zn): Hb–272
Si–Zr . . . . 42 (Zr): Hb–379

Si^+ . . . . Si^+ . . . . 15 (Si): Hb/B–17/9, 23/4, 27, 83/4, 110/1

See the "Instructions for the user"
at the beginning of the volume

Si^{2+} . . . . Si^{2+} . . . . 15 (Si): Hb/B–17/9, 23/4, 27, 46/7, 84/5, 110/1

Si^{3+} . . . . Si^{3+} . . . . 15 (Si): Hb/B–17/9, 23/4, 27, 85/6, 110/1

Si^{4+} . . . . Si^{4+}:
- Diffusion . . . . 15 (Si): Hb/B–62
- Ionic polarization . . . . 15 (Si): Hb/B–111
- Ionic refraction . . . . 15 (Si): Hb/B–78
- Ionic susceptibility . . . . 15 (Si): Hb/B–92/3
- Ionization potential . . . . 15 (Si): Hb/B–110/1
- Parachor . . . . 15 (Si): Hb/B–59
- Properties of the ion . . . . 15 (Si): Hb/B–17/9, 23/7
- Scattering factors . . . . 15 (Si): Hb/B–46/8
- Spectra . . . . 15 (Si): Hb/B–86

Si^{n+} . . . . Si^{n+} (n = 5 to 14) . . . . 15 (Si): Hb/B–17, 19, 86/8, 110/1

Si^{-} . . . . Si^{-} . . . . 15 (Si): Hb/B–27, 110
Si^{2-} . . . . Si^{2-} . . . . 15 (Si): Hb/B–27, 110
Si^{3-} . . . . Si^{3-} . . . . 15 (Si): Hb/B–27, 110
Si^{4-} . . . . Si^{4-} . . . . 15 (Si): Hb/B–17/8, 23, 25/7, 78, 92/3

$SiTa_2$ . . . . Ta_2Si . . . . 50 (Ta): Hb/A2–514
Hb/B1–251/2, 256/7

–. . . . . Ta_2Si solid solutions:
- $Ta_2(Si,Al)$ . . . . 50 (Ta): Hb/B2–17

$SiTa_3$ . . . . Ta_3Si . . . . 50 (Ta): Hb/B1–251/2, 256
$SiTa_4$ . . . . Ta_4Si . . . . 50 (Ta): Hb/A2–514
Hb/B1–252, 254

–. . . . . Ta_4Si solid solutions:
- Ta_4Si-Nb_4Si . . . . 50 (Ta): Hb/B2–104

$SiTa_{4.5}$ . . . . $Ta_{4.5}Si$ . . . . 50 (Ta): Hb/B1–251/2, 255/6
SiTe . . . . SiTe . . . . 15 (Si): Hb/B–756/7
$SiTi_{0.3}V_{2.7}$ . . . . $(V_{0.9}Ti_{0.1})_3Si$ . . . . 48 (V): Hb/B2–660/1
SiTi . . . . TiSi . . . . 41 (Ti): Hb–375
$SiTi_2$ . . . . Ti_2Si . . . . 41 (Ti): Hb–375
SiV_2 . . . . V_2Si . . . . 48 (V): Hb/B1–355
$SiV_{2.7}Zr_{0.3}$ . . . . $(V_{0.9}Zr_{0.1})_3Si$ . . . . 48 (V): Hb/B2–674
SiV_3 . . . . V_3Si . . . . 48 (V): Hb/A2–381
Hb/B1–352/3, 354/5

–. . . . . V_3Si solid solutions:
- $(V,Nb)_3Si$ . . . . 49 (Nb): Hb/B2–268/70
- V_3Si-V_3Ga . . . . 48 (V): Hb/B2–592
- $V_3(Si,Ge)$ . . . . 48 (V): Hb/B2–683/4

–. . . . . V_3Si systems:
- V_3Si-Cr_3Si . . . . 52 (Cr): Hb/B–941
- V_3Si-Nb_3Sn . . . . 49 (Nb): Hb/B2–307

SiZr . . . . ZrSi . . . . 42 (Zr): Hb–379

Hb = Main Series Volume; GD = Gmelin-Durrer; Ab = Appended Volume to Main Series; Eb = Supplement Volume to Main Series; Erg. W. = New Supplement Series Volume

Si_2	Si_2	15 (Si):	Hb/B–33/4, 91
Si_2Sr	$SrSi_2$	29 (Sr):	Hb–209 Eb–280
Si_2Sr_2	Sr_2Si_2	29 (Sr):	Eb–280
Si_2Ta	$TaSi_2$	50 (Ta):	Hb/A2–513/4 Hb/B1–251/2, 259/63
–	$TaSi_2$ solid solutions:		
	$TaSi_2–NbSi_2$	50 (Ta):	Hb/B2–104
	$Ta(Si,Al)_2$	50 (Ta):	Hb/B2–16/7
	$Ta(Si,Be)_2$	50 (Ta):	Hb/B2–10
	$Ta(Ge,Si)_2$	50 (Ta):	Hb/B2–65
	$(Ta,V)Si_2$	50 (Ta):	Hb/B2–77/8
–	$TaSi_2$ systems:		
	$TaSi_2–CrSi_2$	52 (Cr):	Hb/B–942
	$TaSi_2–CrSi_2–TiSi_2$	52 (Cr):	Hb/B–942
Si_2Th	$ThSi_2$	44 (Th):	Hb–312/3
Si_2Ti	$TiSi_2$	41 (Ti):	Hb–375/6
–	$TiSi_2$ systems:		
	$TiSi_2–CrSi_2$	52 (Cr):	Hb/B–905
	$TiSi_2–CrSi_2–TaSi_2$	52 (Cr):	Hb/B–942
	$TiSi_2–VSi_2$	48 (V):	Hb/B2–562
Si_2U	USi_2	55 (U):	Hb–179
Si_2V	VSi_2	48 (V):	Hb/A2–381, 517, 709 Hb/B1–352/3, 356/7
–	VSi_2 solid solutions:		
	$(V,Nb)Si_2$	49 (Nb):	Hb/B2–268/70
	$(V,Ta)Si_2$	50 (Ta):	Hb/B2–77/8
	$V(Si,Ge)_2$	48 (V):	Hb/B2–683/4
–	VSi_2 systems:		
	$VSi_2–TiSi_2$	48 (V):	Hb/B2–562
	$VSi_2–ZrSi_2$	48 (V):	Hb/B2–563
Si_2V_3	V_3Si_2	48 (V):	Hb/B1–352/3, 356
Si_2W	WSi_2	54 (W):	Hb–205
Si_2Zr	$ZrSi_2$	42 (Zr):	Hb–379/81
–	$ZrSi_2$ systems:		
	$ZrSi_2–CrSi_2$	52 (Cr):	Hb/B–906
	$ZrSi_2–VSi_2$	48 (V):	Hb/B2–563
Si_3Ta_5	Ta_5Si_3	50 (Ta):	Hb/A2–514 Hb/B1–251/2, 257/9
–	Ta_5Si_3 solid solutions:		
	$Ta_5Si_3–Nb_5Si_3$	50 (Ta):	Hb/B2–104
	$Ta_5(Si,Al)_3$	50 (Ta):	Hb/B2–16/7
	$(Ta,Ti)_5Si_3$	50 (Ta):	Hb/B2–39
	$(Ta,V)_5Si_3$	50 (Ta):	Hb/B2–77/8
Si_3Ti_2	Ti_2Si_3	41 (Ti):	Hb–375
Si_3Ti_5	Ti_5Si_3 solid solutions:		
	$(Ta,Ti)_5Si_3$	50 (Ta):	Hb/B2–39

*See the "Instructions for the user"
at the beginning of the volume*

Si_3V_5	V_5Si_3	48 (V):	Hb/A2–381 Hb/B1–352/3, 356
–	V_5Si_3 solid solutions:		
	$(V,Nb)_5Si_3$	49 (Nb):	Hb/B2–268/70
	$(V,Ta)_5Si_3$	50 (Ta):	Hb/B2–77/8
	$V_5(Si,Ge)_3$	48 (V):	Hb/B2–683
–	V_5Si_3 systems:		
	V_5Si_3–V_5B_3	48 (V):	Hb/B1–357
	V_5Si_3–Zr_5Si_3	48 (V):	Hb/B2–563
Si_3W	WSi_3	54 (W):	Hb–205
Si_3W_2	W_2Si_3	54 (W):	Hb–204/5
Si_3Zr_5	Zr_5Si_3 systems:		
	Zr_5Si_3–V_5Si_3	48 (V):	Hb/B2–563
$Si_{12}Ta$	$TaSi_{12}$ solid solutions:		
	$Ta(Be,Si)_{12}$	50 (Ta):	Hb/B2–10
Sm	Samarium:		
	Abundance	39 (SE):	Hb/A1–34/5
	Band model	39 (SE):	Hb/B3–213/5
	Boiling point	39 (SE):	Hb/B3–105/6
	Characteristic electron energy losses	39 (SE):	Hb/B3–225/6
	Charge carrier concentration	39 (SE):	Hb/B3–228
	Cosmochemistry	39 (SE):	Hb/A1–23/7
	Crystal chemistry	39 (SE):	Hb/A1–28/32
	Crystallographic properties	39 (SE):	Hb/B3–61, 66/70, 76/9
	Crystal structure	39 (SE):	Hb/B3–66/70
	Debye temperature	39 (SE):	Hb/B3–187
	Density	39 (SE):	Hb/B3–85/7
	Diffusion	39 (SE):	Hb/B3–77/8
	Disorder	39 (SE):	Hb/B3–76
	Economic deposits	39 (SE):	Hb/A1–110/22
	Elasticity	39 (SE):	Hb/B3–92/102
	Electrical properties	39 (SE):	Hb/B3–209/83
	Electrical resistance	39 (SE):	Hb/B3–236/7
	Electron density	39 (SE):	Hb/B3–216
	Electron emission	39 (SE):	Hb/B3–278/83
	Enthalpy of fusion	39 (SE):	Hb/B3–120
	– of sublimation	39 (SE):	Hb/B3–110, 112/3
	– of transformation	39 (SE):	Hb/B3–118
	– of vaporization	39 (SE):	Hb/B3–110, 112/3
	Equation of state	39 (SE):	Hb/B3–83/4
	Geochemistry	39 (SE):	Hb/A1–28/54
	Grüneisen constant	39 (SE):	Hb/B3–84/5
	Hall effect	39 (SE):	Hb/B3–266
	Hardness	39 (SE):	Hb/B3–103
	History	39 (SE):	Hb/A1–4/22

Hb = Main Series Volume; GD = Gmelin-Durrer; Ab = Appended Volume to Main Series; Eb = Supplement Volume to Main Series; Erg. W. = New Supplement Series Volume

Sm	Samarium:		
	Hyperfine interactions	39 (SE):	Hb/B3–221
	Isomer shift	39 (SE):	Hb/B3–216
	Magnetic properties	39 (SE):	Hb/B3–294/6
	Mechanical properties	39 (SE):	Hb/B3–85/103, 193/5, 207/9
	Melting point	39 (SE):	Hb/B3–118/9
	Mineralogy	39 (SE):	Hb/A1–54/110
	Molar volume	39 (SE):	Hb/B3–85/7
	Occurrence, extraterrestrial	39 (SE):	Hb/A1–22/7
	–, terrestrial	39 (SE):	Hb/A1–27/54
	– in the biosphere	39 (SE):	Hb/A1–53/4
	– in the lithosphere	39 (SE):	Hb/A1–27/54
	Optical properties	39 (SE):	Hb/B3–339
	Polymorphism	39 (SE):	Hb/B3–61
	Position in the periodic system	39 (SE):	Hb/A1–2/4
	Radioactivity	39 (SE):	Hb/A1–22
	Strength	39 (SE):	Hb/B3–102
	Structure	39 (SE):	Hb/B3–79
	Surface tension	39 (SE):	Hb/B3–208/9
	Susceptibility	39 (SE):	Hb/B3–294/5
	Thermal conductivity	39 (SE):	Hb/B3–198/9
	Thermal expansion	39 (SE):	Hb/B3–88/92
	Thermal properties	39 (SE):	Hb/B3–83/92, 103/6, 140/6, 187, 198/9
	Thermodynamic functions	39 (SE):	Hb/B3–140/5
	Thermoelectric power	39 (SE):	Hb/B3–274
	Twinning	39 (SE):	Hb/B3–78
	Ultrasound absorption	39 (SE):	Hb/B3–193
	Valence	39 (SE):	Hb/A1–4
	Vapor pressure	39 (SE):	Hb/B3–103/6
–	Samarium ions	39 (SE):	Hb/A1–3
			Hb/B3–284
		50 (Ta):	Hb/B2–204
–	Samarium systems:		
	Sm–Ag	61 (Ag):	Hb/C–220/1
	Sm–H	39 (SE):	Hb/C1–2/3, 58/9
	Sm–D	39 (SE):	Hb/C1–34, 59
	Sm–N–O	39 (SE):	Hb/C2–180
	Sm–O	39 (SE):	Hb/C1–313/4
	Sm–Pb	47 (Pb):	Hb/C4–1367
	Sm–Pb–Zn	47 (Pb):	Hb/C4–1363, 1367
	Sm–Sm_2O_3	39 (SE):	Hb/C1–313/4
	Sm–Sn	46 (Sn):	Hb/D–312/3
	Sm–V	48 (V):	Hb/B2–600/1
Sm^{2+}	Sm^{2+}	39 (SE):	Hb/A1–3
Sm^{3+}	Sm^{3+}	39 (SE):	Hb/A1–3
			Hb/B3–284
		50 (Ta):	Hb/B2–204

See the "Instructions for the user"
at the beginning of the volume

$SmSn_3$ . . . . $SmSn_3$ . . . . 46 (Sn): Hb/D–312/5
Sm_2Sn_3 . . . . Sm_2Sn_3 . . . . 46 (Sn): Hb/D–312/5
Sm_4Sn_3 . . . . Sm_4Sn_3 . . . . 46 (Sn): Hb/D–312/5
Sm_5Sn_3 . . . . Sm_5Sn_3 . . . . 46 (Sn): Hb/D–312/5
Sm_5Sn_4 . . . . Sm_5Sn_4 . . . . 46 (Sn): Hb/D–312/3
$Sm_{11}Sn_{10}$ . . . . $Sm_{11}Sn_{10}$ . . . . 46 (Sn): Hb/D–312/5

$Sn_{0.55}Zr$ . . . . $ZrSn_{0.55}$ . . . . 46 (Sn): Hb/D–342, 346
Sn . . . . Tin:
Abundance . . . . 46 (Sn): Hb/A–214/5
Anodic polishing . . . . 46 (Sn): Hb/B–293
Atom . . . . 46 (Sn): Hb/B–62/129
Atomic nucleus . . . . 46 (Sn): Hb/B–55/62
Atomic radius . . . . 46 (Sn): Hb/B–100/1
Atomic refraction . . . . 46 (Sn): Hb/B–129
Atomic scattering factors . . . . 46 (Sn): Hb/B–71/2
Atomic susceptibility . . . . 46 (Sn): Hb/B–101
Atomic volume . . . . 46 (Sn): Hb/B–100/1
Atomic weight . . . . 46 (Sn): Hb/B–1/4
Behavior as anode . . . . 46 (Sn): Hb/B–285/93
– as cathode . . . . 46 (Sn): Hb/B–281/5
Boiling point . . . . 46 (Sn): Hb/B–177
Bond properties . . . . 46 (Sn): Hb/C1–1/4
Characteristic electron
energy losses . . . . 46 (Sn): Hb/B–196/7
Chemical reactions . . . . 46 (Sn): Hb/B–347/82
– with acids . . . . 46 (Sn): Hb/B–365/75
– with alkaline solutions . . . . 46 (Sn): Hb/B–364/5
– with aqueous salt solutions . . . . . 46 (Sn): Hb/B–375/81
– with metal compounds . . . . 46 (Sn): Hb/B–360/4
– with metals . . . . 46 (Sn): Hb/B–355
– with nonmetal compounds . . . . . 46 (Sn): Hb/B–355/60
– with nonmetals . . . . 46 (Sn): Hb/B–347/55
– with organic compounds . . . . 46 (Sn): Hb/B–374/5, 381/2
Colloidal forms . . . . 46 (Sn): Hb/B–41/3
Compressibility . . . . 46 (Sn): Hb/B–167/8
Condensation . . . . 46 (Sn): Hb/B–177
Conduction mechanism . . . . 46 (Sn): Hb/B–202/4
Constitution of vapor . . . . 46 (Sn): Hb/B–176
Cosmochemistry . . . . 46 (Sn): Hb/A–175/82
Critical constants . . . . 46 (Sn): Hb/B–177/8
Crystal chemistry . . . . 46 (Sn): Hb/A–182/214
Crystal form . . . . 46 (Sn): Hb/B–134/7
Crystallographic properties . . . . . . 46 (Sn): Hb/B–130/60
Crystal potential . . . . 46 (Sn): Hb/B–149
Debye temperature . . . . 46 (Sn): Hb/B–185/6
Deformation . . . . 46 (Sn): Hb/B–159/60
Density . . . . 46 (Sn): Hb/B–161/3

Sn. . . . . Tin:		
Deposition from melts	46 (Sn):	Hb/B–343/7
– from nonaqueous solutions	46 (Sn):	Hb/B–342/3
Detection	46 (Sn):	Hb/B–415/8
– in iron	59 (Fe):	Hb/FII1–43/4
– in special substances	46 (Sn):	Hb/B–417/8
Determination	46 (Sn):	Hb/B–418/23
– in iron	59 (Fe):	Hb/FII1–44/53 Hb/FII2–300, 350
– in special substances	46 (Sn):	Hb/B–421/3
Diadochy	46 (Sn):	Hb/A–186/92
Diffusion, self-diffusion	46 (Sn):	Hb/B–152/4
– in aluminum	46 (Sn):	Hb/D–208
– in antimony	46 (Sn):	Hb/D–1
– in bismuth	46 (Sn):	Hb/D–23/5
– in cadmium	46 (Sn):	Hb/D–158
– in gallium	46 (Sn):	Hb/D–223
– in indium	46 (Sn):	Hb/D–234/5, 263
– in lithium	46 (Sn):	Hb/D–61
– in magnesium	46 (Sn):	Hb/D–81
– in mercury	46 (Sn):	Hb/D–186/7
– in Pb–Sn alloys	47 (Pb):	Hb/C4–1482/6, 1518
– in sodium	46 (Sn):	Hb/D–67
– in titanium	46 (Sn):	Hb/D–325
– in zinc	46 (Sn):	Hb/D–129/30
– in zirconium	46 (Sn):	Hb/D–337/9
Disorder	46 (Sn):	Hb/B–151/2
Dressing and treatment of ores	46 (Sn):	Hb/B–6/20
Economic deposits:		
Africa	46 (Sn):	Hb/A–374/91
Asia	46 (Sn):	Hb/A–393/426
Australia and Oceania	46 (Sn):	Hb/A–426/35
Europe	46 (Sn):	Hb/A–348/74, 392/3
North America	46 (Sn):	Hb/A–435/9
South America	46 (Sn):	Hb/A–439/51
Soviet Union	46 (Sn):	Hb/A–391/408
Economics	46 (Sn):	Hb/A–342/8 Hb/B–5/6
Elasticity	46 (Sn):	Hb/B–166/7
Electrical conductivity	46 (Sn):	Hb/B–205/16
Electrical properties	46 (Sn):	Hb/B–191/242
Electrochemical behavior	46 (Sn):	Hb/B–250/347
Electrodeposition	46 (Sn):	Hb/B–293/347
Electrokinetic phenomena	46 (Sn):	Hb/B–273
Electrolytic refining	46 (Sn):	Hb/B–31/7
Electron affinity	46 (Sn):	Hb/B–100
Electron configuration	46 (Sn):	Hb/B–62/71
Electron density	46 (Sn):	Hb/B–147/9

*See the "Instructions for the user"
at the beginning of the volume*

Sn . . . . Tin:

Electronegativity	46 (Sn):	Hb/B–100
Electron emission	46 (Sn):	Hb/B–240/2
Energy band structure	46 (Sn):	Hb/B–191/6
Epitaxy	46 (Sn):	Hb/B–158/9
Equation of state	46 (Sn):	Hb/B–161
Galvanomagnetic effects	46 (Sn):	Hb/B–229/35
Geochemical cycle	46 (Sn):	Hb/A–308/11
Geochemistry	46 (Sn):	Hb/A–182/311
Growing of crystals	46 (Sn):	Hb/B–137/42
Hall effect	46 (Sn):	Hb/B–229/31
Heat of fusion	46 (Sn):	Hb/B–180
– of sublimation	46 (Sn):	Hb/B–177
– of vaporization	46 (Sn):	Hb/B–177
History	46 (Sn):	Hb/A–1/174
Interaction with electrons	46 (Sn):	Hb/B–98/100
Interfacial tension	46 (Sn):	Hb/B–174/6
Internal friction	46 (Sn):	Hb/B–168/71
Ionization	46 (Sn):	Hb/B–98/9
Lattice defects	46 (Sn):	Hb/B–151/2
Lattice energy	46 (Sn):	Hb/B–151
Lattice structure	46 (Sn):	Hb/B–142/4
Lattice vibrations	46 (Sn):	Hb/B–149/51
Lilienfeld radiation	46 (Sn):	Hb/B–249/50
Magnetic properties	46 (Sn):	Hb/B–191/242
Magnetoreflection	46 (Sn):	Hb/B–250
Mechanical properties	46 (Sn):	Hb/B–161/91
Melting point	46 (Sn):	Hb/B–178/80
Mineralogy	46 (Sn):	Hb/A–311/41
Mobility of charge carriers	46 (Sn):	Hb/B–202/4
Modifications	46 (Sn):	Hb/B–130/4
Molecule	46 (Sn):	Hb/B–129/30
Names and symbols	46 (Sn):	Hb/A–3/9
Nature of bonding	46 (Sn):	Hb/B–147/9
Nuclear magnetic resonance	46 (Sn):	Hb/B–197/9
Nuclear properties	46 (Sn):	Hb/B–55/62
Occurrence, extraterrestrial	46 (Sn):	Hb/A–175/82
–, terrestrial	46 (Sn):	Hb/A–182/311
– in meteorites	46 (Sn):	Hb/A–176/82
– in minerals	46 (Sn):	Hb/A–193/214
– in tektites and impactites	46 (Sn):	Hb/A–182
– in the atmosphere	46 (Sn):	Hb/A–308
– in the hydrosphere	46 (Sn):	Hb/A–302/8
– in the lithosphere	46 (Sn):	Hb/A–215/302
– in the magmatic cycle	46 (Sn):	Hb/A–215/79
– in the metamorphic cycle	46 (Sn):	Hb/A–293/302
– in the sedimentary cycle	46 (Sn):	Hb/A–279/93
– on the moon	46 (Sn):	Hb/A–182
Optical properties	46 (Sn):	Hb/B–242/50

Sn . . . . Tin:

Optical terms . . . . 46 (Sn): Hb/B–64/71
Orientation of single crystals . . . . . 46 (Sn): Hb/B–154
Overvoltage . . . . 46 (Sn): Hb/B–273/81
Paramagnetic resonance absorption . 46 (Sn): Hb/B–199
Photoelectric electron emission . . . . 46 (Sn): Hb/B–240/2
Polymorphism . . . . 46 (Sn): Hb/B–130/4
Position in the electrochemical
series . . . . 46 (Sn): Hb/B–250
– in the periodic system . . . . 46 (Sn): Hb/B–1
Potentials . . . . 46 (Sn): Hb/B–255/73
Preparation, industrial . . . . 46 (Sn): Hb/B–4/41
–, –, historical data . . . . 46 (Sn): Hb/A–53/84
–, –, hydrometallurgy. . . . . 46 (Sn): Hb/B–18/22
–, –, ore preparation . . . . 46 (Sn): Hb/B–6/7
–, –, purification of crude tin . . . . . . 46 (Sn): Hb/B–28/38
–, –, recovery . . . . 46 (Sn): Hb/B–22/8
–, –, smelting . . . . 46 (Sn): Hb/B–7/14
–, –, volatization methods . . . . 46 (Sn): Hb/B–14/7
– in pure state. . . . . 46 (Sn): Hb/B–37/8
– of special forms. . . . . 46 (Sn): Hb/B–41/3
Purification . . . . 46 (Sn): Hb/B–28/37
Purity testing . . . . 46 (Sn): Hb/B–38/41
Recovery . . . . 46 (Sn): Hb/B–22/8
Recrystallization . . . . 46 (Sn): Hb/B–160
Reflectivity . . . . 46 (Sn): Hb/B–242/9
Refractive index. . . . . 46 (Sn): Hb/B–242/9
Self-diffusion . . . . 46 (Sn): Hb/B–152/4
Separation, analytical . . . . 46 (Sn): Hb/B–420/1
Slip. . . . . 46 (Sn): Hb/B–159/60
Sorption . . . . 46 (Sn): Hb/C1–5
Sound absorption. . . . . 46 (Sn): Hb/B–168/71
Special forms . . . . 46 (Sn): Hb/B–41/3
Spectra:
Arc spectrum . . . . 46 (Sn): Hb/B–72/87
Spark spectra
see Sn^+ ... Sn^+, *etc.*
Terms . . . . 46 (Sn): Hb/B–64/71
X-ray spectrum . . . . 46 (Sn): Hb/B–87/98
Standard potentials. . . . . 46 (Sn): Hb/B–250/5
Structure of the melt . . . . 46 (Sn): Hb/B–144/7
– of thin films . . . . 46 (Sn): Hb/B–155/8
Superconductivity . . . . 46 (Sn): Hb/B–216/29
Surface structure . . . . 46 (Sn): Hb/B–154/5
Surface tension . . . . 46 (Sn): Hb/B–174/6
Susceptibility . . . . 46 (Sn): Hb/B–199/202
Thermal conductivity . . . . 46 (Sn): Hb/B–186/91
Thermal diffusivity . . . . 46 (Sn): Hb/B–191

See the "Instructions for the user"
at the beginning of the volume

Sn	Tin:		
	Thermal expansion	46 (Sn):	Hb/B–163/6
	Thermal properties	46 (Sn):	Hb/B–161/91
	Thermal radiation	46 (Sn):	Hb/B–249
	Thermodynamic functions	46 (Sn):	Hb/B–180/5
	Thermoelectric force	46 (Sn):	Hb/B–235/40
	Thermomagnetic effects	46 (Sn):	Hb/B–240
	Toxicity	46 (Sn):	Hb/B–411/5
	Uses, historical data	46 (Sn):	Hb/A–101/74
	Vaporization	46 (Sn):	Hb/B–176/7
	Vapor pressure	46 (Sn):	Hb/B–176
	Viscosity	46 (Sn):	Hb/B–171/3
	Wave functions	46 (Sn):	Hb/B–62/4
–	Tin ions		
	see Sn^+ ... Sn^+, *etc.*		
–	Tin isotopes:		
	Abundance	46 (Sn):	Hb/A–175, 181, 215
			Hb/B–3, 43
	Detection and determination	46 (Sn):	Hb/B–53/5
	Handling	46 (Sn):	Hb/B–43/4
	Nuclear properties	46 (Sn):	Hb/B–55/62
	Preparation and separation	46 (Sn):	Hb/B–43/53
–	Tin systems:		
	Sn–Ag	61 (Ag):	Hb/C–266/70
	Sn–Ag–AgCl–$SnCl_2$	61 (Ag):	Hb/B4–247
	Sn–Ag–Al	61 (Ag):	Hb/C–290/1
	Sn–Ag–Cd	61 (Ag):	Hb/C–286/9
	Sn–Ag–Cd–Cu	61 (Ag):	Hb/C–406
	Sn–Ag–Cu	61 (Ag):	Hb/C–403/6
	Sn–Ag–Hg	61 (Ag):	Hb/C–289/90
	Sn–Ag–Mg	27 (Mg):	Hb/A4–751
		61 (Ag):	Hb/C–286
	Sn–Ag–Mn	61 (Ag):	Hb/C–351
	Sn–Ag–Sb	61 (Ag):	Hb/B4–252
	Sn–Al	35 (Al):	Hb/A5–844/5
		46 (Sn):	Hb/D–208/9
	Sn–Al–Bi	35 (Al):	Hb/A5–854/5
		46 (Sn):	Hb/D–218
	Sn–Al–Cd	35 (Al):	Hb/A5–858
		46 (Sn):	Hb/D–222
	Sn–Al–Cu–Sb	60 (Cu):	Hb/B3–1196
	Sn–Al–Ga–Ge–Zn	46 (Sn):	Hb/D–397
	Sn–Al–In	46 (Sn):	Hb/D–273
	Sn–Al–Mg	46 (Sn):	Hb/D–218/9
	Sn–Al–Mn	35 (Al):	Hb/A6–947/8
	Sn–Al–Nd	46 (Sn):	Hb/D–312
	Sn–Al–Ni	35 (Al):	Hb/A6–1004

Sn . . . . Tin systems:

Sn–Al–Pb	35 (Al):	Hb/A5–863/4
	47 (Pb):	Hb/C4–1529
Sn–Al–Sb	35 (Al):	Hb/A5–854
	46 (Sn):	Hb/D–218
Sn–Al–Si	35 (Al):	Hb/A5–853
	46 (Sn):	Hb/D–217/8
Sn–Al–Ti	46 (Sn):	Hb/D–333/4
Sn–Al–Ti–Zr	46 (Sn):	Hb/D–382/3
Sn–Al–Zn	35 (Al):	Hb/A5–855/7
	46 (Sn):	Hb/D–220/1
Sn–Al–Zr	46 (Sn):	Hb/D–376
Sn–As–Ga–Ge	46 (Sn):	Hb/D–397
Sn–As–Ga–In	46 (Sn):	Hb/D–277
Sn–Au	62 (Au):	Hb/3–840/1
Sn–B–Ni	57 (Ni):	Hb/B3–1193
Sn–Ba	46 (Sn):	Hb/D–126
Sn–Be	46 (Sn):	Hb/D–79
Sn–Bi	46 (Sn):	Hb/D–23/8
Sn–Bi–Bi_2Mg_3	46 (Sn):	Hb/D–115
Sn–Bi–Cd	46 (Sn):	Hb/D–178/9
Sn–Bi–Cd–Pb	47 (Pb):	Hb/C4–1525/6
Sn–Bi–Cd–Pb–Zn	47 (Pb):	Hb/C4–1528/9
Sn–Bi–Ga–Pb	47 (Pb):	Hb/C4–1529
Sn–Bi–InBi	46 (Sn):	Hb/D–267/9
Sn–Bi–Mg	46 (Sn):	Hb/D–115
Sn–Bi–Pb	47 (Pb):	Hb/C4–1518/9
Sn–Bi–Sb	46 (Sn):	Hb/D–58/9
Sn–Bi–Zn	46 (Sn):	Hb/D–155
Sn–Bi_2Mg_3–Mg_2Sn	46 (Sn):	Hb/D–115
Sn–Br–Pb	47 (Pb):	Hb/C4–1584
Sn–C–Nb	49 (Nb):	Hb/B4–330
Sn–Ca	46 (Sn):	Hb/D–117
Sn–Cd	46 (Sn):	Hb/D–159/60
Sn–Cd–Hg	46 (Sn):	Hb/D–206/7
Sn–Cd–In	46 (Sn):	Hb/D–270
Sn–Cd–In–Zn	46 (Sn):	Hb/D–271
Sn–Cd–Pb	47 (Pb):	Hb/C4–1524/5
Sn–Cd–Pb–Zn	47 (Pb):	Hb/C4–1528/9
Sn–Cd–Sb	46 (Sn):	Hb/D–177/8
Sn–Cd–Tl	46 (Sn):	Hb/D–291
Sn–Cd–Zn	46 (Sn):	Hb/D–181/2
Sn–Ce	46 (Sn):	Hb/D–302/4
Sn–Cl–Pb	47 (Pb):	Hb/C4–1583/4
Sn–Cl–Pb–S	47 (Pb):	Hb/C4–1586
Sn–Co	58 (Co):	Hb/A1–194/6
		Eb/A–409/15
Sn–Cr	52 (Cr):	Hb/A2–726/7

See the "Instructions for the user"
at the beginning of the volume

Sn . . . . Tin systems:

Sn–Cu	60 (Cu):	Hb/B3–1193
Sn–Cu–Fe–S	60 (Cu):	Hb/B3–1327
Sn–Cu–H	60 (Cu):	Hb/B3–1193/4
Sn–Cu–Mg	27 (Mg):	Hb/A4–722
Sn–Cu–P	60 (Cu):	Hb/B3–1196
Sn–Cu–S	60 (Cu):	Hb/B3–1195/6
Sn–Cu–Sb	60 (Cu):	Hb/B3–1196
Sn–Er	46 (Sn):	Hb/D–321/2
Sn–Eu	46 (Sn):	Hb/D–315
Sn–Fe	59 (Fe):	Hb/A9–1903/5
Sn–Ga	36 (Ga):	Hb–62/3
	46 (Sn):	Hb/D–223/4
Sn–Ga–Ge–In–Zn	46 (Sn):	Hb/D–397
Sn–Ga–Ge–Sb	46 (Sn):	Hb/D–397
Sn–Ga–In	46 (Sn):	Hb/D–274
Sn–Ga–In–Sb	46 (Sn):	Hb/D–277
Sn–Ga–In–Zn	46 (Sn):	Hb/D–278
Sn–Ga–Pb–Zn	47 (Pb):	Hb/C4–1529
Sn–Ga–Tl	46 (Sn):	Hb/D–291
Sn–Ga–Zn	46 (Sn):	Hb/D–231/3
Sn–GaSb	46 (Sn):	Hb/D–231
Sn–Gd	46 (Sn):	Hb/D–316
Sn–Ge	46 (Sn):	Hb/D–390/1
Sn–Ge–Mg	46 (Sn):	Hb/D–395
Sn–Ge–Ni–Pb	57 (Ni):	Hb/B1–153
Sn–H	46 (Sn):	Hb/C1–5/6
Sn–Hf	46 (Sn):	Hb/D–386
Sn–Hg	46 (Sn):	Hb/D–187/90
Sn–Hg–In	46 (Sn):	Hb/D–271/3
Sn–I	46 (Sn):	Hb/C1–444
Sn–In	46 (Sn):	Hb/D–235/6
Sn–In–Pb	47 (Pb):	Hb/C4–1529/30
Sn–In–Sb	46 (Sn):	Hb/D–264/5
Sn–Ir	68 (Pt):	Hb/A6–734
Sn–K	46 (Sn):	Hb/D–76
Sn–La	46 (Sn):	Hb/D–296
Sn–Li	46 (Sn):	Hb/D–61/2
Sn–Li–Mg	46 (Sn):	Hb/D–116
Sn–Mg	27 (Mg):	Hb/A4–617/9
	46 (Sn):	Hb/D–81/2
Sn–Mg–Pb	27 (Mg):	Hb/A4–656/7
	47 (Pb):	Hb/C4–1522
Sn–Mg–Zn	27 (Mg):	Hb/A4–635
Sn–Mn–Ni	57 (Ni):	Hb/B1–312
Sn–Mo–Ni	57 (Ni):	Hb/B1–251
Sn–Na	46 (Sn):	Hb/D–67/8
Sn–Nb	49 (Nb):	Hb/B2–208/9

Sn	Tin systems:		
	Sn–Nb–Zr	49 (Nb):	Hb/B2–235/6
	Sn–Nd	46 (Sn):	Hb/D–310
	Sn–Ni	57 (Ni):	Hb/B1–128/31
	Sn–O	46 (Sn):	Hb/C1–24/5
	Sn–Pb	47 (Pb):	Hb/C4–1487/9
	Sn–Pb–PbS–SnS	47 (Pb):	Hb/C4–1584
	Sn–Pb–S	47 (Pb):	Hb/C4–1584
	Sn–Pb–Sb	47 (Pb):	Hb/C4–1515
	Sn–Pb–Se–Te	47 (Pb):	Hb/C4–1594/5
	Sn–Pb–Te	47 (Pb):	Hb/C4–1588
	Sn–Pb–Tl	47 (Pb):	Hb/C4–1530
	Sn–Pb–Zn	47 (Pb):	Hb/C4–1522/4
	Sn–Pd	68 (Pt):	Hb/A5–586/8
	Sn–Pr	46 (Sn):	Hb/D–307
	Sn–Pt	68 (Pt):	Hb/A6–758/63
	Sn–Rh	68 (Pt):	Hb/A5–555/6
	Sn–Ru	63 (Ru):	Eb–193/4
		68 (Pt):	Hb/A5–538/9
	Sn–Sb	46 (Sn):	Hb/D–1/4
	Sn–Sb–Zn	46 (Sn):	Hb/D–153
	Sn–Sm	46 (Sn):	Hb/D–312/3
	Sn–SnO_2	46 (Sn):	Hb/C1–24/5
	Sn–Sr	46 (Sn):	Hb/D–123
	Sn–Ta	50 (Ta):	Hb/B2–65
	Sn–Tb	46 (Sn):	Hb/D–318
	Sn–Th	46 (Sn):	Hb/D–388
	Sn–Ti	46 (Sn):	Hb/D–325/7
	Sn–Ti–V	48 (V):	Hb/B2–686/7
	Sn–Ti–Zr	46 (Sn):	Hb/D–378
	Sn–Tl	46 (Sn):	Hb/D–279/81
	Sn–Tl–Zn	46 (Sn):	Hb/D–290
	Sn–V	48 (V):	Hb/B2–684/7
	Sn–Y	46 (Sn):	Hb/D–293/4
	Sn–Zn	46 (Sn):	Hb/D–130/2
	Sn–Zr	46 (Sn):	Hb/D–339/42
Sn^+	Sn^+:		
	Electrochemical behavior	46 (Sn):	Hb/B–271
	Heat capacity	46 (Sn):	Hb/B–185
	Ionic entropy	46 (Sn):	Hb/B–185
	Ionic radius	46 (Sn):	Hb/B–128
	Ionization potential	46 (Sn):	Hb/B–114
	Optical terms	46 (Sn):	Hb/B–102/4
	Spectra	46 (Sn):	Hb/B–106/9
Sn^{2+}	Sn^{2+}:		
	Bond properties	46 (Sn):	Hb/C1–1/4
	Catalytic properties	46 (Sn):	Hb/B–390/1
	Chemical reactions	46 (Sn):	Hb/B–383/91

See the "Instructions for the user"
at the beginning of the volume

Sn^{2+} . . . . Sn^{2+}:
Electrochemical behavior . . . . 46 (Sn): Hb/B–252, 271/2, 295/331, 342/7
Exchange reactions . . . . 46 (Sn): Hb/B–389/90
Heat capacity . . . . 46 (Sn): Hb/B–185
Hydration . . . . 46 (Sn): Hb/B–383/4
Hydrolysis . . . . 46 (Sn): Hb/B–384/5
Inhibitor effect . . . . 46 (Sn): Hb/B–391
Ionic activity . . . . 46 (Sn): Hb/B–384
Ionic entropy . . . . 46 (Sn): Hb/B–185
Ionic refraction . . . . 46 (Sn): Hb/B–129
Ionic susceptibility . . . . 46 (Sn): Hb/B–129
Ionization potential . . . . 46 (Sn): Hb/B–114
Optical terms . . . . 46 (Sn): Hb/B–104
Properties of the free ion . . . . 46 (Sn): Hb/B–104, 128
– of the ion in crystals . . . . 46 (Sn): Hb/B–115/23, 128/9
– – in solutions . . . . 46 (Sn): Hb/B–124/7
Redox reactions . . . . 46 (Sn): Hb/B–395/409
Sorption . . . . 46 (Sn): Hb/B–390
Spectra:
Crystal spectra . . . . 21 (Na): Eb/6–333/4, 347
46 (Sn): Hb/B–115/23
Spark spectrum . . . . 46 (Sn): Hb/B–110/1
Spectra in solutions . . . . 46 (Sn): Hb/B–124/7

Sn^{3+} . . . . Sn^{3+}:
Chemical reactions . . . . 46 (Sn): Hb/B–409
Ionization potential . . . . 46 (Sn): Hb/B–114
Optical terms . . . . 46 (Sn): Hb/B–104/5
Properties of the free ion . . . . 46 (Sn): Hb/B–104/5
– of the ion in crystals . . . . 46 (Sn): Hb/B–123/4
Spectra:
Crystal spectra . . . . 46 (Sn): Hb/B–123/4
Spark spectrum . . . . 46 (Sn): Hb/B–111/3

Sn^{4+} . . . . Sn^{4+}:
Bond properties . . . . 46 (Sn): Hb/C1–2/4
Chemical reactions . . . . 46 (Sn): Hb/B–391/5
Electrochemical behavior . . . . 46 (Sn): Hb/B–295, 331/42
Faraday effect . . . . 46 (Sn): Hb/B–129
Inhibitor effect . . . . 46 (Sn): Hb/B–394/5
Ionic activity . . . . 46 (Sn): Hb/B–392
Ionic entropy . . . . 46 (Sn): Hb/B–185
Ionic susceptibility . . . . 46 (Sn): Hb/B–129
Ionization potential . . . . 46 (Sn): Hb/B–114
Optical terms . . . . 46 (Sn): Hb/B–105
Properties of the free ion . . . . 46 (Sn): Hb/B–105, 128
– of the ion in crystals . . . . 46 (Sn): Hb/B–128/9
– – in solutions . . . . 46 (Sn): Hb/B–127/9
Redox reactions . . . . 46 (Sn): Hb/B–410/1

Sn^{4+}	Sn^{4+}:		
	Sorption	46 (Sn):	Hb/B–394
	Spectra:		
	Spark spectrum	46 (Sn):	Hb/B–113
	Spectra in solutions	46 (Sn):	Hb/B–127/8
Sn^{5+}	Sn^{5+}	46 (Sn):	Hb/B–105, 113/4
Sn^{n+}	Sn^{n+} (n = 6 to 49)	46 (Sn):	Hb/B–114
Sn^{-}	Sn^{-}	46 (Sn):	Hb/B–128
Sn^{4-}	Sn^{4-}	46 (Sn):	Hb/B–128
SnSr	SrSn	46 (Sn):	Hb/D–123/4
$SnSr_2$	Sr_2Sn	46 (Sn):	Hb/D–123/4
$SnTaV_2$	TaV_2Sn	50 (Ta):	Hb/B2–85
$SnTa_2V$	Ta_2VSn	50 (Ta):	Hb/B2–85
$SnTa_3$	Ta_3Sn	50 (Ta):	Hb/B2–65/7
–	Ta_3Sn solid solutions:		
	$(Ta,Nb)_3Sn$	50 (Ta):	Hb/B2–118
SnTe	SnTe solid solutions:		
	$SnTe-AgBiTe_2$	61 (Ag):	Hb/B4–252
	$SnTe-AgSbTe_2$	61 (Ag):	Hb/B4–251
	$SnTe-AgSbTe_2-PbTe$	61 (Ag):	Hb/B4–272
	SnTe–PbS–PbTe	47 (Pb):	Hb/C4–1594
	SnTe–PbSe	47 (Pb):	Hb/C4–1595
	SnTe–PbTe	47 (Pb):	Hb/C4–1589/94
	SnTe–PbTe–SnSe	47 (Pb):	Hb/C4–1595
–	SnTe systems:		
	$SnTe-Ag_2Te-Te$	61 (Ag):	Hb/B4–251
	SnTe–PbTe–SnSe	47 (Pb):	Hb/C4–1594/5
$SnTi_2$	Ti_2Sn	46 (Sn):	Hb/D–325, 328
$SnTi_3$	Ti_3Sn	46 (Sn):	Hb/D–325, 328
–	Ti_3Sn solid solutions:		
	$(Ti,Nb)_3Sn$	49 (Nb):	Hb/B2–234/5
–	Ti_3Sn systems:		
	Ti_3Sn-Ti_3Al-Zr	46 (Sn):	Hb/D–382/3
SnV_3	V_3Sn	48 (V):	Hb/A2–392 Hb/B2–588/91, 684/6
SnYb	YbSn	46 (Sn):	Hb/D–323/4
SnZr	ZrSn	46 (Sn):	Hb/D–342, 347
$SnZr_3$	Zr_3Sn	46 (Sn):	Hb/D–342, 346/7
–	Zr_3Sn solid solutions:		
	$(Zr,Nb)_3Sn$	49 (Nb):	Hb/B2–236/40
$SnZr_4$	Zr_4Sn	46 (Sn):	Hb/D–340, 346/7, 356, 362
Sn_2	Sn_2	46 (Sn):	Hb/B–129/30
Sn_2^{+}	Sn_2^{+}	46 (Sn):	Hb/B–176 Hb/C1–22
Sn_2Tb	$TbSn_2$	46 (Sn):	Hb/B–318/9

See the "Instructions for the user"
at the beginning of the volume

Sn_2Tl . . . . Sn_2Tl . . . . 46 (Sn): Hb/D–279/81
Sn_2Tm . . . . $TmSn_2$ . . . . 46 (Sn): Hb/D–323
Sn_2Y . . . . YSn_2 . . . . 46 (Sn): Hb/D–293/5
–. . . . . YSn_2 systems:
YSn_2–YAl_2 . . . . 46 (Sn): Hb/D–295

Sn_2Zr . . . . $ZrSn_2$ . . . . 46 (Sn): Hb/D–340, 342, 347
Sn_2Zr_3 . . . . Zr_3Sn_2 . . . . 46 (Sn): Hb/D–340
Sn_3^+ . . . . Sn_3^+ . . . . 46 (Sn): Hb/B–176
Sn_3Sr . . . . $SrSn_3$ . . . . 46 (Sn): Hb/D–123/4
Sn_3Ta_2 . . . . Ta_2Sn_3 . . . . 50 (Ta): Hb/B2–65/6
Sn_3Tb . . . . $TbSn_3$ . . . . 46 (Sn): Hb/D–318/9
Sn_3Tb_5 . . . . Tb_5Sn_3 . . . . 46 (Sn): Hb/D–318/9
Sn_3Th . . . . $ThSn_3$ . . . . 46 (Sn): Hb/D–388
Sn_3Ti_5 . . . . Ti_5Sn_3 . . . . 46 (Sn): Hb/D–325, 328
–. . . . . Ti_5Sn_3 solid solutions:
$(Ti,Zr)_5Sn_3$ . . . . 46 (Sn): Hb/D–378, 380

Sn_3Tm_5 . . . . Tm_5Sn_3 . . . . 46 (Sn): Hb/D–323
Sn_3Y . . . . YSn_3 . . . . 46 (Sn): Hb/D–293/5
–. . . . . YSn_3 systems:
YSn_3–YIn_3 . . . . 46 (Sn): Hb/D–295

Sn_3Y_5 . . . . Y_5Sn_3 . . . . 46 (Sn): Hb/D–293/5
Sn_3Yb . . . . $YbSn_3$ . . . . 46 (Sn): Hb/D–324
Sn_3Yb_5 . . . . Yb_5Sn_3 . . . . 46 (Sn): Hb/D–292
Sn_3Zr_5 . . . . Zr_5Sn_3 . . . . 46 (Sn): Hb/D–340, 342, 347
–. . . . . Zr_5Sn_3 solid solutions:
$(Zr,Ti)_5Sn_3$ . . . . 46 (Sn): Hb/D–378, 380

Sn_4^+ . . . . Sn_4^+ . . . . 46 (Sn): Hb/B–176
Sn_4Sr . . . . $SrSn_4$ . . . . 46 (Sn): Hb/D–124
Sn_4Tb_5 . . . . Tb_5Sn_4 . . . . 46 (Sn): Hb/D–318/9
Sn_4Y_5 . . . . Y_5Sn_4 . . . . 46 (Sn): Hb/D–294/5
Sn_4Zr_5 . . . . Zr_5Sn_4 . . . . 46 (Sn): Hb/D–342, 347
Sn_5^+ . . . . Sn_5^+ . . . . 46 (Sn): Hb/B–176
Sn_5Sr . . . . $SrSn_5$ . . . . 46 (Sn): Hb/D–123, 125
Sn_5Ti_6 . . . . Ti_6Sn_5 . . . . 46 (Sn): Hb/D–325, 328
Sn_6^+ . . . . Sn_6^+ . . . . 46 (Sn): Hb/B–176
$Sn_{10}Tb_{11}$ . . . . . . $Tb_{11}Sn_{10}$ . . . . 46 (Sn): Hb/D–318/9
$Sn_{10}Tm_{11}$ . . . . . . $Tm_{11}Sn_{10}$ . . . . 46 (Sn): Hb/D–323
$Sn_{10}Y_{11}$ . . . . $Y_{11}Sn_{10}$ . . . . 46 (Sn): Hb/D–293/5

Sr . . . . Strontium:
Absorption of radiation. . . . . 29 (Sr): Hb–59/60
Eb–125
Abundance . . . . 29 (Sr): Eb–13/4
Affinity. . . . . 29 (Sr): Hb–41
Atom. . . . . 29 (Sr): Hb–38/41
Eb–92/103
Atomic nucleus . . . . 29 (Sr): Hb–38
Eb–91/2

Hb = Main Series Volume; GD = Gmelin-Durrer; Ab = Appended Volume to Main Series;
Eb = Supplement Volume to Main Series; Erg. W. = New Supplement Series Volume

Sr	Strontium:		
	Chemical reactions	29 (Sr):	Hb–64/6 Eb–131/2
	Cosmochemistry	29 (Sr):	Hb–3 Eb–1/4
	Crystal chemistry	29 (Sr):	Eb–4/13
	Crystallographic properties	29 (Sr):	Hb–42 Eb–103/6
	Detection	28 (Ca):	Hb/B3–1490/8
	– in iron	59 (Fe):	Hb/FII1–6
	– in special substances	28 (Ca):	Hb/B3–1499/500
	Determination.	28 (Ca):	Hb/B3–1500/6
	– in presence of other ions	28 (Ca):	Hb/B3–1507/12
	– in special substances	28 (Ca):	Hb/B3–1512/8
	Economic deposits:		
	Africa	29 (Sr):	Hb–7/8 Eb–66
	America	29 (Sr):	Hb–8/9 Eb–66/75
	Asia	29 (Sr):	Eb–65/6
	Australia	29 (Sr):	Eb–66
	Europe	29 (Sr):	Hb–4/8 Eb–61/4
	Soviet Union	29 (Sr):	Hb–7/8 Eb–64/5
	Economics.	29 (Sr):	Eb–60/1
	Electrical properties	29 (Sr):	Hb–61/2 Eb–127/9
	Electrochemical behavior	29 (Sr):	Hb–62/4 Eb–129/31
	Electronegativity	29 (Sr):	Eb–103
	Formation	29 (Sr):	Eb–88
	Geochemical cycle	29 (Sr):	Eb–57/9
	Geochemistry	29 (Sr):	Hb–3/4 Eb–4/59
	History.	29 (Sr):	Hb–1/2 Eb–1
	Ionization potentials	29 (Sr):	Hb–62 Eb–128/9
	Magnetic properties	29 (Sr):	Hb–60/1 Eb–126/7
	Mechanical properties	29 (Sr):	Hb–42/3 Eb–106/10
	Mineralogy	29 (Sr):	Hb–9/16 Eb–75/82
	Molecule.	29 (Sr):	Hb–41 Eb–103
	Nuclear properties	29 (Sr):	Hb–38 Eb–91/2

See the "Instructions for the user"
at the beginning of the volume

Sr . . . . Strontium:

Occurrence, extraterrestrial	29 (Sr):	Hb–3 Eb–1/4
–, terrestrial	29 (Sr):	Hb–3/4 Eb–4/59
– in meteorites and tektites	29 (Sr):	Hb–3 Eb–3/4
– in soils	29 (Sr):	Hb–4 Eb–39/41
– in the atmosphere	29 (Sr):	Eb–49
– in the biosphere	29 (Sr):	Hb–3/4 Eb–49/57
– in the hydrosphere	29 (Sr):	Hb–3/4 Eb–43/9
– in the lithosphere	29 (Sr):	Hb–3/4 Eb–14/43
Optical properties	29 (Sr):	Hb–44/60 Eb–115/25
Position in the electrochemical series	29 (Sr):	Hb–63 Eb–129/30
– in the periodic system	29 (Sr):	Hb–41 Eb–102
Preparation	29 (Sr):	Hb–35/8 Eb–84, 88
–, industrial	29 (Sr):	Eb–83/4
Purification	29 (Sr):	Hb–37
Scattering of X-rays	29 (Sr):	Eb–125
Separation, analytical	28 (Ca):	Hb/B3–1507/12
Sorption	29 (Sr):	Eb–109/10
Special forms	29 (Sr):	Hb–37/8 Eb–88
Spectra:		
Arc spectrum	29 (Sr):	Hb–47/50, 54/6 Eb–116/8
Band spectrum	29 (Sr):	Hb–45 Eb–120
Spark spectra	29 (Sr):	Hb–50/4, 56/7 Eb–118/20
X-ray spectrum	29 (Sr):	Hb–58/9 Eb–120/5
Storage	29 (Sr):	Hb–37
Thermal properties	29 (Sr):	Hb–43/4 Eb–110/4
Toxicity	29 (Sr):	Eb–146
Treatment of raw materials	29 (Sr):	Hb–16/24 Eb–83/4
Uses	29 (Sr):	Eb–84

Sr . . . . Strontium:
Valence . . . . 29 (Sr): Hb–41
Eb–102/3

–. . . . . Strontium ions
see Sr^{+} ... Sr^{+}, etc.

–. . . . . Strontium isotopes:
Detection and determination. . . . . . 28 (Ca): Hb/B3–1518/26
Erg. W. 7b–355/6
Enrichment and separation . . . . . . 29 (Sr): Eb–88/91
Nuclear properties . . . . 29 (Sr): Hb–38
Eb–91/2
Toxicity . . . . 29 (Sr): Eb–146
Uses . . . . Erg. W. 8–207

–. . . . . Strontium systems:
Sr–Ag . . . . 61 (Ag): Hb/C–35/6
Sr–Ba . . . . 30 (Ba): Eb–275
Sr–Ca . . . . 29 (Sr): Eb–147
Sr–Cd . . . . 33 (Cd): Eb–382
Sr–Cu . . . . 60 (Cu): Hb/B3–1136
Sr–Hg . . . . 34 (Hg): Hb/A2–1039/40
Sr–Mg. . . . . 27 (Mg): Hb/A3–477
Sr–NH_3 . . . . 29 (Sr): Hb–91
Eb–171
Sr–O–S . . . . 9 (S): Hb/B1–183
Sr–Pb . . . . 47 (Pb): Hb/C3–1193/4
Sr–Pb–Zn. . . . . 47 (Pb): Hb/C4–1225
Sr–Sn . . . . 46 (Sn): Hb/D–123
Sr–$SrBr_2$. . . . . 29 (Sr): Eb–226
Sr–SrI_2 . . . . 29 (Sr): Eb–237
Sr–Zn . . . . 32 (Zn): Eb–627

Sr^{+} . . . . Sr^{+}:
Ionization potential . . . . 29 (Sr): Hb–62
Eb–128/9
Properties of the ion . . . . 29 (Sr): Eb–92/4, 96, 101
Spectra . . . . 29 (Sr): Hb–50/7
Eb–118/9

Sr^{2+}. . . . . Sr^{2+}:
Chemical reactions . . . . 29 (Sr): Hb–66/8
Eb–132/46
– with inorganic compounds. . . . . . 29 (Sr): Hb–66/8
Eb–134/8
– with organic compounds. . . . . . . 29 (Sr): Hb–68
Eb–138/46
Electrochemical behavior . . . . 29 (Sr): Hb–62/4
Eb–129/31
Ionic entropy . . . . 29 (Sr): Hb–44
Eb–114

See the "Instructions for the user" at the beginning of the volume

Sr^{2+}	Sr^{2+}:		
	Ionic polarization	29 (Sr):	Eb–127
	Ionic refraction	29 (Sr):	Eb–115
	Ionic susceptibility	29 (Sr):	Hb–61 Eb–126/7
	Ionization potential	29 (Sr):	Hb–62 Eb–128/9
	Lyotropic behavior	29 (Sr):	Eb–133/4
	Properties of the ion	29 (Sr):	Hb–39/41 Eb–92/102
	Solvation	29 (Sr):	Hb–66 Eb–132/3
	Sorption	29 (Sr):	Hb–43 Eb–109/10
	Spectra	29 (Sr):	Hb–50/7 Eb–119
Sr^{n+}	Sr^{n+} (n = 3 to 38)	29 (Sr):	Eb–92, 119/20, 128/9
SrTe	SrTe	29 (Sr):	Hb–187 Eb–261/2
$SrZn_{13}$	$Zn_{13}Sr$	32 (Zn):	Eb–627
Sr_2	Sr_2	29 (Sr):	Eb–103
T	Tritium *see H ... Hydrogen isotopes*		
Ta	Tantalum:		
	Absorption of X- and gamma rays	50 (Ta):	Hb/A2–320/6
	Abundance	50 (Ta):	Hb/A1–9, 49/50
	Atom	50 (Ta):	Hb/A2–300/26
	Atomic nucleus	50 (Ta):	Hb/A2–277/86
	Atomic radius	50 (Ta):	Hb/A2–304/6
	Atomic scattering factors	50 (Ta):	Hb/A2–302/3
	Atomic weight	50 (Ta):	Hb/A2–278/9
	Behavior as anode	50 (Ta):	Hb/A2–446/87
	– as cathode	50 (Ta):	Hb/A2–432/46
	Boiling point	50 (Ta):	Hb/A2–373
	Bond properties	48 (V):	Hb/A2–389/92
	Cathodic sputtering	50 (Ta):	Hb/A2–404/6
	Chemical constant	50 (Ta):	Hb/A2–379
	Chemical reactions	50 (Ta):	Hb/A2–495/547
	– on heating	50 (Ta):	Hb/A2–495
	– with acids	50 (Ta):	Hb/A2–535/43
	– with alkaline solutions	50 (Ta):	Hb/A2–533/5
	– with alloys	50 (Ta):	Hb/A2–524/9
	– with aqueous solutions	50 (Ta):	Hb/A2–543/5
	– with metal compounds	50 (Ta):	Hb/A2–529/33
	– with metals	50 (Ta):	Hb/A2–519/24

Ta	Tantalum:		
	Chemical reactions		
	– with nonmetal compounds	50 (Ta):	Hb/A2–514/9
	– with nonmetals	50 (Ta):	Hb/A2–495/514
	– with organic compounds	50 (Ta):	Hb/A2–545/7
	Cleavage	50 (Ta):	Hb/A2–328
	Colloidal forms	50 (Ta):	Hb/A1–266
	Compressibility	50 (Ta):	Hb/A2–350
	Cosmochemistry	50 (Ta):	Hb/A1–9/12
	Critical temperature	50 (Ta):	Hb/A2–374
	Crystal chemistry	50 (Ta):	Hb/A1–12/49
	Crystallographic properties	50 (Ta):	Hb/A2–326/46
	Debye temperature	50 (Ta):	Hb/A2–379
	Deformation	50 (Ta):	Hb/A2–338/41
	Density	50 (Ta):	Hb/A2–346/7
	Deposition of pure tantalum compounds	50 (Ta):	Hb/A1–246/51
	Detection	50 (Ta):	Hb/A2–547/51
	– in iron	59 (Fe):	Hb/FII1–81/3
	– in special substances	50 (Ta):	Hb/A2–551/2
	Determination	50 (Ta):	Hb/A2–547/8, 552/7
	– in iron	59 (Fe):	Hb/FII1–83/5 Hb/FII2–301
	– in presence of other ions	50 (Ta):	Hb/A2–557/62
	– in special substances	50 (Ta):	Hb/A2–563/74
	Diffusion	50 (Ta):	Hb/A2–346
	–, self-diffusion	50 (Ta):	Hb/A2–345/6
	– of boron	50 (Ta):	Hb/B1–193
	– of caesium	50 (Ta):	Hb/B2–2
	– of carbon	50 (Ta):	Hb/B1–211/2
	– of hydrogen	50 (Ta):	Hb/B1–3/4, 16/7
	– of niobium	50 (Ta):	Hb/B2–85
	– of nitrogen	50 (Ta):	Hb/B1–66
	– of oxygen	50 (Ta):	Hb/B1–17
	– of silicon	50 (Ta):	Hb/B1–251
	– of titanium	50 (Ta):	Hb/B2–24
	Disorder	50 (Ta):	Hb/A2–333/7, 341/2
	Dressing and treatment of ores	50 (Ta):	Hb/A1–212/31
	Economic deposits	50 (Ta):	Hb/A1–107/25
	Economics	50 (Ta):	Hb/A1–110/3
	Elasticity	50 (Ta):	Hb/A2–347/50
	Electrical conductivity	50 (Ta):	Hb/A2–386/9
	Electrical properties	50 (Ta):	Hb/A2–383/408
	Electrochemical behavior	50 (Ta):	Hb/A2–410/95
	Electrolytic deposition	50 (Ta):	Hb/A2–487/95
	Electrolytic polishing	50 (Ta):	Hb/A2–338, 487
	Electron configuration	50 (Ta):	Hb/A2–300
	Electronegativity	48 (V):	Hb/A2–392
	Electron emission	50 (Ta):	Hb/A2–397/404

See the "Instructions for the user" at the beginning of the volume

Ta . . . . Tantalum:

Electron structure . . . . 50 (Ta): Hb/A2–331/2
Energy band structure . . . . 50 (Ta): Hb/A2–383/5
Epitaxy. . . . . 50 (Ta): Hb/A2–328
Equation of state . . . . 50 (Ta): Hb/A2–370
Expansion coefficient. . . . . 50 (Ta): Hb/A2–370/2
g factor . . . . 50 (Ta): Hb/A2–301, 385
Galvanomagnetic effects . . . . 50 (Ta): Hb/A2–396/7
Geochemical character . . . . 50 (Ta): Hb/A1–49/50
Geochemical cycle . . . . 50 (Ta): Hb/A1–106/7
Geochemistry . . . . 50 (Ta): Hb/A1–12/107
de Haas-van Alphen effect . . . . 50 (Ta): Hb/A2–386
Hall effect . . . . 50 (Ta): Hb/A2–396/7
Hardness . . . . 50 (Ta): Hb/A2–354/61
Heat of fusion . . . . 50 (Ta): Hb/A2–374
– of sublimation. . . . . 50 (Ta): Hb/A2–373
– of vaporization . . . . 50 (Ta): Hb/A2–373/4
History. . . . . 50 (Ta): Hb/A1–1/9
Increments . . . . 50 (Ta): Hb/A2–326
Ion emission . . . . 50 (Ta): Hb/A2–404/8
Ionization potentials . . . . 50 (Ta): Hb/A2–302
Lattice defects. . . . . 50 (Ta): Hb/A2–333/7, 341/2
Lattice energy . . . . 50 (Ta): Hb/A2–332/3
Lattice structure. . . . . 50 (Ta): Hb/A2–328/31
Magnetic moment . . . . 50 (Ta): Hb/A2–301
Magnetic properties . . . . 50 (Ta): Hb/A2–385/6, 396/7
Mechanical properties . . . . 50 (Ta): Hb/A2–346/70
Melting point . . . . 50 (Ta): Hb/A2–374
Mineralogy . . . . 50 (Ta): Hb/A1–125/211
Modifications . . . . 50 (Ta): Hb/A2–326/7
Nuclear magnetic resonance. . . . . . 50 (Ta): Hb/A2–385
Nuclear moments. . . . . 50 (Ta): Hb/A2–279/81
Nuclear properties . . . . 50 (Ta): Hb/A2–277/86
Occurrence, extraterrestrial . . . . . . 50 (Ta): Hb/A1–9/12
–, terrestrial . . . . 50 (Ta): Hb/A1–12/107
– in meteorites and tektites . . . . . . 50 (Ta): Hb/A1–10/2
– in minerals . . . . 50 (Ta): Hb/A1–22/49, 125/211
– in soils . . . . 50 (Ta): Hb/A1–100/2
– in the atmosphere . . . . 50 (Ta): Hb/A1–105/6
– in the hydrosphere . . . . 50 (Ta): Hb/A1–104/5
– in the lithosphere . . . . 50 (Ta): Hb/A1–50/104
– in the magmatic cycle . . . . 50 (Ta): Hb/A1–50/91
– in the metamorphic cycle . . . . . . 50 (Ta): Hb/A1–102/4
– in the sedimentary cycle . . . . 50 (Ta): Hb/A1–91/100
Optical properties. . . . . 50 (Ta): Hb/A2–408/10
Overvoltage phenomena . . . . 50 (Ta): Hb/A2–422/32
Oxidation . . . . 50 (Ta): Hb/A2–495/505
Paramagnetic resonance . . . . 50 (Ta): Hb/A2–301, 385

Ta	Tantalum:		
	Passivation	50 (Ta):	Hb/A2–446/76
	Polarizability	50 (Ta):	Hb/A2–302
	Polarography	50 (Ta):	Hb/A2–421/2
	Polymorphism	50 (Ta):	Hb/A2–326/7
	Position in the electrochemical series	50 (Ta):	Hb/A2–410
	– in the periodic system	50 (Ta):	Hb/A1–7
	Potentials	50 (Ta):	Hb/A2–410/21
	Preparation, industrial	50 (Ta):	Hb/A1–211/70
	– in pure state	50 (Ta):	Hb/A1–248/51, 262
	– of metallic tantalum	50 (Ta):	Hb/A1–251/8
	– of special forms	50 (Ta):	Hb/A1–263/6
	Purification of metallic tantalum	50 (Ta):	Hb/A1–258/62
	– of solutions and concentrates	50 (Ta):	Hb/A1–231/5
	Purity testing	50 (Ta):	Hb/A1–266/70
	Recovery	50 (Ta):	Hb/A1–262/3
	Recrystallization	50 (Ta):	Hb/A2–342/5
	Scattering of X- and gamma rays	50 (Ta):	Hb/A2–320/6
	Self-diffusion	50 (Ta):	Hb/A2–345/6
	Separation, analytical	50 (Ta):	Hb/A2–557/62
	– from niobium	49 (Nb):	Hb/A–27
		50 (Ta):	Hb/A1–235/46
			Hb/A2–558/60
	Sorption of carbon monoxide	50 (Ta):	Hb/A2–517/8
	– of hydrogen	50 (Ta):	Hb/A2–506/7
			Hb/B1–2/3
	– of nitrogen	50 (Ta):	Hb/A2–508
			Hb/B1–61/6
	– of oxygen	50 (Ta):	Hb/A2–503/4, 505
	– of rare gases	50 (Ta):	Hb/A2–505
			Hb/B1–1
	Sound absorption and velocity	50 (Ta):	Hb/A2–350/1
	Special forms	50 (Ta):	Hb/A1–263/6
	Spectra:		
	Arc spectrum	50 (Ta):	Hb/A2–308/9
	Spark spectra	50 (Ta):	Hb/A2–310/1
	Terms	50 (Ta):	Hb/A2–300, 306/11
	X-ray spectrum	50 (Ta):	Hb/A2–312/20
	Standard potential	50 (Ta):	Hb/A2–410/1
	Strength	50 (Ta):	Hb/A2–361/9
	Structure	50 (Ta):	Hb/A2–330/1, 337/8
	Superconductivity	50 (Ta):	Hb/A2–390/5
	Surface tension	50 (Ta):	Hb/A2–369/70
	Susceptibility	50 (Ta):	Hb/A2–385/6
	Technology	50 (Ta):	Hb/A1–211/66
	Temperature radiation	50 (Ta):	Hb/A2–409/10
	Thermal conductivity	50 (Ta):	Hb/A2–380/3

See the "Instructions for the user"
at the beginning of the volume

Ta . . . . Tantalum:

Thermal properties . . . . 50 (Ta): Hb/A2–370/83
Thermodynamic functions . . . . 50 (Ta): Hb/A2–375/8
Thermoelectric effects . . . . 50 (Ta): Hb/A2–395/6
Thermomagnetic effects . . . . 50 (Ta): Hb/A2–396/7
Toxicity . . . . 50 (Ta): Hb/A1–276
Twinning . . . . 50 (Ta): Hb/A2–327/8
Uses . . . . 50 (Ta): Hb/A1–271/6
Vapor pressure . . . . 50 (Ta): Hb/A2–372/3
Viscosity . . . . 50 (Ta): Hb/A2–351/3
Wave functions . . . . 50 (Ta): Hb/A2–301/2

–. . . . . Tantalum ions
see Ta^{+} ... Ta^{+}*, etc.*

–. . . . . Tantalum isotopes:

Abundance . . . . 50 (Ta): Hb/A1–50
Hb/A2–277
Detection and determination. . . . . . 50 (Ta): Hb/A2–574
Enrichment and separation . . . . . . 50 (Ta): Hb/A2–286/300
Nuclear decay. . . . . 50 (Ta): Hb/A2–281/6
Nuclear properties . . . . 50 (Ta): Hb/A2–277/81
Preparation . . . . 50 (Ta): Hb/A2–290/300

–. . . . . Tantalum systems:

Ta–Al . . . . 50 (Ta): Hb/B2–11/2
Ta–Al–B . . . . 50 (Ta): Hb/B2–320
Ta–Al–C . . . . 50 (Ta): Hb/B2–323
Ta–Al–C–V . . . . 50 (Ta): Hb/B2–343
Ta–Al–O . . . . 50 (Ta): Hb/B2–197
Ta–Al–Si . . . . 50 (Ta): Hb/B2–16/7
Ta–Al–Ti . . . . 50 (Ta): Hb/B2–40
Ta–Al–V . . . . 50 (Ta): Hb/B2–78/9
Ta–As . . . . 50 (Ta): Hb/B1–272
Ta–B. 50 (Ta): Hb/B1–192/8
Ta–B–C . . . . 50 (Ta): Hb/B1–250
Ta–B–C–N . . . . 50 (Ta): Hb/B1–250
Ta–B–Cs . . . . 50 (Ta): Hb/B2–320
Ta–B–Ga . . . . 50 (Ta): Hb/B2–320
Ta–B–Ge . . . . 50 (Ta): Hb/B2–322
Ta–B–Li. . . . . 50 (Ta): Hb/B2–320
Ta–B–Nb . . . . 50 (Ta): Hb/B2–323
Ta–B–Rb . . . . 50 (Ta): Hb/B2–320
Ta–B–Si. . . . . 50 (Ta): Hb/B1–264
Ta–B–V . . . . 50 (Ta): Hb/B2–322
Ta–B–Zr . . . . 50 (Ta): Hb/B2–321/2
Ta–Be . . . . 50 (Ta): Hb/B2–2
Ta–Be–O . . . . 50 (Ta): Hb/B2–172
Ta–Bi . . . . 50 (Ta): Hb/B1–274
Ta–Bi–O . . . . 50 (Ta): Hb/B1–275

Hb = Main Series Volume; GD = Gmelin-Durrer; Ab = Appended Volume to Main Series;
Eb = Supplement Volume to Main Series; Erg. W. = New Supplement Series Volume

Ta . . . . Tantalum systems:

Ta–Br	50 (Ta):	Hb/B1–141/2
Ta–C.	50 (Ta):	Hb/B1–211/6
Ta–C–Ga	50 (Ta):	Hb/B2–323
Ta–C–Ge	50 (Ta):	Hb/B2–341
Ta–C–Hf	50 (Ta):	Hb/B2–333/40
Ta–C–Hf–N.	50 (Ta):	Hb/B2–340
Ta–C–Hf–Ti	50 (Ta):	Hb/B2–341
Ta–C–Hf–Zr	50 (Ta):	Hb/B2–341
Ta–C–N	50 (Ta):	Hb/B1–248/50
Ta–C–N–O	50 (Ta):	Hb/B1–250
Ta–C–N–Zr.	50 (Ta):	Hb/B2–333
Ta–C–Nb	50 (Ta):	Hb/B2–344/6
Ta–C–Nb–Ti	50 (Ta):	Hb/B2–346/7
Ta–C–Ni	57 (Ni):	Hb/B3–1206
Ta–C–O	50 (Ta):	Hb/B1–248
Ta–C–Si.	50 (Ta):	Hb/B1–264/5
Ta–C–Ti.	50 (Ta):	Hb/B2–324/6
Ta–C–Ti–V	50 (Ta):	Hb/B2–343
Ta–C–V	50 (Ta):	Hb/B2–341/3
Ta–C–V–Zr.	50 (Ta):	Hb/B2–343/4
Ta–C–Zr	50 (Ta):	Hb/B2–327/33
Ta–Cl	50 (Ta):	Hb/B1–91
Ta–Cl–O–S.	50 (Ta):	Hb/B1–178
Ta–Cl–S.	50 (Ta):	Hb/B1–178
Ta–Co	58 (Co):	Eb/A–418/20
Ta–Cr	52 (Cr):	Hb/A2–728/30
Ta–Cr–N	52 (Cr):	Hb/B–942
Ta–Cr–Ni	57 (Ni):	Hb/B1–227/8
Ta–Cu	60 (Cu):	Hb/B3–1214
Ta–Dy	50 (Ta):	Hb/B2–22/3
Ta–Er	50 (Ta):	Hb/B2–23
Ta–Fe	59 (Fe):	Hb/A9–1944/5
Ta–Ga	50 (Ta):	Hb/B2–17/8
Ta–Gd	50 (Ta):	Hb/B2–22
Ta–Ge	50 (Ta):	Hb/B2–64
Ta–Ge–N	50 (Ta):	Hb/B2–267
Ta–H.	50 (Ta):	Hb/B1–1/16
Ta–D.	50 (Ta):	Hb/B1–16
Ta–T.	50 (Ta):	Hb/B1–16/7
Ta–Hf	50 (Ta):	Hb/B2–52/3
Ta–Hf–N	50 (Ta):	Hb/B2–267
Ta–Hf–Nb.	50 (Ta):	Hb/B2–117
Ta–Hf–Ti	50 (Ta):	Hb/B2–61
Ta–Hf–V	50 (Ta):	Hb/B2–84
Ta–Hf–Zr	50 (Ta):	Hb/B2–61
Ta–Hg	50 (Ta):	Hb/B2–11
Ta–In	50 (Ta):	Hb/B2–18

See the "Instructions for the user"
at the beginning of the volume

Ta . . . . Tantalum systems:

System	Volume	Reference
Ta–Li–O	50 (Ta):	Hb/B2–126
Ta–Mg–O	50 (Ta):	Hb/B2–172
Ta–N	50 (Ta):	Hb/B1–61/9
Ta–N–Si	50 (Ta):	Hb/B1–263/4
Ta–N–V	50 (Ta):	Hb/B2–267
Ta–Na–O	50 (Ta):	Hb/B2–141
Ta–Nb	50 (Ta):	Hb/B2–85/6
Ta–Nb–Ni–Ti	57 (Ni):	Hb/B1–173
Ta–Nb–O	50 (Ta):	Hb/B2–245/6
Ta–Nb–Si	49 (Nb):	Hb/B2–104
Ta–Nb–Ti	50 (Ta):	Hb/B2–104
Ta–Nb–V	50 (Ta):	Hb/B2–118
Ta–Nb–Zr	50 (Ta):	Hb/B2–110
Ta–Ni	57 (Ni):	Hb/B1–166/7
Ta–Ni–V	57 (Ni):	Hb/B1–172
Ta–O	50 (Ta):	Hb/B1–17/23
Ta–O–Pb	50 (Ta):	Hb/B2–232
Ta–O–Si	50 (Ta):	Hb/B1–263
Ta–O–Th	50 (Ta):	Hb/B2–229
Ta–O–Ti	50 (Ta):	Hb/B2–221
Ta–O–Ti–Zr	50 (Ta):	Hb/B2–229
Ta–O–V	50 (Ta):	Hb/B2–244
Ta–O–Y	50 (Ta):	Hb/B2–205
Ta–O–Zr	50 (Ta):	Hb/B2–228
Ta–Ru	63 (Ru):	Eb–198/9
Ta–S	50 (Ta):	Hb/B1–169/70
Ta–Sc	50 (Ta):	Hb/B2–18/9
Ta–Se	50 (Ta):	Hb/B1–179/80
Ta–Si	50 (Ta):	Hb/B1–250/5
Ta–Si–Ti	50 (Ta):	Hb/B2–39
Ta–Si–V	50 (Ta):	Hb/B2–77/8
Ta–Sn	50 (Ta):	Hb/B2–65
Ta–TaC–TaN	50 (Ta):	Hb/B1–248/9
Ta–$TaCl_5$–TaS_3	50 (Ta):	Hb/B1–178
Ta–TaI_5	50 (Ta):	Hb/B1–161
Ta–Te	50 (Ta):	Hb/B1–189/90
Ta–Th	50 (Ta):	Hb/B2–62
Ta–Ti	50 (Ta):	Hb/B2–24/5
Ta–Ti–V	50 (Ta):	Hb/B2–79
Ta–Ti–Zr	50 (Ta):	Hb/B2–51
Ta–V	48 (V):	Hb/A2–402
	50 (Ta):	Hb/B2–68/9
Ta–V–Zr	50 (Ta):	Hb/B2–83
Ta–Y	50 (Ta):	Hb/B2–19/20
Ta–Y_2O_3	50 (Ta):	Hb/B2–205
Ta–Zn	50 (Ta):	Hb/B2–10/1
Ta–Zr	50 (Ta):	Hb/B2–41/3

Ta^+ . . . . Ta^+:
Electrochemical behavior . . . . 50 (Ta): Hb/A2–487/8
Form factors. . . . . 50 (Ta): Hb/A2–303
Ion emission . . . . 50 (Ta): Hb/A2–404, 407/8
Ionization potential . . . . 50 (Ta): Hb/A2–302
Properties of the ion . . . . 50 (Ta): Hb/A2–300, 305
Spectra . . . . 50 (Ta): Hb/A2–310/1

Ta^{2+} . . . . Ta^{2+}:
Electrochemical behavior . . . . 50 (Ta): Hb/A2–420, 487/8
Form factors. . . . . 50 (Ta): Hb/A2–303
Ionization potential . . . . 50 (Ta): Hb/A2–302
Properties of the ion . . . . 50 (Ta): Hb/A2–300, 305
Spectra . . . . 50 (Ta): Hb/A2–311

Ta^{3+} . . . . Ta^{3+}:
Electrochemical behavior . . . . 50 (Ta): Hb/A2–420, 487/8
Form factors. . . . . 50 (Ta): Hb/A2–303
Ionization potential . . . . 50 (Ta): Hb/A2–302
Properties of the ion . . . . 50 (Ta): Hb/A2–301, 305

Ta^{4+} . . . . Ta^{4+}:
Electrochemical behavior . . . . 50 (Ta): Hb/A2–420, 487/8
Form factors. . . . . 50 (Ta): Hb/A2–303
Ionization potential . . . . 50 (Ta): Hb/A2–302
Properties of the ion . . . . 50 (Ta): Hb/A2–301, 305

Ta^{5+} . . . . Ta^{5+}:
Electrochemical behavior . . . . 50 (Ta): Hb/A2–420, 487/95
Ionic refraction . . . . 50 (Ta): Hb/A2–326
Ionic susceptibility . . . . 50 (Ta): Hb/A2–326
Ionization potential . . . . 50 (Ta): Hb/A2–302
Properties of the ion . . . . 50 (Ta): Hb/A2–301, 305/6

TaTe . . . . TaTe . . . . 50 (Ta): Hb/B1–189
–. . . . . TaTe solid solutions:
(Ta,Nb)Te . . . . 50 (Ta): Hb/B2–319/20

$TaTe_2$ . . . . $TaTe_2$ . . . . 50 (Ta): Hb/B1–189/91
–. . . . . $TaTe_2$ solid solutions:
$(Ta,Nb)Te_2$. . . . . 50 (Ta): Hb/B2–319/20

$TaTe_3$ . . . . $TaTe_3$ . . . . 50 (Ta): Hb/B1–190, 192
$TaTe_4$ . . . . $TaTe_4$ . . . . 50 (Ta): Hb/B1–189/90, 191/2
TaTi . . . . TaTi solid solutions:
Ta(Ti,Al) . . . . 50 (Ta): Hb/B2–40

TaV_2. . . . . TaV_2 . . . . 50 (Ta): Hb/B2–68/72
–. . . . . TaV_2 systems:
TaV_2–Nb . . . . 50 (Ta): Hb/B2–118

Ta_2W_5. . . . . Ta_2W_5 . . . . 50 (Ta): Hb/B1–51
Ta_3Te_2 . . . . Ta_3Te_2 . . . . 50 (Ta): Hb/B1–190
$Ta_{31}Te_8$ . . . . $Ta_{31}Te_8$ . . . . 50 (Ta): Hb/B1–190

See the "Instructions for the user" at the beginning of the volume

Tb. . . . . Terbium:

Abundance	39 (SE):	Hb/A1-34/5
Band model	39 (SE):	Hb/B3-213/5
Boiling point	39 (SE):	Hb/B3-105, 107
Charge carrier concentration	39 (SE):	Hb/B3-228
Cosmochemistry	39 (SE):	Hb/A1-23/7
Crystal chemistry	39 (SE):	Hb/A1-28/32
Crystal field parameters	39 (SE):	Hb/B3-219
Crystallographic properties	39 (SE):	Hb/B3-62, 66/72, 76/83
Crystal structure	39 (SE):	Hb/B3-66/7, 71/2
Debye temperature	39 (SE):	Hb/B3-190
Density	39 (SE):	Hb/B3-85/7
Diffusion	39 (SE):	Hb/B3-77/8
Disorder	39 (SE):	Hb/B3-76
Domain structure	39 (SE):	Hb/B3-332
Economic deposits	39 (SE):	Hb/A1-110/22
Elasticity	39 (SE):	Hb/B3-92/102
Electrical properties	39 (SE):	Hb/B3-209/83
Electrical resistance	39 (SE):	Hb/B3-241/3, 260
Electron emission	39 (SE):	Hb/B3-281/3
Enthalpy of fusion	39 (SE):	Hb/B3-120
– of sublimation	39 (SE):	Hb/B3-110, 114
– of transformation	39 (SE):	Hb/B3-118
– of vaporization	39 (SE):	Hb/B3-110, 114
Ettingshausen effect	39 (SE):	Hb/B3-273
Exchange interaction	39 (SE):	Hb/B3-298/303
Geochemistry	39 (SE):	Hb/A1-28/54
Grüneisen constant	39 (SE):	Hb/B3-84/5
Hall effect	39 (SE):	Hb/B3-268/9
History	39 (SE):	Hb/A1-4/22
Hyperfine interactions	39 (SE):	Hb/B3-222
Hysteresis	39 (SE):	Hb/B3-333
Internal friction	39 (SE):	Hb/B3-195
Lattice dynamics	39 (SE):	Hb/B3-82/3
Magnetic anisotropy	39 (SE):	Hb/B3-303/6
Magnetic form factor	39 (SE):	Hb/B3-315/6
Magnetic lag	39 (SE):	Hb/B3-334
Magnetic moment	39 (SE):	Hb/B3-334
Magnetic properties	39 (SE):	Hb/B3-298/336
Magnetic resonance	39 (SE):	Hb/B3-331/2
Magnetic structure	39 (SE):	Hb/B3-307/12
Magnetization	39 (SE):	Hb/B3-316/8
Magnetostriction	39 (SE):	Hb/B3-323/6
Mechanical properties	39 (SE):	Hb/B3-85/103, 193/5, 207/9
Melting point	39 (SE):	Hb/B3-118/9
Mineralogy	39 (SE):	Hb/A1-54/110
Molar volume	39 (SE):	Hb/B3-85/7

Tb . . . . Terbium:

Nuclear magnetic resonance . . . . . . 39 (SE): Hb/B3–227/8
Occurrence, extraterrestrial . . . . . . 39 (SE): Hb/A1–22/7
–, terrestrial . . . . 39 (SE): Hb/A1–27/54
– in meteorites . . . . 39 (SE): Hb/A1–25/7
– in minerals . . . . 39 (SE): Hb/A1–35/53
– in the biosphere . . . . 39 (SE): Hb/A1–53/4
– in the lithosphere . . . . 39 (SE): Hb/A1–27/54
Optical properties . . . . 39 (SE): Hb/B3–341
Polymorphism . . . . 39 (SE): Hb/B3–62
Position in the periodic system . . . . 39 (SE): Hb/A1–2/4
Spin waves . . . . 39 (SE): Hb/B3–328/30
Strength . . . . 39 (SE): Hb/B3–102/3
Structure . . . . 39 (SE): Hb/B3–80
Surface tension . . . . 39 (SE): Hb/B3–208/9
Susceptibility . . . . 39 (SE): Hb/B3–334/6
Thermal conductivity . . . . 39 (SE): Hb/B3–200/4
Thermal expansion . . . . 39 (SE): Hb/B3–88/92
Thermal properties . . . . 39 (SE): Hb/B3–83/92, 103/14, 156/60, 190, 200/5
Thermodynamic functions . . . . 39 (SE): Hb/B3–156/60
Thermoelectric power . . . . 39 (SE): Hb/B3–275
Ultrasound absorption . . . . 39 (SE): Hb/B3–193/4
Valence . . . . 39 (SE): Hb/A1–4
Vapor pressure . . . . 39 (SE): Hb/B3–103/5, 107

–. . . . . Terbium ions . . . . 39 (SE): Hb/A1–3
Hb/B3–317
Hb/C1–242, 399
49 (Nb): Hb/B3–134, 135

–. . . . . Terbium systems:

Tb–Ag . . . . 61 (Ag): Hb/C–227/8
Tb–H . . . . 39 (SE): Hb/C1–2/3, 67
Tb–Pb . . . . 47 (Pb): Hb/C4–1369
Tb–Sn . . . . 46 (Sn): Hb/D–318

Tb^{3+} . . . . Tb^{3+} . . . . 39 (SE): Hb/A1–3
Hb/B3–317
49 (Nb): Hb/B3–134, 135

Tb^{4+} . . . . Tb^{4+} . . . . 39 (SE): Hb/C1–242, 399

Tc . . . . Technetium:

Chemical reactions . . . . 69 (Ma): Hb–9/10
History . . . . 69 (Ma): Hb–1
70 (Re): Hb–1/10
Occurrence . . . . 69 (Ma): Hb–1/2
70 (Re): Hb–12/3
Physical properties . . . . 69 (Ma): Hb–2/8
Preparation . . . . 69 (Ma): Hb–2

–. . . . . Technetium isotopes . . . . 69 (Ma): Hb–2/3, 9

See the "Instructions for the user"
at the beginning of the volume

Te . . . . Tellurium:

Absorption of X-rays . . . . 11 (Te): Hb–203
Abundance . . . . 11 (Te): Hb–9
Affinity . . . . 11 (Te): Hb–161
Atom . . . . 11 (Te): Hb–151/62
Atomic nucleus . . . . 11 (Te): Hb–140/51
Chemical reactions . . . . 11 (Te): Hb–228/38
– with elements . . . . 11 (Te): Hb–228/31
– with inorganic compounds . . . . . 11 (Te): Hb–232/7
– with organic compounds . . . . . . 11 (Te): Hb–238
Cosmochemistry . . . . 11 (Te): Hb–6
Crystallographic properties . . . . . . 11 (Te): Hb–165/71
Detection . . . . 11 (Te): Hb–240/7
– in iron . . . . 59 (Fe): Hb/FII1–31
Determination . . . . 11 (Te): Hb–247/58
– in iron . . . . 59 (Fe): Hb/FII1–31
– in special substances . . . . 11 (Te): Hb–262
Economic deposits:
Africa . . . . 11 (Te): Hb–53
America . . . . 11 (Te): Hb–40/9, 53/65
Asia . . . . 11 (Te): Hb–50/2
Australia and Oceania . . . . 11 (Te): Hb–33/40, 52/3
Europe . . . . 11 (Te): Hb–19/33, 49/50
Electrical conductivity . . . . 11 (Te): Hb–209/16
Electrical properties . . . . 11 (Te): Hb–209/22
Electrochemical behavior . . . . 11 (Te): Hb–223/8
Formation . . . . 11 (Te): Hb–125
Geochemical character . . . . 11 (Te): Hb–7/8
Geochemistry . . . . 11 (Te): Hb–6/124
History . . . . 11 (Te): Hb–1/5
Ionization potentials . . . . 11 (Te): Hb–222
Magnetic properties . . . . 11 (Te): Hb–204/9
Mechanical properties . . . . 11 (Te): Hb–171/6
Mineralogy . . . . 11 (Te): Hb–65/124
Molecule . . . . 11 (Te): Hb–162/5
Occurrence, extraterrestrial . . . . . . 11 (Te): Hb–6
–, terrestrial . . . . 11 (Te): Hb–6/124
– in industrial products . . . . 11 (Te): Hb–16/8
– in the biosphere . . . . 11 (Te): Hb–15
– in the hydrosphere . . . . 11 (Te): Hb–15
– in the lithosphere . . . . 11 (Te): Hb–7/15
Optical properties . . . . 11 (Te): Hb–182/203
Position in the periodic system . . . . 11 (Te): Hb–161/2
Preparation . . . . 11 (Te): Hb–125/38
–, industrial . . . . 11 (Te): Hb–125/30
– in pure state . . . . 11 (Te): Hb–130/6
– of special forms . . . . 11 (Te): Hb–136/8
Radioactivity . . . . 11 (Te): Hb–151

Te	Tellurium:		
	Scattering of X-rays	11 (Te):	Hb–203
	Separation, analytical	11 (Te):	Hb–258/62
	Spectra:		
	Arc spectrum	11 (Te):	Hb–184/9, 191
	Band spectrum	11 (Te):	Hb–191/9
	Continuous spectrum	11 (Te):	Hb–199
	Spark spectra	11 (Te):	Hb–189/91
	Terms	11 (Te):	Hb–187/9, 197
	X-ray spectrum	11 (Te):	Hb–200/3
	Thermal properties	11 (Te):	Hb–176/82
	Toxicity	11 (Te):	Hb–238/9
	Uses	11 (Te):	Hb–139/40, 282
	Valence	11 (Te):	Hb–160/1
–.	Tellurium ions *see Te^+ ... Te^+, etc.*		
–.	Tellurium isotopes	11 (Te):	Hb–140/2
–.	Tellurium solutions:		
	Te–H_2Te	11 (Te):	Hb–272
–.	Tellurium systems:		
	Te–Ag	61 (Ag):	Hb/B3–199/206
	Te–Ag–Au	62 (Au):	Hb/3–766
	Te–Ag–Cu	61 (Ag):	Hb/B4–477/8
	Te–Ag–Fe.	61 (Ag):	Hb/B4–415
	Te–Ag–Ge	61 (Ag):	Hb/B4–245
	Te–Ag–Pb	61 (Ag):	Hb/B4–269
	Te–Ag_2Te–AuTe_2.	62 (Au):	Hb/3–766
	Te–Ag_2Te–SnTe	61 (Ag):	Hb/B4–251
	Te–Al	35 (Al):	Hb/B1–281/2
	Te–As	17 (As):	Hb–464
	Te–As–Bi	19 (Bi):	Eb–858
	Te–Au	62 (Au):	Hb/3–720/1
	Te–Bi	19 (Bi):	Hb–175/7
			Eb–775/85
	Te–Bi–O	19 (Bi):	Eb–817
	Te–Bi–Pb	47 (Pb):	Hb/C3–973/5
	Te–Bi–Sb	19 (Bi):	Eb–861/5
	Te–Bi–Sb–Se	19 (Bi):	Eb–865/6
	Te–Br	11 (Te):	Hb–335/7
	Te–Co	58 (Co):	Eb/A–662/3
	Te–Cr	52 (Cr):	Hb/B–342
	Te–Cr–Se	52 (Cr):	Hb/B–343
	Te–Cu	60 (Cu):	Hb/B1–619/20
	Te–Cu–Pb	60 (Cu):	Hb/B3–1202
	Te–Fe	59 (Fe):	Hb/A6–1174
	Te–Ga	36 (Ga):	Hb–90
	Te–Ge	45 (Ge):	Eb–543/4

Te . . . . Tellurium systems:

	Te–Hg	34 (Hg):	Hb/B3–1111/3
	Te–I	11 (Te):	Hb–344/5
	Te–In	37 (In):	Hb–102
	Te–Na	21 (Na):	Hb–643 Eb/3–1202
	Te–Nb	49 (Nb):	Hb/B1–291/2
	Te–Ni	57 (Ni):	Hb/B2–756/7
	Te–O–Pb	47 (Pb):	Hb/C2–678/9
	Te–Pb	47 (Pb):	Hb/C2–640/3
	Te–Pb–Sb	47 (Pb):	Hb/C3–940/1
	Te–Pb–Se–Sn	47 (Pb):	Hb/C4–1594/5
	Te–Pb–Sn	47 (Pb):	Hb/C4–1588
	Te–S	11 (Te):	Hb–350/1
	Te–S–Se	11 (Te):	Hb–363
	Te–Sb	18 (Sb):	Hb/B3–554/5
	Te–Se	11 (Te):	Hb–361/2
	Te–Ta	50 (Ta):	Hb/B1–189/90
	Te–$TeBr_4$	11 (Te):	Hb–335/7
	Te–$TeCl_4$	11 (Te):	Hb–323/5
	Te–Ti	41 (Ti):	Hb–353/4
	Te–Tl	38 (Tl):	Hb/3–376
	Te–V	48 (V):	Hb/B1–315/6
	Te–Zn	32 (Zn):	Eb–969
Te^+	Te^+	11 (Te):	Hb–155, 189, 221/2
Te^{2+}	Te^{2+}	11 (Te):	Hb–155, 189, 221/2
Te^{3+}	Te^{3+}	11 (Te):	Hb–155, 189/90, 221/2
Te^{4+}	Te^{4+}:		
	Electrochemical behavior	11 (Te):	Hb–223/4, 226/7, 286
	Ionic susceptibility	11 (Te):	Hb–205/6
	Ionization potential	11 (Te):	Hb–222
	Properties of the ion	11 (Te):	Hb–155/6
	Spectra	11 (Te):	Hb–190
Te^{5+}	Te^{5+}	11 (Te):	Hb–190, 222
Te^{6+}	Te^{6+}:		
	Ionic refraction	11 (Te):	Hb–183
	Ionic susceptibility	11 (Te):	Hb–205/6
	Ionization potential	11 (Te):	Hb–222
	Properties of the ion	11 (Te):	Hb–155/6
	Spectra	11 (Te):	Hb–190/1
Te^{7+}	Te^{7+}	11 (Te):	Hb–222
Te^-	Te^-	11 (Te):	Hb–154/5, 226/7
–	Te_2^{2-}	11 (Te):	Hb–223/6, 272
Te^{2-}	Te^{2-}:		
	Chemical reactions	11 (Te):	Hb–270/2
	Electrochemical behavior	11 (Te):	Hb–223/7

Hb = Main Series Volume; GD = Gmelin-Durrer; Ab = Appended Volume to Main Series;
Eb = Supplement Volume to Main Series; Erg. W. = New Supplement Series Volume

Te^{2-}	Te^{2-}:		
	Ionic refraction	11 (Te):	Hb–183
	Ionic susceptibility	11 (Te):	Hb–206
	Properties of the ion	11 (Te):	Hb–153/4, 156
TeTh	ThTe	44 (Th):	Hb–295
$TeTh_3$	Th_3Te	44 (Th):	Hb–295
TeTi	TiTe	41 (Ti):	Hb–353/4
TeTl	TlTe	38 (Tl):	Hb/3–377
$TeTl_2$	Tl_2Te	38 (Tl):	Hb/3–376/7
–	Tl_2Te systems:		
	Tl_2Te–PbTe	47 (Pb):	Hb/C4–1353/4
TeU	UTe	55 (U):	Hb–159
TeV	VTe	48 (V):	Hb/B1–315/8
TeZn	ZnTe	32 (Zn):	Hb–247/8 Eb–969/70
–	ZnTe · H_2O	32 (Zn):	Hb–248
–	ZnTe solid solutions:		
	ZnTe–HgTe	34 (Hg):	Hb/B4–1502/7
$TeZr_2$	Zr_2Te	42 (Zr):	Hb–355
$Te_{1.4}V$	$VTe_{1.4}$	48 (V):	Hb/B1–316, 318/9
$Te_{1.77}V$	$VTe_{1.77}$	48 (V):	Hb/B1–316, 319
Te_2	Te_2	11 (Te):	Hb–162/5, 191/9
Te_2^{2-}	Te_2^{2-}	11 (Te):	Hb–223/6, 272
Te_2Th	$ThTe_2$	44 (Th):	Hb–295
Te_2Ti	$TiTe_2$	41 (Ti):	Hb–353/4
Te_2Ti_x	Ti_xTe_2	41 (Ti):	Hb–354
Te_2Tl_3	Tl_3Te_2	38 (Tl):	Hb/3–377
Te_2U	UTe_2	55 (U):	Hb–159
Te_2V	VTe_2	48 (V):	Hb/B1–316, 319/20
Te_2Zr	$ZrTe_2$	42 (Zr):	Hb–355
$Te_{2.66}Th$	$ThTe_{2.66}$	44 (Th):	Hb–295
Te_3Tl_2	Tl_2Te_3 systems:		
	Tl_2Te_3–Ag_2Te	61 (Ag):	Hb/B4–228/9
Te_3Tl_5	Tl_5Te_3	38 (Tl):	Hb/3–376, 377
Te_3U_2	U_2Te_3	55 (U):	Hb–159
Te_3V_2	V_2Te_3	48 (V):	Hb/B1–316, 319
Te_4TiV_2	TiV_2Te_4	48 (V):	Hb/B2–561
Te_4Ti_2V	Ti_2VTe_4	48 (V):	Hb/B2–561
Te_4V_3	V_3Te_4	48 (V):	Hb/B1–315/6, 318
Te_4V_5	V_5Te_4	48 (V):	Hb/B1–315/6, 317
Te_5Tl_2	Tl_2Te_5	38 (Tl):	Hb/3–376, 377
Te_x^{2-}	Te_x^{2-}	11 (Te):	Hb–272
Te_xTi	$TiTe_x$	41 (Ti):	Hb–353/4
Th	Thorium:		
	Absorption of X-rays	44 (Th):	Hb–161
	Abundance	44 (Th):	Hb–10

See the "Instructions for the user"
at the beginning of the volume

Th . . . . Thorium:

Age determination . . . . 44 (Th): Hb–17
Atom . . . . 44 (Th): Hb–120/7
Erg. W. 8–1/28
Atomic nucleus . . . . 44 (Th): Hb–93/112, 359, 370, 383, 393, 402
Chemical reactions . . . . 44 (Th): Hb–180/4, 364, 377, 390, 396, 406
Cosmochemistry . . . . 44 (Th): Hb–3/4
Crystal chemistry . . . . 44 (Th): Hb–5/10
Crystallographic properties . . . . . . 44 (Th): Hb–128/32
Detection . . . . 44 (Th): Hb–197, 376/7
Determination . . . . 44 (Th): Hb–197, 376/7
Dressing and treatment of ores . . . . 44 (Th): Hb–36/40, 67/72
Economic deposits . . . . 44 (Th): Hb–19/27
Economics . . . . 44 (Th): Hb–27/8
Electrical properties . . . . 44 (Th): Hb–162/78
Electrochemical behavior . . . . 44 (Th): Hb–178/80
Electronegativity . . . . 44 (Th): Hb–128
Electron emission . . . . 44 (Th): Hb–165/76
Formation . . . . 44 (Th): Hb–83, 356, 366, 379, 391, 397
Geochemical character . . . . 44 (Th): Hb–10
Geochemical cycle . . . . 44 (Th): Hb–19
Geochemistry . . . . 44 (Th): Hb–5/19, 356, 365/6, 378, 391, 396
History:
^{227}Th (RdAc) . . . . 44 (Th): Hb–355
^{228}Th (RdTh) . . . . 44 (Th): Hb–364/5
^{230}Th (Io) . . . . 44 (Th): Hb–377/8
^{231}Th (UY) . . . . 44 (Th): Hb–390/1
^{232}Th (Th) . . . . 44 (Th): Hb–1/2
^{234}Th (UX_1) . . . . 44 (Th): Hb–396
Ionization potentials . . . . 44 (Th): Hb–178
Erg. W. 8–23
Magnetic properties . . . . 44 (Th): Hb–161/2
Mechanical properties . . . . 44 (Th): Hb–133/45
Mineralogy . . . . 44 (Th): Hb–29/32
Nuclear decay,
alpha radioactivity . . . . 44 (Th): Hb–115/7, 360/1, 370/2, 384/6, 395, 405
Erg. W. 7a–85/94
–, beta radioactivity . . . . 44 (Th): Hb–360, 370, 393/4, 402/4
–, decay scheme . . . . 44 (Th): Hb–112/5, 363, 373, 389, 395
Erg. W. 7a–97

Th . . . . Thorium:

Nuclear decay,
gamma radioactivity . . . . 44 (Th): Hb–117/8, 361/3, 372/4, 387/9, 395, 404/5
Erg. W. 7a–80

Nuclear fission,
fission by charged particles . . . . . 44 (Th): Hb–104/9
–, – by gamma rays . . . . 44 (Th): Hb–109/12, 384
–, – by neutrons . . . . 44 (Th): Hb–101/4, 359/60, 370, 383/4, 402
Erg. W. 7a–140
–, spontaneous fission . . . . 44 (Th): Hb–118/9, 390
Erg. W. 7a–134/8

Nuclear properties . . . . 44 (Th): Hb–93/112, 359, 370, 383, 393, 402

Occurrence, extraterrestrial . . . . . . 44 (Th): Hb–3/4
–, terrestrial . . . . 44 (Th): Hb–5/19, 356, 365/6, 378, 391, 396
– in the atmosphere . . . . 44 (Th): Hb–18
– in the biosphere . . . . 44 (Th): Hb–19
– in the hydrosphere . . . . 44 (Th): Hb–17/8, 365/6, 378
– in the lithosphere . . . . 44 (Th): Hb–10/7, 356, 365/6, 378, 391, 396

Optical properties . . . . 44 (Th): Hb–151/61
Oxidation states . . . . Erg. W. 8–8/11
Position in the periodic system . . . . 44 (Th): Hb–127/8
Preparation . . . . 44 (Th): Hb–83/8
–, industrial, of thorium compounds . 44 (Th): Hb–66/74
– in pure state:
^{227}Th (RdAc) . . . . 44 (Th): Hb–356/9
^{228}Th (RdTh) . . . . 44 (Th): Hb–366/9
^{230}Th (Io) . . . . 44 (Th): Hb–379/82
^{231}Th(UY) . . . . 44 (Th): Hb–391/3
^{232}Th (Th) . . . . 44 (Th): Hb–64/5
^{234}Th (UX_1) . . . . 44 (Th): Hb–397/401

– of pure thorium preparations . . . . 44 (Th): Hb–33/66
– of special forms . . . . 44 (Th): Hb–90/2
Purification of thorium preparations . 44 (Th): Hb–63/5, 72/3
Purity testing . . . . 44 (Th): Hb–63
Radioactivity . . . . 44 (Th): Hb–112/20
Separation . . . . 44 (Th): Hb–40/63
Sorption . . . . 44 (Th): Hb–142/5, 406
Spectra:
Arc spectrum . . . . 44 (Th): Hb–152, 154
Spark spectra . . . . 44 (Th): Hb–152/6
X-ray spectrum . . . . 44 (Th): Hb–156/61

Th . . . . Thorium:
Thermal properties . . . . 44 (Th): Hb–145/51
Toxicity . . . . 44 (Th): Hb–196/7
Uses . . . . 44 (Th): Hb–74/7
Erg. W. 8–224
Valence . . . . 44 (Th): Hb–127/8

–. . . . . Thorium ions
see Th^+ ... Th^+, etc.

–. . . . . Thorium isotopes
see Th ... Thorium

–. . . . . Thorium systems:
Th–Ag . . . . 61 (Ag): Hb/C–254/6
Th–Al . . . . 35 (Al): Hb/A5–840/2
Th–B. . . . . 44 (Th): Hb–295
Th–Bi–Pb . . . . 47 (Pb): Hb/C4–1470
Th–C. . . . . 44 (Th): Hb–297/8
Th–C–Pu . . . . Erg. W. 4–210/1
Th–Co . . . . 58 (Co): Eb/A–406
Th–Cu . . . . 60 (Cu): Hb/B3–1192
Th–Cu–Si. . . . . 60 (Cu): Hb/B3–1192
Th–H . . . . 44 (Th): Hb–200
Th–La–Ru. . . . . 63 (Ru): Eb–192/3
Th–Mg . . . . 27 (Mg): Hb/A4–613/4
Th–Na . . . . 44 (Th): Hb–198
Th–Nb. . . . . 49 (Nb): Hb/B2–196/7
Th–Nb–Zr. . . . . 49 (Nb): Hb/B2–200
Th–Ni . . . . 57 (Ni): Hb/B1–118/20
Th–O–Ta . . . . 50 (Ta): Hb/B2–229
Th–Pb . . . . 47 (Pb): Hb/C4–1468/70
Th–Ru . . . . 63 (Ru): Eb–191/2
Th–S. . . . . 44 (Th): Hb–277/8
Th–Se . . . . 44 (Th): Hb–294
Th–Sn . . . . 46 (Sn): Hb/D–388
Th–Ta . . . . 50 (Ta): Hb/B2–62
Th–V. . . . . 48 (V): Hb/B2–677/80
Th–W . . . . 54 (W): Hb–103

Th^+ . . . . Th^+ . . . . 44 (Th): Hb–123, 154/6, 178
Erg. W. 8–23

Th^{2+} . . . . Th^{2+} . . . . 44 (Th): Hb–155, 178
Erg. W. 8–11/23

Th^{3+} . . . . Th^{3+} . . . . 44 (Th): Hb–155, 178
Erg. W. 8–5, 11/23

Th^{4+} . . . . Th^{4+}:
Chemical reactions
with inorganic compounds . . . . . 44 (Th): Hb–184/9
– with organic compounds. . . . . . . 44 (Th): Hb–189/96
Electrochemical behavior . . . . 44 (Th): Hb–178/80

Hb = Main Series Volume; GD = Gmelin-Durrer; Ab = Appended Volume to Main Series;
Eb = Supplement Volume to Main Series; Erg. W. = New Supplement Series Volume

Th^{4+} . . . . Th^{4+}:
Ionization potential . . . . 44 (Th): Hb–178
Erg. W. 8–23
Properties of the ion . . . . 44 (Th): Hb–121/5
Erg. W. 8–5
Sorption . . . . 44 (Th): Hb–143/5
Susceptibility . . . . 44 (Th): Hb–162

Th^{n+} . . . . Th^{n+} (n = 5 to 10) . . . . 44 (Th): Hb–178
Erg. W. 8–23

ThA . . . . Thorium A = ^{216}Po
see Po ... Polonium

ThB . . . . Thorium B = ^{212}Pb
see Pb ... Lead isotopes

ThC . . . . Thorium C = ^{212}Bi
see Bi ... Bismuth isotopes

ThC′ . . . . Thorium C′ = ^{212}Po
see Po ... Polonium

ThC″ . . . . Thorium C″ = ^{208}Tl
see Tl ... Thallium isotopes

ThD . . . . Thorium D = ^{208}Pb
see Pb ... Lead isotopes

ThEm . . . . Thorium emanation = ^{220}Rn
see Rn ... Radon

ThX . . . . Thorium X = ^{224}Ra
see Ra ... Radium

Ti . . . . Titanium:
Absorption of X-rays . . . . 41 (Ti): Hb–179
Atom . . . . 41 (Ti): Hb–137/41
Atomic nucleus . . . . 41 (Ti): Hb–132/7
Chemical reactions . . . . 41 (Ti): Hb–189/99
Colloidal forms . . . . 41 (Ti): Hb–131
Coordination number . . . . 41 (Ti): Hb–142
Cosmochemistry . . . . 41 (Ti): Hb–6/10
Crystal chemistry . . . . 41 (Ti): Hb–11/21
Crystallographic properties . . . . . . 41 (Ti): Hb–142/7
Detection in iron . . . . 59 (Fe): Hb/FII2–166/7
Determination in iron . . . . 59 (Fe): Hb/FII2–167/82, 305/7, 358
Dressing and treatment of ores . . . . 41 (Ti): Hb–81/117
Economic deposits:
Africa . . . . 41 (Ti): Hb–55/9
America . . . . 41 (Ti): Hb–59/69
Asia . . . . 41 (Ti): Hb–49/52
Australia . . . . 41 (Ti): Hb–52/5
Europe . . . . 41 (Ti): Hb–43/9
Soviet Union . . . . 41 (Ti): Hb–47/9

See the "Instructions for the user" at the beginning of the volume

Ti . . . . Titanium:

Economics	41 (Ti):	Hb–69/70
Electrical properties	41 (Ti):	Hb–180/6
Electrochemical behavior	41 (Ti):	Hb–186/9
Electronegativity	41 (Ti):	Hb–141/2
Formation	41 (Ti):	Hb–129
Geochemistry	41 (Ti):	Hb–11/43
History	41 (Ti):	Hb–1/5
Magnetic properties	41 (Ti):	Hb–179/80
Mechanical properties	41 (Ti):	Hb–147/56
Mineralogy	41 (Ti):	Hb–70/9
Nuclear properties	41 (Ti):	Hb–132/7
Occurrence, extraterrestrial	41 (Ti):	Hb–6/10
–, terrestrial	41 (Ti):	Hb–11/43
– in meteorites	41 (Ti):	Hb–7/10
– in minerals	41 (Ti):	Hb–11/21
– in the atmosphere	41 (Ti):	Hb–41
– in the biosphere	41 (Ti):	Hb–41/3
– in the hydrosphere	41 (Ti):	Hb–38/41
– in the lithosphere	41 (Ti):	Hb–21/38
Optical properties	41 (Ti):	Hb–160/79
Physiological action	41 (Ti):	Hb–199
Position in the periodic system	41 (Ti):	Hb–141
Preparation	41 (Ti):	Hb–129
–, industrial	41 (Ti):	Hb–117/27
– in pure state	41 (Ti):	Hb–129/30
– of dense titanium	41 (Ti):	Hb–125/7
– of special forms	41 (Ti):	Hb–130/2
Scattering of X-rays	41 (Ti):	Hb–179
Sorption of hydrogen	41 (Ti):	Hb–203/8
Spectra:		
Arc spectrum	41 (Ti):	Hb–162/7
Spark spectra	41 (Ti):	Hb–167/70
X-ray spectrum	41 (Ti):	Hb–172/9
Thermal properties	41 (Ti):	Hb–156/60
Toxicity	41 (Ti):	Hb–199
Uses	41 (Ti):	Hb–127/8
Valence	41 (Ti):	Hb–141

–. . . . . Titanium ions
see Ti^{+} ... Ti^{+}, *etc.*

–. . . . . Titanium isotopes:

Abundance	41 (Ti):	Hb–132/3
Enrichment and separation	41 (Ti):	Hb–132
Nuclear properties	41 (Ti):	Hb–133/4

–. . . . . Titanium systems:

Ti–Ag	61 (Ag):	Hb/C–237/41
Ti–Ag–Cu	61 (Ag):	Hb/C–400/2

Hb = Main Series Volume; GD = Gmelin-Durrer; Ab = Appended Volume to Main Series; Eb = Supplement Volume to Main Series; Erg. W. = New Supplement Series Volume

Ti . . . . Titanium systems:

Ti–Al	35 (Al):	Hb/A5–827/8
Ti–Al–Nb	49 (Nb):	Hb/B2–118/9
Ti–Al–Sn	46 (Sn):	Hb/D–333/4
Ti–Al–Sn–Zr	46 (Sn):	Hb/D–382/3
Ti–Al–Ta	50 (Ta):	Hb/B2–40
Ti–Au	62 (Au):	Hb/3–833/4
Ti–B	41 (Ti):	Hb–354/5
Ti–B–Nb	49 (Nb):	Hb/B4–307/8
Ti–B–V	48 (V):	Hb/B2–659
Ti–Be	41 (Ti):	Hb–200
Ti–Be–V	48 (V):	Hb/B2–661/2
Ti–C	41 (Ti):	Hb–357/8
Ti–C–Fe	59 (Fe):	Hb/A9–1874/8
Ti–C–Hf–Ta	50 (Ta):	Hb/B2–341
Ti–C–N	41 (Ti):	Hb–369
Ti–C–Nb	49 (Nb):	Hb/B4–315/20
Ti–C–Nb–Ta	50 (Ta):	Hb/B2–346/7
Ti–C–Ni	57 (Ni):	Hb/B3–1182
Ti–C–O	41 (Ti):	Hb–368
Ti–C–Ta	50 (Ta):	Hb/B2–324/6
Ti–C–Ta–V	50 (Ta):	Hb/B2–343
Ti–C–V	48 (V):	Hb/B2–659
Ti–Ce–V	48 (V):	Hb/B2–662
Ti–Co	58 (Co):	Hb/A1–193/4 Eb/A–400/4
Ti–Cr	52 (Cr):	Hb/A2–721/4
Ti–Cr–Ni	57 (Ni):	Hb/B1–223/5
Ti–Cu	60 (Cu):	Hb/B3–1189
Ti–Cu–Si	60 (Cu):	Hb/B3–1190
Ti–Fe	59 (Fe):	Hb/A9–1867/70
Ti–Fe–Si	59 (Fe):	Hb/A9–1888/93
Ti–Gd–V	48 (V):	Hb/B2–662
Ti–H	41 (Ti):	Hb–203/8
Ti–H–Mn	56 (Mn):	Hb/C1–6
Ti–H–Nb	49 (Nb):	Hb/B4–196/7
Ti–H–V	48 (V):	Hb/B2–657/8
Ti–Hf–Nb	49 (Nb):	Hb/B2–189
Ti–Hf–Nb–Zr	49 (Nb):	Hb/B2–194
Ti–Hf–Ta	50 (Ta):	Hb/B2–61
Ti–La–V	48 (V):	Hb/B2–662
Ti–Mg	27 (Mg):	Hb/A4–604
Ti–Mo–Ni	57 (Ni):	Hb/B1–251
Ti–N	41 (Ti):	Hb–271/2
Ti–N–V	48 (V):	Hb/B2–658
Ti–Nb	49 (Nb):	Hb/B2–62/4
Ti–Nb–Ni–Ta	57 (Ni):	Hb/B1–173
Ti–Nb–O	49 (Nb):	Hb/B3–178/9

See the "Instructions for the user" at the beginning of the volume

Ti	Titanium systems:		
	Ti–Nb–Ta	50 (Ta):	Hb/B2–104
	Ti–Nb–V	49 (Nb):	Hb/B2–276/7
	Ti–Nb–Zr	49 (Nb):	Hb/B2–174/6
	Ti–Ni	57 (Ni):	Hb/B1–79/82
	Ti–Ni–O	57 (Ni):	Hb/B1–97
	Ti–Ni–V	57 (Ni):	Hb/B1–159
	Ti–O	41 (Ti):	Hb–211/5
	Ti–O–S	41 (Ti):	Hb–343
	Ti–O–Ta	50 (Ta):	Hb/B2–221
	Ti–O–Ta–Zr	50 (Ta):	Hb/B2–229
	Ti–O–V	48 (V):	Hb/B2–658
	Ti–Os	68 (Pt):	Hb/A6–721
	Ti–P	41 (Ti):	Hb–378
	Ti–Pb	47 (Pb):	Hb/C4–1377/8
	Ti–Pb–Zr	47 (Pb):	Hb/C4–1423
	Ti–Pd	68 (Pt):	Hb/A5–585
	Ti–Ru	63 (Ru):	Eb–187/9
		68 (Pt):	Hb/A5–538
	Ti–S	41 (Ti):	Hb–337/9
	Ti–Se	41 (Ti):	Hb–352
	Ti–Si	41 (Ti):	Hb–375
	Ti–Si–Ta	50 (Ta):	Hb/B2–39
	Ti–Si–V	48 (V):	Hb/B2–659/61
	Ti–Sn	46 (Sn):	Hb/D–325/7
	Ti–Sn–V	48 (V):	Hb/B2–686/7
	Ti–Sn–Zr	46 (Sn):	Hb/D–378
	Ti–Ta	50 (Ta):	Hb/B2–24/5
	Ti–Ta–V	50 (Ta):	Hb/B2–79
	Ti–Ta–Zr	50 (Ta):	Hb/B2–51
	Ti–Te	41 (Ti):	Hb–353/4
	Ti–V	48 (V):	Hb/B2–602/4, 648/57
	Ti–V–Zr	48 (V):	Hb/B2–674/5
	Ti–Zn	41 (Ti):	Hb–201
	Ti–Zr	42 (Zr):	Hb–194
Ti^{+}	Ti^{+}	41 (Ti):	Hb–137/9, 167/8, 186
Ti^{2+}	Ti^{2+}:		
	Chemical reactions	41 (Ti):	Hb–468/9, 475
	Ionization potential	41 (Ti):	Hb–186
	Properties of the ion	41 (Ti):	Hb–137/40
	Spectra	41 (Ti):	Hb–168/9
Ti^{3+}	Ti^{3+}:		
	Chemical reactions	41 (Ti):	Hb–469/71, 475/81
	Ionization potential	41 (Ti):	Hb–186
	Properties of the ion	41 (Ti):	Hb–137/40
	Spectra	41 (Ti):	Hb–169/70

Hb = Main Series Volume; GD = Gmelin-Durrer; Ab = Appended Volume to Main Series; Eb = Supplement Volume to Main Series; Erg. W. = New Supplement Series Volume

Ti^{4+}	Ti^{4+}:		
	Chemical reactions	41 (Ti):	Hb–471/5, 475/81
	Ionization potential	41 (Ti):	Hb–186
	Properties of the ion	41 (Ti):	Hb–137/40
	Spectra	41 (Ti):	Hb–170
Ti^{n+}	Ti^{n+} (n = 5 to 12)	41 (Ti):	Hb–137, 170, 186
$TiZn_3$	$TiZn_3$	41 (Ti):	Hb–201/2
–.	$TiZn_3$ solid solutions:		
	$(Ti,Nb)Zn_3$	49 (Nb):	Hb/B2–118
Tl	Thallium:		
	Abundance	38 (Tl):	Hb/1–4/5
	Affinity.	38 (Tl):	Hb/1–43
	Atom.	38 (Tl):	Hb/1–36/42
	Atomic nucleus	38 (Tl):	Hb/1–30/6
	Behavior as electrode	38 (Tl):	Hb/1–128/9
	Cells	38 (Tl):	Hb/1–117/25
	Chemical reactions	38 (Tl):	Hb/1–132/7
	Cosmochemistry	38 (Tl):	Hb/1–2
	Crystallographic properties	38 (Tl):	Hb/1–44/8
	Detection	38 (Tl):	Hb/1–144/58
	Determination.	38 (Tl):	Hb/1–159/81, 185/6
	Diffusion.	38 (Tl):	Hb/1–54/5
		46 (Sn):	Hb/D–279
		47 (Pb):	Hb/C4–1327/9
	Economics.	38 (Tl):	Hb/1–18/9
	Electrical properties	38 (Tl):	Hb/1–102/13
	Electrochemical behavior	38 (Tl):	Hb/1–114/31
	Formation	38 (Tl):	Hb/1–20/7
	Geochemistry	38 (Tl):	Hb/1–3/12
	History.	38 (Tl):	Hb/1–1
			Hb/3–497/526
	Magnetic properties	38 (Tl):	Hb/1–98/101
	Mechanical properties	38 (Tl):	Hb/1–49/56
	Mineralogy	38 (Tl):	Hb/1–14/8
	Molecule.	38 (Tl):	Hb/1–43
	Nuclear properties	38 (Tl):	Hb/1–30/6
	Occurrence, extraterrestrial	38 (Tl):	Hb/1–2
	–, terrestrial	38 (Tl):	Hb/1–3/14
	Optical properties.	38 (Tl):	Hb/1–72/98
	Overvoltage phenomena	38 (Tl):	Hb/1–127/8
	Position in the electrochemical series	38 (Tl):	Hb/1–115
	– in the periodic system	38 (Tl):	Hb/1–43
	Potentials	38 (Tl):	Hb/1–114/7, 127, 130/1
	Preparation	38 (Tl):	Hb/1–20/7
	– in pure state.	38 (Tl):	Hb/1–23, 25/6

See the "Instructions for the user"
at the beginning of the volume

Tl	Thallium:		
	Separation, analytical	38 (Tl):	Hb/1–181/5
	Sorption	38 (Tl):	Hb/1–55/6
	Special forms	38 (Tl):	Hb/1–26/7
	Spectra:		
	Arc spectrum	38 (Tl):	Hb/1–74/88
	Band spectrum	38 (Tl):	Hb/1–94
	Spark spectra	38 (Tl):	Hb/1–88/93
	X-ray spectrum	38 (Tl):	Hb/1–94/8
	Storage	38 (Tl):	Hb/1–26
	Thermal properties	38 (Tl):	Hb/1–56/72
	Toxicity	38 (Tl):	Hb/1–143
	Uses	38 (Tl):	Hb/1–28/9
	Valence	38 (Tl):	Hb/1–42/3
–	Thallium ions *see* Tl^{+} ... Tl^{+}, *etc.*		
–	Thallium isotopes	38 (Tl):	Hb/1–30/1
			Hb/3–497/526
–	Thallium systems:		
	Tl–Ag	61 (Ag):	Hb/C–205/6
	Tl–Ag–Al	35 (Al):	Hb/A6–1087
	Tl–Ag–Mg	27 (Mg):	Hb/A4–748/9
	Tl–Ag–S	61 (Ag):	Hb/B4–225
	Tl–Al	35 (Al):	Hb/A5–814/5
	Tl–As	38 (Tl):	Hb/3–433
	Tl–Au	62 (Au):	Hb/3–832/3
	Tl–Au–Zn	62 (Au):	Hb/3–798
	Tl–Bi	38 (Tl):	Hb/2–191/4
	Tl–Bi–Cd	38 (Tl):	Hb/2–215/6
	Tl–Bi–Pb	47 (Pb):	Hb/C4–1347
	Tl–Ca	38 (Tl):	Hb/2–209/11
	Tl–Cd	38 (Tl):	Hb/2–213/4
	Tl–Cd–$CdCl_2$–TlCl	38 (Tl):	Hb/3–482/4
	Tl–Cd–Mg	27 (Mg):	Hb/A4–582/4
	Tl–Cd–Pb	47 (Pb):	Hb/C4–1347
	Tl–Cd–Sn	46 (Sn):	Hb/D–291
	Tl–Cu	60 (Cu):	Hb/B3–1183
	Tl–Ga–Sn	46 (Sn):	Hb/D–291
	Tl–Ge	45 (Ge):	Eb–480
	Tl–Hg	34 (Hg):	Hb/A2–1146/8
	Tl–In	37 (In):	Hb–59/60
		38 (Tl):	Hb/2–216/7
	Tl–In–Pb	47 (Pb):	Hb/C4–1347
	Tl–K	38 (Tl):	Hb/2–208/9
	Tl–Li	38 (Tl):	Hb/2–203/5
	Tl–Mg	27 (Mg):	Hb/A4–576/7
	Tl–Mg–Zn	27 (Mg):	Hb/A4–581
	Tl–Na	38 (Tl):	Hb/2–205/8

Tl	Thallium systems:		
	Tl–Na–NaBr	38 (Tl):	Hb/3–443
	Tl–Ni	57 (Ni):	Hb/B1–64/5
	Tl–P	38 (Tl):	Hb/3–425/6
	Tl–Pb	47 (Pb):	Hb/C4–1329/32
	Tl–Pb–PbI_2–TlI	47 (Pb):	Hb/C4–1358
	Tl–Pb–Pt	68 (Pt):	Hb/A6–767
	Tl–Pb–Sn	47 (Pb):	Hb/C4–1530
	Tl–Pt	68 (Pt):	Hb/A6–755/7
	Tl–S	38 (Tl):	Hb/3–339
	Tl–Sb	38 (Tl):	Hb/2–187/91
	Tl–Se	38 (Tl):	Hb/3–370/1
	Tl–Si	38 (Tl):	Hb/3–424
	Tl–Sn	46 (Sn):	Hb/D–279/81
	Tl–Sn–Zn	46 (Sn):	Hb/D–290
	Tl–Te	38 (Tl):	Hb/3–376
	Tl–Zn	38 (Tl):	Hb/2–212/3
Tl^+	Tl^+:		
	Chemical reactions	38 (Tl):	Hb/1–137/40
	Diffusion in sodium halides	21 (Na):	Eb/6–291
	Ionic entropy	38 (Tl):	Hb/1–71/2
	Ionic mobility	38 (Tl):	Hb/1–126
	Ionic refraction	38 (Tl):	Hb/1–73
	Ionization	38 (Tl):	Hb/1–113
	Properties of the ion	38 (Tl):	Hb/1–38/9, 41
	Redox cells	38 (Tl):	Hb/1–130/1
	Spectra:		
	Crystal spectra	21 (Na):	Eb/6–333/4
	Luminescence spectra	21 (Na):	Eb/6–348/9
	Spark spectrum	38 (Tl):	Hb/1–88/91
Tl^{2+}	Tl^{2+}	21 (Na):	Eb/6–337
		38 (Tl):	Hb/1–39, 41, 91/3, 113
Tl^{3+}	Tl^{3+}:		
	Chemical reactions	38 (Tl):	Hb/1–140/3
	Ionization	38 (Tl):	Hb/1–113
	Properties of the ion	38 (Tl):	Hb/1–39/41
	Redox cells	38 (Tl):	Hb/1–130/1
	Spectra	38 (Tl):	Hb/1–93
Tl^{n+}	Tl^{n+} (n = 4 to 21)	38 (Tl):	Hb/1–39
Tl_2^{2+}	Tl_2^{2+}	38 (Tl):	Hb/1–43
Tm	Thulium:		
	Abundance	39 (SE):	Hb/A1–34/5
	Band model	39 (SE):	Hb/B3–213/5
	Boiling point	39 (SE):	Hb/B3–105, 108
	Charge carrier concentration	39 (SE):	Hb/B3–228
	Cosmochemistry	39 (SE):	Hb/A1–23/7
	Crystal chemistry	39 (SE):	Hb/A1–28/32

See the "Instructions for the user"
at the beginning of the volume

Tm . . . . Thulium:
Crystal field parameters . . . . 39 (SE): Hb/B3–218/9
Crystallographic properties . . . . . . 39 (SE): Hb/B3–62/3, 66/7, 73, 77/80
Crystal structure . . . . 39 (SE): Hb/B3–66/7, 73
Debye temperature . . . . 39 (SE): Hb/B3–192
Density . . . . 39 (SE): Hb/B3–85/7
Diffusion . . . . 39 (SE): Hb/B3–77/8
Economic deposits . . . . 39 (SE): Hb/A1–110/22
Elasticity . . . . 39 (SE): Hb/B3–92/8
Electrical properties . . . . 39 (SE): Hb/B3–209/83
Electrical resistance . . . . 39 (SE): Hb/B3–248/9, 262
Electron density . . . . 39 (SE): Hb/B3–217
Electron emission . . . . 39 (SE): Hb/B3–278/83
Enthalpy of fusion . . . . 39 (SE): Hb/B3–120
– of sublimation . . . . 39 (SE): Hb/B3–110, 115
– of vaporization . . . . 39 (SE): Hb/B3–110, 115
Exchange interaction . . . . 39 (SE): Hb/B3–298/303
Geochemistry . . . . 39 (SE): Hb/A1–28/54
Grüneisen constant . . . . 39 (SE): Hb/B3–84/5
Hall effect . . . . 39 (SE): Hb/B3–271/2
History . . . . 39 (SE): Hb/A1–4/22
Hyperfine interactions . . . . 39 (SE): Hb/B3–224
Isomer shift . . . . 39 (SE): Hb/B3–217
Magnetic anisotropy . . . . 39 (SE): Hb/B3–305/6
Magnetic form factor . . . . 39 (SE): Hb/B3–315/6
Magnetic moment . . . . 39 (SE): Hb/B3–334
Magnetic properties . . . . 39 (SE): Hb/B3–298/336
Magnetic structure . . . . 39 (SE): Hb/B3–307/9, 315
Magnetization . . . . 39 (SE): Hb/B3–317, 320/2
Mechanical properties . . . . 39 (SE): Hb/B3–85/103, 193/5, 207/9
Melting point . . . . 39 (SE): Hb/B3–118/9
Mineralogy . . . . 39 (SE): Hb/A1–54/110
Molar volume . . . . 39 (SE): Hb/B3–85/7
Occurrence, extraterrestrial . . . . . . 39 (SE): Hb/A1–22/7
–, terrestrial . . . . 39 (SE): Hb/A1–27/54
– in meteorites . . . . 39 (SE): Hb/A1–25/7
– in minerals . . . . 39 (SE): Hb/A1–35/53
– in the biosphere . . . . 39 (SE): Hb/A1–53/4
– in the lithosphere . . . . 39 (SE): Hb/A1–27/54
Optical properties . . . . 39 (SE): Hb/B3–342
Polymorphism . . . . 39 (SE): Hb/B3–62/3
Position in the periodic system . . . . 39 (SE): Hb/A1–2/4
Spectra . . . . 49 (Nb): Hb/B3–133
Structure . . . . 39 (SE): Hb/B3–80
Surface tension . . . . 39 (SE): Hb/B3–208/9
Susceptibility . . . . 39 (SE): Hb/B3–334/6
Thermal conductivity . . . . 39 (SE): Hb/B3–200/4

Tm . . . . Thulium:

Thermal expansion . . . . 39 (SE): Hb/B3–88/92
Thermal properties . . . . 39 (SE): Hb/B3–83/92, 103/20, 175/9, 192, 200/6
Thermodynamic functions . . . . 39 (SE): Hb/B3–175/9
Thermoelectric power . . . . 39 (SE): Hb/B3–277
Valence . . . . 39 (SE): Hb/A1–4
Vapor pressure . . . . 39 (SE): Hb/B3–103/5, 108

–. . . . . Thulium ions
see Tm^{3+} ... Tm^{3+}

–. . . . . Thulium isotopes:

Nuclear decay . . . . Erg. W. 7a–86
Uses . . . . Erg. W. 8–207

–. . . . . Thulium systems:

Tm–Ag . . . . 61 (Ag): Hb/C–234/5
Tm–H . . . . 39 (SE): Hb/C1–2/3, 75
Tm–Pb . . . . 47 (Pb): Hb/C4–1371

Tm^{3+} . . . . Tm^{3+} . . . . 39 (SE): Hb/A1–3
Hb/B3–317
49 (Nb): Hb/B3–133

Tn . . . . Thoron = ^{220}Rn
see Rn ... Radon

Tu . . . . Thulium
see Tm ... Thulium

U . . . . Uranium:

Atom . . . . 55 (U): Hb–45/6, 52/4
Erg. W. 8–1/34
Atomic nucleus . . . . 55 (U): Hb–44/52, 271/9
Atomic radius . . . . 55 (U): Hb–52
Erg. W. 8–4/5
Catalytic properties . . . . 55 (U): Hb–66
Chemical reactions . . . . 55 (U): Hb–64/6
Cosmochemistry . . . . 55 (U): Hb–2
Crystallographic properties . . . . . . 55 (U): Hb–54
Detection . . . . 55 (U): Hb–67/9
– in iron . . . . 59 (Fe): Hb/FII1–62/3
– of uranium isotopes
in soils and fallout . . . . Erg. W. 7b–353/8
Determination . . . . 55 (U): Hb–69/75
– in iron . . . . 59 (Fe): Hb/FII1–63/6
– in special substances . . . . 55 (U): Hb–80
Dressing and treatment of ores . . . . 55 (U): Hb–38/9
Economic deposits . . . . 55 (U): Hb–3/12
Economics . . . . 55 (U): Hb–37/8

See the "Instructions for the user"
at the beginning of the volume

U . . . . Uranium:
Electrical properties . . . . 55 (U): Hb–63/4
Electrochemical behavior . . . . 55 (U): Hb–64
Electron configuration . . . . 55 (U): Hb–45/6
Erg. W. 8–3/23
Formation . . . . 55 (U): Hb–39/41, 271
Geochemistry . . . . 55 (U): Hb–1/12
History . . . . 55 (U): Hb–1, 272/9
Ionization potentials . . . . Erg. W. 8–23
Magnetic properties . . . . 55 (U): Hb–62
Mechanical properties . . . . 55 (U): Hb–54/5
Mineralogy . . . . 55 (U): Hb–12/37
Nuclear decay,
alpha radioactivity . . . . 55 (U): Hb–46/50, 271/2
Erg. W. 7a–82/96
–, decay scheme . . . . Erg. W. 7a–97, 102
Nuclear fission,
fission by charged particles . . . . . 55 (U): Hb–45
Erg. W. 7a–151/2
–, – by neutrons . . . . Erg. W. 7a–140, 143/5
–, spontaneous fission . . . . Erg. W. 7a–134/8, 140
–, theoretical . . . . Erg. W. 7a–163/70
Nuclear properties . . . . 55 (U): Hb–44/52, 271/9
Nuclear structure . . . . 55 (U): Hb–44, 271
Occurrence, extraterrestrial . . . . . . 55 (U): Hb–2
–, terrestrial . . . . 55 (U): Hb–1/12
Optical properties . . . . 55 (U): Hb–56/62
Oxidation states . . . . Erg. W. 8–8/11
Position in the periodic system . . . . 55 (U): Hb–54
Preparation . . . . 55 (U): Hb–39/43
–, industrial . . . . 55 (U): Hb–38/42
– in pure state . . . . 55 (U): Hb–42/3
Production in charged particle
reactions . . . . Erg. W. 7b–145
Purification . . . . 55 (U): Hb–51
Radioactivity . . . . 55 (U): Hb–46/52
Recoil reactions . . . . Erg. W. 8–315/9
Safety measures . . . . Erg. W. 8–385/93
Separation, analytical . . . . 55 (U): Hb–75/9
Special forms . . . . 55 (U): Hb–43
Spectra:
Absorption spectra . . . . 55 (U): Hb–58
Erg. W. 8–52, 63/4
Emission spectra . . . . 55 (U): Hb–56/8
Mössbauer spectra . . . . Erg. W. 8–89/90
X-ray spectrum . . . . 55 (U): Hb–58/61
Thermal properties . . . . 55 (U): Hb–55
Uses . . . . 55 (U): Hb–43/4
Erg. W. 8–206

Hb = Main Series Volume; GD = Gmelin-Durrer; Ab = Appended Volume to Main Series;
Eb = Supplement Volume to Main Series; Erg. W. = New Supplement Series Volume

U . . . . Uranium ions
see U^+ ... U^+, etc.

–. . . . . Uranium isotopes
see U ... Uranium

–. . . . . Uranium systems:
U–Ag . . . . 61 (Ag): Hb/C–315/6
U–Bi–Ni. . . . . 57 (Ni): Hb/B1–277
U–C–Ni . . . . 57 (Ni): Hb/B1–278
U–C–Pu. . . . . Erg. W. 4–211/3
U–Co . . . . 58 (Co): Eb/A–452/5
U–Co–Ru . . . . 63 (Ru): Eb–224
U–Cr–Ni . . . . 57 (Ni): Hb/B1–277
U–Cu . . . . 60 (Cu): Hb/B3–1240
U–Cu–O . . . . 60 (Cu): Hb/B3–1240
U–Mn–Ni . . . . 57 (Ni): Hb/B1–314
U–Nb . . . . 49 (Nb): Hb/A–173/4
U–Ni. . . . . 57 (Ni): Hb/B1–272/3
U–Ni–Si. . . . . 57 (Ni): Hb/B1–277
U–O . . . . 55 (U): Hb–82
U–O–Pu. . . . . Erg. W. 4–59/60
U–Ru . . . . 63 (Ru): Eb–213/6

U^+ . . . . U^+ . . . . 55 (U): Hb–45
Erg. W. 8–23

U^{2+} . . . . U^{2+} . . . . 55 (U): Hb–45
Erg. W. 8–11/23

U^{3+} . . . . U^{3+}:
Crystal field effects . . . . Erg. W. 8–24/5
Ionic susceptibility . . . . 55 (U): Hb–62
Ionization potential . . . . Erg. W. 8–23
Magnetic moment . . . . Erg. W. 8–26/7
Properties of the ion . . . . 55 (U): Hb–45
Erg. W. 8–5, 11/22
Spectra . . . . Erg. W. 8–52

U^{4+} . . . . U^{4+}:
Crystal field effects . . . . Erg. W. 8–24/5
Electrochemical behavior . . . . 55 (U): Hb–255/6
Ionic susceptibility . . . . 55 (U): Hb–62, 136, 254/5
Ionization potential . . . . Erg. W. 8–23
Properties of the ion . . . . 55 (U): Hb–45, 52
Erg. W. 8–5, 11/22
Spectra . . . . 55 (U): Hb–245/6
Erg. W. 8–63/4

U^{5+} . . . . U^{5+} . . . . 55 (U): Hb–45, 246
Erg. W. 8–5, 11/25

U^{6+} . . . . U^{6+} . . . . 21 (Na): Eb/6–354/5
55 (U): Hb–254
Erg. W. 8–5, 23

see also O_2U^{2+} ... $UO_2{}^{2+}$

See the "Instructions for the user" at the beginning of the volume

U^{n+} . . . . U^{n+} (n = 7 to 10) . . . . Erg. W. 8–23

UI . . . . Uranium I = ^{238}U
see U ... Uranium

UII . . . . Uranium II = ^{234}U
see U ... Uranium

UX_1 . . . . Uranium X_1 = ^{234}Th
see Th ... Thorium

UX_2 . . . . Uranium X_2 = ^{234}Pa
see Pa ... Protactinium

UY . . . . Uranium Y = ^{231}Th
see Th ... Thorium

UZ. . . . . Uranium Z = ^{234}Pa
see Pa ... Protactinium

V . . . . Vanadium:
Absorption of X- and
gamma rays . . . . 48 (V): Hb/A2–384/9
Abundance . . . . 48 (V): Hb/A1–11/2, 39
Atom. . . . . 48 (V): Hb/A2–332/48
Atomic nucleus . . . . 48 (V): Hb/A2–321/9
Atomic radius . . . . 48 (V): Hb/A2–343/6
Atomic scattering factors. . . . . 48 (V): Hb/A2–340/3
Atomic weight . . . . 48 (V): Hb/A2–323
Behavior as electrode . . . . 48 (V): Hb/A2–489/93
Boiling point . . . . 48 (V): Hb/A2–437
Bond properties. . . . . 48 (V): Hb/A2–389/92
Cells . . . . 48 (V): Hb/A2–471/4
Characteristic electron
energy losses. . . . . 48 (V): Hb/A2–451
Chemical constant . . . . 48 (V): Hb/A2–449
Chemical reactions . . . . 48 (V): Hb/A2–514/24
– by action of electricity . . . . 48 (V): Hb/A2–514
– on heating. . . . . 48 (V): Hb/A2–514
– with compounds . . . . 48 (V): Hb/A2–519/24
– with elements. . . . . 48 (V): Hb/A2–514/9
Cleavage. . . . . 48 (V): Hb/A2–396
Colloidal forms . . . . 48 (V): Hb/A1–317
Compressibility . . . . 48 (V): Hb/A2–419
Coordination number. . . . . 48 (V): Hb/A2–321
Cosmochemistry . . . . 48 (V): Hb/A1–11/8
Creep properties . . . . 48 (V): Hb/A2–436
Crystal chemistry . . . . 48 (V): Hb/A1–18/39
Crystal form . . . . 48 (V): Hb/A2–396
Crystal formation . . . . 48 (V): Hb/A2–397
Crystallographic properties . . . . . . 48 (V): Hb/A2–394/414
Debye temperature . . . . 48 (V): Hb/A2–446/7
Deformation. . . . . 48 (V): Hb/A2–408/9

Hb = Main Series Volume; GD = Gmelin-Durrer; Ab = Appended Volume to Main Series;
Eb = Supplement Volume to Main Series; Erg. W. = New Supplement Series Volume

V . . . . Vanadium:

Density	48 (V):	Hb/A2–415
Detection	48 (V):	Hb/A2–640/52
– in iron	59 (Fe):	Hb/FII1–135/8 Hb/FII2–336/7, 340
– in special substances	48 (V):	Hb/A2–652/7
Determination	48 (V):	Hb/A2–658/76
– in iron	59 (Fe):	Hb/FII1–138/64 Hb/FII2–302/5, 336/7, 357/8
– in presence of other ions	48 (V):	Hb/A2–676/92
– in special substances	48 (V):	Hb/A2–692/724
Diadochy	48 (V):	Hb/A1–25/9
Diffusion	48 (V):	Hb/A2–407/8 Hb/B1–325 Hb/B2–602
Disorder	48 (V):	Hb/A2–406/7
Dressing and treatment of raw materials:		
Chlorination	48 (V):	Hb/A1–216/21
Enrichment as by-product	48 (V):	Hb/A1–196/202
– of ores	48 (V):	Hb/A1–187/96
Heating under reducing conditions	48 (V):	Hb/A1–216
Melt dressing	48 (V):	Hb/A1–223/6
Preparation of solution	48 (V):	Hb/A1–226/42
Roasting	48 (V):	Hb/A1–203/16
Sintering methods	48 (V):	Hb/A1–221/2
Treatment of solution	48 (V):	Hb/A1–242/80
Economic deposits:		
Africa	48 (V):	Hb/A1–129/34
America	48 (V):	Hb/A1–134/46
Asia	48 (V):	Hb/A1–127/9
Australia and Oceania	48 (V):	Hb/A1–129
Europe	48 (V):	Hb/A1–125/7
Soviet Union	48 (V):	Hb/A1–126/8
Economics	48 (V):	Hb/A1–123/5
Elasticity	48 (V):	Hb/A2–417/9
Electrical conductivity	48 (V):	Hb/A2–454/6
Electrical properties	48 (V):	Hb/A2–450/63
Electrochemical behavior	48 (V):	Hb/A2–464/514
Electrode sputtering	48 (V):	Hb/A2–462/3
Electrolytic deposition	48 (V):	Hb/A1–302/6 Hb/A2–493/8
Electrolytic polishing	48 (V):	Hb/A2–406, 492
Electron affinity	48 (V):	Hb/A2–393
Electron configuration	48 (V):	Hb/A2–332
Electronegativity	48 (V):	Hb/A2–392

V . . . . Vanadium:

Electron emission	48 (V):	Hb/A2–461/2
Electron energies	48 (V):	Hb/A2–332/4
Energy band structure	48 (V):	Hb/A2–450/1
Epitaxy	48 (V):	Hb/A2–403
Equation of state	48 (V):	Hb/A2–414/5
Expansion coefficient	48 (V):	Hb/A2–416/7
Fermi surface	48 (V):	Hb/A2–451
Formation	48 (V):	Hb/A2–321
Galvanomagnetic effects	48 (V):	Hb/A2–460/1
Geochemical character	48 (V):	Hb/A1–39
Geochemical cycle	48 (V):	Hb/A1–122/3
Geochemistry	48 (V):	Hb/A1–18/123
Hall effect	48 (V):	Hb/A2–461
Hardness	48 (V):	Hb/A2–422/8
Heat of fusion	48 (V):	Hb/A2–439
– of sublimation	48 (V):	Hb/A2–437/8
– of vaporization	48 (V):	Hb/A2–437/8
History	48 (V):	Hb/A1–1/10
Increments	48 (V):	Hb/A2–394
Interaction with electrons	48 (V):	Hb/A2–392/3
Internal friction	48 (V):	Hb/A2–420/1
Ionization	48 (V):	Hb/A2–392/3
Isomorphism	48 (V):	Hb/A2–402
Lattice defects	48 (V):	Hb/A2–406/7
Lattice energy	48 (V):	Hb/A2–412/4
Lattice structure	48 (V):	Hb/A2–397/402
Lattice vibrations	48 (V):	Hb/A2–447/9
Magnetic moment	48 (V):	Hb/A2–334/6
Magnetic properties	48 (V):	Hb/A2–450/61
Magnetostriction	48 (V):	Hb/A2–454
Mechanical properties	48 (V):	Hb/A2–414/37
Melting point	48 (V):	Hb/A2–438/9
Mesonic atom	48 (V):	Hb/A2–383/4
Mineralogy	48 (V):	Hb/A1–36/9, 146/87
Modifications	48 (V):	Hb/A2–395/6
Nuclear magnetic resonance and relaxation	48 (V):	Hb/A2–451/3
Nuclear properties	48 (V):	Hb/A2–321/9
Occurrence, extraterrestrial	48 (V):	Hb/A1–11/8
–, terrestrial	48 (V):	Hb/A1–18/123
– in meteorites	48 (V):	Hb/A1–15/7
– in minerals	48 (V):	Hb/A1–18/25, 30/5
– in soils	48 (V):	Hb/A1–97/104
– in tektites and impactites	48 (V):	Hb/A1–17/8
– in the atmosphere	48 (V):	Hb/A1–114
– in the biosphere	48 (V):	Hb/A1–105, 116/22
– in the hydrosphere	48 (V):	Hb/A1–110/5

V . . . . Vanadium:

Occurrence in the lithosphäre . . . . . 48 (V): Hb/A1–40/110
– in the magmatic cycle . . . . 48 (V): Hb/A1–40/52
– in the metamorphic cycle . . . . . . 48 (V): Hb/A1–104/10
– in the sedimentary cycle . . . . 48 (V): Hb/A1–52/104
Optical properties. . . . . 48 (V): Hb/A2–463/4
Overvoltage phenomena . . . . 48 (V): Hb/A2–489
Paramagnetic resonance absorption . 48 (V): Hb/A2–336/40
Polarography . . . . 48 (V): Hb/A2–474/87
Polymorphism . . . . 48 (V): Hb/A2–395/6
Position in the electrochemical
series . . . . 48 (V): Hb/A2–469
– in the periodic system . . . . 48 (V): Hb/A2–321
Potentials . . . . 48 (V): Hb/A2–470/1, 502/6
Preparation . . . . 48 (V): Hb/A2–321
–, industrial . . . . 48 (V): Hb/A1–187/320
see also Dressing
– in pure state. . . . . 48 (V): Hb/A1–311/5
– of metallic vanadium . . . . 48 (V): Hb/A1–285/320
– of special forms. . . . . 48 (V): Hb/A1–315/7
Purification . . . . 48 (V): Hb/A1–306/15
Purity testing . . . . 48 (V): Hb/A1–317/20
Radiation of the mesonic atom . . . . 48 (V): Hb/A2–383/4
Recrystallization . . . . 48 (V): Hb/A2–409/12
Redox potentials and cells . . . . 48 (V): Hb/A2–498/506
Scattering of electrons . . . . 48 (V): Hb/A2–393/4
– of X- and gamma rays . . . . 48 (V): Hb/A2–384/9
Separation, analytical . . . . 48 (V): Hb/A2–676/92
Sorption of hydrogen. . . . . 48 (V): Hb/B1–1/5
Sound absorption. . . . . 48 (V): Hb/A2–420/1
Sound velocity . . . . 48 (V): Hb/A2–419/20
Spectra:
Arc spectrum . . . . 48 (V): Hb/A2–352/4
Excitation . . . . 48 (V): Hb/A2–348/52
Spark spectra . . . . 48 (V): Hb/A2–354/71
Terms . . . . 48 (V): Hb/A2–332, 348/50
X-ray spectrum . . . . 48 (V): Hb/A2–371/83
Standard potential . . . . 48 (V): Hb/A2–464/9
Strength . . . . 48 (V): Hb/A2–428/36
Structure . . . . 48 (V): Hb/A2–403/6
– of thin films . . . . 48 (V): Hb/A2–402/3
Superconductivity . . . . 48 (V): Hb/A2–457/60
Superlattice . . . . 48 (V): Hb/A2–403
Surface tension . . . . 48 (V): Hb/A2–437
Susceptibility . . . . 48 (V): Hb/A2–453/4
Thermal conductivity. . . . . 48 (V): Hb/A2–449/50
Thermal expansion . . . . 48 (V): Hb/A2–416/7
Thermal properties . . . . 48 (V): Hb/A2–414/50

*See the "Instructions for the user"
at the beginning of the volume*

V . . . . Vanadium:

Thermal radiation	48 (V):	Hb/A2–464
Thermodynamic functions	48 (V):	Hb/A2–439/46
Thermoelectric effects	48 (V):	Hb/A2–461
Toxicity	48 (V):	Hb/A2–725/8
Twinning	48 (V):	Hb/A2–396/7
Valence	48 (V):	Hb/A2–321
Vapor pressure	48 (V):	Hb/A2–437
Wave functions	48 (V):	Hb/A2–332/4
Workability	48 (V):	Hb/A2–421/2

–. . . . . Vanadium ions
see V^+ ... V^+, etc.

–. . . . . Vanadium isotopes:

Abundance	48 (V):	Hb/A1–39 Hb/A2–322
Detection and determination	48 (V):	Hb/A2–724/5
Nuclear decay	48 (V):	Hb/A2–327/9
Nuclear properties	48 (V):	Hb/A2–321/6
Preparation	48 (V):	Hb/A2–329/32
Separation	48 (V):	Hb/A2–332

–. . . . . Vanadium systems:

V–Al	35 (Al):	Hb/A5–864
V–Al–C–Ta	50 (Ta):	Hb/B2–343
V–Al–Nb	49 (Nb):	Hb/B2–272
V–Al–Ta	50 (Ta):	Hb/B2–78/9
V–As	48 (V):	Hb/B1–364
V–As–C	48 (V):	Hb/B1–366/7
V–B	48 (V):	Hb/B1–321
V–B–C	48 (V):	Hb/B1–352
V–B–Nb	49 (Nb):	Hb/B4–312
V–B–Si	48 (V):	Hb/B1–357
V–B–Ta	50 (Ta):	Hb/B2–322
V–B–Ti	48 (V):	Hb/B2–659
V–Be	48 (V):	Hb/B2–578/9
V–Be–Ti	48 (V):	Hb/B2–661/2
V–Bi	48 (V):	Hb/B2–577
V–C	48 (V):	Hb/B1–325/6
V–C–Fe–Si	59 (Fe):	Hb/A9–1936
V–C–Ga	48 (V):	Hb/B2–592
V–C–Ge–Si	48 (V):	Hb/B2–684
V–C–H	48 (V):	Hb/B1–335
V–C–Hf	48 (V):	Hb/B2–677
V–C–N	48 (V):	Hb/B1–350
V–C–Nb	49 (Nb):	Hb/B4–330/1
V–C–Nb–Si	49 (Nb):	Hb/B2–271
V–C–O	48 (V):	Hb/B1–335/6
V–C–Ta	50 (Ta):	Hb/B2–341/3

V . . . . Vanadium systems:

System	Volume	Reference
V–C–Ta–Ti	50 (Ta):	Hb/B2–343
V–C–Ta–Zr	50 (Ta):	Hb/B2–343/4
V–C–Ti	48 (V):	Hb/B2–659
V–C–Zr	48 (V):	Hb/B2–674
V–Cd	48 (V):	Hb/B2–581
V–Ce	48 (V):	Hb/B2–598
V–Ce–Ti	48 (V):	Hb/B2–662
V–Co	58 (Co):	Eb/A–416/7
V–Co–O	58 (Co):	Eb/A–873
V–Cr	52 (Cr):	Hb/A2–727/8
V–Cu	60 (Cu):	Hb/B3–1209
V–Dy	48 (V):	Hb/B2–600/1
V–Er	48 (V):	Hb/B2–599/600
V–Fe	59 (Fe):	Hb/A9–1920/2
V–Fe–FeS–VS	59 (Fe):	Hb/A9–1924/5
V–Fe–Fe_3C–V_4C_3	59 (Fe):	Hb/A9–1926/31
V–Ga	48 (V):	Hb/B2–581/3
V–Ga–N	48 (V):	Hb/B2–592
V–Ga–O	48 (V):	Hb/B2–591/2
V–Ga–O–Si	48 (V):	Hb/B2–592
V–Ga–Si	48 (V):	Hb/B2–592
V–Gd	48 (V):	Hb/B2–600/1
V–Gd–Ti	48 (V):	Hb/B2–662
V–Ge	48 (V):	Hb/B2–680/1
V–Ge–Si	48 (V):	Hb/B2–683/4
V–H	48 (V):	Hb/B1–1/7
V–H–Ti	48 (V):	Hb/B2–657/8
V–H–Zr	48 (V):	Hb/B2–673
V–Hf	48 (V):	Hb/B2–675/7
V–Hf–Ta	50 (Ta):	Hb/B2–84
V–Hg	48 (V):	Hb/B2–581
V–La	48 (V):	Hb/B2–596/7
V–La–Ti	48 (V):	Hb/B2–662
V–Lu	48 (V):	Hb/B2–600/1
V–Mo	48 (V):	Hb/A2–402
V–N	48 (V):	Hb/B1–167/8
V–N–Ta	50 (Ta):	Hb/B2–267
V–N–Ti	48 (V):	Hb/B2–658
V–N–Zn	48 (V):	Hb/B2–581
V–Nb	49 (Nb):	Hb/B2–243/4
V–Nb–Si	49 (Nb):	Hb/B2–268
V–Nb–Ta	50 (Ta):	Hb/B2–118
V–Nb–Ti	49 (Nb):	Hb/B2–276/7
V–Nb–Zr	49 (Nb):	Hb/B2–295/6
V–Nd	48 (V):	Hb/B2–600
V–Ni	57 (Ni):	Hb/B1–153/5
V–Ni–O	57 (Ni):	Hb/B1–159
		Hb/B3–1198

See the "Instructions for the user" at the beginning of the volume

V . . . . Vanadium systems:
V–Ni–Si . . . . 57 (Ni): Hb/B1–159
V–Ni–Ta . . . . 57 (Ni): Hb/B1–172
V–Ni–Ti . . . . 57 (Ni): Hb/B1–159
V–Ni–Zr . . . . 57 (Ni): Hb/B1–159
V–O . . . . 48 (V): Hb/B1–11/2, 18/9
V–O–Ta . . . . 50 (Ta): Hb/B2–244
V–O–Ti . . . . 48 (V): Hb/B2–658
V–P . . . . 48 (V): Hb/B1–358
V–Pb . . . . 48 (V): Hb/B2–687
V–Pr . . . . 48 (V): Hb/B2–600
V–Rb . . . . 48 (V): Hb/B2–578
V–Ru . . . . 63 (Ru): Eb–194/5
V–S . . . . 48 (V): Hb/B1–267/8
V–Sb . . . . 48 (V): Hb/B2–575
V–Sc . . . . 48 (V): Hb/B2–592/3
V–Se . . . . 48 (V): Hb/B1–308/9
V–Si . . . . 48 (V): Hb/B1–352/4
V–Si–Ta . . . . 50 (Ta): Hb/B2–77/8
V–Si–Ti . . . . 48 (V): Hb/B2–659
V–Sm . . . . 48 (V): Hb/B2–600
V–Sn . . . . 48 (V): Hb/B2–684
V–Sn–Ti . . . . 48 (V): Hb/B2–686
V–Ta . . . . 48 (V): Hb/A2–402
50 (Ta): Hb/B2–68/9
V–Ta–Ti . . . . 50 (Ta): Hb/B2–79
V–Ta–Zn . . . . 50 (Ta): Hb/B2–83
V–Te . . . . 48 (V): Hb/B1–315/6
V–Th . . . . 48 (V): Hb/B2–677/8
V–Ti . . . . 48 (V): Hb/B2–602/4, 648/57
V–Ti–Zr . . . . 48 (V): Hb/B2–674/5
V–VC–VN . . . . 48 (V): Hb/B1–350
V–VN . . . . 48 (V): Hb/B1–176
V–VO . . . . 48 (V): Hb/B1–12/3
V–VSe . . . . 48 (V): Hb/B1–308
V–Y . . . . 48 (V): Hb/B2–593
V–Yb . . . . 48 (V): Hb/B2–600
V–Zn . . . . 48 (V): Hb/B2–579/80
V–Zr . . . . 48 (V): Hb/B2–662/3
V^+ . . . . V^+:
Compton scattering . . . . 48 (V): Hb/A2–393
Form factors . . . . 48 (V): Hb/A2–341/3
Ionization potential . . . . 48 (V): Hb/A2–392/3
Properties of the ion . . . . 48 (V): Hb/A2–332/4, 346/8, 389
Spectra . . . . 48 (V): Hb/A2–354/7
V^{2+} . . . . V^{2+}:
Chemical reactions . . . . 48 (V): Hb/A2–524/7

Hb = Main Series Volume; GD = Gmelin-Durrer; Ab = Appended Volume to Main Series;
Eb = Supplement Volume to Main Series; Erg. W. = New Supplement Series Volume

V^{2+} . . . . V^{2+}:

Compton scattering . . . . 48 (V): Hb/A2–393
Detection . . . . 48 (V): Hb/A2–646
Determination . . . . 48 (V): Hb/A2–677
Electrochemical behavior . . . . 48 (V): Hb/A2–464/7, 476, 498/504, 510/3
Form factors . . . . 48 (V): Hb/A2–341/3
Increments . . . . 48 (V): Hb/A2–394
Ionization potential . . . . 48 (V): Hb/A2–392/3
Paramagnetic resonance . . . . 48 (V): Hb/A2–336/8
Properties of the ion . . . . 48 (V): Hb/A2–332/5, 346/8, 389
Spectra . . . . 21 (Na): Eb/6–336
48 (V): Hb/A2–357/61
Transition reactions . . . . 48 (V): Hb/A2–550/640

–. . . . . $^{50}V^{2+}$ . . . . 48 (V): Hb/A2–337
–. . . . . $^{51}V^{2+}$ . . . . 48 (V): Hb/A2–337
–. . . . . $[V(H_2O)_6]^{2+}$ . . . . 48 (V): Hb/A2–337, 359/60, 525

V^{3+} . . . . V^{3+}:

see also OV^+ ... VO^+

Chemical reactions . . . . 48 (V): Hb/A2–527/33
Compton scattering . . . . 48 (V): Hb/A2–393
Detection . . . . 48 (V): Hb/A2–646
Determination . . . . 48 (V): Hb/A2–676/7
Electrochemical behavior . . . . 48 (V): Hb/A2–464/9, 475/6, 498/505, 510/3
Form factors . . . . 48 (V): Hb/A2–342/3
Increments . . . . 48 (V): Hb/A2–394
Ionization potential . . . . 48 (V): Hb/A2–392/3
Paramagnetic resonance . . . . 48 (V): Hb/A2–337/9
Properties of the ion . . . . 48 (V): Hb/A2–332/5, 346/8
Spectra . . . . 48 (V): Hb/A2–361/7
Transition reactions . . . . 48 (V): Hb/A2–550/640

–. . . . . $[V(H_2O)_6]^{3+}$ . . . . 48 (V): Hb/A2–339, 363, 365/6, 527/8

V^{4+} . . . . V^{4+}:

see also OV^{2+} ... VO^{2+}

Form factors . . . . 48 (V): Hb/A2–342/3
Increments . . . . 48 (V): Hb/A2–394
Ionization potential . . . . 48 (V): Hb/A2–392/3
Paramagnetic resonance . . . . 48 (V): Hb/A2–339/40
Properties of the ion . . . . 48 (V): Hb/A2–332/6, 346/8
Spectra . . . . 48 (V): Hb/A2–367/8

V^{5+} . . . . V^{5+}:

see also OV^{3+} ... VO^{3+},
O_2V^+ ... $VO_2{}^+$

See the "Instructions for the user"
at the beginning of the volume

V^{5+}	V^{5+}:		
	Form factors	48 (V):	Hb/A2–343
	Increments	48 (V):	Hb/A2–394
	Ionization potential	48 (V):	Hb/A2–392/3
	Properties of the ion	48 (V):	Hb/A2–332/4, 346/8
	Spectra	48 (V):	Hb/A2–368/9
V^{n+}	V^{n+} (n = 6 to 20)	48 (V):	Hb/A2–332, 334, 369/71, 392/3
V^-	V^-	48 (V):	Hb/A2–333/4
VZn_3	VZn_3	48 (V):	Hb/B2–579/81
V_2Zr	V_2Zr	48 (V):	Hb/B2–662/5, 669
–	ZrV_2 solid solutions:		
	$Zr(V,Nb)_2$	49 (Nb):	Hb/B2–295
V_4Zn_5	V_4Zn_5	48 (V):	Hb/B2–579/81
W	Tungsten:		
	Atom	54 (W):	Hb–41/3, 63
	Atomic nucleus	54 (W):	Hb–40/1
	Chemical reactions	54 (W):	Hb–86/90
	Colloidal forms	54 (W):	Hb–33
	Cosmochemistry	54 (W):	Hb–4
	Crystallographic properties	54 (W):	Hb–43/5
	Detection	54 (W):	Hb–90/3
	– in iron	59 (Fe):	Hb/FII1–86/7 Hb/FII2–335/6
	Determination	54 (W):	Hb–93/8
	– in iron	59 (Fe):	Hb/FII1–87/98 Hb/FII2–301, 335/6, 351/2
	Dressing and treatment of ores	54 (W):	Hb–19/23
	Economic deposits	54 (W):	Hb–4/11
	Economics	54 (W):	Hb–16/9
	Electrical properties	54 (W):	Hb–68/84
	Electrochemical behavior	54 (W):	Hb–85/6
	Formation	54 (W):	Hb–25/9
	History	54 (W):	Hb–1/3
	Magnetic properties	54 (W):	Hb–67/8
	Mechanical properties	54 (W):	Hb–46/51
	Mineralogy	54 (W):	Hb–12/6
	Occurrence	54 (W):	Hb–3/16
	Optical properties	54 (W):	Hb–58/67
	Preparation	54 (W):	Hb–25/32
	–, industrial	54 (W):	Hb–19/23
	– of special forms	54 (W):	Hb–32/7
	Purification	54 (W):	Hb–31/2
	Separation, analytical	54 (W):	Hb–99/100
	Special forms	54 (W):	Hb–32/7
	Spectra	54 (W):	Hb–61/7
	Thermal properties	54 (W):	Hb–51/8

W . . . . Tungsten:
Uses . . . . 54 (W): Hb–24/5
Working . . . . 54 (W): Hb–37/40
–. . . . . Tungsten ions . . . . 54 (W): Hb–61/3
–. . . . . Tungsten isotopes. . . . . 54 (W): Hb–40
–. . . . . Tungsten systems:
W–Ag . . . . 61 (Ag): Hb/C–314/5
W–C. 54 (W): Hb–187/9
W–C–Fe. . . . . 59 (Fe): Hb/A5–1116
W–C–Ni. . . . . 57 (Ni): Hb/B3–1230
W–Co . . . . 58 (Co): Hb/A1–201/2
Eb/A–433/7, 443/4
W–Co–O . . . . 58 (Co): Eb/A–877
W–Cr . . . . 54 (W): Hb–105
W–Cr–Ni . . . . 57 (Ni): Hb/B1–269/70
W–Cu . . . . 60 (Cu): Hb/B3–1235
W–Cu–O . . . . 60 (Cu): Hb/B3–1235
W–Fe . . . . 59 (Fe): Hb/A5–1115/6
W–Mo. . . . . 54 (W): Hb–105
W–Nb . . . . 49 (Nb): Hb/A–161
W–Ni . . . . 57 (Ni): Hb/B1–255/6
W–Ni–O . . . . 57 (Ni): Hb/B3–1225
W–Ni–Si . . . . 57 (Ni): Hb/B1–269
W–Pb . . . . 54 (W): Hb–104
W–Pt . . . . 68 (Pt): Hb/A6–774/5
W–Re . . . . 70 (Re): Hb–84/5
W–Ru . . . . 63 (Ru): Eb–210/3
W–Th . . . . 54 (W): Hb–103

Xe. . . . . Xenon:
Affinity. . . . . 1 (EG): Hb–223/4
Atom. . . . . 1 (EG): Hb–182/5, 190
Atomic nucleus . . . . 1 (EG): Hb–182
Chemical reactions . . . . 1 (EG): Hb–223/4, 229
Cosmochemistry . . . . 1 (EG): Hb–6
Detection . . . . 1 (EG): Hb–240/1
Determination. . . . . 1 (EG): Hb–240/1
Electrical properties . . . . 1 (EG): Hb–191/2
Geochemistry . . . . 1 (EG): Hb–7/35
History. . . . . 1 (EG): Hb–3/6
Ionization . . . . 1 (EG): Hb–192
Mechanical properties . . . . 1 (EG): Hb–185/8
Molecule. . . . . 1 (EG): Hb–185
Occurrence, extraterrestrial . . . . . . 1 (EG): Hb–6
–, terrestrial . . . . 1 (EG): Hb–7/35
– in the atmosphere . . . . 1 (EG): Hb–7/17
– in the hydrosphere . . . . 1 (EG): Hb–17/28
– in the lithosphere . . . . 1 (EG): Hb–28/35

See the "Instructions for the user"
at the beginning of the volume

Formula	Entry	Volume	Pages
Xe	Xenon:		
	Optical properties	1 (EG):	Hb–189/91
	Position in the periodic system	1 (EG):	Hb–223
	Recovery from air	1 (EG):	Hb–169/70
		3 (O):	Hb/2–197/8
	Solubility	1 (EG):	Hb–188
	Sorption	1 (EG):	Hb–188
	– on nickel	57 (Ni):	Hb/B2–316
	– on platinum	68 (Pt):	Hb/C1–1, 3
	Spectra	1 (EG):	Hb–189/91
	Thermal properties	1 (EG):	Hb–185/8
	Uses	1 (EG):	Hb–249/51
	Valence	1 (EG):	Hb–223/4
–	$Xe \cdot 5.75\,H_2O$	Erg. W. 1–156/7	
–	$Xe \cdot 5.75\,D_2O$	Erg. W. 1–157/8	
–	$Xe \cdot nH_2O$	1 (EG):	Hb–225
–	Xenon compounds	Erg. W. 1–24/129, 154/60	
–	Xenon isotopes	1 (EG):	Hb–182
–	Xenon mixtures:		
	$Xe-CO_2$	14 (C):	Hb/C1–509
–	Xenon systems:		
	Xe–Ag	61 (Ag):	Hb/B1–6/8
	$Xe-B_2H_6$	13 (B):	Eb–115
	Xe–O	3 (O):	Hb/5–1190
Xe^+	Xe^+	1 (EG):	Hb–192
Xe_2	Xe_2	1 (EG):	Hb–185
Y	Yttrium:		
	Abundance	39 (SE):	Hb/A1–34/5
	Cosmochemistry	39 (SE):	Hb/A1–23/7
	Crystal chemistry	39 (SE):	Hb/A1–28/32
	Crystallographic properties	39 (SE):	Hb/B3–28/32
	Economic deposits	39 (SE):	Hb/A1–110/22
	Electrical properties	39 (SE):	Hb/B3–48/55
	Geochemistry	39 (SE):	Hb/A1–28/54
	History	39 (SE):	Hb/A1–4/22
	Magnetic properties	39 (SE):	Hb/B3–48/55
	Mechanical properties	39 (SE):	Hb/B3–32/7
	Mineralogy	39 (SE):	Hb/A1–54/110
	Occurrence, extraterrestrial	39 (SE):	Hb/A1–22/7
	–, terrestrial	39 (SE):	Hb/A1–27/54
	– in meteorites	39 (SE):	Hb/A1–25/7
	– in the biosphere	39 (SE):	Hb/A1–53/4
	– in the lithosphere	39 (SE):	Hb/A1–27/54
	Optical properties	39 (SE):	Hb/B3–56
	Position in the periodic system	39 (SE):	Hb/A1–2/4
	Thermal properties	39 (SE):	Hb/B3–32/48
	Valence	39 (SE):	Hb/A1–4

Hb = Main Series Volume; GD = Gmelin-Durrer; Ab = Appended Volume to Main Series;
Eb = Supplement Volume to Main Series; Erg. W. = New Supplement Series Volume

Y	Yttrium ions	39 (SE):	Hb/A1–3
			Hb/B3–45
–	Yttrium systems:		
	Y–Ag	61 (Ag):	Hb/C–211/2
	Y–H	39 (SE):	Hb/C1–2/3, 21/3
	Y–H–Sc	39 (SE):	Hb/C1–31/3
	Y–Nb	49 (Nb):	Hb/B2–54/5
	Y–Ni	57 (Ni):	Hb/B1–65/7
	Y–O	39 (SE):	Hb/C1–188/9
	Y–O–Ta	50 (Ta):	Hb/B2–205
	Y–Pb	47 (Pb):	Hb/C4–1360/1
	Y–Pb–Zn	47 (Pb):	Hb/C4–1361, 1363
	Y–Sn	46 (Sn):	Hb/D–293/4
	Y–Ta	50 (Ta):	Hb/B2–19/20
	Y–V	48 (V):	Hb/B2–593
Y^{+}	Y^{+}	39 (SE):	Hb/B3–45
Y^{2+}	Y^{2+}	39 (SE):	Hb/B3–45
Y^{3+}	Y^{3+}	21 (Na):	Eb/6–293
		39 (SE):	Hb/A1–3
			Hb/B3–45
Yb	Ytterbium:		
	Abundance	39 (SE):	Hb/A1–34/5
	Band model	39 (SE):	Hb/B3–211
	Boiling point	39 (SE):	Hb/B3–105, 108
	Characteristic electron energy losses	39 (SE):	Hb/B3–225/6
	Charge carrier concentration	39 (SE):	Hb/B3–228
	Cosmochemistry	39 (SE):	Hb/A1–23/7
	Crystal chemistry	39 (SE):	Hb/A1–28/32
	Crystallographic properties	39 (SE):	Hb/B3–63/7, 73/80
	Crystal structure	39 (SE):	Hb/B3–66/7, 73/4
	Debye temperature	39 (SE):	Hb/B3–192/3
	Density	39 (SE):	Hb/B3–85/7
	Diffusion	39 (SE):	Hb/B3–77/8
			Hb/C2–178
	Disorder	39 (SE):	Hb/B3–76
	Economic deposits	39 (SE):	Hb/A1–110/22
	Elasticity	39 (SE):	Hb/B3–92/102
	Electrical properties	39 (SE):	Hb/B3–209/83
	Electrical resistance	39 (SE):	Hb/B3–250/3, 263
	Electron density	39 (SE):	Hb/B3–217
	Electron emission	39 (SE):	Hb/B3–280/3
	Enthalpy of fusion	39 (SE):	Hb/B3–120
	– of sublimation	39 (SE):	Hb/B3–110, 115/6
	– of transformation	39 (SE):	Hb/B3–118
	– of vaporization	39 (SE):	Hb/B3–110, 115/6
	Epitaxy	39 (SE):	Hb/B3–80

See the "Instructions for the user"
at the beginning of the volume

Yb. . . . . Ytterbium:
Equation of state . . . . 39 (SE): Hb/B3–83/4
Geochemistry . . . . 39 (SE): Hb/A1–28/54
Grüneisen constant. . . . . 39 (SE): Hb/B3–84/5
Hall effect . . . . 39 (SE): Hb/B3–272
Hardness . . . . 39 (SE): Hb/B3–103
History. . . . . 39 (SE): Hb/A1–4/22
Isomer shift . . . . 39 (SE): Hb/B3–217
Magnetic properties . . . . 39 (SE): Hb/B3–336
Mechanical properties . . . . 39 (SE): Hb/B3–85/103, 193/5, 207/9
Melting point . . . . 39 (SE): Hb/B3–118/9
Mineralogy . . . . 39 (SE): Hb/A1–54/110
Molar volume . . . . 39 (SE): Hb/B3–85/7
Nuclear magnetic resonance. . . . . . 39 (SE): Hb/B3–227/8
Occurrence, extraterrestrial . . . . . . 39 (SE): Hb/A1–22/7
–, terrestrial . . . . 39 (SE): Hb/A1–27/54
– in meteorites . . . . 39 (SE): Hb/A1–25/7
– in minerals . . . . 39 (SE): Hb/A1–35/53
– in the biosphere. . . . . 39 (SE): Hb/A1–53/4
– in the lithosphere . . . . 39 (SE): Hb/A1–27/54
Optical properties. . . . . 39 (SE): Hb/B3–342
Polymorphism . . . . 39 (SE): Hb/B3–57, 63/4
Position in the periodic system . . . . 39 (SE): Hb/A1–2/4
Strength . . . . 39 (SE): Hb/B3–102
Structure . . . . 39 (SE): Hb/B3–80
Surface tension . . . . 39 (SE): Hb/B3–208/9
Susceptibility . . . . 39 (SE): Hb/B3–336
Thermal conductivity . . . . 39 (SE): Hb/B3–205
Thermal expansion . . . . 39 (SE): Hb/B3–91
Thermal properties . . . . 39 (SE): Hb/B3–83/92, 103/20, 179/81, 192, 205
Thermodynamic functions . . . . 39 (SE): Hb/B3–179/81
Thermoelectric power . . . . 39 (SE): Hb/B3–277
Valence . . . . 39 (SE): Hb/A1–4
Vapor pressure . . . . 39 (SE): Hb/B3–103/5, 108
Viscosity. . . . . 39 (SE): Hb/B3–207/8
–. . . . . Ytterbium ions. . . . . 39 (SE): Hb/A1–3
Hb/C1–125/6, 241
50 (Ta): Hb/B2–204
–. . . . . Ytterbium isotopes . . . . Erg. W. 7a–87
–. . . . . Ytterbium systems:
Yb–Ag. . . . . 61 (Ag): Hb/C–235/6
Yb–Gd–N. . . . . 39 (SE): Hb/C2–179
Yb–H . . . . 39 (SE): Hb/C1–2, 77/8
Yb–O . . . . 39 (SE): Hb/C1–426
Yb–Pb. . . . . 47 (Pb): Hb/C4–1371/2
Yb–V . . . . 48 (V): Hb/B2–600

Yb^{2+} . . . . Yb^{2+} . . . . 39 (SE): Hb/A1–3
Yb^{3+} . . . . Yb^{3+} . . . . 39 (SE): Hb/A1–3
Hb/C1–125/6, 241
50 (Ta): Hb/B2–204

Zn . . . . Zinc:
Absorption of light . . . . 32 (Zn): Hb–39/40
Eb–366/7
– of X-rays . . . . 32 (Zn): Hb–40
Abundance . . . . 32 (Zn): Eb–15
Atom . . . . 32 (Zn): Hb–11/3
Eb–278/9
Atomic nucleus . . . . 32 (Zn): Hb–11
Eb–277
Atomic scattering factors . . . . 32 (Zn): Eb–286/7
Behavior as anode . . . . 32 (Zn): Hb–105
Eb–477/86
– as cathode . . . . 32 (Zn): Hb–105
Eb–472/7
– as electrode . . . . 32 (Zn): Hb–48, 105
Eb–463, 472/86
– at the dropping mercury electrode . 32 (Zn): Eb–465/8
– at the dropping
zinc amalgam electrode . . . . 32 (Zn): Eb–469
– in alternating current electrolysis . . 32 (Zn): Hb–105/6
Eb–487/8
– in glowlight electrolysis . . . . 32 (Zn): Eb–488
Birefringence . . . . 32 (Zn): Eb–368
Boiling point . . . . 32 (Zn): Hb–35
Eb–359
Catalytic properties . . . . 32 (Zn): Eb–604/5
Cells . . . . 32 (Zn): Hb–50/4
Eb–416/30
Chemical constant . . . . 32 (Zn): Hb–38
Eb–365
Chemical reactions . . . . 32 (Zn): Hb–56/73
Eb–538/605
– in soils . . . . 32 (Zn): Eb–604
– with acids . . . . 32 (Zn): Hb–60/6
Eb–559/91
– with air . . . . 32 (Zn): Hb–56
Eb–542/5
– with alkaline solutions . . . . 32 (Zn): Hb–60
Eb–558/9
– with aqueous salt solutions . . . . . 32 (Zn): Hb–66/70
Eb–591/8
– with metal compounds . . . . 32 (Zn): Hb–70
Eb–555/8

See the "Instructions for the user"
at the beginning of the volume

Zn	Zinc:		
	Chemical reactions with metals	32 (Zn):	Hb–59 Eb–552
	– with nonmetal compounds	32 (Zn):	Hb–57/9, 70 Eb–545/55
	– with nonmetals	32 (Zn):	Hb–56, 59 Eb–538/45, 552/3
	– with organic substances	32 (Zn):	Hb–70/2 Eb–584/91, 598/604
	– with oxygen	32 (Zn):	Hb–56 Eb–538/41
	– with water	32 (Zn):	Hb–57/9 Eb–545/52
	Colloidal forms	32 (Zn):	Hb–18 Eb–273/4
	Commercial products.	32 (Zn):	Hb–93/5 Eb–252/3
	Compressibility	32 (Zn):	Hb–24 Eb–343/5
	Coordination number.	32 (Zn):	Eb–279
	Cosmochemistry	32 (Zn):	Hb–5 Eb–1/2
	Crystal chemistry	32 (Zn):	Eb–2/14
	Crystallization	32 (Zn):	Hb–14/5 Eb–281/2
	Crystallographic properties	32 (Zn):	Hb–13/21 Eb–280/302
	Debye temperature	32 (Zn):	Eb–364
	Density	32 (Zn):	Hb–21/2 Eb–303/6
	Detection	32 (Zn):	Hb–108/11 Eb–727/37
	– in iron	59 (Fe):	Hb/FII1–58/9
	– in special substances	32 (Zn):	Hb–122/4
	Determination.	32 (Zn):	Hb–112/7 Eb–737/53
	– in iron	59 (Fe):	Hb/FII1–59/61 Hb/FII2–316
	– in special substances	32 (Zn):	Hb–122/4 Eb–760/70
	Dielectric constant	32 (Zn):	Eb–379
	Difference effect	32 (Zn):	Eb–486/7
	Diffusion.	32 (Zn):	Hb–32 Eb–348/53
		46 (Sn):	Hb/D–129/30
	Discharge phenomena	32 (Zn):	Hb–54
	Dressing and treatment of ores	32 (Zn):	Hb–74, 89/92 Eb–137/80

Hb = Main Series Volume; GD = Gmelin-Durrer; Ab = Appended Volume to Main Series;
Eb = Supplement Volume to Main Series; Erg. W. = New Supplement Series Volume

Zn . . . . Zinc:
Economic deposits:
Africa . . . . 32 (Zn): Hb–4
Eb–100/3
America . . . . 32 (Zn): Hb–3/4
Eb–103/18
Arctic . . . . 32 (Zn): Eb–103
Asia . . . . 32 (Zn): Hb–4
Eb–90/7
Australia and Oceania . . . . 32 (Zn): Hb–4
Eb–97/100
Europe . . . . 32 (Zn): Hb–3
Eb–80/90
Soviet Union . . . . 32 (Zn): Eb–88/94
Economics. . . . . 32 (Zn): Hb–4/5
Eb–78/80, 267/8
Elasticity. . . . . 32 (Zn): Hb–22/7
Eb–317/21
Electrical conductivity . . . . 32 (Zn): Hb–44/6
Eb–380/92
Electrical properties . . . . 32 (Zn): Hb–44/8, 54/6
Eb–379/400
Electrocapillarity . . . . 32 (Zn): Eb–464
Electrochemical behavior . . . . 32 (Zn): Hb–48/54, 72, 99/108
Eb–401/537
Electrokinetic properties . . . . 32 (Zn): Eb–463/4
Electrolytic deposition . . . . 32 (Zn): Hb–92/3, 99, 102/8
Eb–219/52, 488/537
Electrolytic solution pressure . . . . . 32 (Zn): Hb–48/9
Eb–406
Electron emission. . . . . 32 (Zn): Hb–55
Eb–396/400
Enthalpy . . . . 32 (Zn): Eb–364/5
Entropy . . . . 32 (Zn): Eb–365
Equation of state . . . . 32 (Zn): Eb–353
Estimation. . . . . 32 (Zn): Hb–120/2
Eb–262/4
Flow under pressure . . . . 32 (Zn): Hb–32
Eb–345
Formation . . . . 32 (Zn): Hb–73
Free energy . . . . 32 (Zn): Eb–364/5
Frictional electricity. . . . . 32 (Zn): Hb–46
Galvanomagnetic effects . . . . 32 (Zn): Hb–44
Eb–376/9
Geochemical character. . . . . 32 (Zn): Eb–14/5
Geochemical cycle . . . . 32 (Zn): Eb–51/2, 77/8
Geochemistry . . . . 32 (Zn): Hb–9/11
Eb–2/78

See the "Instructions for the user"
at the beginning of the volume

Zn . . . . Zinc:

Hall effect . . . . 32 (Zn): Hb–44
Eb–377/9

Hardness . . . . 32 (Zn): Hb–30/1
Eb–306/17

Heat of fusion . . . . 32 (Zn): Hb–36
Eb–360

– of sublimation. . . . . 32 (Zn): Eb–360

– of vaporization . . . . 32 (Zn): Hb–36
Eb–360

History. . . . . 32 (Zn): Hb–1/2

Industrial hazards. . . . . 32 (Zn): Hb–11
Eb–262, 724/7

Internal energy . . . . 32 (Zn): Eb–364/5

Internal friction . . . . 32 (Zn): Hb–32
Eb–347

Internal pressure . . . . 32 (Zn): Eb–345

Ionic mobility . . . . 32 (Zn): Hb–99
Eb–406

Ionization potentials . . . . 32 (Zn): Hb–54/5
Eb–400

Lattice structure. . . . . 32 (Zn): Hb–15
Eb–282/7

Lattice vibrations . . . . 32 (Zn): Eb–363/4

Magnetic properties . . . . 32 (Zn): Hb–44
Eb–374/9

Mechanical properties . . . . 32 (Zn): Hb–21/32
Eb–303/53

Melting point . . . . 32 (Zn): Hb–35/6
Eb–359

Mineralogy . . . . 32 (Zn): Hb–5/10
Eb–118/36

Modifications . . . . 32 (Zn): Hb–13/4
Eb–280/1

Molecule. . . . . 32 (Zn): Hb–12
Eb–279/80

Nuclear properties . . . . 32 (Zn): Hb–11
Eb–277

Occupational hazards. . . . . 32 (Zn): Hb–11
Eb–262

Occurrence, extraterrestrial . . . . . . 32 (Zn): Hb–5
Eb–1/2

–, terrestrial . . . . 32 (Zn): Hb–5/11
Eb–2/78

– in meteorites . . . . 32 (Zn): Eb–2

– in soils . . . . 32 (Zn): Hb–10
Eb–52/8

– in the atmosphere . . . . 32 (Zn): Eb–51

Hb = Main Series Volume; GD = Gmelin-Durrer; Ab = Appended Volume to Main Series;
Eb = Supplement Volume to Main Series; Erg. W. = New Supplement Series Volume

Zn . . . . Zinc:
Occurrence in the biosphere . . . . . . 32 (Zn): Hb–10/1
Eb–58/76
– in the hydrosphere . . . . 32 (Zn): Hb–10
Eb–47/51
– in the lithosphere . . . . 32 (Zn): Hb–9/10
Eb–15/47
– in the magmatic cycle . . . . 32 (Zn): Eb–15/32
– in the metamorphic cycle . . . . . . 32 (Zn): Eb–45/7
– in the sedimentary cycle . . . . 32 (Zn): Hb–9
Eb–32/45
Optical properties . . . . 32 (Zn): Hb–39/43
Eb–366/74
Orientation of single crystals . . . . . 32 (Zn): Eb–287/8
– of thin films . . . . 32 (Zn): Eb–288/9
Overvoltage phenomena . . . . 32 (Zn): Hb–101
Eb–469/72, 493/5, 531
Passivity . . . . 32 (Zn): Hb–72, 101
Eb–477/86
Photoelectric effect, electrochemical . 32 (Zn): Eb–464/5
Physiological action . . . . 32 (Zn): Eb–63/6, 75/6
Plasticity . . . . 32 (Zn): Hb–25/7
Eb–321/2
Polarization . . . . 32 (Zn): Hb–99/100
Eb–439, 488, 493/5, 509
Polymorphism . . . . 32 (Zn): Hb–13/4
Eb–280/1
Position in the electrochemical series . . . . 32 (Zn): Eb–404/5
– in the periodic system . . . . 32 (Zn): Eb–278/9
Potentials . . . . 32 (Zn): Hb–49
Eb–406/16
Preparation, industrial:
Dressing of ores . . . . 32 (Zn): Hb–74, 89/92
Eb–137/59
Electrolytic deposition . . . . 32 (Zn): Hb–92/3, 102/8
Eb–219/52
Leaching . . . . 32 (Zn): Hb–89/92
Eb–178/80
Reduction of zinc oxide . . . . 32 (Zn): Hb–77/86
Eb–180/206
Refining . . . . 32 (Zn): Hb–93/5
Eb–252/62
Roasting of ores . . . . 32 (Zn): Hb–74/7
Eb–159/78
Zinc oxide process . . . . 32 (Zn): Hb–87/9
Eb–206/19

See the "Instructions for the user" at the beginning of the volume

Zn . . . . Zinc:

Preparation in pure state	32 (Zn):	Hb–95/6 Eb–252
– of special forms	32 (Zn):	Hb–16/8, 96/8, 103/4 Eb–251/2, 269/74, 508
Primary cells	32 (Zn):	Eb–434/63
Purity testing	32 (Zn):	Hb–120/2 Eb–262/4
Recrystallization	32 (Zn):	Hb–19/21 Eb–299/302
Rectifier effect	32 (Zn):	Hb–102 Eb–486
Refining	32 (Zn):	Hb–93/5 Eb–252/62
Reflection of light	32 (Zn):	Hb–39 Eb–366/8
– of X-rays	32 (Zn):	Eb–374
Refraction	32 (Zn):	Hb–39/40 Eb–366/7
Scattering of X-rays	32 (Zn):	Eb–374
Separation, analytical	32 (Zn):	Hb–117/9 Eb–753/60
Sorption	32 (Zn):	Hb–32
Sound velocity	32 (Zn):	Hb–24 Eb–353
Specific heat	32 (Zn):	Hb–36/8 Eb–361/3
Spectra:		
Arc spectrum	32 (Zn):	Hb–41/3 Eb–368/9
Band spectrum	32 (Zn):	Hb–42 Eb–280
Spark spectra	32 (Zn):	Hb–41/3 Eb–369/70
X-ray spectrum	32 (Zn):	Hb–43 Eb–370/4
Standard cells	32 (Zn):	Hb–49 Eb–431/2
Standard potential	32 (Zn):	Hb–49 Eb–401/4
Storage batteries	32 (Zn):	Eb–432/4
Strength	32 (Zn):	Hb–27/30 Eb–322/43
Structure of liquid and gaseous zinc .	32 (Zn):	Eb–290
– of solid zinc	32 (Zn):	Hb–18/21 Eb–290/302
– of thin films	32 (Zn):	Eb–288/9
Surface tension	32 (Zn):	Hb–32 Eb–345/6

Hb = Main Series Volume; GD = Gmelin-Durrer; Ab = Appended Volume to Main Series;
Eb = Supplement Volume to Main Series; Erg. W. = New Supplement Series Volume

Zn . . . . Zinc:
Surface treatment:
Aftertreatment . . . . 32 (Zn): Eb–635/6
Dyeing . . . . 32 (Zn): Eb–667/72
Etching . . . . 32 (Zn): Eb–634/5
Metallic coatings . . . . 32 (Zn): Eb–673/707
Organic coatings . . . . 32 (Zn): Eb–707/17
Paints . . . . 32 (Zn): Eb–707/17
Polishing . . . . 32 (Zn): Eb–631/4
Purification . . . . 32 (Zn): Eb–628/31
Surface protection . . . . 32 (Zn): Eb–672
Treatment with solutions . . . . . . 32 (Zn): Eb–636/66
Susceptibility . . . . 32 (Zn): Hb–44
Eb–374/6
Thermal conductivity . . . . 32 (Zn): Hb–38/9
Eb–365/6
Thermal expansion . . . . 32 (Zn): Hb–32/3
Eb–353/7
Thermal properties . . . . 32 (Zn): Hb–32/9
Eb–353/66
Thermal radiation . . . . 32 (Zn): Hb–40/1
Eb–368
Thermoelectric effects . . . . 32 (Zn): Hb 46/8
Eb–392/6
Thermomagnetic effects . . . . 32 (Zn): Eb–376/9
Toxicity . . . . 32 (Zn): Hb–11
Eb–262, 724/7
Uses . . . . 32 (Zn): Hb–98
Eb–264/7
Valence . . . . 32 (Zn): Hb–13
Eb–279
Vaporization . . . . 32 (Zn): Hb–33/4
Eb–358/9
Vapor pressure . . . . 32 (Zn): Hb–34/5
Eb–357/8
Viscosity . . . . 32 (Zn): Hb–32
Eb–347/8
Wettability . . . . 32 (Zn): Eb–346/7
Zinc coatings,
electrolytic deposition . . . . 32 (Zn): Eb–490/537
–. . . . . Zinc ions
see Zn^{+} *...* Zn^{+}*, etc.*
–. . . . . Zinc isotopes:
Enrichment and separation . . . . . . 32 (Zn): Eb–274/7
Nuclear properties . . . . 32 (Zn): Hb–11
Eb–277
–. . . . . Zinc systems:
Zn–Ag . . . . 61 (Ag): Hb/C–39/47

See the "Instructions for the user"
at the beginning of the volume

Zn . . . . Zinc systems:

Zn–Ag–Al	35 (Al):	Hb/A6–1081/3
	61 (Ag):	Hb/C–172/3
Zn–Ag–Bi	61 (Ag):	Hb/B4–143/4
Zn–Ag–Bi–Sb	61 (Ag):	Hb/B4–144
Zn–Ag–Cd	61 (Ag):	Hb/C–119/20
Zn–Ag–Cd–Cu	61 (Ag):	Hb/C–396/8
Zn–Ag–Cu	61 (Ag):	Hb/C–389/92
Zn–Ag–Mg	27 (Mg):	Hb/A4–745
	61 (Ag):	Hb/C–76/9
Zn–Ag–Mn	61 (Ag):	Hb/C–345/6
Zn–Ag–Pb	61 (Ag):	Hb/C–308/10
Zn–Ag–Sb	61 (Ag):	Hb/B4–142/3
Zn–Al	35 (Al):	Hb/A5–683/706
Zn–Al–Cd	35 (Al):	Hb/A5–791/3
Zn–Al–Ga–Ge–Sn	46 (Sn):	Hb/D–397
Zn–Al–Mg	35 (Al):	Hb/A5–762/73
Zn–Al–Mg–Si	35 (Al):	Hb/A5–783
Zn–Al–Ni	35 (Al):	Hb/A6–1003
Zn–Al–Pb	47 (Pb):	Hb/C4–1290/1
Zn–Al–Si	35 (Al):	Hb/A5–760/1
Zn–Al–Sn	35 (Al):	Hb/A5–855/7
	46 (Sn):	Hb/D–220/1
Zn–As	32 (Zn):	Hb–282/3
Zn–As–Pb	47 (Pb):	Hb/C4–1223
Zn–Au	62 (Au):	Hb/3–786/8
Zn–Au–Bi	62 (Au):	Hb/3–798
Zn–Au–Ge	62 (Au):	Hb/3–839
Zn–Au–Ni	62 (Au):	Hb/3–877/8
Zn–Au–Pb	62 (Au):	Hb/3–798
Zn–Au–Tl	62 (Au):	Hb/3–798
Zn–B–Ni	57 (Ni):	Hb/B3–1137
Zn–Ba	32 (Zn):	Eb–627
Zn–Bi	32 (Zn):	Hb–291/2
		Eb–615/6
Zn–Bi–Cd	33 (Cd):	Hb–213
Zn–Bi–Cd–Pb	47 (Pb):	Hb/C4–1251
Zn–Bi–Cd–Pb–Sn	47 (Pb):	Hb/C4–1528/9
Zn–Bi–Cu	60 (Cu):	Hb/B3–1152
Zn–Bi–Mg	27 (Mg):	Hb/A4–524/6
Zn–Bi–Pb	47 (Pb):	Hb/C4–1224/5
Zn–Bi–Sb	32 (Zn):	Hb–293
		Eb–617
Zn–Bi–Sn	46 (Sn):	Hb/D–155
Zn–C–Ni	57 (Ni):	Hb/B1–48
Zn–C–O	32 (Zn):	Eb–977
Zn–Ca	32 (Zn):	Hb–325
		Eb–623/4
Zn–Ca–Mg	27 (Mg):	Hb/A4–527/9

Zn . . . . Zinc systems:

Zn–Cd	33 (Cd):	Hb–210/3
		Eb–383/6
Zn–Cd–Cu–H.	60 (Cu):	Hb/B3–1157
Zn–Cd–In–Sn	46 (Sn):	Hb/D–271
Zn–Cd–Mg	27 (Mg):	Hb/A4–555/6
	33 (Cd):	Hb–213/4
Zn–Cd–Pb	47 (Pb):	Hb/C4–1248/9
Zn–Cd–Pb–Sn	47 (Pb):	Hb/C4–1528/9
Zn–Cd–Sb	33 (Cd):	Hb–213
Zn–Cd–Sn	46 (Sn):	Hb/D–181/2
Zn–Ce–Pb	47 (Pb):	Hb/C4–1363, 1365
Zn–Co	58 (Co):	Hb/A1–188/9
		Eb/A–391/3
Zn–Cr	52 (Cr):	Hb/A2–717/9
Zn–Cs–Pb.	47 (Pb):	Hb/C4–1225
Zn–Cu	60 (Cu):	Hb/B3–1145
Zn–Cu–Fe–S	60 (Cu):	Hb/B3–1325/6
Zn–Cu–H	60 (Cu):	Hb/B3–1145
Zn–Cu–H–Mg	60 (Cu):	Hb/B3–1153
Zn–Cu–Mg	27 (Mg):	Hb/A4–719/22
Zn–Cu–O	60 (Cu):	Hb/B3–1145/6
Zn–Cu–P	60 (Cu):	Hb/B3–1151
Zn–Cu–S	60 (Cu):	Hb/B3–1148
Zn–Cu–Si.	60 (Cu):	Hb/B3–1150
Zn–Dy–Pb	47 (Pb):	Hb/C4–1363, 1370
Zn–Er–Pb.	47 (Pb):	Hb/C4–1371
Zn–Eu–Pb	47 (Pb):	Hb/C4–1363, 1368
Zn–Fe	59 (Fe):	Hb/A9–1827/33
Zn–Ga.	36 (Ga):	Hb–60
Zn–Ga–Ge–In–Sn.	46 (Sn):	Hb/D–397
Zn–Ga–In–Sn	46 (Sn):	Hb/D–278
Zn–Ga–Pb	47 (Pb):	Hb/C4–1305
Zn–Ga–Pb–Sn	47 (Pb):	Hb/C4–1529
Zn–Ga–Sn	46 (Sn):	Hb/D–231/3
Zn–Gd–Pb	47 (Pb):	Hb/C4–1363, 1369
Zn–Ge.	45 (Ge):	Eb–478/9
Zn–H–Ni	57 (Ni):	Hb/B1–46, 48
Zn–Hg.	34 (Hg):	Hb/A2–1062/4
Zn–Ho–Pb	47 (Pb):	Hb/C4–1363, 1370
Zn–In	37 (In):	Hb–58
Zn–K.	32 (Zn):	Eb–622
Zn–La–Pb.	47 (Pb):	Hb/C4–1363
Zn–Li	32 (Zn):	Eb–617/9
Zn–Lu–Pb.	47 (Pb):	Hb/C4–1363, 1373
Zn–Mg	27 (Mg):	Hb/A4–483/7
	32 (Zn):	Hb–321
Zn–Mg–Mn.	27 (Mg):	Hb/A4–673
Zn–Mg–Pb	47 (Pb):	Hb/C4–1225

See the "Instructions for the user"
at the beginning of the volume

Zn . . . . Zinc systems:

Zn–Mg–Si . . . . 27 (Mg): Hb/A4–520/1
Zn–Mg–Sn . . . . 27 (Mg): Hb/A4–635
Zn–Mg–Tl . . . . 27 (Mg): Hb/A4–581
Zn–N–V . . . . 48 (V): Hb/B2–581
Zn–Na . . . . 32 (Zn): Hb–294
Eb–621
Zn–Na–Pb . . . . 47 (Pb): Hb/C4–1225
Zn–Nb . . . . 49 (Nb): Hb/B2–18
Zn–Nb–O . . . . 49 (Nb): Hb/B3–94
Zn–Nd–Pb . . . . 47 (Pb): Hb/C4–1363, 1367
Zn–Ni . . . . 57 (Ni): Hb/B1–32/4
Zn–O . . . . 32 (Zn): Eb–809
Zn–Pb . . . . 47 (Pb): Hb/C4–1213/4
Zn–Pb–Pr . . . . 47 (Pb): Hb/C4–1363, 1366
Zn–Pb–Sb . . . . 47 (Pb): Hb/C4–1223/4
Zn–Pb–Sm . . . . 47 (Pb): Hb/C4–1363, 1367
Zn–Pb–Sn . . . . 47 (Pb): Hb/C4–1522/4
Zn–Pb–Sr . . . . 47 (Pb): Hb/C4–1225
Zn–Pb–Y . . . . 47 (Pb): Hb/C4–1361, 1363
Zn–Sb . . . . 32 (Zn): Hb–286/90
Eb–606/8
Zn–Sb–Sn . . . . 46 (Sn): Hb/D–153
Zn–Se . . . . 32 (Zn): Eb–967
Zn–Si . . . . 32 (Zn): Hb–272
Zn–Sn . . . . 46 (Sn): Hb/D–130/2
Zn–Sn–Tl . . . . 46 (Sn): Hb/D–290
Zn–Sr . . . . 32 (Zn): Eb–627
Zn–Ta . . . . 50 (Ta): Hb/B2–10/1
Zn–Te . . . . 32 (Zn): Eb–969
Zn–Ti . . . . 41 (Ti): Hb–201
Zn–Tl . . . . 38 (Tl): Hb/2–212/3
Zn–V . . . . 48 (V): Hb/B2–579/80
Zn–Zr . . . . 42 (Zr): Hb–193

Zn^+ . . . . Zn^+:

Ionization potential . . . . 32 (Zn): Hb–55
Eb–400
Spectra . . . . 32 (Zn): Hb–41/3
Eb–369/70

Zn^{2+} . . . . Zn^{2+}:

Chemical reactions . . . . 32 (Zn): Hb–108/12
Eb–719/20
Diffusion in sodium halides . . . . . . 21 (Na): Eb/6–293
Electrochemical behavior . . . . 32 (Zn): Eb–719
Increments . . . . 32 (Zn): Eb–718
Ionic mobility . . . . 32 (Zn): Hb–99
Eb–406
Ionization potential . . . . 32 (Zn): Eb–400

Formula	Entry	Volume	Reference
Zn^{2+}	Zn^{2+}:		
	Properties of the ion	32 (Zn):	Eb–278, 718
	Scattering factors	32 (Zn):	Eb–286
	Spectra	21 (Na):	Eb/6–337
		32 (Zn):	Hb–41/3
			Eb–370
–	$[Zn(H_2O)_6]^{2+}$	32 (Zn):	Hb–111
			Eb–717, 867
–	$[Zn(H_4O_2)_6]^{2+}$	32 (Zn):	Hb–111
–	$[Zn(H_2O)_x]^{2+}$	32 (Zn):	Eb–717/8
Zn^{3+}	Zn^{3+}	32 (Zn):	Eb–370, 400
Zn^{n+}	Zn^{n+} (n = 4 to 30)	32 (Zn):	Eb–400
Zn^{n-}	Zn^{n-} (n = 1 to 4)	32 (Zn):	Eb–400
Zn_2	Zn_2	32 (Zn):	Hb–12
			Eb–279/80
Zn_2Zr	$ZrZn_2$ solid solutions:		
	$(Zr,Nb)Zn_2$	49 (Nb):	Hb/B2–174
Zr	Zirconium:		
	Absorption of X-rays	42 (Zr):	Hb–147
	Abundance	42 (Zr):	Hb–9
	Atom	42 (Zr):	Hb–103/11
	Atomic nucleus	42 (Zr):	Hb–102/3
	Chemical reactions	42 (Zr):	Hb–158/72
	Cosmochemistry	42 (Zr):	Hb–2/4
	Crystal chemistry	42 (Zr):	Hb–4/9
	Crystallographic properties	42 (Zr):	Hb–112/5
	Detection	42 (Zr):	Hb–192
	– in iron	59 (Fe):	Hb/FII1–20/1
	Determination	42 (Zr):	Hb–192
	– in iron	59 (Fe):	Hb/FII1–21/6
			Hb/FII2–300
	Diffusion	42 (Zr):	Hb–133
		46 (Sn):	Hb/D–337/9
	Dressing and treatment of ores	42 (Zr):	Hb–50/66
	Economic deposits:		
	Africa	42 (Zr):	Hb–34/6
	America	42 (Zr):	Hb–36/43
	Asia	42 (Zr):	Hb–25/8
	Australia	42 (Zr):	Hb–28/34
	Europe	42 (Zr):	Hb–23/5
	Soviet Union	42 (Zr):	Hb–24/6
	Economics	42 (Zr):	Hb–23
	Electrical properties	42 (Zr):	Hb–148/56
	Electrochemical behavior	42 (Zr):	Hb–156/8
	Electronegativity	42 (Zr):	Hb–111
	Formation	42 (Zr):	Hb–84

See the "Instructions for the user" at the beginning of the volume

Zr . . . . Zirconium:

Geochemical character . . . . 42 (Zr): Hb–9
Geochemical cycle . . . . 42 (Zr): Hb–22/3
Geochemistry . . . . 42 (Zr): Hb–4/23
History . . . . 42 (Zr): Hb–1/2
Magnetic properties . . . . 42 (Zr): Hb–147/8
Mechanical properties . . . . 42 (Zr): Hb–115/34
Mineralogy . . . . 42 (Zr): Hb–43/50
Nuclear properties . . . . 42 (Zr): Hb–102/3
Occurrence, extraterrestrial . . . . . . 42 (Zr): Hb–2/4
–, terrestrial . . . . 42 (Zr): Hb–4/23
– in minerals . . . . 42 (Zr): Hb–4/9
– in the biosphere . . . . 42 (Zr): Hb–22
– in the hydrosphere . . . . 42 (Zr): Hb–22
– in the lithosphere . . . . 42 (Zr): Hb–10/21
Optical properties . . . . 42 (Zr): Hb–140/7
Position in the periodic system . . . . 42 (Zr): Hb–111
Preparation by electrolysis . . . . 42 (Zr): Hb–92/4
– by reduction . . . . 42 (Zr): Hb–85/92
– by thermal decomposition . . . . . . 42 (Zr): Hb–94/6
– of special forms . . . . 42 (Zr): Hb–99/101
Purification . . . . 42 (Zr): Hb–96
Purity testing . . . . 42 (Zr): Hb–96/7
Scattering of X-rays . . . . 42 (Zr): Hb–147
Separation methods . . . . 42 (Zr): Hb–66/76
Sorption . . . . 42 (Zr): Hb–133/4
Spectra:
Arc spectrum . . . . 42 (Zr): Hb–141/2
Spark spectra . . . . 42 (Zr): Hb–142/4
X-ray spectrum . . . . 42 (Zr): Hb–144/7
Subsequent treatment . . . . 42 (Zr): Hb–97/9
Surface treatment . . . . 42 (Zr): Hb–99
Thermal properties . . . . 42 (Zr): Hb–134/40
Toxicity . . . . 42 (Zr): Hb–191
Uses . . . . 42 (Zr): Hb–78/84
Valence . . . . 42 (Zr): Hb–111
Workability . . . . 42 (Zr): Hb–125/7

–. . . . . Zirconium ions
see Zr^{+} *...* Zr^{+}*, etc.*

–. . . . . Zirconium isotopes:

Detection in fallout . . . . Erg. W. 7b–355/6
Enrichment and separation . . . . . . 42 (Zr): Hb–101/2
Nuclear properties . . . . 42 (Zr): Hb–102/3

–. . . . . Zirconium systems:

Zr–Ag . . . . 61 (Ag): Hb/C–247/9
Zr–Ag–C . . . . 61 (Ag): Hb/B4–239
Zr–Al . . . . 35 (Al): Hb/A5–836/7

Hb = Main Series Volume; GD = Gmelin-Durrer; Ab = Appended Volume to Main Series;
Eb = Supplement Volume to Main Series; Erg. W. = New Supplement Series Volume

Zr . . . . Zirconium systems:

Zr–Al–Nb	49 (Nb):	Hb/B2–174
Zr–Al–Sn	46 (Sn):	Hb/D–376
Zr–Al–Sn–Ti	46 (Sn):	Hb/D–382/3
Zr–Au	62 (Au):	Hb/3–835
Zr–B	42 (Zr):	Hb–356
Zr–B–Nb	49 (Nb):	Hb/B4–310/1
Zr–B–Ta	50 (Ta):	Hb/B2–321/2
Zr–Be–Nb	49 (Nb):	Hb/B2–173
Zr–C	42 (Zr):	Hb–361
Zr–C–Fe	59 (Fe):	Hb/A9–1897/901
Zr–C–Hf–Ta	50 (Ta):	Hb/B2–341
Zr–C–N–Ta	50 (Ta):	Hb/B2–333
Zr–C–Nb	49 (Nb):	Hb/B4–321/5
Zr–C–Ta	50 (Ta):	Hb/B2–327/33
Zr–C–Ta–V	50 (Ta):	Hb/B2–343/4
Zr–C–V	48 (V):	Hb/B2–674
Zr–Co	58 (Co):	Eb/A–404/6
Zr–Cr	52 (Cr):	Hb/A2–724/5
Zr–Cr–Ni	57 (Ni):	Hb/B1–226
Zr–Cu	60 (Cu):	Hb/B3–1191
Zr–Cu–Si	60 (Cu):	Hb/B3–1191
Zr–Fe	59 (Fe):	Hb/A9–1894/6
Zr–H	42 (Zr):	Hb–197/203
Zr–H–Ni	57 (Ni):	Hb/B1–111, 114
Zr–H–V	48 (V):	Hb/B2–673
Zr–Hf–Nb	49 (Nb):	Hb/B2–192
Zr–Hf–Nb–Ti	49 (Nb):	Hb/B2–194
Zr–Hf–Ta	50 (Ta):	Hb/B2–61
Zr–Mg	27 (Mg):	Hb/A4–604/5
Zr–Mn–Ni	57 (Ni):	Hb/B1–312
Zr–N	42 (Zr):	Hb–263/5
Zr–Nb	49 (Nb):	Hb/B2–126/8
Zr–Nb–$NbBe_2$–$ZrBe_2$	49 (Nb):	Hb/B2–173
Zr–Nb–O	49 (Nb):	Hb/B2–128/9, 171/2 Hb/B3–209
Zr–Nb–Sn	49 (Nb):	Hb/B2–235/6
Zr–Nb–Ta	50 (Ta):	Hb/B2–110
Zr–Nb–Th	49 (Nb):	Hb/B2–200
Zr–Nb–Ti	49 (Nb):	Hb/B2–174/6
Zr–Nb–V	49 (Nb):	Hb/B2–295/6
Zr–Ni	57 (Ni):	Hb/B1–102/4
Zr–Ni–O	57 (Ni):	Hb/B1–114
Zr–Ni–V	57 (Ni):	Hb/B1–159
Zr–O	42 (Zr):	Hb–208/12
Zr–O–Ta	50 (Ta):	Hb/B2–228
Zr–O–Ta–Ti	50 (Ta):	Hb/B2–229
Zr–P	42 (Zr):	Hb–389/90

See the "Instructions for the user"
at the beginning of the volume

Zr	Zirconium systems:		
	Zr–Pb	47 (Pb):	Hb/C4–1420/2
	Zr–Pb–Ti	47 (Pb):	Hb/C4–1423
	Zr–Ru	63 (Ru):	Eb–189/91
	Zr–S	42 (Zr):	Hb–326
	Zr–Si	42 (Zr):	Hb–379
	Zr–Sn	46 (Sn):	Hb/D–339/42
	Zr–Sn–Ti	46 (Sn):	Hb/D–378
	Zr–Ta	50 (Ta):	Hb/B2–41/3
	Zr–Ta–Ti	50 (Ta):	Hb/B2–51
	Zr–Ta–V	50 (Ta):	Hb/B2–83
	Zr–Ti.	42 (Zr):	Hb–194
	Zr–Ti–V	48 (V):	Hb/B2–674/5
	Zr–Ti_3Al–Ti_3Sn	46 (Sn):	Hb/D–382/3
	Zr–V	48 (V):	Hb/B2–662/3
	Zr–Zn	42 (Zr):	Hb–193
Zr^+	Zr^+	42 (Zr):	Hb–104/5, 142/4, 156
Zr^{2+}	Zr^{2+}	42 (Zr):	Hb–104/5, 144, 156
Zr^{3+}	Zr^{3+}	42 (Zr):	Hb–104/5, 144, 156
Zr^{4+}	Zr^{4+}:		
	see also OZr^{2+} ... ZrO^{2+}		
	Catalytic properties	42 (Zr):	Hb–191
	Ionic polarization	42 (Zr):	Hb–148
	Ionic refraction	42 (Zr):	Hb–140
	Ionic susceptibility	42 (Zr):	Hb–148
	Ionization potential	42 (Zr):	Hb–156
	Properties of the ion	42 (Zr):	Hb–104/8
	Sorption	42 (Zr):	Hb–134
	Spectra	42 (Zr):	Hb–144
Zr^{5+}	Zr^{5+}	42 (Zr):	Hb–104, 144, 156
Zr^{n+}	Zr^{n+} (n = 6 to 40)	42 (Zr):	Hb–156
Zr^{n-}	Zr^{n-} (n = 1 to 3)	42 (Zr):	Hb–156

Hb = Main Series Volume; GD = Gmelin-Durrer; Ab = Appended Volume to Main Series;
Eb = Supplement Volume to Main Series; Erg. W. = New Supplement Series Volume

See the "Instructions for the user"
at the beginning of the volume

Appendix

The following transuranium elements which are described in the Gmelin Handbook have no name accepted by the IUPAC. They are listed according to their atomic numbers.

104 . . . . Element 104 (Kurchatovium, Rutherfordium):

Electron configuration . . . . Erg. W. 8–28/30
Formation . . . . Erg. W. 7b–97
History . . . . Erg. W. 7a–8/9
Nuclear decay . . . . Erg. W. 7a–51, 120, 123/6
Nuclear fission . . . . Erg. W. 7a–51, 136
Nuclear properties . . . . Erg. W. 7a–132/3
Separation . . . . Erg. W. 7b–349/53
Spectra . . . . Erg. W. 8–83

105 . . . . Element 105 (Hahnium, Nielsbohrium):

Electron configuration . . . . Erg. W. 8–28/30
Formation . . . . Erg. W. 7b–97
History . . . . Erg. W. 7a–8/9
Nuclear decay . . . . Erg. W. 7a–51, 123/6
Nuclear fission . . . . Erg. W. 7a–136
Nuclear properties . . . . Erg. W. 7a–133
Separation . . . . Erg. W. 7b–349/53
Spectra . . . . Erg. W. 8–84

106 . . . . Element 106:

Electron configuration . . . . Erg. W. 8–28/30
Existence . . . . Erg. W. 7a–174/5
Nuclear properties . . . . Erg. W. 7a–133
Separation . . . . Erg. W. 7b–350/1
Spectra . . . . Erg. W. 8–84

107 . . . . Element 107:

Electron configuration . . . . Erg. W. 8–28/30
Existence . . . . Erg. W. 7a–174/5
Separation . . . . Erg. W. 7b–350/1
Spectra . . . . Erg. W. 8–84

108 . . . . Element 108:

Electron configuration . . . . Erg. W. 8–28/30
Existence . . . . Erg. W. 7a–174/5
Separation . . . . Erg. W. 7b–350/1
Spectra . . . . Erg. W. 8–84

109 . . . . Element 109:

Electron configuration . . . . Erg. W. 8–28/30
Existence . . . . Erg. W. 7a–174/5
Separation . . . . Erg. W. 7b–350/1
Spectra . . . . Erg. W. 8–84

See the "Instructions for the user" at the beginning of the volume

110 . . . . Element 110:
Electron configuration . . . . Erg. W. 8–28/30
Existence . . . . Erg. W. 7a–174/5
Separation. . . . . Erg. W. 7b–350/1
Spectra . . . . Erg. W. 8–84

111 . . . . Element 111:
Electron configuration . . . . Erg. W. 8–28/30
Existence . . . . Erg. W. 7a–174/5
Separation. . . . . Erg. W. 7b–350/1
Spectra . . . . Erg. W. 8–84

112 . . . . Element 112:
Electron configuration . . . . Erg. W. 8–28/30
Existence . . . . Erg. W. 7a–9, 174/5
Separation. . . . . Erg. W. 7b–350/1
Spectra . . . . Erg. W. 8–84

113 . . . . Element 113:
Electron configuration . . . . Erg. W. 8–28/30
Existence . . . . Erg. W. 7a–174/5
Physical properties . . . . Erg. W. 8–29
Separation. . . . . Erg. W. 7b–350/1
Spectra . . . . Erg. W. 8–84

114 . . . . Element 114:
Electron configuration . . . . Erg. W. 8–28/30
Existence . . . . Erg. W. 7a–170/6
Physical properties . . . . Erg. W. 8–29
Separation. . . . . Erg. W. 7b–350/1
Spectra . . . . Erg. W. 8–84

115 . . . . Element 115:
Electron configuration . . . . Erg. W. 8–28/30
Existence . . . . Erg. W. 7a–174/5
Separation. . . . . Erg. W. 7b–350/1
Spectra . . . . Erg. W. 8–84

116 . . . . Element 116:
Electron configuration . . . . Erg. W. 8–28/30
Existence . . . . Erg. W. 7a–174/5
Spectra . . . . Erg. W. 8–84

117 to 120 . . . . . . Elements 117 to 120:
Electron configuration . . . . Erg. W. 8–28/30
Physical properties . . . . Erg. W. 8–29
Spectra . . . . Erg. W. 8–84

121 to 132 . . . . . . Elements 121 to 132. . . . . Erg. W. 8–28/30